Craftsman 2020
NATIONAL
PLUMBING & HVAC
ESTIMATOR

By James A. Thomson

Includes Free Estimating Software Download

Includes inside the back cover:

Inside the back cover of this book you'll find a software download certificate. To access the download, follow the instructions printed there. The download includes the National Estimator, an easy to-use estimating program with all the cost estimates in this book. The software will run on PCs using Windows XP, Vista, 7, 8, or 10 operating systems.

Quarterly price updates on the Web are free and automatic all during 2020. You'll be prompted when it's time to collect the next update. A connection to the Web is required.

Download all of Craftsman's most popular costbooks for one low price with the Craftsman Site License. http://CraftsmanSiteLicense.com

- Turn your estimate into a bid.
- Turn your bid into a contract.
- ConstructionContractWriter.com

Craftsman Book Company
6058 Corte del Cedro, Carlsbad, CA 92011

Acknowledgments

The sample "Standard Form Subcontract" and "Subcontract Change Order" forms used in the final section of this book are reprinted with the permission of the publisher, the Associated General Contractors of America (National Office), 1957 E Street NW, Washington, District of Columbia 20006.

Looking for other construction reference manuals?
Craftsman has the books to fill your needs. **Call toll-free 1-800-829-8123**
Visit our Web site: http://www.craftsman-book.com

Cover design by: *Jennifer Johnson*
Photos*: iStock by Getty Images™*

© 2019 Craftsman Book Company
ISBN 978-1-57218-358-2
Published September 2019 for the year 2020.

Contents

How to Use This Book

This 2020 National Plumbing & HVAC Estimator is a guide to estimating labor and material costs for plumbing, heating, ventilating and air conditioning systems in residential, commercial and industrial buildings.

 Inside the back cover of this book you'll find a software download certificate. To access the download, follow the instructions printed there. The download includes an easy to-use estimating program with all the cost estimates in this book. The software will run on PCs using Windows XP, Vista, 7, 8, or 10 operating systems.

When the *National Estimator* program has been installed, click Help on the menu bar to see a list of topics that will get you up and running. Or, go online to www.craftsman-book.com and click on Support, then Tutorials, to view an interactive tutorial for *National Estimator*.

Costs in This Manual will apply within a few percent on a wide variety of projects. Using the information given on the pages that follow will explain how to use these costs and suggest procedures to follow when compiling estimates. Reading the remainder of this section will help you produce more reliable estimates for plumbing and HVAC work.

 Manhour Estimates in This Book will be accurate for some jobs and inaccurate for others. No manhour estimate fits all jobs because every construction project is unique. Expect installation times to vary widely from job to job, from crew to crew, and even for the same crew from day to day.

There's no way to eliminate all errors when making manhour estimates. But you can minimize the risk of a major error by:

1. Understanding what's included in the manhour estimates in this book, and

2. Adjusting the manhour estimates in this book for unusual job conditions.

The Craft@Hrs Column. Manhour estimates in this book are listed in the column headed *Craft@Hrs*. For example, on page 19 you'll see an estimate for installing a 6 gallon hot water heater. In the *Craft@Hrs* column opposite 6 gallon you'll see:

P1 @ .500

To the left of the @ symbol you see an abbreviation for the recommended work crew.

Page 7 shows the wage rates and craft codes used in this book.

To the right of the @ symbol you see a number. The number is the estimated manhours (not crew hours) required to install each unit of material listed. In the case of a 6 gallon hot water heater, P1 @ .500 means that .500 manhours are required to install 1 hot water heater.

Costs in the Labor $ Column are based on manhour estimates in the *Craft@Hrs* column. Multiply the manhour estimate by the assumed hourly labor cost to find the installation cost in the *Labor $* column. For example, .500 manhours times $36.70 (the average wage for crew P1) is $18.35.

Quarterly price updates on the Web are free and automatic all during 2020. You'll be prompted by Craftsman Software Update when it's time to collect the next update. A connection to the Web is required.

Manhour Estimates include all productive labor normally associated with installing the materials described. These estimates assume normal conditions: experienced craftsmen working on reasonably well planned and managed new construction with fair to good productivity. Labor estimates also assume that materials are standard grade, appropriate tools are on hand, work done by other crafts is adequate, layout and installation are relatively uncomplicated, and working conditions don't slow progress.

All manhour estimates include tasks such as:

■ Unloading and storing construction materials, tools and equipment on site.

■ Working no more than two floors above or below ground level.

■ Working no more than 10 feet above an uncluttered floor.

■ Normal time lost due to work breaks.

■ Moving tools and equipment from a storage area or truck not more than 200 feet from the work area.

■ Returning tools and equipment to the storage area or truck at the end of the day.

■ Planning and discussing the work to be performed.

■ Normal handling, measuring, cutting and fitting.

■ Regular cleanup of construction debris.

■ Infrequent correction or repairs required because of faulty installation.

If the work you're estimating won't be done under these conditions, you need to apply a correction factor to adjust the manhour estimates in this book to fit your job.

Applying Correction Factors. Analyze your job carefully to determine whether a labor correction factor is needed. Failure to consider job conditions is probably the most common reason for inaccurate estimates.

Use one or more of the recommended correction factors in Table 1 to adjust for unusual job conditions. To make the adjustment, multiply the manhour estimate by the appropriate conversion factor. On some jobs, several correction factors may be needed. A correction factor less than 1.00 means that favorable working conditions will reduce the manhours required.

 **Supervision Expense** to the installing contractor is not included in the labor cost. The cost of supervision and non-productive labor varies widely from job to job. Calculate the cost of supervision and non-productive labor and add this to the estimate.

Hourly Labor Costs also vary from job to job. This book assumes an average manhour labor cost of $43.25 for plumbers and $41.94 for sheet metal workers. If these hourly labor costs are not accurate for your jobs, adjust the labor costs up or down by an appropriate percentage. Instructions on the next page explain how to make these adjustments. If you're using the National Estimator disk, it's easy to set your own wage rates.

Hourly labor costs in this book include the basic wage, fringe benefits, the employer's contribution to welfare, pension, vacation and apprentice funds, and all tax and insurance charges based on wages. Table 2 at the top of the next page shows how hourly labor costs in this book were calculated. It's important that you understand what's included in the figures in each of the six columns in Table 2. Here's an explanation:

Column 1, the base wage per hour, is the craftsman's hourly wage. These figures are representative of what many contractors are paying plumbers, sheet metal workers and helpers in 2020.

Column 2, taxable fringe benefits, includes vacation pay, sick leave and other taxable benefits. These fringe benefits average about 5.53% of the base wage for many plumbing and HVAC contractors. This benefit is in addition to the base wage.

Condition	Correction Factor
Work in large open areas, no partitions	.85
Prefabrication under ideal conditions, bench work	.90
Large quantities of repetitive work	.90
Very capable tradesmen	.95
Work 300' from storage area	1.03
Work 400' from storage area	1.05
Work 500' from storage area	1.07
Work on 3rd through 5th floors	1.05
Work on 6th through 9th floors	1.10
Work on 10th through 13th floors	1.15
Work on 14th through 17th floors	1.20
Work on 18th through 21st floors	1.25
Work over 21 floors	1.35
Work in cramped shafts	1.30
Work in commercial kitchens	1.10
Work above a sloped floor	1.25
Work in attic space	1.50
Work in crawl space	1.20
Work in a congested equipment room	1.20
Work 15' above floor level	1.10
Work 20' above floor level	1.20
Work 25' above floor level	1.30
Work 30' above floor level	1.40
Work 35' to 40' above floor level	1.50

Table 1 Recommended Correction Factors

 Column 3, insurance and employer-paid taxes in percent, shows the insurance and tax rate for the craft workers. The cost of insurance in this column includes workers' compensation and contractor's casualty and liability coverage. Insurance rates vary widely from state to state and depend on a contractor's loss experience. Note that taxes and insurance increase the hourly labor cost by approximately 30%. There is no legal way to avoid these costs.

Column 4, insurance and employer taxes in dollars, shows the hourly cost of taxes and insurance. Insurance and taxes are paid on the costs in both columns 1 and 2.

Column 5, non-taxable fringe benefits, includes employer paid non-taxable benefits such as medical coverage and tax-deferred pension and profit sharing plans. These fringe benefits average 4.88% of the base wage for many plumbing and HVAC contractors.

Column Number	1	2	3	4	5	6
Craft	Base wage per hour	Taxable fringe benefits (at 5.53% of base wage)	Insurance and employer taxes (%)	Insurance and employer taxes ($)	Non-taxable fringe benefits (at 4.88% of base wage)	Total hourly cost used in this book
Laborer	20.89	1.15	32.14%	7.08	1.02	30.14
Plumber	31.85	1.76	24.03%	8.08	1.56	43.25
Sheet Metal Worker	30.50	1.69	25.66%	8.26	1.49	41.94
Operating Engineer	31.09	1.72	24.87%	8.16	1.52	42.49
Sprinkler Fitter	31.30	1.73	24.79%	8.19	1.53	42.75
Electrician	30.81	1.70	19.67%	6.39	1.50	40.40
Cement Mason	26.26	1.45	22.92%	6.35	1.28	35.34

Craft Code	Crew Composition	Average Hourly Cost per Manhour
ER	4 building plumbers, 2 building laborers, 1 operating engineer	39.40
SN	4 building sheet metal workers, 2 building laborers, 1 operating engineer	38.65
P1	1 building plumber and 1 building laborer	36.70
ST	1 sprinkler fitter	42.75
SK	4 sprinkler fitters, 2 building laborers, 1 operating engineer	39.11
SL	1 sprinkler fitter and 1 laborer	36.45
S2	1 building sheet metal worker, 1 building laborer	36.04
BE	1 electrician	40.40
CF	1 cement mason	35.34
SW	1 sheet metal worker	41.94

Table 2 Labor Costs Used in This Book

The employer pays no taxes or insurance on these benefits.

Column 6, the total hourly cost in dollars, is the sum of columns 1, 2, 4, and 5. The labor costs in Column 6 were used to compute costs in the Labor $ column of this book.

Adjusting Costs in the Labor $ Column. The hourly labor costs used in this book may apply within a few percent on many of your jobs. But wage rates may be much higher or lower in some areas. If the hourly costs shown in Column 6 of Table 2 are not accurate for your work, adjust labor costs to fit your jobs.

For example, suppose your hourly labor costs are as follows:

Plumber	$19.00
Laborer	$16.00
Total hourly crew cost	$35.00

Your average cost per manhour would be $17.50 ($35.00 per crew hour divided by 2 because this is a crew of two).

A labor cost of $17.50 is about 48% of the $36.70 labor cost used for crew P1. Multiply costs in the Labor $ column by .477 to find your estimated cost.

For example, notice on page 19 that the labor cost for installing a 6 gallon hot water heater is $18.35 each. If installed by your plumbing crew working at an average cost of $17.50 per manhour, your estimated cost would be 48% of $18.35 or $8.75 per heater.

Adjusting the labor costs in this book will make your estimates much more accurate. Making adjustments to labor costs is both quick and easy if you use the National Estimator program.

Equipment Cost will vary according to need and application. It is typically $31.00 per day for a 2-ton chain hoist.

Material Costs in this manual are intended to reflect what medium- to low-volume contractors will be paying in 2020 after applying normal discounts. These costs include charges for delivery to within 25 to 30 miles of the supplier.

Overhead and Profit for the installing contractor are not included in the costs in this manual unless specifically identified in the text. Markup can vary widely with local economic conditions, competition and the installing contractor's operating expenses. Add the markup that's appropriate for your company, the job and the competitive environment.

How Accurate Are These Figures? As accurate as possible considering that the editors don't know your material suppliers, haven't seen the plans or specifications, don't know what building code applies or where the job is, had to project material costs at least six months into the future, and had no record of how much work the crew that will be assigned to the job can handle.

You wouldn't bid a job under those conditions. And I don't claim that all plumbing and HVAC work is done at these prices.

Estimating Is an Art, not a science. There is no one price that applies on all jobs. On many jobs the range between high and low bid will be 10% or more. There's room for legitimate disagreement on what the correct costs are, even when complete plans and specifications are available, the date and site are established, and labor and material costs are identical for all bidders.

No estimate fits all jobs. Good estimates are custom made for a particular project and a single contractor through judgment, analysis and experience. This book is not intended as a substitute for judgment, analysis and sound estimating practice. It's an aid in developing an informed opinion of cost, not an answer book.

Additional Costs to Consider

Here's a checklist of additional costs to consider before submitting any bid.

1. Sales taxes

2. Mobilization costs

3. Payment and performance bond costs

4. Permits and fees

5. Storage container rental costs

6. Utility costs

7. Tool costs

8. Callback costs during warranty period

9. Demobilization costs

Exclusions and Clarifications

Neither the job specifications nor the contract may identify exactly what work should be included in the plumbing and HVAC bid. Obviously, you have to identify what work is included in the job.

The most efficient way to define the scope of the work is to prepare a list of tasks not normally performed by your company and attach that list to each bid submitted. Here's a good list of work that should be excluded from your bid.

Your Bid Should Exclude

Final cleaning of plumbing fixtures

Backings for plumbing fixtures

Toilet room accessories

Electrical work, including motor starters

Electrical wiring and conduit over 100 volts

Temporary utilities

Painting, priming and surface preparation

Structural cutting, patching or repairing

Fire protection and landscape sprinklers

Equipment supports

Surveying and layout of control lines

Removal or stockpiling of excess soil

Concrete work, including forming and rebar

Setting of equipment furnished by others

Equipment, unless shown, and personnel hoisting

Wall and floor blockouts

Pitch pockets

The costs of performance or payment bonds

Site utilities

Asbestos removal or disposal

Contaminated soil removal or disposal

Major increases in copper material prices

Fire dampers not shown on the plans

Your Bid Should Include

Trash sweep-up only. Others haul it away

Site utilities from building to property line only

Piping to 5 feet outside the building only

Plumbing & HVAC permits for your work only

Beware of Price Changes

There's no way to be sure what prices will be in three to six months. All labor, equipment, material and subcontract prices in a bid should be based on costs anticipated when the project is expected to be built, not when the estimate is compiled. That presents a problem. Except for the installation of underground utilities, most plumbing and HVAC work is done six months to a year after the bid is submitted. When possible, get price protection in writing from your suppliers and subcontractors. If your suppliers and subs won't guarantee prices, include an escalation allowance in your bid to cover anticipated price increases.

Material Pricing Conditions

All equipment and material prices quoted by your vendors will be conditional. They usually don't include sales tax and are subject to specific payment and shipping terms. Every estimator should understand the meaning of common shipping terms. They define who pays the freight and who has responsibility for processing freight-damage claims. Here's a summary of important conditions you should understand.

F.O.B. Factory (Free On Board at the Factory): Title passes to the buyer when the goods are delivered by the seller to the freight carrier. The buyer pays the freight and is responsible for freight-damage claims.

F.O.B. Factory F.F.A. (Free On Board at the Factory, Full Freight Allowed): The title passes to the buyer when the goods are delivered by the seller to the freight carrier. The seller pays the freight charges, but the buyer is responsible for freight-damage claims.

F.O.B. (city of destination) (Free On Board to your city): The title passes to the buyer when the goods are delivered by the seller to the freight terminal in the city, or nearest city, of destination. The seller pays the freight and is responsible for freight-damage claims to the terminal. The buyer pays the freight charge and is responsible for freight-damage claims from the terminal to the final destination.

F.O.B. Job Site (Free On Board at job site, or contractor's shop): The title passes to the buyer when the goods are delivered to the job site (or shop). The seller pays the freight and is responsible for freight-damage claims.

F.A.S. Port [of a specific city] (Free Alongside Ship at the nearest port): The title passes to the buyer when goods are delivered to the ship dock or port terminal. The seller pays the freight and is responsible for freight-damage claims to the ship dock or port terminal only. The buyer pays the freight and is responsible for freight-damage claims from the ship dock or port terminal to the designated delivery point.

Obviously, it's to your advantage to instruct all vendors to quote costs F.O.B. the job site or your shop.

Reducing Costs

Most construction specifications allow the use of alternative equipment and materials. It's the estimator's responsibility to select the most cost-effective products. Research and compare your costs before making any decisions. Avoid selecting any material or equipment simply because that's what you've always done.

Don't recommend plastic products such as ABS, PVC, or polypropylene pipe or corrugated flexible ducts until you've checked local code requirements. Most building codes prohibit use of these materials inside public buildings such as schools, care centers and hospitals.

It's wise to select 100% factory-packaged equipment. Beware of equipment labeled "Some assembly required." Field labor costs for mounting loose coils, motors and similar equipment are very high.

Value Engineering

Let's suppose you've submitted a combined plumbing and HVAC bid for $233,000. Your cutthroat competitor put in a bid at $4,000 less, $229,000. Obviously there's no way you're going to get the job. Right?

Not so fast! Maybe value engineering can help you win that contract — while fattening your profit margin.

Suppose the proposal you submitted had two parts. Part I is the bid for $233,000, based entirely on job plans and specs, just the way they were written. But appended to your proposal is Part II, a list of suggestions for saving money without sacrificing any of the capacity or quality designed into the system. Here's an example of what might be in Part II:

1. Deduct for providing pipe hanger spacings per UPC in lieu of specified spacings: $1,750.00

2. Deduct for reducing heating hot water pipe sizes by using 40 degrees F Delta T in lieu of specified 20 degrees F Delta T: $4,600.00

3. Deduct for providing pressure/temperature taps at air handling units, pumps and chillers in lieu of specified thermometers and pressure gauges: $875.00

4. Deduct for eliminating water treatment in closed piping systems: $1,800.00

5. Deduct for piping chilled and heating hot water pumps in parallel in lieu of providing 100% standby pumps: $2,900.00

Total deductions: **$11,925.00**

Adopting these suggestions would make you low bidder by nearly $8,000. A saving like that will be tempting to most owners, especially if the owner understands that your suggestions result in a system that is every bit as good and maybe better than the system as originally designed.

You're not offering to undercut the competition. Far from it. You're using knowledge and experience to create better value for the owner. That's called value engineering and it's likely to win the respect of nearly all cost-conscious owners.

Notice that reducing costs is only part of what value engineering is all about. You don't cut costs at the expense of system quality, integrity, capacity or performance.

Don't waste your time, and your client's, by offering to substitute cheaper or lower-quality fixtures or equipment. Any cutthroat contractor with a price list can do that. Recommend the use of inferior materials and you'll be associated with the inferior goods you promote. Some owners consider even the suggestion to be insulting.

The recommendations you make (like most of those in the example) will require design changes. You can expect to be examined (or even challenged) on these points. Be ready to explain and defend each of your suggestions. Convince the client (or the design engineer) that your ideas are based on sound engineering principles and you're well on the way to winning the owner's confidence and the contract.

Now, let's go back to the list and see how we might justify the five value engineering recommendations.

1. **Pipe Hanger Spacing.** The pipe hanger spacings recommended in the Uniform Plumbing Code (UPC) are calculated by experienced, professional structural engineers. The safety factors used in these calculations are very conservative. They've been widely used for many years and have proved to be more than adequate. There's no need for more hangers than the UPC requires.

2. **Changing HHW Delta T.** In hydronic heating systems, heat measured in Btus is pumped to terminal units. The proposed change of the Delta T, from 20 degrees F to 40 degrees F, has no effect whatsoever on how many Btus the system delivers. You're not changing anything but the volume of water being pumped. At lower volume levels, the size of the pump, the pipe and the pipe insulation can all be reduced. Not one of these changes will affect the system's ability to transmit heat. Furthermore, operating costs will also drop, since less pump horsepower will be needed to run the smaller pump.

3. **Thermometers/Pressure Gauges.** Thermometers and pressure gauges installed on or near vibrating machinery have a very short life expectancy. Gauges quickly lose accuracy under harsh conditions. Readings will become less and less reliable. That's potentially dangerous. You can avoid this problem by using insertion-type pressure/temperature taps instead. Store these sensitive gauges in a desk drawer or a tool crib when not in use. Safely stored, they're protected from damage. They'll give accurate readings longer and won't need to be replaced as often. And they're simple to use. Just insert a gauge in one of the conveniently located taps. Make the reading, then remove the gauge and put it away.

4. **Water Treatment.** ITT Bell & Gossett has done studies on corrosion in closed hydronic systems that have a make-up water rate of no more than 5% per year. These studies show that corrosion virtually stops when entrained air is either removed or depleted. No water treatment is needed in this closed system.

5. **100% Standby Pumps.** Two pumps piped and operated in parallel are more economical. Even if one pump fails, the other pump can maintain delivery at 75 to 80% of the designed flow rate. That's usually adequate for emergency operation.

These cost-saving ideas are small, but could tip the balance in your favor. I hope they demonstrate the potential that value engineering has when bidding jobs. Any time you're compiling an estimate, keep an eye out for ways to save money or reduce the owner's cost. Jot a note to yourself about each potential saving you identify. Before submitting the bid, make a list of your alternate suggestions. Maybe best of all, markup on your value engineering suggestions can be higher than your normal markup. If value engineering can cut costs by $10,000, maybe as much as $4,000 of that should end up in your pocket!

Value Engineering: Surplus Materials

Value engineering doesn't begin and end with job plans and specs. Value engineering means getting the most value at the least cost, no matter whether it's value to the owner or value to the contractor. Smart mechanical contractors learn to build extra value into their jobs by controlling shrinkage of materials. Nearly every significant plumbing and HVAC job ends with at least some surplus material on hand. Material left over when the job is done tends to be discarded as waste or hauled off the job in the back of a truck that doesn't have your company name on the door. And why not? It's surplus — not needed. The owner didn't need it. So now it's up for grabs.

Not quite. Let's consider who actually owns that surplus material. When your company has been paid, every piece of material your crew installed belongs to the building owner. But what about those fittings, hangers and valves delivered to the job site but never actually used? Almost certainly, those materials were included in your bid. So aren't they the property of the owner? Not in my opinion. The owner contracted for a mechanical system and (presumably) has one. Unless it's a cost-plus job or a labor-only job, the owner didn't buy materials delivered to the job site. The owner bought a mechanical system and has one — completely separate and apart from any surplus materials. In my mind, the property owner has no more claim to left-over materials than the same owner would have claim to labor hours not expended or equipment not used on the same job.

Unless there's some provision in your contract to the contrary, surplus material belongs to the installing contractor. But your right to that material and the chance of actually getting it back to your shop are two very different propositions. I see recovery of surplus material as a training issue. As a matter of company policy, make it clear to your crews that surplus material belongs to your company. The supervisor on every job should be accountable for recovery of excess material. Every significant job will have at least some surplus. Accounting for that surplus should be part of your routine close-out procedure. Fortunately, it's not difficult. I'll explain.

Control of surplus materials begins with a good checklist, or form. I recommend the Materials, Equipment and Tool form, "MET" for short. A blank MET form appears following this section. Your MET should show both what's delivered to the job site (material, equipment and tools) and surplus "drops" returned to your shop at project close-out. A MET ensures that the estimator, the shop inventory manager and your field supervisor are on the same page. Your MET establishes accountability. Nothing falls through the cracks. Job input equals job output plus returns. Everything delivered to the job and not expended should be returned to your shop.

Here's how it works:

1. Based on the estimate that won you the job, the items needed are purchased for the job and staged for delivery to the job site.

2. As materials, equipment and tools are delivered to the job site, your supervisor completes the first three columns of the MET form: Description, Quantity and Date.

3. As work is completed, the same supervisor completes the four columns under Returned to Inventory: Quantity Returned, Date, Status Code and Value. The status code will be either "RS" (Returned and Salvaged) or "RN" (Returned New).

4. Back at your shop, both RS and RN materials should be restored to inventory.

5. If your company has an inventory manager, have that manager assign the return value to each item returned. If you're using QuickBooks Pro, the "Adjust Inventory" feature can handle this task quite easily. Add two new categories under "Inventory Stock on Hand by Vendor." The first new category is Returned Salvage. The second is Returned New. Be sure the value of RS materials includes the cost of any reconditioning done to restore salvaged materials (such as pumps and boilers) to serviceable condition.

6. Comparing MET deployed to the job site with MET returned to inventory yields MET actually used on the job. That's a very important number to every plumbing and HVAC estimator. Be sure actual usage gets entered on the Project Summary form.

7. When the take-off on your next estimate is complete, compare that materials list with a summary of RS and RN materials on hand from prior jobs.

8. Evaluate which returned materials can be redeployed on the new job.

9. It's a management decision to either (1) charge the new job for the cost of RS and RN materials already on hand, or (2) consider materials on hand as "free" and a competitive advantage in winning the new bid. Either way, RN and RS materials are an asset to your company.

Plumbing and HVAC materials are expensive. Every mechanical contractor has an interest in MET tracking. Everyone in your company should be aware of the need for good materials management. Used correctly, the MET form in this book can help engineer more value into your jobs.

Maximizing the Value of Old Estimates

There should be two profits in every job. The first is money in the bank — a return on time and expenses. The second is what you learn from the job — primarily by comparing the estimate you made with what turns out to be your actual cost. On some jobs, the value of lessons learned may outweigh net revenue.

Every plumbing and HVAC contractor has marginal jobs. That's normal. What *shouldn't* be normal is repeating mistakes. The best way to avoid trouble in your future is to keep track of your past. Keeping old estimates available for reference can help prevent errors on new estimates.

As your file of completed estimates grows, organization becomes more important. You need an easy way to find similar projects with the same components and comparable scope of work. If your estimating file is in QuickBooks Pro, searching by keyword may be enough. Otherwise, I recommend creating a short summary for each completed job, and an index that references all summaries available for comparison. You'll find a blank Project Summary form at the end of this section. To make reference easier, create an index by type of job and equipment used. You may choose to use an alphabetical index based on client name or project ID.

How to complete the Project Summary form is obvious. The many ways to use this form may not be so obvious, so here are a few pointers.

1. Use your index of Project Summary forms to find completed jobs most similar to the job you're bidding. Believe it or not, Project Summary forms with the widest margin of error will be most useful. Ask yourself: Who worked on those projects? Who was the field superintendent? Who were the vendors? Did the errors result from poor estimating or the poor performance of vendors, supervisors or crews? The most common estimating errors occur when (a) inspecting the job site, (b) examining the plans or (c) reading the specifications. What did you miss and why? Look for pitfalls to avoid in the job now being estimated. Identify the biggest two or three mistakes made when bidding that job. Make a notation about each on the Project Summary form.

2. Now look at your bid for the current job. Which mistakes made on a prior job might you expect on this job? Concentrate on the big three oversights to avoid: Inspecting the job site; examining the plans; and reading the specifications.

3. Unless there's a major error in take-off, your estimate of material costs should be within about 5 percent of the actual costs of materials. However, it's common for labor cost estimates to vary 20 percent or more from actual labor costs. This is precisely where data from old jobs comes in handy. If your Project Summary files show that some project types are consistent money-losers, either shift your company's focus to another class of work, factor more contingency into your bids, or find some way to wring inefficiencies out of the labor component. Poor staging, delivery and retrieval procedures drag down labor productivity on any job.

4. Use your file of Project Summary forms to spot any common thread that runs through either money-making jobs or money-losing jobs. For example, if the names of certain subcontractors or vendors are prominent on low-margin jobs, maybe there's a relationship between your profit margin and choice of subs and suppliers. Even the best and most reliable vendors can become complacent if not challenged occasionally.

5. Project Summary forms should note changes and extras identified after the contract was signed — both for which your company was paid and changes done without additional compensation. Projects with changes and extras that exceed about 4 percent of the contract price deserve special scrutiny. Jobs with changes beyond about 4 percent aren't good for business, at least in my opinion. Nearly all changes have a negative impact on your job schedule and require a disproportionate investment of management resources. Too many changes can antagonize the owner and design staff, even if they were responsible for the altered plans. You may know of a mechanical contractor with a reputation for capitalizing on change orders. But I've rarely seen a job plagued with changes that turned into a money-maker for anyone — except the attorneys. Your file of Project Summary forms will show job types that carry change order risk. Before finalizing and submitting any bid, consider whether the job will get mired in disputes over changes and extras. If similar jobs have ended on the courthouse steps, factor that risk into your estimate.

Utility of a Project Summary forms file is limited only by your ingenuity. The important point is to keep and organize the source of your second profit available on every job. What you learn can be more valuable than what you earn.

The Estimating Procedure

Every plumbing and HVAC estimator works under deadline pressure. You'll seldom have the luxury of spending as much time as you would like on an estimate. Estimators who aren't organized waste valuable time and tend to make careless errors. Try to be well-organized and consistent in your approach to estimating. For most projects, I recommend that you follow the procedures listed below and in the order listed:

1. Get a second set of project drawings and specifications for use by your suppliers and subcontractors. Remember that your subs and suppliers need access to the plans and specs and time to prepare their quotes.

2. Study the plans and specs carefully. Highlight important items. Make a list of specific tasks that require labor unit correction factors. The estimate is never complete until you're totally familiar with the project and the applicable construction codes.

3. Get the general contractor or owner to identify the proposed construction schedule and subcontractor lay-down (storage) area. Work schedule and site conditions always affect your costs.

4. Contact all potential suppliers and subcontractors as early as possible. Set a time when each can come to your office to make their take-offs from the spare set of contract documents.

When this important preliminary work is done, or in progress, it's time to begin your detailed take-off.

Guidelines for Good Estimating

You can compile estimates on a legal pad, a printed estimating form or on a computer. Regardless of the method, these guidelines will apply:

List Each Cost Separately on your take-off sheet. Don't combine system estimates, even if the materials are the same type. A combined system estimate may have to be completely redone if materials for one system are changed at a later date. Use the Estimate Detail Sheet on page 16 if you don't already have a good material take-off form.

Use Engineer's Identification Numbers when listing equipment. The word pump without any other description is ambiguous when there are several pumps included in the project.

Don't Forget Labor Adjustment factors if your labor costs are significantly higher or lower than the costs used in this book. See instructions on page 7 for adjusting labor costs.

Use Colored Pencils or highlighters to mark the items you've taken off and listed. Use a different color for each piping or ducting system.

Log Telephone Quotes and other important phone conversations on a telephone quote form. See the sample on page 18.

Project Estimated Costs for labor, material and equipment to the time when the work is expected to be done, not when the job is being estimated.

The only good estimate is a complete estimate. You've probably heard this saying, "He who makes the most mistakes is likely to be low bidder, and live to regret it."

Preparing the Proposal

It's both common courtesy and good business practice to deliver an unpriced copy of your bid or proposal letter to the general contractor three or four days before the bid deadline date. This gives the contractor time to study your proposal and obtain alternate pricing for items you may have excluded. To avoid misunderstandings, make sure your proposals include, as a minimum, the following elements:

1. The complete name and address of the proposed project.
2. Specification title and issue date.
3. A complete listing of drawings and their issue or revision date.
4. A complete list of addenda and their dates of issue.
5. A list of specification section numbers covered by your proposal.
6. A list of exclusions, clarifications and assumptions.

Your final bid can be phoned in or sent by fax, but it should reach the general contractor or owner no more than five or ten minutes before the bid deadline. Prices submitted too early may have to be revised because of last-minute price changes by subcontractors or suppliers.

MET Worksheet
Material, Equipment and Tool Delivery and Surplus Return Record

Project ID _____ Job Location _____

Supervisor _____ Start Date _____

Description of Material, Equipment or Tool Delivered or Returned	Delivered to Job Site		Returned to Inventory			
	Quantity Delivered	Date Delivered	Quantity Returned	Date Returned	Status Code RN or RS	Value at Return

PROJECT SUMMARY

Project ID _____

Job Location _____

Short description _____

Supervisor _____

Index ID _____

Start Date _____

Estimator _____

Client _____

Major vendors _____

Subcontractors _____

Sources of cost deviation _____

_____

_____

Related Projects by ID Number _____

Thumbnail Summary	Labor	Material	Equipment	Subcontract	Deployed RN/RS	Total
Actual cost						
Estimate						
Over/(Under)						
Full Summary						
Bid amount						
Estimated cost						
Projected profit						
Cost overrun						
Bid profit						
Change orders						
Cost of changes						
Total profit						
Total profit with RN/RS						
Redeployment						

Estimate Detail Sheet

Data carried forward from Take-Off Quantity Survey Sheet(s)

Company/Department _____ Estimator _____ Date _____

Project _____ Checked by _____ Date _____

Address _____ Notes: _____

Job description _____ Estimate # _____

CSI Division/Account _____ Estimate due _____

Item Description	Quantity	Unit	Crew @ MH/Unit	Manhours Ext.	Materials Unit $	Materials Ext. $	Labor Unit $	Labor Ext. $	Equipment Unit $	Equipment Ext. $	Subcontract Unit $	Subcontract Ext. $	Total $
Totals This Sheet				Manhours	Material $		Labor $		Equipment $		Subcontract $		Total $

Carry totals forward to Estimate Summary Sheet   Estimate # _____   Estimate Detail Sheet _____ of _____

Quotation Sheet

Job: _____

Supplier: _____

Salesperson: _____ **Phone No:** _____

Per Plans/Specs: _____ **Freight:** _____ **Terms:** _____

Description	Delivery Time	Price

By: _____

Record of Telephone Conversation

Date:_____ Time:_____ Project: _____

Telecon with:_____

Company: _____ Phone No:_____

Subject: _____

Details of Conversation: _____

_____

_____

_____

_____

_____

_____

_____

_____

_____

_____

_____

_____

_____

_____

_____

_____

_____

_____

_____

_____

_____

_____

By:_____

Description	Craft@Hrs	Unit	Material $	Labor $	Equipment $	Total $

Electric domestic hot water heater (residential). Set in place only (floor models). Make additional allowances for pipe and electrical connections. (See below)

Description	Craft@Hrs	Unit	Material $	Labor $	Equipment $	Total $
6 gallon						
1.5 KW/110V	P1@.500	Ea	422.00	18.40	—	440.40
10 gallon						
1.5 KW/110V	P1@.500	Ea	472.00	18.40	—	490.40
15 gallon						
1.5 KW/110V	P1@.750	Ea	497.00	27.50	—	524.50
20 gallon						
1.5 KW/110V	P1@.750	Ea	513.00	27.50	—	540.50
30 gallon						
1.5 KW/110V	P1@1.00	Ea	457.00	36.70	—	493.70
40 gallon						
1.5 KW/110V	P1@1.20	Ea	479.00	44.00	—	523.00
50 gallon						
3 KW/110V	P1@1.30	Ea	516.00	47.70	—	563.70
12 gallon						
3 KW/220V	P1@.500	Ea	416.00	18.40	—	434.40
20 gallon						
3 KW/220V	P1@.750	Ea	455.00	27.50	—	482.50
30 gallon						
3 KW/220V	P1@1.00	Ea	497.00	36.70	—	533.70
40 gallon						
3 KW/220V	P1@1.20	Ea	541.00	44.00	—	585.00
50 gallon						
3 KW/220V	P1@1.30	Ea	579.00	47.70	—	626.70

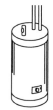

Electric domestic hot water heater (commercial), 208/240 volt. Set in place only. Make additional allowances for pipe and electrical connections. (See below)

Description	Craft@Hrs	Unit	Material $	Labor $	Equipment $	Total $
96 gallon, 12 kw	P1@1.50	Ea	2,390.00	55.10	—	2,445.10
96 gallon, 18 kw	P1@1.50	Ea	3,240.00	55.10	—	3,295.10
96 gallon, 36 kw	P1@1.50	Ea	3,360.00	55.10	—	3,415.10
120 gallon, 18 kw	P1@2.00	Ea	3,440.00	73.40	—	3,513.40
120 gallon, 36 kw	P1@2.00	Ea	3,550.00	73.40	—	3,623.40
120 gallon, 54 kw	P1@2.00	Ea	4,200.00	73.40	—	4,273.40
120 gallon, 63 kw	P1@2.00	Ea	4,530.00	73.40	—	4,603.40

Gas-fired domestic hot water heater (residential). Set in place only, Make additional allowances for pipe and combustion venting connections. (See below)

Description	Craft@Hrs	Unit	Material $	Labor $	Equipment $	Total $
30 gallon	P1@1.00	Ea	490.00	36.70	—	526.70
40 gallon	P1@1.00	Ea	792.00	36.70	—	828.70
50 gallon	P1@1.50	Ea	900.00	55.10	—	955.10

Domestic Hot Water Heaters

Description	Craft@Hrs	Unit	Material $	Labor $	Equipment $	Total $

Gas-fired domestic hot water heater (commercial), standard efficiency. Set in place only, Make additional allowances for pipe and combustion venting connections. (See below)

Description	Craft@Hrs	Unit	Material $	Labor $	Equipment $	Total $
50 gal./ 95 gph	P1@2.00	Ea	2,380.00	73.40	—	2,453.40
67 gal./106 gph	P1@2.00	Ea	2,810.00	73.40	—	2,883.40
76 gal./175 gph	P1@2.00	Ea	3,770.00	73.40	—	3,843.40
91 gal./291 gph	P1@2.00	Ea	4,550.00	73.40	—	4,623.40

Gas-fired domestic hot water heater (commercial), energy miser. Set in place only, Make additional allowances for pipe and combustion venting connections. (See below)

Description	Craft@Hrs	Unit	Material $	Labor $	Equipment $	Total $
50 gal./ 95 gph	P1@2.00	Ea	5,600.00	73.40	—	5,673.40
67 gal./106 gph	P1@2.00	Ea	5,860.00	73.40	—	5,933.40
76 gal./175 gph	P1@2.00	Ea	7,260.00	73.40	—	7,333.40
91 gal./291 gph	P1@2.00	Ea	8,630.00	73.40	—	8,703.40

Tankless natural gas water heaters. Ambient pressure. DOE and UL rated. For residential, multi-dwelling and light commercial potable water applications. Add the cost of piping, tempering valve, circulating pump, controls, and electrical connection, post-installation inspection by both the fire marshal and the mechanical inspector to validate federal, state and local energy tax credits or energy tax credit offsets. For larger arrays (laundries, institutional facilities, food processing plants), develop an estimate based on the required capacity and multiply these costs by the number of heaters required. Rated in Btus and gallons per minute capacity. (1 Mbh = 1,000 Btus)

Description	Craft@Hrs	Unit	Material $	Labor $	Equipment $	Total $
19.5-140 Mbh, .75-5.8 Gpm	P1@16.0	Ea	1,880.00	587.00	—	2,467.00
11-199 Mbh, .5-7 Gpm	P1@20.0	Ea	2,220.00	734.00	—	2,954.00
25-235 Mbh, .75-9.6 Gpm	P1@20.0	Ea	2,220.00	734.00	—	2,954.00

Tankless electric point-of-use water heaters. Ambient pressure, DOE and UL rated. For residential, multi-dwelling and light commercial potable water applications. Cost does not include piping, tempering valve, circulating pump, controls, storage tank, electrical connection. Add the cost of post-installation inspection by the mechanical inspector to validate federal, state and local energy tax credits or energy tax credit offsets. In rated gallons per minute capacity.

Description	Craft@Hrs	Unit	Material $	Labor $	Equipment $	Total $
5.5 Kw/40 Amp, .75-2 Gpm	P1@4.00	Ea	475.00	147.00	—	622.00
9.5 Kw/50 Amp, .75-2.5 Gpm	P1@4.25	Ea	563.00	156.00	—	719.00
19 Kw/100 Amp, 1-3.5 Gpm	P1@4.50	Ea	936.00	165.00	—	1,101.00
28 Kw/120 Amp, 1.5-5 Gpm	P1@4.75	Ea	1,710.00	174.00	—	1,884.00

Description	Craft@Hrs	Unit	Material $	Labor $	Equipment $	Total $

Domestic hot water heater connection assembly. Includes supply, return, recirculation and relief piping and fittings (copper), relief and isolation valves. Make additional allowances for gas and venting connections where applicable.

Description	Craft@Hrs	Unit	Material $	Labor $	Equipment $	Total $
¾" residential	P1@1.75	Ea	313.00	64.20	—	377.20
¾" commercial	P1@2.25	Ea	420.00	82.60	—	502.60
1" commercial	P1@2.75	Ea	736.00	101.00	—	837.00
1¼" commercial	P1@3.50	Ea	901.00	128.00	—	1,029.00
1½" commercial	P1@3.75	Ea	938.00	138.00	—	1,076.00
2" commercial	P1@4.50	Ea	1,000.00	165.00	—	1,165.00
2½" commercial	P1@5.75	Ea	2,080.00	211.00	—	2,291.00
3" commercial	P1@6.50	Ea	3,190.00	239.00	—	3,429.00

Domestic water heater combustion vent connection. Make additional allowances for piping distances greater than 25'.

Description	Craft@Hrs	Unit	Material $	Labor $	Equipment $	Total $
2" B-vent	P1@.090	LF	6.47	3.30	—	9.77
3" B-vent	P1@.100	LF	7.99	3.67	—	11.66
4" B-vent	P1@.110	LF	10.60	4.04	—	14.64
6" B-vent	P1@.130	LF	13.00	4.77	—	17.77
Tankless heater vent kit	P1@2.50	Ea	454.00	91.80	—	545.80
Power vent kit	P1@2.00	Ea	1,160.00	73.40	—	1,233.40

Water Softeners and Controllers

Description	Craft@Hrs	Unit	Material $	Labor $	Equipment $	Total $

Water softener, time clock controller. Including brine tank, brine well and pick-up tube. Labor includes setting in place, connecting the unit to an existing domestic water distribution system, start up and testing.

Description	Craft@Hrs	Unit	Material $	Labor $	Equipment $	Total $
20,000 grain water softener, TCC	P1@4.50	Ea	587.00	165.00	—	752.00
30,000 grain water softener, TCC	P1@4.50	Ea	626.00	165.00	—	791.00
45,000 grain water softener, TCC	P1@4.50	Ea	696.00	165.00	—	861.00
50,000 grain water softener, TCC	P1@4.75	Ea	785.00	174.00	—	959.00
60,000 grain water softener, TCC	P1@4.75	Ea	928.00	174.00	—	1,102.00
75,000 grain water softener, TCC	P1@5.00	Ea	996.00	184.00	—	1,180.00
90,000 grain water softener, TCC	P1@5.50	Ea	1,350.00	202.00	—	1,552.00
120,000 grain water softener, TCC	P1@5.75	Ea	1,450.00	211.00	—	1,661.00

Water softener, mechanically-metered controller. Including brine tank, brine well and pick up tube. Labor includes setting in place, connecting the unit to an existing domestic water distribution system, start up and testing.

Description	Craft@Hrs	Unit	Material $	Labor $	Equipment $	Total $
20,000 grain water softener, MMC	P1@4.50	Ea	763.00	165.00	—	928.00
30,000 grain water softener, MMC	P1@4.50	Ea	797.00	165.00	—	962.00
45,000 grain water softener, MMC	P1@4.50	Ea	867.00	165.00	—	1,032.00
50,000 grain water softener, MMC	P1@4.75	Ea	955.00	174.00	—	1,129.00
60,000 grain water softener, MMC	P1@4.75	Ea	1,120.00	174.00	—	1,294.00
75,000 grain water softener, MMC	P1@5.00	Ea	1,190.00	184.00	—	1,374.00
90,000 grain water softener, MMC	P1@5.50	Ea	1,530.00	202.00	—	1,732.00
120,000 grain water softener, MMC	P1@5.75	Ea	1,630.00	211.00	—	1,841.00

Description	Craft@Hrs	Unit	Material $	Labor $	Equipment $	Total $

Water softener, electronically-metered controller. Including brine tank, brine well and pick up tube. Labor includes setting in place, connecting the unit to an existing domestic water distribution system, start up and testing.

Description	Craft@Hrs	Unit	Material $	Labor $	Equipment $	Total $
20,000 grain water softener, EMC	P1@4.50	Ea	809.00	165.00	—	974.00
30,000 grain water softener, EMC	P1@4.50	Ea	833.00	165.00	—	998.00
45,000 grain water softener, EMC	P1@4.50	Ea	914.00	165.00	—	1,079.00
50,000 grain water softener, EMC	P1@4.75	Ea	1,000.00	174.00	—	1,174.00
60,000 grain water softener, EMC	P1@4.75	Ea	1,170.00	174.00	—	1,344.00
75,000 grain water softener, EMC	P1@5.00	Ea	1,230.00	184.00	—	1,414.00
90,000 grain water softener, EMC	P1@5.50	Ea	1,570.00	202.00	—	1,772.00
120,000 grain water softener, EMC	P1@5.75	Ea	1,680.00	211.00	—	1,891.00

Water softener accessories

Description	Craft@Hrs	Unit	Material $	Labor $	Equipment $	Total $
By-pass valve	P1@.400	Ea	77.90	14.70	—	92.60
Manifold adapter kit	P1@.200	Ea	21.00	7.34	—	28.34
Turbulator	P1@.400	Ea	38.40	14.70	—	53.10

Iron filter, electronically-metered controller. Manganese green sand filter.
Labor includes setting in place, connecting the unit to an existing domestic water distribution system, start-up and testing.

Description	Craft@Hrs	Unit	Material $	Labor $	Equipment $	Total $
42,000 iron filter (1.5 cf media), 5 gpm	P1@4.00	Ea	765.00	147.00	—	912.00
65,000 iron filter (2.0 cf media), 6 gpm	P1@4.50	Ea	905.00	165.00	—	1,070.00
84,000 iron filter (2.5 cf media), 8 gpm	P1@4.75	Ea	966.00	174.00	—	1,140.00
Replacement green sand media	P1@1.20	CF	44.30	44.00	—	88.30

Iron filter accessories

Description	Craft@Hrs	Unit	Material $	Labor $	Equipment $	Total $
By-pass valve	P1@.400	Ea	77.90	14.70	—	92.60
Air vent	P1@.200	Ea	61.80	7.34	—	69.14
Air controller	P1@.400	Ea	69.80	14.70	—	84.50

Water Softener Accessories

Description	Craft@Hrs	Unit	Material $	Labor $	Equipment $	Total $

Combination iron filter/water softener. Zeolite resins soften water and remove iron and manganese. Controller automatically controls PH level. Labor includes setting in place, connecting the unit to an existing domestic water distribution system, start-up and testing.

Description	Craft@Hrs	Unit	Material $	Labor $	Equipment $	Total $
40,000 iron filter, 1.3 cf media	P1@4.00	Ea	1,480.00	147.00	—	1,627.00
60,000 iron filter, 1.7 cf media	P1@4.50	Ea	1,600.00	165.00	—	1,765.00
80,000 iron filter, 2.5 cf media	P1@4.75	Ea	2,320.00	174.00	—	2,494.00

Hot water softener, time clock controller. Brass valve construction. Designed for 150 F. maximum operating temperature. Includes brine tank, brine well and pick-up tube. Labor includes setting in place, connecting the unit to an existing domestic water distribution system, start-up and testing.

Description	Craft@Hrs	Unit	Material $	Labor $	Equipment $	Total $
20,000 grain hot water softener	P1@4.50	Ea	1,860.00	165.00	—	2,025.00
30,000 grain hot water softener	P1@4.50	Ea	1,970.00	165.00	—	2,135.00
40,000 grain hot water softener	P1@4.50	Ea	2,060.00	165.00	—	2,225.00
60,000 grain hot water softener	P1@4.75	Ea	2,440.00	174.00	—	2,614.00

Pressure tank, fiberglass wound. Labor includes setting in place, connecting the tank to a domestic water distribution system and testing.

Description	Craft@Hrs	Unit	Material $	Labor $	Equipment $	Total $
Fiberglass pressure tank, 20 gallon	P1@2.00	Ea	255.00	73.40	—	328.40
Fiberglass pressure tank, 30 gallon	P1@2.00	Ea	287.00	73.40	—	360.40
Fiberglass pressure tank, 80 gallon	P1@2.75	Ea	465.00	101.00	—	566.00
Fiberglass pressure tank, 120 gallon	P1@3.50	Ea	613.00	128.00	—	741.00
Brass tank tee assembly, ¾"	P1@3.50	Ea	35.70	128.00	—	163.70
Brass tank tee assembly, 1"	P1@3.50	Ea	66.50	128.00	—	194.50
Brass tank tee assembly, 1¼"	P1@3.50	Ea	113.00	128.00	—	241.00

Description	Craft@Hrs	Unit	Material $	Labor $	Equipment $	Total $

Ultra-violet water disinfection unit. Stainless steel reactor, audible and visible alarm, lamp end-of-life indicator and 7-day override. Gpm rating at 30,000 mj/m2 output. Labor includes setting in place, connecting to the water system and testing.

Description	Craft@Hrs	Unit	Material $	Labor $	Equipment $	Total $
UV system, 1 gpm, ¼" in/out	P1@3.00	Ea	246.00	110.00	—	356.00
UV system, 6 gpm, ½" in/out	P1@3.00	Ea	478.00	110.00	—	588.00
UV system, 8 gpm, ¾" in/out	P1@4.00	Ea	553.00	147.00	—	700.00
UV system, 12 gpm, ¾" in/out	P1@4.00	Ea	709.00	147.00	—	856.00
UV replacement lamp, 20W, 1 gpm	P1@.750	Ea	55.00	27.50	—	82.50
UV replacement lamp, 32W, 6 gpm	P1@.750	Ea	62.40	27.50	—	89.90
UV replacement lamp, 39W, 8-12 gpm	P1@.750	Ea	79.90	27.50	—	107.40
UV replacement ballast, 420 Mv/110V	P1@1.00	Ea	241.00	36.70	—	277.70

Kitchen equipment booster heater

Description	Craft@Hrs	Unit	Material $	Labor $	Equipment $	Total $
1,000 watt	P1@4.00	Ea	609.00	147.00	—	756.00

Dishwasher

Description	Craft@Hrs	Unit	Material $	Labor $	Equipment $	Total $
Built-in	P1@5.00	Ea	907.00	184.00	—	1,091.00

Garbage disposal

Description	Craft@Hrs	Unit	Material $	Labor $	Equipment $	Total $
½ HP	P1@2.00	Ea	181.00	73.40	—	254.40
¾ HP	P1@2.00	Ea	302.00	73.40	—	375.40

Grease and oil interceptor

Description	Craft@Hrs	Unit	Material $	Labor $	Equipment $	Total $
4 GPM	P1@4.00	Ea	337.00	147.00	—	484.00
10 GPM	P1@5.00	Ea	549.00	184.00	—	733.00
15 GPM	P1@7.00	Ea	820.00	257.00	—	1,077.00
20 GPM	P1@8.00	Ea	992.00	294.00	—	1,286.00

Hair and lint interceptor

Description	Craft@Hrs	Unit	Material $	Labor $	Equipment $	Total $
1½"	P1@.650	Ea	211.00	23.90	—	234.90
2"	P1@.750	Ea	301.00	27.50	—	328.50

All bronze ¾" to 1½" in-line NPT pump

Description	Craft@Hrs	Unit	Material $	Labor $	Equipment $	Total $
1/12 HP	P1@1.50	Ea	551.00	55.10	—	606.10
1/6 HP	P1@1.50	Ea	823.00	55.10	—	878.10
1/4 HP	P1@1.50	Ea	965.00	55.10	—	1,020.10

Kitchen Equipment Connections

Description	Craft@Hrs	Unit	Material $	Labor $	Equipment $	Total $
Kitchen appliance gas trim						
½"	P1@1.15	Ea	40.60	42.20	—	82.80
¾"	P1@1.30	Ea	74.30	47.70	—	122.00
1"	P1@1.60	Ea	86.10	58.70	—	144.80
1¼"	P1@2.10	Ea	142.00	77.10	—	219.10
1½"	P1@2.50	Ea	180.00	91.80	—	271.80
2"	P1@3.00	Ea	240.00	110.00	—	350.00
Hot and cold water supply						
½"	P1@1.10	Ea	40.30	40.40	—	80.70
¾"	P1@1.55	Ea	57.00	56.90	—	113.90
1"	P1@1.90	Ea	77.70	69.70	—	147.40
1¼"	P1@2.50	Ea	109.00	91.80	—	200.80
1½"	P1@3.00	Ea	136.00	110.00	—	246.00
Continuous waste						
2-part	P1@.250	Ea	60.80	9.18	—	69.98
3-part	P1@.350	Ea	103.00	12.80	—	115.80
4-part	P1@.450	Ea	132.00	16.50	—	148.50
Indirect waste						
½"	P1@1.05	Ea	13.00	38.50	—	51.50
¾"	P1@1.50	Ea	22.00	55.10	—	77.10
1"	P1@1.90	Ea	35.30	69.70	—	105.00
1¼"	P1@2.15	Ea	52.20	78.90	—	131.10
1½"	P1@2.60	Ea	68.80	95.40	—	164.20
2"	P1@3.00	Ea	105.00	110.00	—	215.00
Kitchen fixture waste tailpiece						
1½"	P1@.100	Ea	13.10	3.67	—	16.77
Kitchen fixture trap with solder bushing						
1½"	P1@.250	Ea	44.00	9.18	—	53.18
2"	P1@.300	Ea	61.00	11.00	—	72.00

Description	Craft@Hrs	Unit	Material $	Labor $	Equipment $	Total $

Water closet, floor-mounted, flush tank, white vitreous china, lined tank. Complete with trim. Make additional allowances for rough-in. Based on American Standard Cadet series. ADA means American Disabilities Act compliant. (Wheelchair accessible)

Description	Craft@Hrs	Unit	Material $	Labor $	Equipment $	Total $
Round bowl	P1@2.10	Ea	265.00	77.10	—	342.10
Elongated bowl	P1@2.10	Ea	320.00	77.10	—	397.10
ADA, 18" high	P1@2.10	Ea	436.00	77.10	—	513.10

Water closet, floor-mounted, flush valve, white vitreous china. Complete with trim. Make additional allowances for rough-in. Based on American Standard. ADA means American Disabilities Act compliant. (Wheelchair accessible)

Description	Craft@Hrs	Unit	Material $	Labor $	Equipment $	Total $
Elongated bowl	P1@2.60	Ea	415.00	95.40	—	510.40
Elongated bowl, ADA 18" high	P1@2.60	Ea	493.00	95.40	—	588.40
Elongated bowl with a bedpan cleanser	P1@4.10	Ea	717.00	150.00	—	867.00
Elongated bowl, ADA 18" high with a bedpan cleanser	P1@4.10	Ea	776.00	150.00	—	926.00

Water closet, wall-hung, flush valve, white vitreous china. Complete with fixture carrier and all trim. Make additional allowances for rough-in. Based on American Standard Afwall series.

Description	Craft@Hrs	Unit	Material $	Labor $	Equipment $	Total $
Elongated bowl	P1@3.55	Ea	658.00	130.00	—	788.00
Elongated bowl with electronic flush valve	P1@3.80	Ea	1,180.00	139.00	—	1,319.00
Elongated bowl with bedpan cleanser	P1@5.05	Ea	953.00	185.00	—	1,138.00
Electronic flush valve, add	P1@.600	Ea	519.00	22.00	—	541.00

Urinal, wall-hung, flush valve, white vitreous china. Complete with trim. Make additional allowances for rough-in.

Description	Craft@Hrs	Unit	Material $	Labor $	Equipment $	Total $
Siphon-jet type	P1@3.15	Ea	658.00	116.00	—	774.00
Wash-out type	P1@3.10	Ea	537.00	114.00	—	651.00
Wash-down type	P1@3.00	Ea	379.00	110.00	—	489.00
Urinal carrier, add	P1@.600	Ea	120.00	22.00	—	142.00
Electronic flush valve, add	P1@.600	Ea	428.00	22.00	—	450.00

Urinal, stall-type, flush valve, white vitreous china. Complete with trim. Make additional allowances for rough-in.

Description	Craft@Hrs	Unit	Material $	Labor $	Equipment $	Total $
Stall urinal	P1@5.00	Ea	1,280.00	184.00	—	1,464.00

Plumbing Fixtures

Description	Craft@Hrs	Unit	Material $	Labor $	Equipment $	Total $

Lavatory, wall-hung, white vitreous china. Complete with trim and fixture carrier. Make additional allowances for rough-in. ADA means American Disabilities Act compliant. (Wheelchair accessible)

Description	Craft@Hrs	Unit	Material $	Labor $	Equipment $	Total $
Wall-hung lav	P1@2.70	Ea	550.00	99.10	—	649.10
Wall-hung, ADA	P1@2.70	Ea	801.00	99.10	—	900.10
Fixture carrier	P1@.500	Ea	114.00	18.40	—	132.40

Countertop lavatory, white. Complete with trim. Make additional allowances for rough-in.

Description	Craft@Hrs	Unit	Material $	Labor $	Equipment $	Total $
Vitreous china	P1@2.00	Ea	407.00	73.40	—	480.40
Enameled steel	P1@2.00	Ea	344.00	73.40	—	417.40
Acrylic	P1@2.00	Ea	250.00	73.40	—	323.40

Bathtub, white, 60" x 32". Complete with trim, including shower head. Make additional allowances for rough-in.

Description	Craft@Hrs	Unit	Material $	Labor $	Equipment $	Total $
Enameled steel	P1@2.50	Ea	573.00	91.80	—	664.80
Cast iron	P1@3.50	Ea	828.00	128.00	—	956.00
Fiberglass	P1@2.50	Ea	529.00	91.80	—	620.80
Acrylic	P1@2.50	Ea	566.00	91.80	—	657.80

Tub and shower combination, fiberglass, white. Complete with trim. Make additional allowances for rough-in.

Description	Craft@Hrs	Unit	Material $	Labor $	Equipment $	Total $
One-piece	P1@4.50	Ea	1,180.00	165.00	—	1,345.00
Two-piece (reno)	P1@5.50	Ea	1,510.00	202.00	—	1,712.00
Four-piece (reno)	P1@6.25	Ea	1,600.00	229.00	—	1,829.00

Shower stall, white, 36" x 36". Complete with trim. Make additional allowances for rough-in.

Description	Craft@Hrs	Unit	Material $	Labor $	Equipment $	Total $
Fiberglass one-piece	P1@3.50	Ea	732.00	128.00	—	860.00
Fiberglass three-piece	P1@4.25	Ea	944.00	156.00	—	1,100.00
Acrylic one-piece	P1@3.50	Ea	1,100.00	128.00	—	1,228.00
Acrylic three-piece	P1@4.25	Ea	1,440.00	156.00	—	1,596.00

Shower basin, 36" x 36". Complete with trim (faucet, shower head and strainer). Make additional allowances for rough-in.

Description	Craft@Hrs	Unit	Material $	Labor $	Equipment $	Total $
Fiberglass	P1@2.50	Ea	499.00	91.80	—	590.80
Acrylic	P1@2.50	Ea	537.00	91.80	—	628.80
Molded stone	P1@2.65	Ea	519.00	97.30	—	616.30

Description	Craft@Hrs	Unit	Material $	Labor $	Equipment $	Total $

Kitchen sink, double compartment. Complete with trim. Make additional allowances for rough-in.

Stainless steel	P1@2.15	Ea	420.00	78.90	—	498.90
Cast iron	P1@2.50	Ea	548.00	91.80	—	639.80
Acrylic	P1@2.15	Ea	497.00	78.90	—	575.90

Kitchen sink, single compartment. Complete with trim. Make additional allowances for rough-in.

Stainless steel	P1@2.00	Ea	355.00	73.40	—	428.40
Cast iron	P1@2.10	Ea	409.00	77.10	—	486.10
Acrylic	P1@2.00	Ea	369.00	73.40	—	442.40

Bar sink. Complete with trim. Make additional allowances for rough-in.

Stainless steel	P1@2.00	Ea	302.00	73.40	—	375.40
Acrylic	P1@2.00	Ea	203.00	73.40	—	276.40

Exam room sink. Complete with trim. Make additional allowances for rough-in.

Stainless steel	P1@2.10	Ea	438.00	77.10	—	515.10
Acrylic	P1@2.10	Ea	374.00	77.10	—	451.10

Laboratory sink. Complete with trim. Make additional allowances for rough-in.

Stainless steel	P1@2.25	Ea	502.00	82.60	—	584.60
Acrylic	P1@2.25	Ea	438.00	82.60	—	520.60

Laundry sink, double compartment. Complete with trim. Make additional allowances for rough-in.

Cast iron	P1@3.50	Ea	587.00	128.00	—	715.00
Acrylic	P1@2.25	Ea	258.00	82.60	—	340.60

Laundry sink, single compartment. Complete with trim. Make additional allowances for rough-in.

Cast iron	P1@2.75	Ea	508.00	101.00	—	609.00
Acrylic	P1@2.00	Ea	178.00	73.40	—	251.40

Mop sink, floor-mounted, 36" x 24". Complete with trim. Make additional allowances for rough-in.

Molded stone	P1@2.65	Ea	715.00	97.30	—	812.30
Terrazzo	P1@2.65	Ea	788.00	97.30	—	885.30
Acrylic	P1@2.35	Ea	551.00	86.20	—	637.20

Plumbing Fixtures

Description	Craft@Hrs	Unit	Material $	Labor $	Equipment $	Total $

Slop sink, enameled cast iron with P-trap, standard. Complete with trim. Make additional allowances for rough-in.

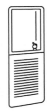

Description	Craft@Hrs	Unit	Material $	Labor $	Equipment $	Total $
Slop sink with P-trap, std.	P1@3.50	Ea	1,230.00	128.00	—	1,358.00

Floor sink, recessed, enameled steel, white. Add 40% to material prices for acid-resisting finish. Complete with strainer. Make additional allowances for rough-in.

Description	Craft@Hrs	Unit	Material $	Labor $	Equipment $	Total $
9" x 9"	P1@1.00	Ea	105.00	36.70	—	141.70
12" x 12"	P1@1.00	Ea	123.00	36.70	—	159.70
15" x 15"	P1@1.15	Ea	83.70	42.20	—	125.90
18" x 18"	P1@1.25	Ea	110.00	45.90	—	155.90
24" x 24"	P1@1.50	Ea	149.00	55.10	—	204.10

Drinking fountain, refrigerated, stainless steel. Complete with trim. Make additional allowances for rough-in. ADA means American Disabilities Act compliant. (Wheelchair accessible)

Description	Craft@Hrs	Unit	Material $	Labor $	Equipment $	Total $
Free-standing	P1@2.00	Ea	1,260.00	73.40	—	1,333.40
Semi-recessed	P1@2.50	Ea	1,670.00	91.80	—	1,761.80
Fully-recessed	P1@2.50	Ea	2,890.00	91.80	—	2,981.80
Wall-hung	P1@2.00	Ea	1,180.00	73.40	—	1,253.40
Wall-hung, ADA	P1@2.50	Ea	2,890.00	91.80	—	2,981.80

Drinking fountain, non-refrigerated. Complete with trim. Make additional allowances for rough-in. ADA means American Disabilities Act compliant. (Wheelchair accessible) S.S. means stainless steel.

Description	Craft@Hrs	Unit	Material $	Labor $	Equipment $	Total $
Recessed, china	P1@2.50	Ea	972.00	91.80	—	1,063.80
Wall-hung, china	P1@2.00	Ea	554.00	73.40	—	627.40
Recessed, S.S.	P1@2.50	Ea	1,110.00	91.80	—	1,201.80
Wall-hung, S.S.	P1@2.00	Ea	590.00	73.40	—	663.40
ADA, S.S.	P1@2.50	Ea	1,010.00	91.80	—	1,101.80

Description	Craft@Hrs	Unit	Material $	Labor $	Equipment $	Total $

Commercial plumbing fixture rough-in. Includes type L copper supply pipe and DWV copper (to 2½") or cast iron (MJ) DWV (over 2½") drain and vent piping. Make additional allowances for plumbing fixtures and trim. Use these costs for preliminary estimates.

Description	Craft@Hrs	Unit	Material $	Labor $	Equipment $	Total $
Water closet, wall-hung, flush valve, with carrier	P1@2.25	Ea	883.00	82.60	—	965.60
Water closet, wall-hung, flush valve, no carrier	P1@1.95	Ea	809.00	71.60	—	880.60
Water closet, floor-mounted, flush valve	P1@2.75	Ea	715.00	101.00	—	816.00
Water closet, floor-mounted, tank type	P1@2.25	Ea	548.00	82.60	—	630.60
Bidet	P1@2.00	Ea	382.00	73.40	—	455.40
Urinal, wall-hung, flush valve, with carrier	P1@3.10	Ea	963.00	114.00	—	1,077.00
Urinal, wall-hung, flush valve, without carrier	P1@2.35	Ea	548.00	86.20	—	634.20
Lavatory, wall-hung, with carrier	P1@2.40	Ea	793.00	88.10	—	881.10
Lavatory	P1@1.90	Ea	382.00	69.70	—	451.70
Sink	P1@1.90	Ea	411.00	69.70	—	480.70
Bath tub	P1@2.35	Ea	587.00	86.20	—	673.20
Shower	P1@2.60	Ea	688.00	95.40	—	783.40
Mop sink	P1@2.40	Ea	487.00	88.10	—	575.10
Slop sink	P1@2.60	Ea	349.00	95.40	—	444.40
Laundry tub	P1@1.95	Ea	414.00	71.60	—	485.60
Wash fountain	P1@2.10	Ea	447.00	77.10	—	524.10
Lab sink, glass drainage	P1@3.80	Ea	1,760.00	139.00	—	1,899.00
Lab sink, acid resistant plastic drainage	P1@2.65	Ea	280.00	97.30	—	377.30
Drinking fountain	P1@2.20	Ea	304.00	80.70	—	384.70
Emergency eyewash and shower	P1@1.75	Ea	115.00	64.20	—	179.20
Washing machine	P1@2.25	Ea	445.00	82.60	—	527.60

Plumbing Fixtures Rough-in

Description	Craft@Hrs	Unit	Material $	Labor $	Equipment $	Total $

Commercial plumbing fixture group rough-in. Includes Type L copper supply pipe and DWV copper (to 2½") or cast iron (MJ) DWV (over 2½") drain and vent piping. Make additional allowances for plumbing fixtures and trim. Use these costs for preliminary estimates.

Description	Craft@Hrs	Unit	Material $	Labor $	Equipment $	Total $
3-piece washroom group	P1@5.50	Ea	1,050.00	202.00	—	1,252.00
3-piece washroom group back to back	P1@9.75	Ea	1,940.00	358.00	—	2,298.00
Kitchen sink, back to back	P1@2.15	Ea	566.00	78.90	—	644.90
Battery of water closets, floor-mounted, tank type, per water closet	P1@1.75	Ea	441.00	64.20	—	505.20
Battery of water closets, floor-mounted, flush valve, per water closet	P1@2.20	Ea	576.00	80.70	—	656.70
Battery of water closets, wall-hung, flush valve, with carrier, per water closet	P1@1.80	Ea	758.00	66.10	—	824.10
Battery of water closets, wall-hung, flush valve, without carrier, per water closet	P1@1.50	Ea	222.00	55.10	—	277.10
Battery of urinals, wall-hung, flush valve with carrier, per urinal	P1@2.45	Ea	911.00	89.90	—	1,000.90
Battery of urinals, wall-hung, flush valve without carrier, per urinal	P1@1.90	Ea	470.00	69.70	—	539.70
Battery of lavatory basins, wall-hung, with carrier, per lavatory	P1@2.00	Ea	709.00	73.40	—	782.40
Battery of lavatory basins, without carrier, per lavatory	P1@1.50	Ea	320.00	55.10	—	375.10

Residential plumbing fixture rough-in. Includes polyethylene (PE) supply pipe and ABS DWV drain and vent piping. Make additional allowances for plumbing fixtures and trim. Use these costs for preliminary estimates.

Description	Craft@Hrs	Unit	Material $	Labor $	Equipment $	Total $
Water closet, floor-mounted, tank type	P1@2.00	Ea	122.00	73.40	—	195.40
Bidet	P1@1.85	Ea	91.70	67.90	—	159.60
Lavatory	P1@1.75	Ea	91.70	64.20	—	155.90
Counter sink	P1@1.75	Ea	101.00	64.20	—	165.20
Bathtub	P1@2.10	Ea	91.70	77.10	—	168.80
Shower	P1@2.45	Ea	134.00	89.90	—	223.90
Laundry tub	P1@1.75	Ea	83.70	64.20	—	147.90
Washing machine	P1@2.00	Ea	104.00	73.40	—	177.40

Copper, Type K with Brazed Joints

Type K hard-drawn copper pipe with wrought copper fittings and brazed joints is used in a wide variety of plumbing and HVAC systems such as potable water, heating hot water, chilled water, compressed air, refrigerant and A.C. condensate.

Brazed joints are those made with silver or other alloy filler metals having melting points at, or above, 1,000 degrees F. Maximum working pressure/temperature relationships for brazed joints are approximately as follows:

Maximum Working Pressure (PSIG)*			
Water Temperature	Nominal Pipe Size (inches)		
(degrees F.)	Up to 1	1¼ to 2	2½ to 4
Up to 350	270	190	150

For copper pipe and solder-type fittings using brazing alloys melting at, or above 1,000 degrees F.

This section has been arranged to save the estimator's time by including all normally-used system components such as pipe, fittings, valves, hanger assemblies, riser clamps and miscellaneous items under one heading. Additional items can be found under "Plumbing and Piping Specialties." The cost estimates in this section are based on the conditions, limitations and wage rates described in the section "How to Use This Book" beginning on page 5.

Description	Craft@Hrs	Unit	Material $	Labor $	Equipment $	Total $

Type K copper pipe with brazed joints, installed horizontally. Complete installation including six to twelve wrought copper tees and six to twelve wrought copper elbows every 100', and hangers spaced to meet plumbing code *(a tee & elbow every 8.3' for ½" dia., a tee and elbow every 16.6' for 4" dia.).*

Description	Craft@Hrs	Unit	Material $	Labor $	Equipment $	Total $
½"	P1@.117	LF	3.42	4.29	—	7.71
¾"	P1@.127	LF	5.98	4.66	—	10.64
1"	P1@.147	LF	9.48	5.39	—	14.87
1¼"	P1@.167	LF	13.00	6.13	—	19.13
1½"	P1@.187	LF	16.10	6.86	—	22.96
2"	P1@.217	LF	27.20	7.96	—	35.16
2½"	P1@.247	LF	44.80	9.06	—	53.86
3"	P1@.277	LF	51.70	10.20	—	61.90
4"	P1@.317	LF	93.90	11.60	—	105.50

Copper, Type K with Brazed Joints

Description	Craft@Hrs	Unit	Material $	Labor $	Equipment $	Total $

Type K copper pipe with brazed joints, installed risers. Complete installation including a wrought copper reducing tee every floor and a riser clamp every other floor.

Description	Craft@Hrs	Unit	Material $	Labor $	Equipment $	Total $
½"	P1@.064	LF	3.08	2.35	—	5.43
¾"	P1@.074	LF	5.57	2.72	—	8.29
1"	P1@.079	LF	8.48	2.90	—	11.38
1¼"	P1@.084	LF	11.70	3.08	—	14.78
1½"	P1@.104	LF	13.90	3.82	—	17.72
2"	P1@.134	LF	23.70	4.92	—	28.62
2½"	P1@.154	LF	35.30	5.65	—	40.95
3"	P1@.174	LF	49.40	6.39	—	55.79
4"	P1@.194	LF	77.50	7.12	—	84.62

Type K copper pipe with brazed joints, pipe only

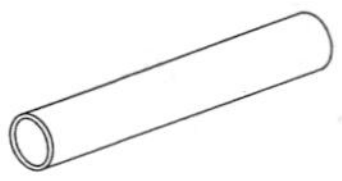

Description	Craft@Hrs	Unit	Material $	Labor $	Equipment $	Total $
½"	P1@.032	LF	2.51	1.17	—	3.68
¾"	P1@.035	LF	4.63	1.28	—	5.91
1"	P1@.038	LF	5.84	1.39	—	7.23
1¼"	P1@.042	LF	7.80	1.54	—	9.34
1½"	P1@.046	LF	8.52	1.69	—	10.21
2"	P1@.053	LF	15.20	1.95	—	17.15
2½"	P1@.060	LF	19.80	2.20	—	22.00
3"	P1@.066	LF	25.90	2.42	—	28.32
4"	P1@.080	LF	46.30	2.94	—	49.24

Type K copper 45-degree ell C x C with brazed joints

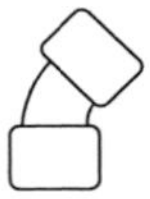

Description	Craft@Hrs	Unit	Material $	Labor $	Equipment $	Total $
½"	P1@.129	Ea	.82	4.73	—	5.55
¾"	P1@.181	Ea	1.56	6.64	—	8.20
1"	P1@.232	Ea	4.60	8.51	—	13.11
1¼"	P1@.284	Ea	6.23	10.40	—	16.63
1½"	P1@.335	Ea	9.28	12.30	—	21.58
2"	P1@.447	Ea	14.70	16.40	—	31.10
2½"	P1@.550	Ea	37.50	20.20	—	57.70
3"	P1@.654	Ea	49.60	24.00	—	73.60
4"	P1@.860	Ea	95.30	31.60	—	126.90

Type K copper 90-degree ell C x C with brazed joints

Description	Craft@Hrs	Unit	Material $	Labor $	Equipment $	Total $
½"	P1@.129	Ea	.48	4.73	—	5.21
¾"	P1@.181	Ea	1.01	6.64	—	7.65
1"	P1@.232	Ea	3.51	8.51	—	12.02
1¼"	P1@.284	Ea	5.67	10.40	—	16.07
1½"	P1@.335	Ea	8.55	12.30	—	20.85
2"	P1@.447	Ea	14.90	16.40	—	31.30
2½"	P1@.550	Ea	31.00	20.20	—	51.20
3"	P1@.654	Ea	39.30	24.00	—	63.30
4"	P1@.860	Ea	83.10	31.60	—	114.70

Description	Craft@Hrs	Unit	Material $	Labor $	Equipment $	Total $

Type K copper 90-degree ell ftg. x C with brazed joints

Description	Craft@Hrs	Unit	Material $	Labor $	Equipment $	Total $
½"	P1@.129	Ea	1.01	4.73	—	5.74
¾"	P1@.181	Ea	3.31	6.64	—	9.95
1"	P1@.232	Ea	4.97	8.51	—	13.48
1¼"	P1@.284	Ea	6.93	10.40	—	17.33
1½"	P1@.335	Ea	10.30	12.30	—	22.60
2"	P1@.447	Ea	20.20	16.40	—	36.60
2½"	P1@.550	Ea	42.20	20.20	—	62.40
3"	P1@.654	Ea	47.50	24.00	—	71.50
4"	P1@.860	Ea	127.00	31.60	—	158.60

Type K copper tee C x C x C with brazed joints

Description	Craft@Hrs	Unit	Material $	Labor $	Equipment $	Total $
½"	P1@.156	Ea	.81	5.73	—	6.54
¾"	P1@.204	Ea	2.19	7.49	—	9.68
1"	P1@.263	Ea	8.23	9.65	—	17.88
1¼"	P1@.322	Ea	13.10	11.80	—	24.90
1½"	P1@.380	Ea	19.20	13.90	—	33.10
2"	P1@.507	Ea	29.60	18.60	—	48.20
2½"	P1@.624	Ea	60.00	22.90	—	82.90
3"	P1@.741	Ea	92.80	27.20	—	120.00
4"	P1@.975	Ea	155.00	35.80	—	190.80

Copper, Type K with Brazed Joints

Description	Craft@Hrs	Unit	Material $	Labor $	Equipment $	Total $

Type K copper branch reducing tee C x C x C with brazed joints

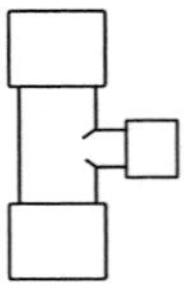

Description	Craft@Hrs	Unit	Material $	Labor $	Equipment $	Total $
½" x 3/8"	P1@.129	Ea	1.65	4.73	—	6.38
¾" x ½"	P1@.182	Ea	3.83	6.68	—	10.51
1" x ½"	P1@.209	Ea	9.77	7.67	—	17.44
1" x ¾"	P1@.234	Ea	9.77	8.59	—	18.36
1¼" x ½"	P1@.241	Ea	14.80	8.84	—	23.64
1¼" x ¾"	P1@.264	Ea	14.80	9.69	—	24.49
1¼" x 1"	P1@.287	Ea	14.80	10.50	—	25.30
1½" x ½"	P1@.291	Ea	18.10	10.70	—	28.80
1½" x ¾"	P1@.307	Ea	18.10	11.30	—	29.40
1½" x 1"	P1@.323	Ea	18.10	11.90	—	30.00
1½" x 1¼"	P1@.339	Ea	18.10	12.40	—	30.50
2" x ½"	P1@.348	Ea	23.70	12.80	—	36.50
2" x ¾"	P1@.373	Ea	23.70	13.70	—	37.40
2" x 1"	P1@.399	Ea	23.70	14.60	—	38.30
2" x 1¼"	P1@.426	Ea	23.70	15.60	—	39.30
2" x 1½"	P1@.452	Ea	23.70	16.60	—	40.30
2½" x ½"	P1@.468	Ea	70.70	17.20	—	87.90
2½" x ¾"	P1@.485	Ea	70.70	17.80	—	88.50
2½" x 1"	P1@.502	Ea	70.70	18.40	—	89.10
2½" x 1½"	P1@.537	Ea	70.70	19.70	—	90.40
2½" x 2"	P1@.555	Ea	70.70	20.40	—	91.10
3" x 1¼"	P1@.562	Ea	80.10	20.60	—	100.70
3" x 2"	P1@.626	Ea	83.10	23.00	—	106.10
3" x 2½"	P1@.660	Ea	83.10	24.20	—	107.30
4" x 1¼"	P1@.669	Ea	145.00	24.60	—	169.60
4" x 1½"	P1@.718	Ea	145.00	26.40	—	171.40
4" x 2"	P1@.768	Ea	145.00	28.20	—	173.20
4" x 2½"	P1@.819	Ea	145.00	30.10	—	175.10
4" x 3"	P1@.868	Ea	145.00	31.90	—	176.90

Description	Craft@Hrs	Unit	Material $	Labor $	Equipment $	Total $

Type K copper reducer with brazed joints

Description	Craft@Hrs	Unit	Material $	Labor $	Equipment $	Total $
½" x 3/8"	P1@.148	Ea	.59	5.43	—	6.02
¾" x ½"	P1@.155	Ea	1.23	5.69	—	6.92
1" x ¾"	P1@.206	Ea	2.00	7.56	—	9.56
1¼" x ½"	P1@.206	Ea	3.36	7.56	—	10.92
1¼" x ¾"	P1@.232	Ea	3.36	8.51	—	11.87
1¼" x 1"	P1@.258	Ea	3.36	9.47	—	12.83
1½" x ½"	P1@.232	Ea	5.14	8.51	—	13.65
1½" x ¾"	P1@.257	Ea	5.14	9.43	—	14.57
1½" x 1"	P1@.283	Ea	5.14	10.40	—	15.54
1½" x 1¼"	P1@.308	Ea	5.14	11.30	—	16.44
2" x ½"	P1@.287	Ea	7.74	10.50	—	18.24
2" x ¾"	P1@.277	Ea	7.74	10.20	—	17.94
2" x 1"	P1@.338	Ea	7.74	12.40	—	20.14
2" x 1¼"	P1@.365	Ea	7.74	13.40	—	21.14
2" x 1½"	P1@.390	Ea	7.74	14.30	—	22.04
2½" x 1"	P1@.390	Ea	21.10	14.30	—	35.40
2½" x 1¼"	P1@.416	Ea	21.10	15.30	—	36.40
2½" x 1½"	P1@.442	Ea	21.10	16.20	—	37.30
2½" x 2"	P1@.497	Ea	21.10	18.20	—	39.30
3" x 1¼"	P1@.468	Ea	24.60	17.20	—	41.80
3" x 1½"	P1@.493	Ea	24.60	18.10	—	42.70
3" x 2"	P1@.548	Ea	24.60	20.10	—	44.70
3" x 2½"	P1@.600	Ea	24.60	22.00	—	46.60
4" x 2"	P1@.652	Ea	47.50	23.90	—	71.40
4" x 2½"	P1@.704	Ea	47.50	25.80	—	73.30
4" x 3"	P1@.755	Ea	47.50	27.70	—	75.20

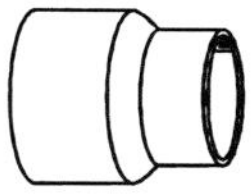

Type K copper adapter C x MPT

Description	Craft@Hrs	Unit	Material $	Labor $	Equipment $	Total $
½"	P1@.090	Ea	1.60	3.30	—	4.90
¾"	P1@.126	Ea	2.29	4.62	—	6.91
1"	P1@.162	Ea	5.67	5.95	—	11.62
1¼"	P1@.198	Ea	10.60	7.27	—	17.87
1½"	P1@.234	Ea	11.90	8.59	—	20.49
2"	P1@.312	Ea	14.90	11.50	—	26.40
2½"	P1@.384	Ea	48.30	14.10	—	62.40
3"	P1@.456	Ea	80.10	16.70	—	96.80

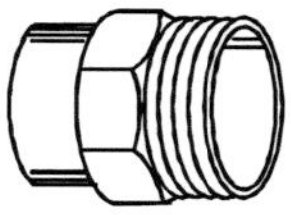

Type K copper adapter C x FPT

Description	Craft@Hrs	Unit	Material $	Labor $	Equipment $	Total $
½"	P1@.090	Ea	2.43	3.30	—	5.73
¾"	P1@.126	Ea	4.41	4.62	—	9.03
1"	P1@.162	Ea	6.66	5.95	—	12.61
1¼"	P1@.198	Ea	14.00	7.27	—	21.27
1½"	P1@.234	Ea	16.90	8.59	—	25.49
2"	P1@.312	Ea	25.20	11.50	—	36.70
2½"	P1@.384	Ea	63.40	14.10	—	77.50
3"	P1@.456	Ea	87.20	16.70	—	103.90

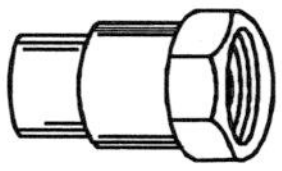

Copper, Type K with Brazed Joints

Description	Craft@Hrs	Unit	Material $	Labor $	Equipment $	Total $

Type K copper flush bushing with brazed joints

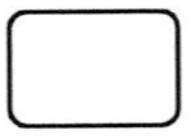

Description	Craft@Hrs	Unit	Material $	Labor $	Equipment $	Total $
½" x 3/8"	P1@.148	Ea	1.48	5.43	—	6.91
¾" x ½"	P1@.155	Ea	2.53	5.69	—	8.22
1" x ¾"	P1@.206	Ea	6.54	7.56	—	14.10
1" x ½"	P1@.206	Ea	7.46	7.56	—	15.02
1¼" x 1"	P1@.258	Ea	7.76	9.47	—	17.23
1½" x 1¼"	P1@.308	Ea	9.87	11.30	—	21.17
2" x 1½"	P1@.390	Ea	18.00	14.30	—	32.30

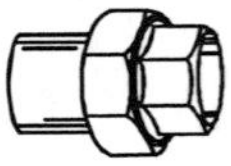

Type K copper union with brazed joints

Description	Craft@Hrs	Unit	Material $	Labor $	Equipment $	Total $
½"	P1@.146	Ea	5.94	5.36	—	11.30
¾"	P1@.205	Ea	7.09	7.52	—	14.61
1"	P1@.263	Ea	15.90	9.65	—	25.55
1¼"	P1@.322	Ea	28.90	11.80	—	40.70
1½"	P1@.380	Ea	35.50	13.90	—	49.40
2"	P1@.507	Ea	51.30	18.60	—	69.90

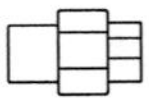

Type K copper dielectric union with brazed joints

Description	Craft@Hrs	Unit	Material $	Labor $	Equipment $	Total $
½"	P1@.146	Ea	5.57	5.36	—	10.93
¾"	P1@.205	Ea	5.57	7.52	—	13.09
1"	P1@.263	Ea	11.50	9.65	—	21.15
1¼"	P1@.322	Ea	19.60	11.80	—	31.40
1½"	P1@.380	Ea	27.00	13.90	—	40.90
2"	P1@.507	Ea	36.10	18.60	—	54.70

Type K copper pressure cap with brazed joints

Description	Craft@Hrs	Unit	Material $	Labor $	Equipment $	Total $
½"	P1@.083	Ea	.49	3.05	—	3.54
¾"	P1@.116	Ea	.99	4.26	—	5.25
1"	P1@.149	Ea	2.23	5.47	—	7.70
1¼"	P1@.182	Ea	3.21	6.68	—	9.89
1½"	P1@.215	Ea	4.68	7.89	—	12.57
2"	P1@.286	Ea	8.14	10.50	—	18.64
2½"	P1@.352	Ea	20.70	12.90	—	33.60
3"	P1@.418	Ea	32.30	15.30	—	47.60
4"	P1@.550	Ea	49.70	20.20	—	69.90

Type K copper coupling with brazed joints

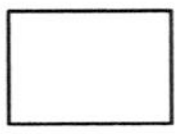

Description	Craft@Hrs	Unit	Material $	Labor $	Equipment $	Total $
½"	P1@.129	Ea	.23	4.73	—	4.96
¾"	P1@.181	Ea	.58	6.64	—	7.22
1"	P1@.232	Ea	2.12	8.51	—	10.63
1¼"	P1@.284	Ea	3.70	10.40	—	14.10
1½"	P1@.335	Ea	5.35	12.30	—	17.65
2"	P1@.447	Ea	7.86	16.40	—	24.26
2½"	P1@.550	Ea	18.60	20.20	—	38.80
3"	P1@.654	Ea	28.30	24.00	—	52.30
4"	P1@.860	Ea	55.40	31.60	—	87.00

Description	Craft@Hrs	Unit	Material $	Labor $	Equipment $	Total $
Class 125 bronze body gate valve, brazed joints						
½"	P1@.240	Ea	18.40	8.81	—	27.21
¾"	P1@.300	Ea	23.00	11.00	—	34.00
1"	P1@.360	Ea	32.40	13.20	—	45.60
1¼"	P1@.480	Ea	40.70	17.60	—	58.30
1½"	P1@.540	Ea	54.90	19.80	—	74.70
2"	P1@.600	Ea	91.80	22.00	—	113.80
2½"	P1@1.00	Ea	152.00	36.70	—	188.70
3"	P1@1.50	Ea	218.00	55.10	—	273.10
Class 125 iron body gate valve, flanged joints						
2"	P1@.500	Ea	351.00	18.40	—	369.40
2½"	P1@.600	Ea	474.00	22.00	—	496.00
3"	P1@.750	Ea	514.00	27.50	—	541.50
4"	P1@1.35	Ea	754.00	49.50	—	803.50
5"	P1@2.00	Ea	1,440.00	73.40	—	1,513.40
6"	P1@2.50	Ea	1,440.00	91.80	—	1,531.80
Class 125 bronze body globe valve, with brazed joints						
½"	P1@.240	Ea	34.10	8.81	—	42.91
¾"	P1@.300	Ea	45.40	11.00	—	56.40
1"	P1@.360	Ea	78.20	13.20	—	91.40
1¼"	P1@.480	Ea	110.00	17.60	—	127.60
1½"	P1@.540	Ea	149.00	19.80	—	168.80
2"	P1@.600	Ea	241.00	22.00	—	263.00
Class 125 iron body globe valve, flanged joints						
2½"	P1@.650	Ea	398.00	23.90	—	421.90
3"	P1@.750	Ea	477.00	27.50	—	504.50
4"	P1@1.35	Ea	644.00	49.50	—	693.50
200 PSIG iron body butterfly valve, lug-type, lever operated						
2"	P1@.450	Ea	156.00	16.50	—	172.50
2½"	P1@.450	Ea	162.00	16.50	—	178.50
3"	P1@.550	Ea	170.00	20.20	—	190.20
4"	P1@.550	Ea	212.00	20.20	—	232.20
200 PSIG iron body butterfly valve, wafer-type, lever operated						
2"	P1@.450	Ea	142.00	16.50	—	158.50
2½"	P1@.450	Ea	145.00	16.50	—	161.50
3"	P1@.550	Ea	156.00	20.20	—	176.20
4"	P1@.550	Ea	189.00	20.20	—	209.20

Copper, Type K with Brazed Joints

Description	Craft@Hrs	Unit	Material $	Labor $	Equipment $	Total $
Class 125 bronze body 2-piece ball valve, brazed joints						
½"	P1@.240	Ea	10.10	8.81	—	18.91
¾"	P1@.300	Ea	13.40	11.00	—	24.40
1"	P1@.360	Ea	23.30	13.20	—	36.50
1¼"	P1@.480	Ea	38.80	17.60	—	56.40
1½"	P1@.540	Ea	52.40	19.80	—	72.20
2"	P1@.600	Ea	66.50	22.00	—	88.50
3"	P1@1.50	Ea	455.00	55.10	—	510.10
4"	P1@1.80	Ea	595.00	66.10	—	661.10
Class 125 bronze body swing check valve, brazed joints						
½"	P1@.240	Ea	20.40	8.81	—	29.21
¾"	P1@.300	Ea	29.20	11.00	—	40.20
1"	P1@.360	Ea	38.10	13.20	—	51.30
1¼"	P1@.480	Ea	54.40	17.60	—	72.00
1½"	P1@.540	Ea	76.70	19.80	—	96.50
2"	P1@.600	Ea	127.00	22.00	—	149.00
2½"	P1@1.00	Ea	256.00	36.70	—	292.70
3"	P1@1.50	Ea	368.00	55.10	—	423.10
Class 125 iron body swing check valve, flanged joints						
2"	P1@.500	Ea	179.00	18.40	—	197.40
2½"	P1@.600	Ea	226.00	22.00	—	248.00
3"	P1@.750	Ea	282.00	27.50	—	309.50
4"	P1@1.35	Ea	414.00	49.50	—	463.50
Class 125 iron body silent check valve, wafer-type						
2"	P1@.500	Ea	131.00	18.40	—	149.40
2½"	P1@.600	Ea	145.00	22.00	—	167.00
3"	P1@.750	Ea	167.00	27.50	—	194.50
4"	P1@1.35	Ea	223.00	49.50	—	272.50
Class 125 bronze body strainer, threaded joints						
½"	P1@.230	Ea	31.40	8.44	—	39.84
¾"	P1@.260	Ea	41.30	9.54	—	50.84
1"	P1@.330	Ea	50.60	12.10	—	62.70
1¼"	P1@.440	Ea	70.70	16.10	—	86.80
1½"	P1@.495	Ea	106.00	18.20	—	124.20
2"	P1@.550	Ea	185.00	20.20	—	205.20
Class 125 iron body strainer, flanged joints						
2"	P1@.500	Ea	141.00	18.40	—	159.40
2½"	P1@.600	Ea	157.00	22.00	—	179.00
3"	P1@.750	Ea	183.00	27.50	—	210.50
4"	P1@1.35	Ea	308.00	49.50	—	357.50

Description	Craft@Hrs	Unit	Material $	Labor $	Equipment $	Total $

Installation of 2-way control valve, threaded joints

½"	P1@.210	Ea	—	7.71	—	7.71
¾"	P1@.275	Ea	—	10.10	—	10.10
1"	P1@.350	Ea	—	12.80	—	12.80
1¼"	P1@.430	Ea	—	15.80	—	15.80
1½"	P1@.505	Ea	—	18.50	—	18.50
2"	P1@.675	Ea	—	24.80	—	24.80
2½"	P1@.830	Ea	—	30.50	—	30.50
3"	P1@.990	Ea	—	36.30	—	36.30

Installation of 3-way control valve, threaded joints

½"	P1@.260	Ea	—	9.54	—	9.54
¾"	P1@.365	Ea	—	13.40	—	13.40
1"	P1@.475	Ea	—	17.40	—	17.40
1¼"	P1@.575	Ea	—	21.10	—	21.10
1½"	P1@.680	Ea	—	25.00	—	25.00
2"	P1@.910	Ea	—	33.40	—	33.40
2½"	P1@1.12	Ea	—	41.10	—	41.10
3"	P1@1.33	Ea	—	48.80	—	48.80

Companion flange, 150 pound cast brass

2"	P1@.290	Ea	308.00	10.60	—	318.60
2½"	P1@.380	Ea	383.00	13.90	—	396.90
3"	P1@.460	Ea	383.00	16.90	—	399.90
4"	P1@.600	Ea	611.00	22.00	—	633.00

Bolt and gasket sets

2"	P1@.500	Ea	4.20	18.40	—	22.60
2½"	P1@.650	Ea	4.90	23.90	—	28.80
3"	P1@.750	Ea	8.32	27.50	—	35.82
4"	P1@1.00	Ea	13.90	36.70	—	50.60

Thermometer with well

7"	P1@.250	Ea	183.00	9.18	—	192.18
9"	P1@.250	Ea	189.00	9.18	—	198.18

Dial-type pressure gauge

2½"	P1@.200	Ea	37.90	7.34	—	45.24
3½"	P1@.200	Ea	50.40	7.34	—	57.74

Copper, Type K with Brazed Joints

Description	Craft@Hrs	Unit	Material $	Labor $	Equipment $	Total $
Pressure/temperature tap						
Tap	P1@.150	Ea	17.40	5.51	—	22.91
Hanger with swivel assembly						
½"	P1@.250	Ea	4.30	9.18	—	13.48
¾"	P1@.250	Ea	4.76	9.18	—	13.94
1"	P1@.250	Ea	4.97	9.18	—	14.15
1¼"	P1@.300	Ea	5.10	11.00	—	16.10
1½"	P1@.300	Ea	5.49	11.00	—	16.49
2"	P1@.300	Ea	5.73	11.00	—	16.73
2½"	P1@.350	Ea	7.79	12.80	—	20.59
3"	P1@.350	Ea	9.63	12.80	—	22.43
4"	P1@.350	Ea	10.60	12.80	—	23.40
Riser clamp						
½"	P1@.100	Ea	2.76	3.67	—	6.43
¾"	P1@.100	Ea	4.09	3.67	—	7.76
1"	P1@.100	Ea	4.13	3.67	—	7.80
1¼"	P1@.105	Ea	4.95	3.85	—	8.80
1½"	P1@.110	Ea	5.22	4.04	—	9.26
2"	P1@.115	Ea	5.56	4.22	—	9.78
2½"	P1@.120	Ea	5.82	4.40	—	10.22
3"	P1@.120	Ea	6.32	4.40	—	10.72
4"	P1@.125	Ea	8.06	4.59	—	12.65

Type K hard-drawn copper pipe with wrought copper fittings and soft-soldered joints is used in a wide variety of plumbing and HVAC systems such as potable water, heating hot water, chilled water and A.C. condensate.

Soft-soldered joints are those made with solders having melting points in the 350 degree F. to 500 degree F. range. Maximum working pressure/temperature relationships for soft-soldered joints are approximately as follows:

Maximum Working Pressures (PSIG)*				
Soft-solder Type	Water Temperature (degrees F.)	Nominal Pipe Size (inches)		
		Up to 1	1¼ to 2	2½ to 4
50-50 tin-lead**	100	200	175	150
	150	150	125	100
	200	100	90	75
	250	85	75	50
95-5 tin-antimony	100	500	400	300
	150	400	350	275
	200	300	250	200
	250	200	175	150

*For copper pipe and solder-type fittings using soft-solders melting at approximately 350 degrees F. to 500 degrees F.
**The use of any solder containing lead is not allowed in potable water systems.

This section has been arranged to save the estimator's time by including all normally-used system components such as pipe, fittings, valves, hanger assemblies, riser clamps and miscellaneous items under one heading. Additional items can be found under "Plumbing and Piping Specialties." The cost estimates in this section are based on the conditions, limitations and wage rates described in the section "How to Use This Book" beginning on page 5.

Description	Craft@Hrs	Unit	Material $	Labor $	Equipment $	Total $

Type K copper pipe with soft-soldered joints installed horizontally.
Complete installation including six to twelve wrought copper tees and six to twelve wrought copper elbows every 100', and hangers spaced to meet plumbing code (*a tee & elbow every 8.3' for ½" dia., a tee and elbow every 16.6' for 4" dia.*)

Description	Craft@Hrs	Unit	Material $	Labor $	Equipment $	Total $
½"	P1@.112	LF	3.42	4.11	—	7.53
¾"	P1@.122	LF	5.98	4.48	—	10.46
1"	P1@.142	LF	9.48	5.21	—	14.69
1¼"	P1@.162	LF	13.00	5.95	—	18.95
1½"	P1@.182	LF	16.10	6.68	—	22.78
2"	P1@.212	LF	27.20	7.78	—	34.98
2½"	P1@.242	LF	44.80	8.88	—	53.68
3"	P1@.272	LF	51.70	9.98	—	61.68
4"	P1@.312	LF	93.90	11.50	—	105.40

Copper, Type K with Soft-Soldered Joints

Description	Craft@Hrs	Unit	Material $	Labor $	Equipment $	Total $

Type K copper pipe with soft-soldered joints, installed risers.
Complete installation including a wrought copper reducing tee every floor and a riser clamp every other floor.

Description	Craft@Hrs	Unit	Material $	Labor $	Equipment $	Total $
½"	P1@.062	LF	3.08	2.28	—	5.36
¾"	P1@.072	LF	5.57	2.64	—	8.21
1"	P1@.077	LF	8.48	2.83	—	11.31
1¼"	P1@.082	LF	11.70	3.01	—	14.71
1½"	P1@.102	LF	13.90	3.74	—	17.64
2"	P1@.132	LF	23.70	4.84	—	28.54
2½"	P1@.152	LF	35.30	5.58	—	40.88
3"	P1@.172	LF	49.40	6.31	—	55.71
4"	P1@.192	LF	77.50	7.05	—	84.55

Type K copper pipe with soft-soldered joints, pipe only

Description	Craft@Hrs	Unit	Material $	Labor $	Equipment $	Total $
½"	P1@.032	LF	2.51	1.17	—	3.68
¾"	P1@.035	LF	4.63	1.28	—	5.91
1"	P1@.038	LF	5.84	1.39	—	7.23
1¼"	P1@.042	LF	7.80	1.54	—	9.34
1½"	P1@.046	LF	8.52	1.69	—	10.21
2"	P1@.053	LF	15.20	1.95	—	17.15
2½"	P1@.060	LF	19.80	2.20	—	22.00
3"	P1@.066	LF	25.90	2.42	—	28.32
4"	P1@.080	LF	46.30	2.94	—	49.24

Type K copper 45-degree ell C x C with soft-soldered joints

Description	Craft@Hrs	Unit	Material $	Labor $	Equipment $	Total $
½"	P1@.107	Ea	.82	3.93	—	4.75
¾"	P1@.150	Ea	1.55	5.51	—	7.06
1"	P1@.193	Ea	4.56	7.08	—	11.64
1¼"	P1@.236	Ea	6.20	8.66	—	14.86
1½"	P1@.278	Ea	9.21	10.20	—	19.41
2"	P1@.371	Ea	14.60	13.60	—	28.20
2½"	P1@.457	Ea	37.50	16.80	—	54.30
3"	P1@.543	Ea	49.40	19.90	—	69.30
4"	P1@.714	Ea	95.30	26.20	—	121.50

Description	Craft@Hrs	Unit	Material $	Labor $	Equipment $	Total $

Type K copper 90-degree ell C x C with soft-soldered joints

Description	Craft@Hrs	Unit	Material $	Labor $	Equipment $	Total $
½"	P1@.107	Ea	.50	3.93	—	4.43
¾"	P1@.150	Ea	1.01	5.51	—	6.52
1"	P1@.193	Ea	3.48	7.08	—	10.56
1¼"	P1@.236	Ea	5.67	8.66	—	14.33
1½"	P1@.278	Ea	8.47	10.20	—	18.67
2"	P1@.371	Ea	14.90	13.60	—	28.50
2½"	P1@.457	Ea	30.80	16.80	—	47.60
3"	P1@.543	Ea	39.00	19.90	—	58.90
4"	P1@.714	Ea	83.10	26.20	—	109.30

Type K copper 90-degree ell ftg. x C with soft-soldered joint

Description	Craft@Hrs	Unit	Material $	Labor $	Equipment $	Total $
½"	P1@.107	Ea	.77	3.93	—	4.70
¾"	P1@.150	Ea	1.66	5.51	—	7.17
1"	P1@.193	Ea	4.93	7.08	—	12.01
1¼"	P1@.236	Ea	6.85	8.66	—	15.51
1½"	P1@.278	Ea	10.20	10.20	—	20.40
2"	P1@.371	Ea	20.20	13.60	—	33.80
2½"	P1@.457	Ea	42.10	16.80	—	58.90
3"	P1@.543	Ea	47.10	19.90	—	67.00
4"	P1@.714	Ea	95.30	26.20	—	121.50

Type K copper tee C x C x C with soft-soldered joints

Description	Craft@Hrs	Unit	Material $	Labor $	Equipment $	Total $
½"	P1@.129	Ea	.81	4.73	—	5.54
¾"	P1@.181	Ea	2.19	6.64	—	8.83
1"	P1@.233	Ea	8.23	8.55	—	16.78
1¼"	P1@.285	Ea	13.10	10.50	—	23.60
1½"	P1@.337	Ea	18.90	12.40	—	31.30
2"	P1@.449	Ea	29.40	16.50	—	45.90
2½"	P1@.552	Ea	60.00	20.30	—	80.30
3"	P1@.656	Ea	92.10	24.10	—	116.20
4"	P1@.863	Ea	153.00	31.70	—	184.70

Copper, Type K with Soft-Soldered Joints

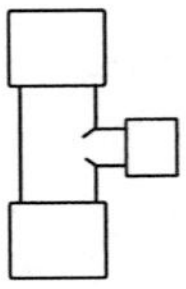

Description	Craft@Hrs	Unit	Material $	Labor $	Equipment $	Total $
Type K copper branch reducing tee C x C x C with soft-soldered joints						
½" x 3/8"	P1@.129	Ea	1.65	4.73	—	6.38
¾" x ½"	P1@.182	Ea	3.50	6.68	—	10.18
1" x ½"	P1@.195	Ea	6.85	7.16	—	14.01
1" x ¾"	P1@.219	Ea	6.85	8.04	—	14.89
1¼" x ½"	P1@.225	Ea	10.30	8.26	—	18.56
1¼" x ¾"	P1@.247	Ea	10.30	9.06	—	19.36
1¼" x 1"	P1@.268	Ea	10.30	9.84	—	20.14
1½" x ½"	P1@.272	Ea	12.70	9.98	—	22.68
1½" x ¾"	P1@.287	Ea	12.70	10.50	—	23.20
1½" x 1"	P1@.302	Ea	12.70	11.10	—	23.80
1½" x 1¼"	P1@.317	Ea	12.70	11.60	—	24.30
2" x ½"	P1@.325	Ea	16.60	11.90	—	28.50
2" x ¾"	P1@.348	Ea	16.60	12.80	—	29.40
2" x 1"	P1@.373	Ea	16.60	13.70	—	30.30
2" x 1¼"	P1@.398	Ea	16.60	14.60	—	31.20
2" x 1½"	P1@.422	Ea	16.60	15.50	—	32.10
2½" x ½"	P1@.437	Ea	50.00	16.00	—	66.00
2½" x ¾"	P1@.453	Ea	50.00	16.60	—	66.60
2½" x 1"	P1@.469	Ea	50.00	17.20	—	67.20
2½" x 1½"	P1@.502	Ea	50.00	18.40	—	68.40
2½" x 2"	P1@.519	Ea	50.00	19.00	—	69.00
3" x 1¼"	P1@.525	Ea	56.50	19.30	—	75.80
3" x 2"	P1@.585	Ea	58.50	21.50	—	80.00
3" x 2½"	P1@.617	Ea	58.50	22.60	—	81.10
4" x 1¼"	P1@.625	Ea	76.60	22.90	—	99.50
4" x 1½"	P1@.671	Ea	76.60	24.60	—	101.20
4" x 2"	P1@.718	Ea	76.60	26.40	—	103.00
4" x 2½"	P1@.765	Ea	76.60	28.10	—	104.70
4" x 3"	P1@.811	Ea	76.60	29.80	—	106.40

Description	Craft@Hrs	Unit	Material $	Labor $	Equipment $	Total $

Type K copper reducer with soft-soldered joints

Description	Craft@Hrs	Unit	Material $	Labor $	Equipment $	Total $
½" x 3/8"	P1@.124	Ea	.62	4.55	—	5.17
¾" x ½"	P1@.129	Ea	1.14	4.73	—	5.87
1" x ¾"	P1@.172	Ea	2.00	6.31	—	8.31
1¼" x ½"	P1@.172	Ea	3.36	6.31	—	9.67
1¼" x ¾"	P1@.193	Ea	3.36	7.08	—	10.44
1¼" x 1"	P1@.215	Ea	3.36	7.89	—	11.25
1½" x ½"	P1@.193	Ea	5.14	7.08	—	12.22
1½" x ¾"	P1@.214	Ea	5.14	7.85	—	12.99
1½" x 1"	P1@.236	Ea	5.14	8.66	—	13.80
1½" x 1¼"	P1@.257	Ea	5.14	9.43	—	14.57
2" x ½"	P1@.239	Ea	7.74	8.77	—	16.51
2" x ¾"	P1@.261	Ea	7.74	9.58	—	17.32
2" x 1"	P1@.282	Ea	7.74	10.30	—	18.04
2" x 1¼"	P1@.304	Ea	7.74	11.20	—	18.94
2" x 1½"	P1@.325	Ea	7.74	11.90	—	19.64
2½" x 1"	P1@.325	Ea	21.10	11.90	—	33.00
2½" x 1¼"	P1@.347	Ea	21.10	12.70	—	33.80
2½" x 1½"	P1@.368	Ea	21.10	13.50	—	34.60
2½" x 2"	P1@.414	Ea	21.10	15.20	—	36.30
3" x 2"	P1@.457	Ea	24.60	16.80	—	41.40
3" x 2½"	P1@.500	Ea	24.60	18.40	—	43.00
4" x 2"	P1@.543	Ea	47.50	19.90	—	67.40
4" x 2½"	P1@.586	Ea	47.50	21.50	—	69.00
4" x 3"	P1@.629	Ea	47.50	23.10	—	70.60

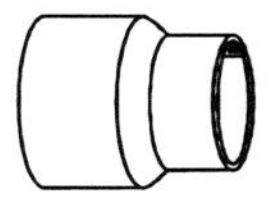

Type K copper adapter C x MPT with soft-soldered joint

Description	Craft@Hrs	Unit	Material $	Labor $	Equipment $	Total $
½"	P1@.075	Ea	1.46	2.75	—	4.21
¾"	P1@.105	Ea	2.29	3.85	—	6.14
1"	P1@.134	Ea	5.67	4.92	—	10.59
1¼"	P1@.164	Ea	10.60	6.02	—	16.62
1½"	P1@.194	Ea	11.90	7.12	—	19.02
2"	P1@.259	Ea	14.90	9.51	—	24.41
2½"	P1@.319	Ea	48.30	11.70	—	60.00
3"	P1@.378	Ea	80.10	13.90	—	94.00

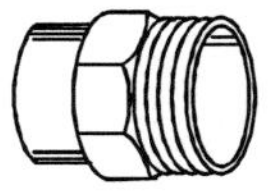

Copper, Type K with Soft-Soldered Joints

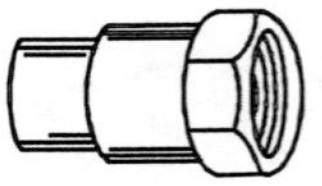

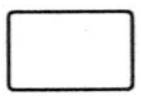

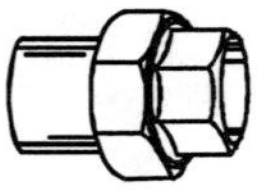

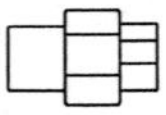

Description	Craft@Hrs	Unit	Material $	Labor $	Equipment $	Total $
Type K copper adapter C x FPT with soft-soldered joint						
½"	P1@.075	Ea	2.43	2.75	—	5.18
¾"	P1@.105	Ea	4.41	3.85	—	8.26
1"	P1@.134	Ea	6.66	4.92	—	11.58
1¼"	P1@.164	Ea	14.00	6.02	—	20.02
1½"	P1@.194	Ea	16.90	7.12	—	24.02
2"	P1@.259	Ea	25.20	9.51	—	34.71
2½"	P1@.319	Ea	63.40	11.70	—	75.10
3"	P1@.378	Ea	87.20	13.90	—	101.10
Type K copper flush bushing with soft-soldered joints						
½" x 3/8"	P1@.124	Ea	1.48	4.55	—	6.03
¾" x ½"	P1@.129	Ea	2.53	4.73	—	7.26
1" x ¾"	P1@.172	Ea	6.54	6.31	—	12.85
1" x ½"	P1@.172	Ea	7.46	6.31	—	13.77
1¼" x 1"	P1@.215	Ea	7.76	7.89	—	15.65
1½" x 1¼"	P1@.257	Ea	9.87	9.43	—	19.30
2" x 1½"	P1@.325	Ea	18.00	11.90	—	29.90
Type K copper union with soft-soldered joint						
½"	P1@.121	Ea	5.94	4.44	—	10.38
¾"	P1@.170	Ea	7.09	6.24	—	13.33
1"	P1@.218	Ea	15.90	8.00	—	23.90
1¼"	P1@.267	Ea	28.90	9.80	—	38.70
1½"	P1@.315	Ea	35.50	11.60	—	47.10
2"	P1@.421	Ea	51.30	15.50	—	66.80
Type K copper dielectric union with soft-soldered joint						
½"	P1@.121	Ea	5.57	4.44	—	10.01
¾"	P1@.170	Ea	5.57	6.24	—	11.81
1"	P1@.218	Ea	11.50	8.00	—	19.50
1¼"	P1@.267	Ea	19.60	9.80	—	29.40
1½"	P1@.315	Ea	27.00	11.60	—	38.60
2"	P1@.421	Ea	36.10	15.50	—	51.60
Type K copper cap with soft-soldered joint						
½"	P1@.069	Ea	.49	2.53	—	3.02
¾"	P1@.096	Ea	.99	3.52	—	4.51
1"	P1@.124	Ea	2.23	4.55	—	6.78
1¼"	P1@.151	Ea	3.21	5.54	—	8.75
1½"	P1@.178	Ea	4.68	6.53	—	11.21
2"	P1@.237	Ea	8.14	8.70	—	16.84
2½"	P1@.292	Ea	20.70	10.70	—	31.40
3"	P1@.347	Ea	32.30	12.70	—	45.00
4"	P1@.457	Ea	49.70	16.80	—	66.50

Description	Craft@Hrs	Unit	Material $	Labor $	Equipment $	Total $

Type K copper coupling with soft-soldered joints

Description	Craft@Hrs	Unit	Material $	Labor $	Equipment $	Total $
½"	P1@.107	Ea	.31	3.93	—	4.24
¾"	P1@.150	Ea	.79	5.51	—	6.30
1"	P1@.193	Ea	2.12	7.08	—	9.20
1¼"	P1@.236	Ea	3.70	8.66	—	12.36
1½"	P1@.278	Ea	5.35	10.20	—	15.55
2"	P1@.371	Ea	7.86	13.60	—	21.46
2½"	P1@.457	Ea	18.60	16.80	—	35.40
3"	P1@.543	Ea	28.30	19.90	—	48.20
4"	P1@.714	Ea	55.40	26.20	—	81.60

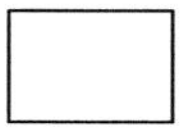

Class 125 bronze body gate valve, soft-soldered joints

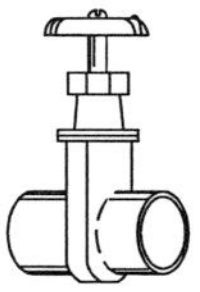

Description	Craft@Hrs	Unit	Material $	Labor $	Equipment $	Total $
½"	P1@.200	Ea	18.20	7.34	—	25.54
¾"	P1@.249	Ea	23.00	9.14	—	32.14
1"	P1@.299	Ea	32.40	11.00	—	43.40
1¼"	P1@.398	Ea	40.70	14.60	—	55.30
1½"	P1@.448	Ea	54.60	16.40	—	71.00
2"	P1@.498	Ea	91.20	18.30	—	109.50
2½"	P1@.830	Ea	151.00	30.50	—	181.50
3"	P1@1.24	Ea	217.00	45.50	—	262.50

Class 125 iron body gate valve, flanged joints

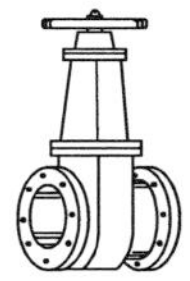

Description	Craft@Hrs	Unit	Material $	Labor $	Equipment $	Total $
2"	P1@.500	Ea	351.00	18.40	—	369.40
2½"	P1@.600	Ea	474.00	22.00	—	496.00
3"	P1@.750	Ea	514.00	27.50	—	541.50
4"	P1@1.35	Ea	754.00	49.50	—	803.50
5"	P1@2.00	Ea	1,440.00	73.40	—	1,513.40
6"	P1@2.50	Ea	1,440.00	91.80	—	1,531.80

Class 125 bronze body globe valve, with soft-soldered joints

Description	Craft@Hrs	Unit	Material $	Labor $	Equipment $	Total $
½"	P1@.200	Ea	34.10	7.34	—	41.44
¾"	P1@.249	Ea	45.40	9.14	—	54.54
1"	P1@.299	Ea	78.20	11.00	—	89.20
1¼"	P1@.398	Ea	110.00	14.60	—	124.60
1½"	P1@.448	Ea	149.00	16.40	—	165.40
2"	P1@.498	Ea	241.00	18.30	—	259.30

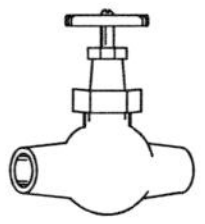

Class 125 iron body globe valve, flanged ends

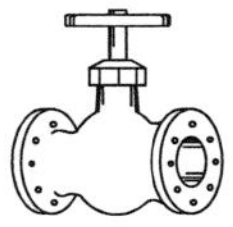

Description	Craft@Hrs	Unit	Material $	Labor $	Equipment $	Total $
2½"	P1@.600	Ea	398.00	22.00	—	420.00
3"	P1@.750	Ea	477.00	27.50	—	504.50
4"	P1@1.35	Ea	640.00	49.50	—	689.50

Copper, Type K with Soft-Soldered Joints

Description	Craft@Hrs	Unit	Material $	Labor $	Equipment $	Total $

200 PSIG iron body butterfly valve, lug-type, lever operated

Description	Craft@Hrs	Unit	Material $	Labor $	Equipment $	Total $
2"	P1@.450	Ea	155.00	16.50	—	171.50
2½"	P1@.450	Ea	161.00	16.50	—	177.50
3"	P1@.550	Ea	169.00	20.20	—	189.20
4"	P1@.550	Ea	211.00	20.20	—	231.20

200 PSIG iron body butterfly valve, wafer-type, lever operated

Description	Craft@Hrs	Unit	Material $	Labor $	Equipment $	Total $
2"	P1@.450	Ea	142.00	16.50	—	158.50
2½"	P1@.450	Ea	145.00	16.50	—	161.50
3"	P1@.550	Ea	156.00	20.20	—	176.20
4"	P1@.550	Ea	189.00	20.20	—	209.20

Class 125 bronze body 2-piece ball valve, soft-soldered joints

Description	Craft@Hrs	Unit	Material $	Labor $	Equipment $	Total $
½"	P1@.200	Ea	10.10	7.34	—	17.44
¾"	P1@.249	Ea	12.10	9.14	—	21.24
1"	P1@.299	Ea	23.30	11.00	—	34.30
1¼"	P1@.398	Ea	38.80	14.60	—	53.40
1½"	P1@.448	Ea	52.40	16.40	—	68.80
2"	P1@.498	Ea	66.50	18.30	—	84.80
3"	P1@1.24	Ea	455.00	45.50	—	500.50
4"	P1@1.45	Ea	595.00	53.20	—	648.20

Class 125 bronze body swing check valve, soft-soldered joints

Description	Craft@Hrs	Unit	Material $	Labor $	Equipment $	Total $
½"	P1@.200	Ea	20.40	7.34	—	27.74
¾"	P1@.249	Ea	29.20	9.14	—	38.34
1"	P1@.299	Ea	38.10	11.00	—	49.10
1¼"	P1@.398	Ea	54.40	14.60	—	69.00
1½"	P1@.448	Ea	76.70	16.40	—	93.10
2"	P1@.498	Ea	127.00	18.30	—	145.30
2½"	P1@.830	Ea	256.00	30.50	—	286.50
3"	P1@1.24	Ea	368.00	45.50	—	413.50

Class 125 iron body swing check valve, flanged ends

Description	Craft@Hrs	Unit	Material $	Labor $	Equipment $	Total $
2"	P1@.500	Ea	179.00	18.40	—	197.40
2½"	P1@.600	Ea	226.00	22.00	—	248.00
3"	P1@.750	Ea	282.00	27.50	—	309.50
4"	P1@1.35	Ea	414.00	49.50	—	463.50

Class 125 iron body silent check valve, wafer-type

Description	Craft@Hrs	Unit	Material $	Labor $	Equipment $	Total $
2"	P1@.500	Ea	131.00	18.40	—	149.40
2½"	P1@.600	Ea	145.00	22.00	—	167.00
3"	P1@.750	Ea	167.00	27.50	—	194.50
4"	P1@1.35	Ea	223.00	49.50	—	272.50

Description	Craft@Hrs	Unit	Material $	Labor $	Equipment $	Total $
Class 125 bronze body strainer, threaded ends						
½"	P1@.230	Ea	31.40	8.44	—	39.84
¾"	P1@.260	Ea	41.30	9.54	—	50.84
1"	P1@.330	Ea	50.60	12.10	—	62.70
1¼"	P1@.440	Ea	70.70	16.10	—	86.80
1½"	P1@.495	Ea	106.00	18.20	—	124.20
2"	P1@.550	Ea	185.00	20.20	—	205.20
Class 125 iron body strainer, flanged ends						
2"	P1@.500	Ea	141.00	18.40	—	159.40
2½"	P1@.600	Ea	157.00	22.00	—	179.00
3"	P1@.750	Ea	183.00	27.50	—	210.50
4"	P1@1.35	Ea	308.00	49.50	—	357.50
Installation of copper 2-way control valve, threaded joints						
½"	P1@.210	Ea	—	7.71	—	7.71
¾"	P1@.275	Ea	—	10.10	—	10.10
1"	P1@.350	Ea	—	12.80	—	12.80
1¼"	P1@.430	Ea	—	15.80	—	15.80
1½"	P1@.505	Ea	—	18.50	—	18.50
2"	P1@.675	Ea	—	24.80	—	24.80
2½"	P1@.830	Ea	—	30.50	—	30.50
3"	P1@.990	Ea	—	36.30	—	36.30
Installation of copper 3-way control valve, threaded joints						
½"	P1@.260	Ea	—	9.54	—	9.54
¾"	P1@.365	Ea	—	13.40	—	13.40
1"	P1@.475	Ea	—	17.40	—	17.40
1¼"	P1@.575	Ea	—	21.10	—	21.10
1½"	P1@.680	Ea	—	25.00	—	25.00
2"	P1@.910	Ea	—	33.40	—	33.40
2½"	P1@1.12	Ea	—	41.10	—	41.10
3"	P1@1.33	Ea	—	48.80	—	48.80
Companion flange, 150 pound cast brass						
2"	P1@.290	Ea	308.00	10.60	—	318.60
2½"	P1@.380	Ea	383.00	13.90	—	396.90
3"	P1@.460	Ea	383.00	16.90	—	399.90
4"	P1@.600	Ea	611.00	22.00	—	633.00

Copper, Type K with Soft-Soldered Joints

Description	Craft@Hrs	Unit	Material $	Labor $	Equipment $	Total $
Bolt and gasket sets						
2"	P1@.500	Ea	4.21	18.40	—	22.61
2½"	P1@.650	Ea	4.94	23.90	—	28.84
3"	P1@.750	Ea	8.41	27.50	—	35.91
4"	P1@1.00	Ea	14.10	36.70	—	50.80
Thermometer with well						
7"	P1@.250	Ea	180.00	9.18	—	189.18
9"	P1@.250	Ea	185.00	9.18	—	194.18
Dial-type pressure gauge						
2½"	P1@.200	Ea	37.20	7.34	—	44.54
3½"	P1@.200	Ea	49.50	7.34	—	56.84
Pressure/temperature tap						
Tap	P1@.150	Ea	17.10	5.51	—	22.61
Hanger with swivel assembly						
½"	P1@.250	Ea	4.22	9.18	—	13.40
¾"	P1@.250	Ea	4.68	9.18	—	13.86
1"	P1@.250	Ea	4.89	9.18	—	14.07
1¼"	P1@.300	Ea	5.02	11.00	—	16.02
1½"	P1@.300	Ea	5.40	11.00	—	16.40
2"	P1@.300	Ea	5.63	11.00	—	16.63
2½"	P1@.350	Ea	7.65	12.80	—	20.45
3"	P1@.350	Ea	9.47	12.80	—	22.27
4"	P1@.350	Ea	10.40	12.80	—	23.20
Riser clamp						
½"	P1@.100	Ea	2.72	3.67	—	6.39
¾"	P1@.100	Ea	4.02	3.67	—	7.69
1"	P1@.100	Ea	4.06	3.67	—	7.73
1¼"	P1@.105	Ea	4.89	3.85	—	8.74
1½"	P1@.110	Ea	5.17	4.04	—	9.21
2"	P1@.115	Ea	5.47	4.22	—	9.69
2½"	P1@.120	Ea	5.76	4.40	—	10.16
3"	P1@.120	Ea	6.23	4.40	—	10.63
4"	P1@.125	Ea	7.71	4.59	—	12.30

Type L hard-drawn copper pipe with wrought copper fittings and brazed joints is used in a wide variety of plumbing and HVAC systems such as potable water, heating hot water, chilled water, compressed air, refrigerant and A.C. condensate.

Brazed joints are those made with silver or other alloy filler metals having melting points at, or above, 1,000 degrees F. Maximum working pressure/temperature relationships for brazed joints are approximately as follows:

Maximum Working Pressure (PSIG)*			
Water Temperature (degrees F.)	Nominal Pipe Size (inches)		
	Up to 1	1¼ to 2	2½ to 4
Up to 350	270	190	150

*For copper pipe and solder-type fittings using brazing alloys melting at, or above, 1,000 degrees F.

This section has been arranged to save the estimator's time by including all normally-used system components such as pipe, fittings, valves, hanger assemblies, riser clamps and miscellaneous items under one heading. Additional items can be found under "Plumbing and Piping Specialties." The cost estimates in this section are based on the conditions, limitations and wage rates described in the section "How to Use This Book" beginning on page 5.

Description	Craft@Hrs	Unit	Material $	Labor $	Equipment $	Total $

Type L copper pipe with brazed joints installed horizontally. Complete installation including six to twelve wrought copper tees and six to twelve wrought copper elbows every 100', and hangers spaced to meet plumbing code *(a tee & elbow every 8.3' for ½" dia., a tee and elbow every 16.6' for 4" dia.)*

Description	Craft@Hrs	Unit	Material $	Labor $	Equipment $	Total $
½"	P1@.116	LF	2.35	4.26	—	6.61
¾"	P1@.126	LF	3.62	4.62	—	8.24
1"	P1@.146	LF	6.50	5.36	—	11.86
1¼"	P1@.166	LF	9.24	6.09	—	15.33
1½"	P1@.186	LF	12.30	6.83	—	19.13
2"	P1@.216	LF	22.30	7.93	—	30.23
2½"	P1@.246	LF	39.50	9.03	—	48.53
3"	P1@.276	LF	54.60	10.10	—	64.70
4"	P1@.316	LF	78.90	11.60	—	90.50

Copper, Type L with Brazed Joints

Description	Craft@Hrs	Unit	Material $	Labor $	Equipment $	Total $

Type L copper pipe with brazed joints, installed risers. Complete installation including a wrought copper reducing tee every floor and a riser clamp every other floor.

Description	Craft@Hrs	Unit	Material $	Labor $	Equipment $	Total $
½"	P1@.063	LF	2.31	2.31	—	4.62
¾"	P1@.073	LF	3.64	2.68	—	6.32
1"	P1@.073	LF	6.17	2.68	—	8.85
1¼"	P1@.083	LF	8.71	3.05	—	11.76
1½"	P1@.103	LF	10.90	3.78	—	14.68
2"	P1@.133	LF	28.50	4.88	—	33.38
2½"	P1@.153	LF	30.80	5.62	—	36.42
3"	P1@.173	LF	39.90	6.35	—	46.25
4"	P1@.193	LF	59.40	7.08	—	66.48

Type L copper pipe with brazed joints, pipe only

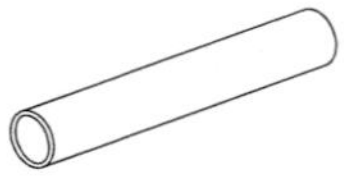

Description	Craft@Hrs	Unit	Material $	Labor $	Equipment $	Total $
½"	P1@.032	LF	1.80	1.17	—	2.97
¾"	P1@.035	LF	2.85	1.28	—	4.13
1"	P1@.038	LF	3.95	1.39	—	5.34
1¼"	P1@.042	LF	5.67	1.54	—	7.21
1½"	P1@.046	LF	7.22	1.69	—	8.91
2"	P1@.053	LF	12.40	1.95	—	14.35
2½"	P1@.060	LF	18.10	2.20	—	20.30
3"	P1@.066	LF	24.50	2.42	—	26.92
4"	P1@.080	LF	38.40	2.94	—	41.34

Type L copper 45-degree ell C x C with brazed joints

Description	Craft@Hrs	Unit	Material $	Labor $	Equipment $	Total $
½"	P1@.129	Ea	.82	4.73	—	5.55
¾"	P1@.181	Ea	1.56	6.64	—	8.20
1"	P1@.232	Ea	4.60	8.51	—	13.11
1¼"	P1@.284	Ea	6.23	10.40	—	16.63
1½"	P1@.335	Ea	9.28	12.30	—	21.58
2"	P1@.447	Ea	14.70	16.40	—	31.10
2½"	P1@.550	Ea	37.50	20.20	—	57.70
3"	P1@.654	Ea	49.60	24.00	—	73.60
4"	P1@.860	Ea	95.30	31.60	—	126.90

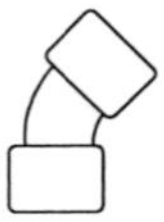

Type L copper 90-degree ell C x C with brazed joints

Description	Craft@Hrs	Unit	Material $	Labor $	Equipment $	Total $
½"	P1@.129	Ea	.48	4.73	—	5.21
¾"	P1@.181	Ea	1.01	6.64	—	7.65
1"	P1@.232	Ea	3.51	8.51	—	12.02
1¼"	P1@.284	Ea	5.67	10.40	—	16.07
1½"	P1@.335	Ea	8.55	12.30	—	20.85
2"	P1@.447	Ea	14.90	16.40	—	31.30
2½"	P1@.550	Ea	31.00	20.20	—	51.20
3"	P1@.654	Ea	39.30	24.00	—	63.30
4"	P1@.860	Ea	83.10	31.60	—	114.70

Description	Craft@Hrs	Unit	Material $	Labor $	Equipment $	Total $

Type L copper 90-degree ell ftg. x C with brazed joint

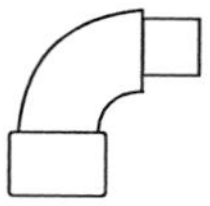

Description	Craft@Hrs	Unit	Material $	Labor $	Equipment $	Total $
½"	P1@.129	Ea	.77	4.73	—	5.50
¾"	P1@.181	Ea	1.66	6.64	—	8.30
1"	P1@.232	Ea	4.97	8.51	—	13.48
1¼"	P1@.284	Ea	6.93	10.40	—	17.33
1½"	P1@.335	Ea	10.30	12.30	—	22.60
2"	P1@.447	Ea	20.20	16.40	—	36.60
2½"	P1@.550	Ea	42.20	20.20	—	62.40
3"	P1@.654	Ea	47.50	24.00	—	71.50
4"	P1@.860	Ea	95.30	31.60	—	126.90

Type L copper tee C x C x C with brazed joints

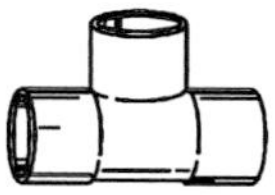

Description	Craft@Hrs	Unit	Material $	Labor $	Equipment $	Total $
½"	P1@.156	Ea	.81	5.73	—	6.54
¾"	P1@.204	Ea	2.19	7.49	—	9.68
1"	P1@.263	Ea	8.23	9.65	—	17.88
1¼"	P1@.322	Ea	13.10	11.80	—	24.90
1½"	P1@.380	Ea	19.20	13.90	—	33.10
2"	P1@.507	Ea	29.60	18.60	—	48.20
2½"	P1@.624	Ea	60.00	22.90	—	82.90
3"	P1@.741	Ea	92.80	27.20	—	120.00
4"	P1@.975	Ea	155.00	35.80	—	190.80

Type L copper branch reducing tee C x C x C with brazed joints

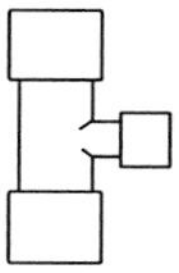

Description	Craft@Hrs	Unit	Material $	Labor $	Equipment $	Total $
½" x 3/8"	P1@.129	Ea	1.65	4.73	—	6.38
¾" x ½"	P1@.182	Ea	3.50	6.68	—	10.18
1" x ½"	P1@.209	Ea	6.85	7.67	—	14.52
1" x ¾"	P1@.234	Ea	6.85	8.59	—	15.44
1¼" x ½"	P1@.241	Ea	10.30	8.84	—	19.14
1¼" x ¾"	P1@.264	Ea	10.30	9.69	—	19.99
1¼" x 1"	P1@.287	Ea	10.30	10.50	—	20.80
1½" x ½"	P1@.291	Ea	12.70	10.70	—	23.40
1½" x ¾"	P1@.307	Ea	12.70	11.30	—	24.00
1½" x 1"	P1@.323	Ea	12.70	11.90	—	24.60
1½" x 1¼"	P1@.339	Ea	12.70	12.40	—	25.10
2" x ½"	P1@.348	Ea	16.60	12.80	—	29.40
2" x ¾"	P1@.373	Ea	16.60	13.70	—	30.30
2" x 1"	P1@.399	Ea	16.60	14.60	—	31.20
2" x 1¼"	P1@.426	Ea	16.60	15.60	—	32.20
2" x 1½"	P1@.452	Ea	16.60	16.60	—	33.20
2½" x ½"	P1@.468	Ea	50.00	17.20	—	67.20
2½" x ¾"	P1@.485	Ea	50.00	17.80	—	67.80
2½" x 1"	P1@.502	Ea	50.00	18.40	—	68.40
2½" x 1½"	P1@.537	Ea	50.00	19.70	—	69.70
2½" x 2"	P1@.555	Ea	50.00	20.40	—	70.40
3" x 1¼"	P1@.562	Ea	56.50	20.60	—	77.10
3" x 2"	P1@.626	Ea	58.50	23.00	—	81.50
3" x 2½"	P1@.660	Ea	58.50	24.20	—	82.70
4" x 1¼"	P1@.669	Ea	102.00	24.60	—	126.60
4" x 1½"	P1@.718	Ea	102.00	26.40	—	128.40
4" x 2"	P1@.768	Ea	102.00	28.20	—	130.20
4" x 2½"	P1@.819	Ea	102.00	30.10	—	132.10
4" x 3"	P1@.868	Ea	102.00	31.90	—	133.90

Copper, Type L with Brazed Joints

Description	Craft@Hrs	Unit	Material $	Labor $	Equipment $	Total $

Type L copper reducer with brazed joints

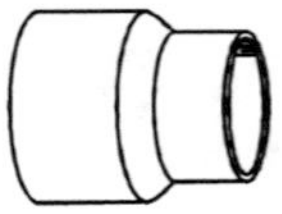

Description	Craft@Hrs	Unit	Material $	Labor $	Equipment $	Total $
½" x 3/8"	P1@.148	Ea	.59	5.43	—	6.02
¾" x ½"	P1@.155	Ea	1.11	5.69	—	6.80
1" x ¾"	P1@.206	Ea	2.00	7.56	—	9.56
1¼" x ½"	P1@.206	Ea	3.36	7.56	—	10.92
1¼" x ¾"	P1@.232	Ea	3.36	8.51	—	11.87
1¼" x 1"	P1@.258	Ea	3.36	9.47	—	12.83
1½" x ½"	P1@.232	Ea	5.14	8.51	—	13.65
1½" x ¾"	P1@.257	Ea	5.14	9.43	—	14.57
1½" x 1"	P1@.283	Ea	5.14	10.40	—	15.54
1½" x 1¼"	P1@.308	Ea	5.14	11.30	—	16.44
2" x ½"	P1@.287	Ea	7.74	10.50	—	18.24
2" x ¾"	P1@.277	Ea	7.74	10.20	—	17.94
2" x 1"	P1@.338	Ea	7.74	12.40	—	20.14
2" x 1¼"	P1@.365	Ea	7.74	13.40	—	21.14
2" x 1½"	P1@.390	Ea	7.74	14.30	—	22.04
2½" x 1"	P1@.390	Ea	21.10	14.30	—	35.40
2½" x 1¼"	P1@.416	Ea	21.10	15.30	—	36.40
2½" x 1½"	P1@.442	Ea	21.10	16.20	—	37.30
2½" x 2"	P1@.497	Ea	21.10	18.20	—	39.30
3" x 1¼"	P1@.468	Ea	24.60	17.20	—	41.80
3" x 1½"	P1@.493	Ea	24.60	18.10	—	42.70
3" x 2"	P1@.548	Ea	24.60	20.10	—	44.70
3" x 2½"	P1@.600	Ea	24.60	22.00	—	46.60
4" x 2"	P1@.652	Ea	47.50	23.90	—	71.40
4" x 2½"	P1@.704	Ea	47.50	25.80	—	73.30
4" x 3"	P1@.755	Ea	47.50	27.70	—	75.20

Type L copper adapter C x MPT with brazed joint

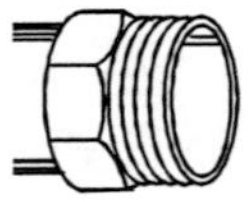

Description	Craft@Hrs	Unit	Material $	Labor $	Equipment $	Total $
½"	P1@.090	Ea	1.46	3.30	—	4.76
¾"	P1@.126	Ea	2.29	4.62	—	6.91
1"	P1@.162	Ea	5.67	5.95	—	11.62
1¼"	P1@.198	Ea	10.60	7.27	—	17.87
1½"	P1@.234	Ea	11.90	8.59	—	20.49
2"	P1@.312	Ea	14.90	11.50	—	26.40
2½"	P1@.384	Ea	48.30	14.10	—	62.40
3"	P1@.456	Ea	80.10	16.70	—	96.80

Type L copper adapter C x FPT with brazed joint

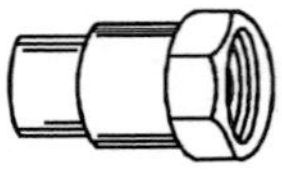

Description	Craft@Hrs	Unit	Material $	Labor $	Equipment $	Total $
½"	P1@.090	Ea	2.43	3.30	—	5.73
¾"	P1@.126	Ea	4.41	4.62	—	9.03
1"	P1@.162	Ea	6.66	5.95	—	12.61
1¼"	P1@.198	Ea	14.00	7.27	—	21.27
1½"	P1@.234	Ea	16.90	8.59	—	25.49
2"	P1@.312	Ea	25.20	11.50	—	36.70
2½"	P1@.384	Ea	63.40	14.10	—	77.50
3"	P1@.456	Ea	87.20	16.70	—	103.90

Description	Craft@Hrs	Unit	Material $	Labor $	Equipment $	Total $

Type L copper flush bushing with brazed joints

Description	Craft@Hrs	Unit	Material $	Labor $	Equipment $	Total $
½" x 3/8"	P1@.148	Ea	1.48	5.43	—	6.91
¾" x ½"	P1@.155	Ea	2.53	5.69	—	8.22
1" x ¾"	P1@.206	Ea	6.54	7.56	—	14.10
1" x ½"	P1@.206	Ea	7.46	7.56	—	15.02
1¼" x 1"	P1@.258	Ea	7.76	9.47	—	17.23
1½" x 1¼"	P1@.308	Ea	9.87	11.30	—	21.17
2" x 1½"	P1@.390	Ea	18.00	14.30	—	32.30

Type L copper union with brazed joints

Description	Craft@Hrs	Unit	Material $	Labor $	Equipment $	Total $
½"	P1@.146	Ea	5.94	5.36	—	11.30
¾"	P1@.205	Ea	7.09	7.52	—	14.61
1"	P1@.263	Ea	15.90	9.65	—	25.55
1¼"	P1@.322	Ea	28.90	11.80	—	40.70
1½"	P1@.380	Ea	35.50	13.90	—	49.40
2"	P1@.507	Ea	51.30	18.60	—	69.90

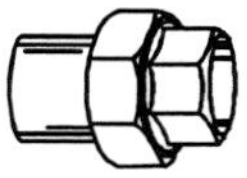

Type L copper dielectric union with brazed joints

Description	Craft@Hrs	Unit	Material $	Labor $	Equipment $	Total $
½"	P1@.146	Ea	5.57	5.36	—	10.93
¾"	P1@.205	Ea	5.57	7.52	—	13.09
1"	P1@.263	Ea	11.50	9.65	—	21.15
1¼"	P1@.322	Ea	19.60	11.80	—	31.40
1½"	P1@.380	Ea	27.00	13.90	—	40.90
2"	P1@.507	Ea	36.10	18.60	—	54.70

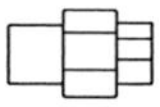

Type L copper cap with brazed joints

Description	Craft@Hrs	Unit	Material $	Labor $	Equipment $	Total $
½"	P1@.083	Ea	.49	3.05	—	3.54
¾"	P1@.116	Ea	.99	4.26	—	5.25
1"	P1@.149	Ea	2.23	5.47	—	7.70
1¼"	P1@.182	Ea	3.21	6.68	—	9.89
1½"	P1@.215	Ea	4.68	7.89	—	12.57
2"	P1@.286	Ea	8.14	10.50	—	18.64
2½"	P1@.352	Ea	20.70	12.90	—	33.60
3"	P1@.418	Ea	32.30	15.30	—	47.60
4"	P1@.550	Ea	49.70	20.20	—	69.90

Type L copper coupling with brazed joints

Description	Craft@Hrs	Unit	Material $	Labor $	Equipment $	Total $
½"	P1@.129	Ea	.34	4.73	—	5.07
¾"	P1@.181	Ea	.79	6.64	—	7.43
1"	P1@.232	Ea	2.12	8.51	—	10.63
1¼"	P1@.284	Ea	3.70	10.40	—	14.10
1½"	P1@.335	Ea	5.35	12.30	—	17.65
2"	P1@.447	Ea	7.86	16.40	—	24.26
2½"	P1@.550	Ea	18.60	20.20	—	38.80
3"	P1@.654	Ea	28.30	24.00	—	52.30
4"	P1@.860	Ea	55.40	31.60	—	87.00

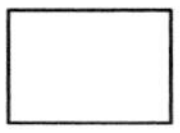

Copper, Type L with Brazed Joints

Description	Craft@Hrs	Unit	Material $	Labor $	Equipment $	Total $

Class 125 bronze body gate valve, brazed joints

Description	Craft@Hrs	Unit	Material $	Labor $	Equipment $	Total $
½"	P1@.240	Ea	18.40	8.81	—	27.21
¾"	P1@.300	Ea	23.00	11.00	—	34.00
1"	P1@.360	Ea	32.40	13.20	—	45.60
1¼"	P1@.480	Ea	40.70	17.60	—	58.30
1½"	P1@.540	Ea	54.90	19.80	—	74.70
2"	P1@.600	Ea	91.80	22.00	—	113.80
2½"	P1@1.00	Ea	152.00	36.70	—	188.70
3"	P1@1.50	Ea	218.00	55.10	—	273.10

Class 125 iron body gate valve, flanged joints

Description	Craft@Hrs	Unit	Material $	Labor $	Equipment $	Total $
2"	P1@.500	Ea	351.00	18.40	—	369.40
2½"	P1@.600	Ea	474.00	22.00	—	496.00
3"	P1@.750	Ea	514.00	27.50	—	541.50
4"	P1@1.35	Ea	754.00	49.50	—	803.50
5"	P1@2.00	Ea	1,440.00	73.40	—	1,513.40
6"	P1@2.50	Ea	1,440.00	91.80	—	1,531.80

Class 125 bronze body globe valve, with brazed joints

Description	Craft@Hrs	Unit	Material $	Labor $	Equipment $	Total $
½"	P1@.240	Ea	34.10	8.81	—	42.91
¾"	P1@.300	Ea	45.40	11.00	—	56.40
1"	P1@.360	Ea	78.20	13.20	—	91.40
1¼"	P1@.480	Ea	110.00	17.60	—	127.60
1½"	P1@.540	Ea	149.00	19.80	—	168.80
2"	P1@.600	Ea	241.00	22.00	—	263.00

Class 125 iron body globe valve, flanged joints

Description	Craft@Hrs	Unit	Material $	Labor $	Equipment $	Total $
2½"	P1@.650	Ea	398.00	23.90	—	421.90
3"	P1@.750	Ea	477.00	27.50	—	504.50
4"	P1@1.35	Ea	644.00	49.50	—	693.50

200 PSIG iron body butterfly valve, lug-type, lever operated

Description	Craft@Hrs	Unit	Material $	Labor $	Equipment $	Total $
2"	P1@.450	Ea	156.00	16.50	—	172.50
2½"	P1@.450	Ea	162.00	16.50	—	178.50
3"	P1@.550	Ea	170.00	20.20	—	190.20
4"	P1@.550	Ea	212.00	20.20	—	232.20

200 PSIG iron body butterfly valve, wafer-type, lever operated

Description	Craft@Hrs	Unit	Material $	Labor $	Equipment $	Total $
2"	P1@.450	Ea	142.00	16.50	—	158.50
2½"	P1@.450	Ea	145.00	16.50	—	161.50
3"	P1@.550	Ea	156.00	20.20	—	176.20
4"	P1@.550	Ea	189.00	20.20	—	209.20

Description	Craft@Hrs	Unit	Material $	Labor $	Equipment $	Total $
Class 125 bronze body 2-piece ball valve, brazed joints						
½"	P1@.240	Ea	10.10	8.81	—	18.91
¾"	P1@.300	Ea	13.40	11.00	—	24.40
1"	P1@.360	Ea	23.30	13.20	—	36.50
1¼"	P1@.480	Ea	38.80	17.60	—	56.40
1½"	P1@.540	Ea	52.40	19.80	—	72.20
2"	P1@.600	Ea	66.50	22.00	—	88.50
3"	P1@1.50	Ea	455.00	55.10	—	510.10
4"	P1@1.80	Ea	595.00	66.10	—	661.10
Class 125 bronze body swing check valve, brazed joints						
½"	P1@.240	Ea	20.40	8.81	—	29.21
¾"	P1@.300	Ea	29.20	11.00	—	40.20
1"	P1@.360	Ea	38.10	13.20	—	51.30
1¼"	P1@.480	Ea	54.40	17.60	—	72.00
1½"	P1@.540	Ea	76.70	19.80	—	96.50
2"	P1@.600	Ea	127.00	22.00	—	149.00
2½"	P1@1.00	Ea	256.00	36.70	—	292.70
3"	P1@1.50	Ea	368.00	55.10	—	423.10
Class 125 iron body swing check valve, flanged joints						
2"	P1@.500	Ea	179.00	18.40	—	197.40
2½"	P1@.600	Ea	226.00	22.00	—	248.00
3"	P1@.750	Ea	282.00	27.50	—	309.50
4"	P1@1.35	Ea	414.00	49.50	—	463.50
Class 125 iron body silent check valve, wafer-type						
2"	P1@.500	Ea	131.00	18.40	—	149.40
2½"	P1@.600	Ea	145.00	22.00	—	167.00
3"	P1@.750	Ea	167.00	27.50	—	194.50
4"	P1@1.35	Ea	223.00	49.50	—	272.50
Class 125 bronze body strainer, threaded joints						
½"	P1@.230	Ea	31.40	8.44	—	39.84
¾"	P1@.260	Ea	41.30	9.54	—	50.84
1"	P1@.330	Ea	50.60	12.10	—	62.70
1¼"	P1@.440	Ea	70.70	16.10	—	86.80
1½"	P1@.495	Ea	106.00	18.20	—	124.20
2"	P1@.550	Ea	185.00	20.20	—	205.20
Class 125 iron body strainer, flanged joints						
2"	P1@.500	Ea	141.00	18.40	—	159.40
2½"	P1@.600	Ea	157.00	22.00	—	179.00
3"	P1@.750	Ea	183.00	27.50	—	210.50
4"	P1@1.35	Ea	308.00	49.50	—	357.50

Copper, Type L with Brazed Joints

Description	Craft@Hrs	Unit	Material $	Labor $	Equipment $	Total $
Installation of 2-way control valve, threaded joints						
½"	P1@.210	Ea	—	7.71	—	7.71
¾"	P1@.275	Ea	—	10.10	—	10.10
1"	P1@.350	Ea	—	12.80	—	12.80
1¼"	P1@.430	Ea	—	15.80	—	15.80
1½"	P1@.505	Ea	—	18.50	—	18.50
2"	P1@.675	Ea	—	24.80	—	24.80
2½"	P1@.830	Ea	—	30.50	—	30.50
3"	P1@.990	Ea	—	36.30	—	36.30
Installation of 3-way control valve, threaded joints						
½"	P1@.260	Ea	—	9.54	—	9.54
¾"	P1@.365	Ea	—	13.40	—	13.40
1"	P1@.475	Ea	—	17.40	—	17.40
1¼"	P1@.575	Ea	—	21.10	—	21.10
1½"	P1@.680	Ea	—	25.00	—	25.00
2"	P1@.910	Ea	—	33.40	—	33.40
2½"	P1@1.12	Ea	—	41.10	—	41.10
3"	P1@1.33	Ea	—	48.80	—	48.80
Companion flange, 150 pound cast brass						
2"	P1@.290	Ea	305.00	10.60	—	315.60
2½"	P1@.380	Ea	376.00	13.90	—	389.90
3"	P1@.460	Ea	376.00	16.90	—	392.90
4"	P1@.600	Ea	599.00	22.00	—	621.00
Bolt and gasket sets						
2"	P1@.500	Ea	4.16	18.40	—	22.56
2½"	P1@.650	Ea	4.86	23.90	—	28.76
3"	P1@.750	Ea	8.25	27.50	—	35.75
4"	P1@1.00	Ea	13.80	36.70	—	50.50
Thermometer with well						
7"	P1@.250	Ea	180.00	9.18	—	189.18
9"	P1@.250	Ea	185.00	9.18	—	194.18
Dial-type pressure gauge						
2½"	P1@.200	Ea	37.70	7.34	—	45.04
3½"	P1@.200	Ea	49.80	7.34	—	57.14

Type L hard-drawn copper pipe with wrought copper fittings and soft-soldered joints is used in a wide variety of plumbing and HVAC systems such as potable water, heating hot water, chilled water, and A.C. condensate.

Soft-soldered joints are those made with solders having melting points in the 350 degree F. to 500 degree F. range. Maximum working temperature/pressure relationships for soft-soldered joints are approximately as follows:

Maximum Working Pressures (PSIG)*				
Soft-solder Type	Water Temperature (degrees F.)	Nominal Pipe Size (inches)		
		Up to 1	1¼ to 2	2½ x 4
50-50 tin-lead**	100	200	175	150
	150	150	125	100
	200	100	90	75
	250	85	75	50
95-5 tin-antimony	100	500	400	300
	150	400	350	275
	200	300	250	200
	250	200	175	150

*For copper pipe and solder-type fittings using soft-solders melting at approximately 350 degrees F. to 500 degrees F.

**The use of any solder containing lead is not allowed in potable water systems.

This section has been arranged to save the estimator's time by including all normally-used system components such as pipe, fittings, valves, hanger assemblies, riser clamps and miscellaneous items under one heading. Additional items can be found under "Plumbing and Piping Specialties." The cost estimates in this section are based on the conditions, limitations and wage rates described in the section "How to Use This Book" beginning on page 5.

Description	Craft@Hrs	Unit	Material $	Labor $	Equipment $	Total $

Type L copper pipe with soft-soldered joints installed horizontally.
Complete installation including eight wrought copper tees and eight wrought copper elbows every 100', and hangers spaced to meet plumbing code. Use these figures for preliminary estimates.

½"	P1@.111	LF	2.16	4.07	—	6.23
¾"	P1@.121	LF	3.35	4.44	—	7.79
1"	P1@.141	LF	6.06	5.17	—	11.23
1¼"	P1@.161	LF	8.64	5.91	—	14.55
1½"	P1@.181	LF	11.70	6.64	—	18.34
2"	P1@.211	LF	21.10	7.74	—	28.84
2½"	P1@.241	LF	36.40	8.84	—	45.24
3"	P1@.271	LF	50.90	9.95	—	60.85
4"	P1@.311	LF	73.50	11.40	—	84.90

Copper, Type L with Soft-Soldered Joints

Description	Craft@Hrs	Unit	Material $	Labor $	Equipment $	Total $

Type L copper pipe with soft-soldered joints, installed risers.

Complete installation including a wrought copper reducing tee every floor and a riser clamp every other floor.

Description	Craft@Hrs	Unit	Material $	Labor $	Equipment $	Total $
½"	P1@.061	LF	2.35	2.24	—	4.59
¾"	P1@.071	LF	3.64	2.61	—	6.25
1"	P1@.076	LF	6.17	2.79	—	8.96
1¼"	P1@.081	LF	8.71	2.97	—	11.68
1½"	P1@.101	LF	10.90	3.71	—	14.61
2"	P1@.131	LF	28.50	4.81	—	33.31
2½"	P1@.151	LF	30.80	5.54	—	36.34
3"	P1@.171	LF	39.90	6.28	—	46.18
4"	P1@.191	LF	59.40	7.01	—	66.41

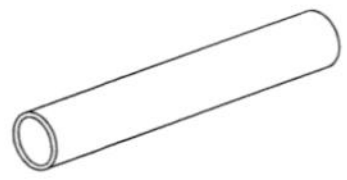

Type L copper pipe with soft-soldered joints, pipe only

Description	Craft@Hrs	Unit	Material $	Labor $	Equipment $	Total $
½"	P1@.032	LF	1.80	1.17	—	2.97
¾"	P1@.035	LF	2.85	1.28	—	4.13
1"	P1@.038	LF	3.95	1.39	—	5.34
1¼"	P1@.042	LF	5.67	1.54	—	7.21
1½"	P1@.046	LF	7.22	1.69	—	8.91
2"	P1@.053	LF	12.40	1.95	—	14.35
2½"	P1@.060	LF	18.10	2.20	—	20.30
3"	P1@.066	LF	24.50	2.42	—	26.92
4"	P1@.080	LF	38.40	2.94	—	41.34

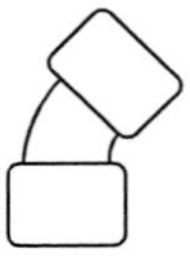

Type L copper 45-degree ell C x C with soft-soldered joints

Description	Craft@Hrs	Unit	Material $	Labor $	Equipment $	Total $
½"	P1@.107	Ea	.82	3.93	—	4.75
¾"	P1@.150	Ea	1.56	5.51	—	7.07
1"	P1@.193	Ea	5.62	7.08	—	12.70
1¼"	P1@.236	Ea	7.62	8.66	—	16.28
1½"	P1@.278	Ea	9.28	10.20	—	19.48
2"	P1@.371	Ea	14.70	13.60	—	28.30
2½"	P1@.457	Ea	37.50	16.80	—	54.30
3"	P1@.543	Ea	49.60	19.90	—	69.50
4"	P1@.714	Ea	95.30	26.20	—	121.50

Type L copper 90-degree ell C x C with soft-soldered joints

Description	Craft@Hrs	Unit	Material $	Labor $	Equipment $	Total $
½"	P1@.107	Ea	.48	3.93	—	4.41
¾"	P1@.150	Ea	1.01	5.51	—	6.52
1"	P1@.193	Ea	3.51	7.08	—	10.59
1¼"	P1@.236	Ea	5.67	8.66	—	14.33
1½"	P1@.278	Ea	8.55	10.20	—	18.75
2"	P1@.371	Ea	14.90	13.60	—	28.50
2½"	P1@.457	Ea	31.00	16.80	—	47.80
3"	P1@.543	Ea	39.30	19.90	—	59.20
4"	P1@.714	Ea	83.10	26.20	—	109.30

Description	Craft@Hrs	Unit	Material $	Labor $	Equipment $	Total $

Type L copper 90-degree ell ftg. x C with soft-soldered joint

Description	Craft@Hrs	Unit	Material $	Labor $	Equipment $	Total $
½"	P1@.107	Ea	1.01	3.93	—	4.94
¾"	P1@.150	Ea	3.31	5.51	—	8.82
1"	P1@.193	Ea	4.97	7.08	—	12.05
1¼"	P1@.236	Ea	6.93	8.66	—	15.59
1½"	P1@.278	Ea	10.30	10.20	—	20.50
2"	P1@.371	Ea	20.20	13.60	—	33.80
2½"	P1@.457	Ea	42.20	16.80	—	59.00
3"	P1@.543	Ea	47.50	19.90	—	67.40
4"	P1@.714	Ea	127.00	26.20	—	153.20

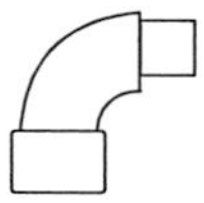

Type L copper tee C x C x C with soft-soldered joints

Description	Craft@Hrs	Unit	Material $	Labor $	Equipment $	Total $
½"	P1@.129	Ea	.81	4.73	—	5.54
¾"	P1@.181	Ea	2.19	6.64	—	8.83
1"	P1@.233	Ea	8.23	8.55	—	16.78
1¼"	P1@.285	Ea	13.10	10.50	—	23.60
1½"	P1@.337	Ea	19.20	12.40	—	31.60
2"	P1@.449	Ea	29.60	16.50	—	46.10
2½"	P1@.552	Ea	60.00	20.30	—	80.30
3"	P1@.656	Ea	92.80	24.10	—	116.90
4"	P1@.863	Ea	155.00	31.70	—	186.70

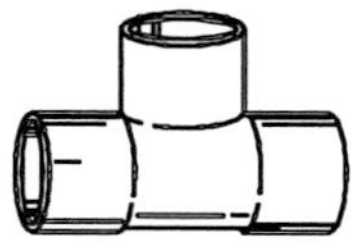

Type L copper branch reducing tee C x C x C with soft-soldered joints

Description	Craft@Hrs	Unit	Material $	Labor $	Equipment $	Total $
½" x 3/8"	P1@.129	Ea	1.65	4.73	—	6.38
¾" x ½"	P1@.182	Ea	3.50	6.68	—	10.18
1" x ½"	P1@.195	Ea	6.85	7.16	—	14.01
1" x ¾"	P1@.219	Ea	6.85	8.04	—	14.89
1¼" x ½"	P1@.225	Ea	10.30	8.26	—	18.56
1¼" x ¾"	P1@.247	Ea	10.30	9.06	—	19.36
1¼" x 1"	P1@.268	Ea	10.30	9.84	—	20.14
1½" x ½"	P1@.272	Ea	12.70	9.98	—	22.68
1½" x ¾"	P1@.287	Ea	12.70	10.50	—	23.20
1½" x 1"	P1@.302	Ea	12.70	11.10	—	23.80
1½" x 1¼"	P1@.317	Ea	12.70	11.60	—	24.30
2" x ½"	P1@.325	Ea	16.60	11.90	—	28.50
2" x ¾"	P1@.348	Ea	16.60	12.80	—	29.40
2" x 1"	P1@.373	Ea	16.60	13.70	—	30.30
2" x 1¼"	P1@.398	Ea	16.60	14.60	—	31.20
2" x 1½"	P1@.422	Ea	16.60	15.50	—	32.10
2½" x ½"	P1@.437	Ea	50.00	16.00	—	66.00
2½" x ¾"	P1@.453	Ea	50.00	16.60	—	66.60
2½" x 1"	P1@.469	Ea	50.00	17.20	—	67.20
2½" x 1½"	P1@.502	Ea	50.00	18.40	—	68.40
2½" x 2"	P1@.519	Ea	50.00	19.00	—	69.00
3" x 1¼"	P1@.525	Ea	56.50	19.30	—	75.80
3" x 2"	P1@.585	Ea	58.50	21.50	—	80.00
3" x 2½"	P1@.617	Ea	58.50	22.60	—	81.10
4" x 1¼"	P1@.625	Ea	102.00	22.90	—	124.90
4" x 1½"	P1@.671	Ea	102.00	24.60	—	126.60
4" x 2"	P1@.718	Ea	102.00	26.40	—	128.40
4" x 2½"	P1@.765	Ea	102.00	28.10	—	130.10
4" x 3"	P1@.811	Ea	102.00	29.80	—	131.80

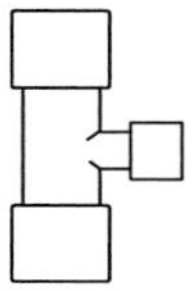

Copper, Type L with Soft-Soldered Joints

Description	Craft@Hrs	Unit	Material $	Labor $	Equipment $	Total $

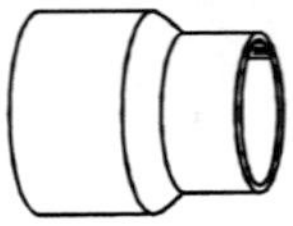

Type L copper reducer with soft-soldered joints

Description	Craft@Hrs	Unit	Material $	Labor $	Equipment $	Total $
½" x 3/8"	P1@.124	Ea	.59	4.55	—	5.14
¾" x ½"	P1@.129	Ea	1.11	4.73	—	5.84
1" x ¾"	P1@.172	Ea	2.00	6.31	—	8.31
1¼" x ½"	P1@.172	Ea	3.36	6.31	—	9.67
1¼" x ¾"	P1@.193	Ea	3.36	7.08	—	10.44
1¼" x 1"	P1@.215	Ea	3.36	7.89	—	11.25
1½" x ½"	P1@.193	Ea	5.14	7.08	—	12.22
1½" x ¾"	P1@.214	Ea	5.14	7.85	—	12.99
1½" x 1"	P1@.236	Ea	5.14	8.66	—	13.80
1½" x 1¼"	P1@.257	Ea	5.14	9.43	—	14.57
2" x ½"	P1@.239	Ea	7.74	8.77	—	16.51
2" x ¾"	P1@.261	Ea	7.74	9.58	—	17.32
2" x 1"	P1@.282	Ea	7.74	10.30	—	18.04
2" x 1¼"	P1@.304	Ea	7.74	11.20	—	18.94
2" x 1½"	P1@.325	Ea	7.74	11.90	—	19.64
2½" x 1"	P1@.325	Ea	21.10	11.90	—	33.00
2½" x 1¼"	P1@.347	Ea	21.10	12.70	—	33.80
2½" x 1½"	P1@.368	Ea	21.10	13.50	—	34.60
2½" x 2"	P1@.414	Ea	21.10	15.20	—	36.30
3" x 2"	P1@.457	Ea	24.60	16.80	—	41.40
3" x 2½"	P1@.500	Ea	24.60	18.40	—	43.00
4" x 2"	P1@.543	Ea	47.50	19.90	—	67.40
4" x 2½"	P1@.586	Ea	47.50	21.50	—	69.00
4" x 3"	P1@.629	Ea	47.50	23.10	—	70.60

Type L copper adapter C x MPT with soft-soldered joint

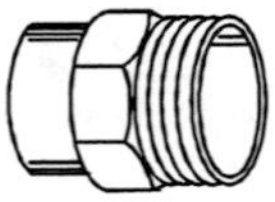

Description	Craft@Hrs	Unit	Material $	Labor $	Equipment $	Total $
½"	P1@.075	Ea	1.46	2.75	—	4.21
¾"	P1@.105	Ea	2.29	3.85	—	6.14
1"	P1@.134	Ea	5.67	4.92	—	10.59
1¼"	P1@.164	Ea	10.60	6.02	—	16.62
1½"	P1@.194	Ea	11.90	7.12	—	19.02
2"	P1@.259	Ea	14.90	9.51	—	24.41
2½"	P1@.319	Ea	48.30	11.70	—	60.00
3"	P1@.378	Ea	80.10	13.90	—	94.00

Type L copper adapter C x FPT with soft-soldered joint

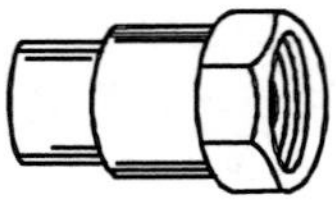

Description	Craft@Hrs	Unit	Material $	Labor $	Equipment $	Total $
½"	P1@.075	Ea	2.43	2.75	—	5.18
¾"	P1@.105	Ea	4.41	3.85	—	8.26
1"	P1@.134	Ea	6.66	4.92	—	11.58
1¼"	P1@.164	Ea	14.00	6.02	—	20.02
1½"	P1@.194	Ea	16.90	7.12	—	24.02
2"	P1@.259	Ea	25.20	9.51	—	34.71
2½"	P1@.319	Ea	63.40	11.70	—	75.10
3"	P1@.378	Ea	87.20	13.90	—	101.10

Description	Craft@Hrs	Unit	Material $	Labor $	Equipment $	Total $

Type L copper flush bushing with soft-soldered joints

Description	Craft@Hrs	Unit	Material $	Labor $	Equipment $	Total $
½" x 3/8"	P1@.124	Ea	1.48	4.55	—	6.03
¾" x ½"	P1@.129	Ea	2.53	4.73	—	7.26
1" x ¾"	P1@.172	Ea	6.54	6.31	—	12.85
1" x ½"	P1@.172	Ea	7.46	6.31	—	13.77
1¼" x 1"	P1@.215	Ea	7.76	7.89	—	15.65
1½" x 1¼"	P1@.257	Ea	9.87	9.43	—	19.30
2" x 1½"	P1@.325	Ea	18.00	11.90	—	29.90

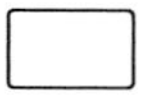

Type L copper union with soft-soldered joint

Description	Craft@Hrs	Unit	Material $	Labor $	Equipment $	Total $
½"	P1@.121	Ea	5.94	4.44	—	10.38
¾"	P1@.170	Ea	7.09	6.24	—	13.33
1"	P1@.218	Ea	15.90	8.00	—	23.90
1¼"	P1@.267	Ea	28.90	9.80	—	38.70
1½"	P1@.315	Ea	35.50	11.60	—	47.10
2"	P1@.421	Ea	51.30	15.50	—	66.80

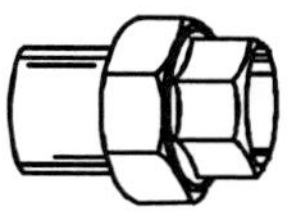

Type L copper dielectric union with soft-soldered joint

Description	Craft@Hrs	Unit	Material $	Labor $	Equipment $	Total $
½"	P1@.121	Ea	5.57	4.44	—	10.01
¾"	P1@.170	Ea	5.57	6.24	—	11.81
1"	P1@.218	Ea	11.50	8.00	—	19.50
1¼"	P1@.267	Ea	19.60	9.80	—	29.40
1½"	P1@.315	Ea	27.00	11.60	—	38.60
2"	P1@.421	Ea	36.10	15.50	—	51.60

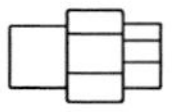

Type L copper cap with soft-soldered joint

Description	Craft@Hrs	Unit	Material $	Labor $	Equipment $	Total $
½"	P1@.069	Ea	.49	2.53	—	3.02
¾"	P1@.096	Ea	.99	3.52	—	4.51
1"	P1@.124	Ea	2.23	4.55	—	6.78
1¼"	P1@.151	Ea	3.21	5.54	—	8.75
1½"	P1@.178	Ea	4.68	6.53	—	11.21
2"	P1@.237	Ea	8.14	8.70	—	16.84
2½"	P1@.292	Ea	20.70	10.70	—	31.40
3"	P1@.347	Ea	32.30	12.70	—	45.00
4"	P1@.457	Ea	49.70	16.80	—	66.50

Type L copper coupling with soft-soldered joints

Description	Craft@Hrs	Unit	Material $	Labor $	Equipment $	Total $
½"	P1@.107	Ea	.34	3.93	—	4.27
¾"	P1@.150	Ea	.79	5.51	—	6.30
1"	P1@.193	Ea	2.12	7.08	—	9.20
1¼"	P1@.236	Ea	3.70	8.66	—	12.36
1½"	P1@.278	Ea	5.35	10.20	—	15.55
2"	P1@.371	Ea	7.86	13.60	—	21.46
2½"	P1@.457	Ea	18.60	16.80	—	35.40
3"	P1@.543	Ea	28.30	19.90	—	48.20
4"	P1@.714	Ea	55.40	26.20	—	81.60

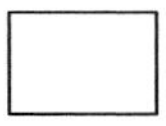

Copper, Type L with Soft-Soldered Joints

Description	Craft@Hrs	Unit	Material $	Labor $	Equipment $	Total $

Class 125 bronze body gate valve, soft-soldered joints

Description	Craft@Hrs	Unit	Material $	Labor $	Equipment $	Total $
½"	P1@.200	Ea	17.50	7.34	—	24.84
¾"	P1@.249	Ea	21.80	9.14	—	30.94
1"	P1@.299	Ea	30.70	11.00	—	41.70
1¼"	P1@.398	Ea	38.90	14.60	—	53.50
1½"	P1@.448	Ea	52.30	16.40	—	68.70
2"	P1@.498	Ea	87.20	18.30	—	105.50
2½"	P1@.830	Ea	146.00	30.50	—	176.50
3"	P1@1.24	Ea	209.00	45.50	—	254.50

Class 125 iron body gate valve, flanged ends

Description	Craft@Hrs	Unit	Material $	Labor $	Equipment $	Total $
2"	P1@.500	Ea	351.00	18.40	—	369.40
2½"	P1@.600	Ea	474.00	22.00	—	496.00
3"	P1@.750	Ea	514.00	27.50	—	541.50
4"	P1@1.35	Ea	754.00	49.50	—	803.50
5"	P1@2.00	Ea	1,440.00	73.40	—	1,513.40
6"	P1@2.50	Ea	1,440.00	91.80	—	1,531.80

Class 125 bronze body globe valve, with soft-soldered joints

Description	Craft@Hrs	Unit	Material $	Labor $	Equipment $	Total $
½"	P1@.200	Ea	34.10	7.34	—	41.44
¾"	P1@.249	Ea	45.40	9.14	—	54.54
1"	P1@.299	Ea	78.20	11.00	—	89.20
1¼"	P1@.398	Ea	110.00	14.60	—	124.60
1½"	P1@.448	Ea	149.00	16.40	—	165.40
2"	P1@.498	Ea	241.00	18.30	—	259.30

Class 125 iron body globe valve, flanged ends

Description	Craft@Hrs	Unit	Material $	Labor $	Equipment $	Total $
2½"	P1@.600	Ea	398.00	22.00	—	420.00
3"	P1@.750	Ea	477.00	27.50	—	504.50
4"	P1@1.35	Ea	640.00	49.50	—	689.50

200 PSIG iron body butterfly valve, lug-type, lever operated

Description	Craft@Hrs	Unit	Material $	Labor $	Equipment $	Total $
2"	P1@.450	Ea	156.00	16.50	—	172.50
2½"	P1@.450	Ea	162.00	16.50	—	178.50
3"	P1@.550	Ea	170.00	20.20	—	190.20
4"	P1@.550	Ea	212.00	20.20	—	232.20

200 PSIG iron body butterfly valve, wafer-type, lever operated

Description	Craft@Hrs	Unit	Material $	Labor $	Equipment $	Total $
2"	P1@.450	Ea	142.00	16.50	—	158.50
2½"	P1@.450	Ea	145.00	16.50	—	161.50
3"	P1@.550	Ea	156.00	20.20	—	176.20
4"	P1@.550	Ea	189.00	20.20	—	209.20

Description	Craft@Hrs	Unit	Material $	Labor $	Equipment $	Total $

Class 125 bronze body 2-piece ball valve, soft-soldered joints

Description	Craft@Hrs	Unit	Material $	Labor $	Equipment $	Total $
½"	P1@.200	Ea	10.10	7.34	—	17.44
¾"	P1@.249	Ea	13.40	9.14	—	22.54
1"	P1@.299	Ea	23.30	11.00	—	34.30
1¼"	P1@.398	Ea	38.80	14.60	—	53.40
1½"	P1@.448	Ea	52.40	16.40	—	68.80
2"	P1@.498	Ea	66.50	18.30	—	84.80
3"	P1@1.24	Ea	455.00	45.50	—	500.50
4"	P1@1.45	Ea	595.00	53.20	—	648.20

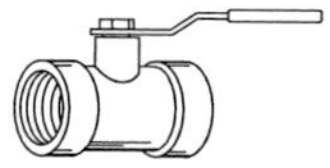

Class 125 bronze body swing check valve, soft-soldered joints

Description	Craft@Hrs	Unit	Material $	Labor $	Equipment $	Total $
½"	P1@.200	Ea	20.40	7.34	—	27.74
¾"	P1@.249	Ea	29.20	9.14	—	38.34
1"	P1@.299	Ea	38.10	11.00	—	49.10
1¼"	P1@.398	Ea	54.40	14.60	—	69.00
1½"	P1@.448	Ea	76.70	16.40	—	93.10
2"	P1@.498	Ea	127.00	18.30	—	145.30
2½"	P1@.830	Ea	256.00	30.50	—	286.50
3"	P1@1.24	Ea	368.00	45.50	—	413.50

Class 125 iron body swing check valve, flanged ends

Description	Craft@Hrs	Unit	Material $	Labor $	Equipment $	Total $
2"	P1@.500	Ea	179.00	18.40	—	197.40
2½"	P1@.600	Ea	226.00	22.00	—	248.00
3"	P1@.750	Ea	282.00	27.50	—	309.50
4"	P1@1.35	Ea	414.00	49.50	—	463.50

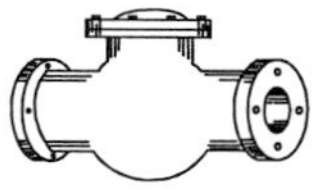

Class 125 iron body silent check valve, wafer-type

Description	Craft@Hrs	Unit	Material $	Labor $	Equipment $	Total $
2"	P1@.500	Ea	131.00	18.40	—	149.40
2½"	P1@.600	Ea	145.00	22.00	—	167.00
3"	P1@.750	Ea	167.00	27.50	—	194.50
4"	P1@1.35	Ea	223.00	49.50	—	272.50

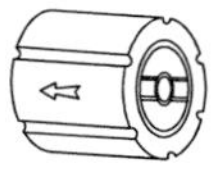

Class 125 bronze body strainer, threaded ends

Description	Craft@Hrs	Unit	Material $	Labor $	Equipment $	Total $
½"	P1@.230	Ea	31.40	8.44	—	39.84
¾"	P1@.260	Ea	41.30	9.54	—	50.84
1"	P1@.330	Ea	50.60	12.10	—	62.70
1¼"	P1@.440	Ea	70.70	16.10	—	86.80
1½"	P1@.495	Ea	106.00	18.20	—	124.20
2"	P1@.550	Ea	185.00	20.20	—	205.20

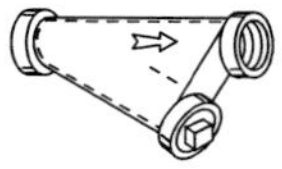

Class 125 iron body strainer, flanged ends

Description	Craft@Hrs	Unit	Material $	Labor $	Equipment $	Total $
2"	P1@.500	Ea	140.00	18.40	—	158.40
2½"	P1@.600	Ea	156.00	22.00	—	178.00
3"	P1@.750	Ea	182.00	27.50	—	209.50
4"	P1@1.35	Ea	306.00	49.50	—	355.50

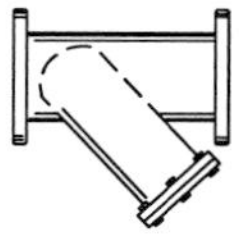

Copper, Type L with Soft-Soldered Joints

Description	Craft@Hrs	Unit	Material $	Labor $	Equipment $	Total $

Installation of 2-way control valve, threaded joints

Description	Craft@Hrs	Unit	Material $	Labor $	Equipment $	Total $
½"	P1@.210	Ea	—	7.71	—	7.71
¾"	P1@.275	Ea	—	10.10	—	10.10
1"	P1@.350	Ea	—	12.80	—	12.80
1¼"	P1@.430	Ea	—	15.80	—	15.80
1½"	P1@.505	Ea	—	18.50	—	18.50
2"	P1@.675	Ea	—	24.80	—	24.80
2½"	P1@.830	Ea	—	30.50	—	30.50
3"	P1@.990	Ea	—	36.30	—	36.30

Installation of 3-way control valve, threaded joints

Description	Craft@Hrs	Unit	Material $	Labor $	Equipment $	Total $
½"	P1@.260	Ea	—	9.54	—	9.54
¾"	P1@.365	Ea	—	13.40	—	13.40
1"	P1@.475	Ea	—	17.40	—	17.40
1¼"	P1@.575	Ea	—	21.10	—	21.10
1½"	P1@.680	Ea	—	25.00	—	25.00
2"	P1@.910	Ea	—	33.40	—	33.40
2½"	P1@1.12	Ea	—	41.10	—	41.10
3"	P1@1.33	Ea	—	48.80	—	48.80

Companion flange, 150 pound cast brass

Description	Craft@Hrs	Unit	Material $	Labor $	Equipment $	Total $
2"	P1@.290	Ea	302.00	10.60	—	312.60
2½"	P1@.380	Ea	372.00	13.90	—	385.90
3"	P1@.460	Ea	372.00	16.90	—	388.90
4"	P1@.600	Ea	592.00	22.00	—	614.00

Bolt and gasket sets

Description	Craft@Hrs	Unit	Material $	Labor $	Equipment $	Total $
2"	P1@.500	Ea	4.16	18.40	—	22.56
2½"	P1@.650	Ea	4.86	23.90	—	28.76
3"	P1@.750	Ea	8.25	27.50	—	35.75
4"	P1@1.00	Ea	13.80	36.70	—	50.50

Thermometer with well

Description	Craft@Hrs	Unit	Material $	Labor $	Equipment $	Total $
7"	P1@.250	Ea	180.00	9.18	—	189.18
9"	P1@.250	Ea	185.00	9.18	—	194.18

Dial-type pressure gauge

Description	Craft@Hrs	Unit	Material $	Labor $	Equipment $	Total $
2½"	P1@.200	Ea	37.70	7.34	—	45.04
3½"	P1@.200	Ea	49.70	7.34	—	57.04

Pressure/temperature tap

Description	Craft@Hrs	Unit	Material $	Labor $	Equipment $	Total $
Tap	P1@.150	Ea	17.30	5.51	—	22.81

Copper, Type L with Soft-Soldered Joints

Description	Craft@Hrs	Unit	Material $	Labor $	Equipment $	Total $
Hanger with swivel assembly						
½"	P1@.250	Ea	4.22	9.18	—	13.40
¾"	P1@.250	Ea	4.68	9.18	—	13.86
1"	P1@.250	Ea	4.89	9.18	—	14.07
1¼"	P1@.300	Ea	5.02	11.00	—	16.02
1½"	P1@.300	Ea	5.40	11.00	—	16.40
2"	P1@.300	Ea	5.63	11.00	—	16.63
2½"	P1@.350	Ea	7.65	12.80	—	20.45
3"	P1@.350	Ea	9.47	12.80	—	22.27
4"	P1@.350	Ea	10.40	12.80	—	23.20
Riser clamp						
½"	P1@.100	Ea	2.72	3.67	—	6.39
¾"	P1@.100	Ea	4.02	3.67	—	7.69
1"	P1@.100	Ea	4.06	3.67	—	7.73
1¼"	P1@.105	Ea	4.89	3.85	—	8.74
1½"	P1@.110	Ea	5.17	4.04	—	9.21
2"	P1@.115	Ea	5.47	4.22	—	9.69
2½"	P1@.120	Ea	5.76	4.40	—	10.16
3"	P1@.120	Ea	6.23	4.40	—	10.63
4"	P1@.125	Ea	7.94	4.59	—	12.53

Copper, Type M with Brazed Joints

Type M hard-drawn copper pipe with wrought copper fittings and brazed joints is used in a wide variety of plumbing and HVAC systems such as potable water, heating hot water and chilled water.

Brazed joints are those made with silver or other alloy filler metals having melting points at, or above, 1,000 degrees F. Maximum working pressure/temperature relationships for brazed joints are approximately as follows:

Maximum Working Pressure (PSIG)*			
Water Temperature (degrees F.)	Nominal Pipe Size (inches)		
	Up to 1	1¼ to 2	2½ to 4
Up to 350	270	190	150

*For copper pipe and solder-type fittings using brazing alloys melting at, or above, 1,000 degrees F.

This section has been arranged to save the estimator's time by including all normally-used system components such as pipe, fittings, valves, hanger assemblies, riser clamps and miscellaneous items under one heading. Additional items can be found under "Plumbing and Piping Specialties." The cost estimates in this section are based on the conditions, limitations and wage rates described in the section "How to Use This Book" beginning on page 5.

Description	Craft@Hrs	Unit	Material $	Labor $	Equipment $	Total $

Type M copper pipe with brazed joints installed horizontally. Complete installation including six to twelve wrought copper tees and six to twelve wrought copper elbows every 100', and hangers spaced to meet plumbing code (a tee & elbow every 8.3' for ½" dia., a tee and elbow every 16.6' for 4" dia.).

Description	Craft@Hrs	Unit	Material $	Labor $	Equipment $	Total $
½"	P1@.113	LF	1.90	4.15	—	6.05
¾"	P1@.123	LF	2.93	4.51	—	7.44
1"	P1@.143	LF	6.08	5.25	—	11.33
1¼"	P1@.163	LF	10.30	5.98	—	16.28
1½"	P1@.183	LF	12.10	6.72	—	18.82
2"	P1@.213	LF	25.50	7.82	—	33.32
2½"	P1@.243	LF	37.30	8.92	—	46.22
3"	P1@.273	LF	46.80	10.00	—	56.80
4"	P1@.313	LF	68.90	11.50	—	80.40

Type M copper pipe with brazed joints, installed risers. Complete installation including a wrought copper reducing tee every floor and a riser clamp every other floor.

Description	Craft@Hrs	Unit	Material $	Labor $	Equipment $	Total $
½"	P1@.062	LF	1.61	2.28	—	3.89
¾"	P1@.072	LF	2.60	2.64	—	5.24
1"	P1@.077	LF	5.09	2.83	—	7.92
1¼"	P1@.082	LF	7.51	3.01	—	10.52
1½"	P1@.102	LF	9.87	3.74	—	13.61
2"	P1@.132	LF	13.10	4.84	—	17.94
2½"	P1@.152	LF	28.50	5.58	—	34.08
3"	P1@.172	LF	35.00	6.31	—	41.31
4"	P1@.192	LF	51.90	7.05	—	58.95

Description	Craft@Hrs	Unit	Material $	Labor $	Equipment $	Total $

Type M copper pipe with brazed joints, pipe only

Description	Craft@Hrs	Unit	Material $	Labor $	Equipment $	Total $
½"	P1@.032	LF	1.16	1.17	—	2.33
¾"	P1@.035	LF	1.90	1.28	—	3.18
1"	P1@.038	LF	2.97	1.39	—	4.36
1¼"	P1@.042	LF	4.45	1.54	—	5.99
1½"	P1@.046	LF	6.24	1.69	—	7.93
2"	P1@.053	LF	11.50	1.95	—	13.45
2½"	P1@.060	LF	15.90	2.20	—	18.10
3"	P1@.066	LF	19.90	2.42	—	22.32
4"	P1@.080	LF	32.00	2.94	—	34.94

Type M copper 45-degree ell C x C with brazed joints

Description	Craft@Hrs	Unit	Material $	Labor $	Equipment $	Total $
½"	P1@.129	Ea	.82	4.73	—	5.55
¾"	P1@.181	Ea	1.56	6.64	—	8.20
1"	P1@.232	Ea	4.60	8.51	—	13.11
1¼"	P1@.284	Ea	6.23	10.40	—	16.63
1½"	P1@.335	Ea	9.28	12.30	—	21.58
2"	P1@.447	Ea	14.70	16.40	—	31.10
2½"	P1@.550	Ea	37.50	20.20	—	57.70
3"	P1@.654	Ea	49.60	24.00	—	73.60
4"	P1@.860	Ea	95.30	31.60	—	126.90

Type M copper 90-degree ell C x C with brazed joints

Description	Craft@Hrs	Unit	Material $	Labor $	Equipment $	Total $
½"	P1@.129	Ea	.47	4.73	—	5.20
¾"	P1@.181	Ea	.98	6.64	—	7.62
1"	P1@.232	Ea	3.51	8.51	—	12.02
1¼"	P1@.284	Ea	5.67	10.40	—	16.07
1½"	P1@.335	Ea	8.55	12.30	—	20.85
2"	P1@.447	Ea	14.90	16.40	—	31.30
2½"	P1@.550	Ea	31.00	20.20	—	51.20
3"	P1@.654	Ea	39.30	24.00	—	63.30
4"	P1@.860	Ea	83.10	31.60	—	114.70

Type M copper 90-degree ell ftg. x C with brazed joints

Description	Craft@Hrs	Unit	Material $	Labor $	Equipment $	Total $
½"	P1@.129	Ea	.77	4.73	—	5.50
¾"	P1@.181	Ea	1.66	6.64	—	8.30
1"	P1@.232	Ea	4.97	8.51	—	13.48
1¼"	P1@.284	Ea	6.93	10.40	—	17.33
1½"	P1@.335	Ea	10.30	12.30	—	22.60
2"	P1@.447	Ea	20.20	16.40	—	36.60
2½"	P1@.550	Ea	42.20	20.20	—	62.40
3"	P1@.654	Ea	47.50	24.00	—	71.50
4"	P1@.860	Ea	95.30	31.60	—	126.90

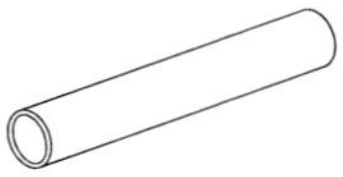

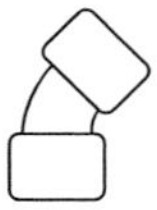

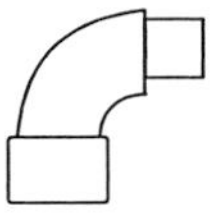

Copper, Type M with Brazed Joints

Description	Craft@Hrs	Unit	Material $	Labor $	Equipment $	Total $

Type M copper tee C x C x C with brazed joints

Description	Craft@Hrs	Unit	Material $	Labor $	Equipment $	Total $
½"	P1@.156	Ea	.81	5.73	—	6.54
¾"	P1@.204	Ea	2.19	7.49	—	9.68
1"	P1@.263	Ea	8.23	9.65	—	17.88
1¼"	P1@.322	Ea	13.10	11.80	—	24.90
1½"	P1@.380	Ea	19.20	13.90	—	33.10
2"	P1@.507	Ea	29.60	18.60	—	48.20
2½"	P1@.624	Ea	60.00	22.90	—	82.90
3"	P1@.741	Ea	92.80	27.20	—	120.00
4"	P1@.975	Ea	155.00	35.80	—	190.80

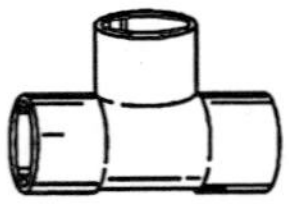

Type M copper reducing branch tee C x C x C with brazed joints

Description	Craft@Hrs	Unit	Material $	Labor $	Equipment $	Total $
½" x 3/8"	P1@.129	Ea	1.65	4.73	—	6.38
¾" x ½"	P1@.182	Ea	3.50	6.68	—	10.18
1" x ½"	P1@.209	Ea	6.85	7.67	—	14.52
1" x ¾"	P1@.234	Ea	6.85	8.59	—	15.44
1¼" x ½"	P1@.241	Ea	10.30	8.84	—	19.14
1¼" x ¾"	P1@.264	Ea	10.30	9.69	—	19.99
1¼" x 1"	P1@.287	Ea	10.30	10.50	—	20.80
1½" x ½"	P1@.291	Ea	12.70	10.70	—	23.40
1½" x ¾"	P1@.307	Ea	12.70	11.30	—	24.00
1½" x 1"	P1@.323	Ea	12.70	11.90	—	24.60
1½" x 1¼"	P1@.339	Ea	12.70	12.40	—	25.10
2" x ½"	P1@.348	Ea	16.60	12.80	—	29.40
2" x ¾"	P1@.373	Ea	16.60	13.70	—	30.30
2" x 1"	P1@.399	Ea	16.60	14.60	—	31.20
2" x 1¼"	P1@.426	Ea	16.60	15.60	—	32.20
2" x 1½"	P1@.452	Ea	16.60	16.60	—	33.20
2½" x ½"	P1@.468	Ea	50.00	17.20	—	67.20
2½" x ¾"	P1@.485	Ea	50.00	17.80	—	67.80
2½" x 1"	P1@.502	Ea	50.00	18.40	—	68.40
2½" x 1½"	P1@.537	Ea	50.00	19.70	—	69.70
2½" x 2"	P1@.555	Ea	50.00	20.40	—	70.40
3" x 1¼"	P1@.562	Ea	56.50	20.60	—	77.10
3" x 2"	P1@.626	Ea	58.50	23.00	—	81.50
3" x 2½"	P1@.660	Ea	58.50	24.20	—	82.70
4" x 1¼"	P1@.669	Ea	102.00	24.60	—	126.60
4" x 1½"	P1@.718	Ea	102.00	26.40	—	128.40
4" x 2"	P1@.768	Ea	102.00	28.20	—	130.20
4" x 2½"	P1@.819	Ea	102.00	30.10	—	132.10
4" x 3"	P1@.868	Ea	102.00	31.90	—	133.90

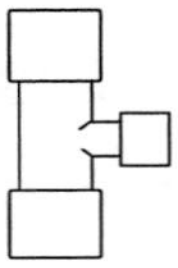

Description	Craft@Hrs	Unit	Material $	Labor $	Equipment $	Total $

Type M copper reducer with brazed joints

Description	Craft@Hrs	Unit	Material $	Labor $	Equipment $	Total $
½" x 3/8"	P1@.148	Ea	.59	5.43	—	6.02
¾" x ½"	P1@.155	Ea	1.11	5.69	—	6.80
1" x ¾"	P1@.206	Ea	2.00	7.56	—	9.56
1¼" x ½"	P1@.206	Ea	3.36	7.56	—	10.92
1¼" x ¾"	P1@.232	Ea	3.36	8.51	—	11.87
1¼" x 1"	P1@.258	Ea	3.36	9.47	—	12.83
1½" x ½"	P1@.232	Ea	5.14	8.51	—	13.65
1½" x ¾"	P1@.257	Ea	5.14	9.43	—	14.57
1½" x 1"	P1@.283	Ea	5.14	10.40	—	15.54
1½" x 1¼"	P1@.308	Ea	5.14	11.30	—	16.44
2" x ½"	P1@.287	Ea	7.74	10.50	—	18.24
2" x ¾"	P1@.277	Ea	7.74	10.20	—	17.94
2" x 1"	P1@.338	Ea	7.74	12.40	—	20.14
2" x 1¼"	P1@.365	Ea	7.74	13.40	—	21.14
2" x 1½"	P1@.390	Ea	7.74	14.30	—	22.04
2½" x 1"	P1@.390	Ea	21.10	14.30	—	35.40
2½" x 1¼"	P1@.416	Ea	21.10	15.30	—	36.40
2½" x 1½"	P1@.442	Ea	21.10	16.20	—	37.30
2½" x 2"	P1@.497	Ea	21.10	18.20	—	39.30
3" x 1¼"	P1@.468	Ea	24.60	17.20	—	41.80
3" x 1½"	P1@.493	Ea	24.60	18.10	—	42.70
3" x 2"	P1@.548	Ea	24.60	20.10	—	44.70
3" x 2½"	P1@.600	Ea	24.60	22.00	—	46.60
4" x 2"	P1@.652	Ea	47.50	23.90	—	71.40
4" x 2½"	P1@.704	Ea	47.50	25.80	—	73.30
4" x 3"	P1@.755	Ea	47.50	27.70	—	75.20

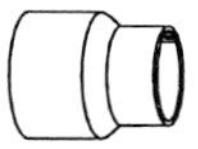

Type M copper adapter C x MPT with brazed joint

Description	Craft@Hrs	Unit	Material $	Labor $	Equipment $	Total $
½"	P1@.090	Ea	1.46	3.30	—	4.76
¾"	P1@.126	Ea	2.29	4.62	—	6.91
1"	P1@.162	Ea	5.67	5.95	—	11.62
1¼"	P1@.198	Ea	10.60	7.27	—	17.87
1½"	P1@.234	Ea	11.90	8.59	—	20.49
2"	P1@.312	Ea	14.90	11.50	—	26.40
2½"	P1@.384	Ea	48.30	14.10	—	62.40
3"	P1@.456	Ea	80.10	16.70	—	96.80

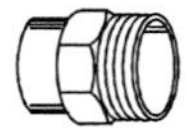

Type M copper adapter C x FPT with brazed joint

Description	Craft@Hrs	Unit	Material $	Labor $	Equipment $	Total $
½"	P1@.090	Ea	2.43	3.30	—	5.73
¾"	P1@.126	Ea	4.41	4.62	—	9.03
1"	P1@.162	Ea	6.66	5.95	—	12.61
1¼"	P1@.198	Ea	14.00	7.27	—	21.27
1½"	P1@.234	Ea	16.90	8.59	—	25.49
2"	P1@.312	Ea	25.20	11.50	—	36.70
2½"	P1@.384	Ea	63.40	14.10	—	77.50
3"	P1@.456	Ea	87.20	16.70	—	103.90

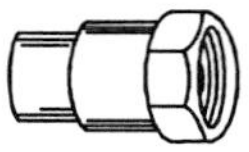

Description	Craft@Hrs	Unit	Material $	Labor $	Equipment $	Total $

Type M copper flush bushing with brazed joints

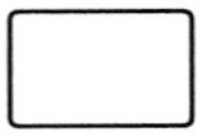

Description	Craft@Hrs	Unit	Material $	Labor $	Equipment $	Total $
½" x 3/8"	P1@.148	Ea	1.48	5.43	—	6.91
¾" x ½"	P1@.155	Ea	2.53	5.69	—	8.22
1" x ¾"	P1@.206	Ea	6.54	7.56	—	14.10
1" x ½"	P1@.206	Ea	7.46	7.56	—	15.02
1¼" x 1"	P1@.258	Ea	7.76	9.47	—	17.23
1½" x 1¼"	P1@.308	Ea	9.87	11.30	—	21.17
2" x 1½"	P1@.390	Ea	18.00	14.30	—	32.30

Type M copper union with brazed joints

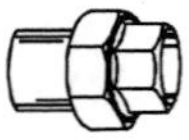

Description	Craft@Hrs	Unit	Material $	Labor $	Equipment $	Total $
½"	P1@.146	Ea	5.94	5.36	—	11.30
¾"	P1@.205	Ea	7.09	7.52	—	14.61
1"	P1@.263	Ea	15.90	9.65	—	25.55
1¼"	P1@.322	Ea	28.90	11.80	—	40.70
1½"	P1@.380	Ea	35.50	13.90	—	49.40
2"	P1@.507	Ea	51.30	18.60	—	69.90

Type M copper dielectric union with brazed joints

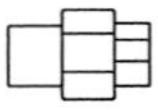

Description	Craft@Hrs	Unit	Material $	Labor $	Equipment $	Total $
½"	P1@.146	Ea	5.57	5.36	—	10.93
¾"	P1@.205	Ea	5.57	7.52	—	13.09
1"	P1@.263	Ea	11.50	9.65	—	21.15
1¼"	P1@.322	Ea	19.60	11.80	—	31.40
1½"	P1@.380	Ea	27.00	13.90	—	40.90
2"	P1@.507	Ea	36.10	18.60	—	54.70

Type M copper cap with brazed joints

Description	Craft@Hrs	Unit	Material $	Labor $	Equipment $	Total $
½"	P1@.083	Ea	.49	3.05	—	3.54
¾"	P1@.116	Ea	.99	4.26	—	5.25
1"	P1@.149	Ea	2.23	5.47	—	7.70
1¼"	P1@.182	Ea	3.21	6.68	—	9.89
1½"	P1@.215	Ea	4.68	7.89	—	12.57
2"	P1@.286	Ea	8.14	10.50	—	18.64
2½"	P1@.352	Ea	20.70	12.90	—	33.60
3"	P1@.418	Ea	32.30	15.30	—	47.60
4"	P1@.550	Ea	49.70	20.20	—	69.90

Type M copper coupling with brazed joints

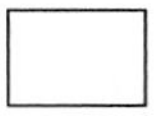

Description	Craft@Hrs	Unit	Material $	Labor $	Equipment $	Total $
½"	P1@.129	Ea	.31	4.73	—	5.04
¾"	P1@.181	Ea	.79	6.64	—	7.43
1"	P1@.232	Ea	2.12	8.51	—	10.63
1¼"	P1@.284	Ea	3.70	10.40	—	14.10
1½"	P1@.335	Ea	5.35	12.30	—	17.65
2"	P1@.447	Ea	7.86	16.40	—	24.26
2½"	P1@.550	Ea	18.60	20.20	—	38.80
3"	P1@.654	Ea	28.30	24.00	—	52.30
4"	P1@.860	Ea	55.40	31.60	—	87.00

Description	Craft@Hrs	Unit	Material $	Labor $	Equipment $	Total $

Class 125 bronze body gate valve, brazed joints

Description	Craft@Hrs	Unit	Material $	Labor $	Equipment $	Total $
½"	P1@.240	Ea	18.40	8.81	—	27.21
¾"	P1@.300	Ea	23.00	11.00	—	34.00
1"	P1@.360	Ea	32.40	13.20	—	45.60
1¼"	P1@.480	Ea	40.70	17.60	—	58.30
1½"	P1@.540	Ea	54.90	19.80	—	74.70
2"	P1@.600	Ea	91.80	22.00	—	113.80
2½"	P1@1.00	Ea	152.00	36.70	—	188.70
3"	P1@1.50	Ea	218.00	55.10	—	273.10

Class 125 iron body gate valve, flanged ends

Description	Craft@Hrs	Unit	Material $	Labor $	Equipment $	Total $
2"	P1@.500	Ea	351.00	18.40	—	369.40
2½"	P1@.600	Ea	474.00	22.00	—	496.00
3"	P1@.750	Ea	514.00	27.50	—	541.50
4"	P1@1.35	Ea	754.00	49.50	—	803.50
5"	P1@2.00	Ea	1,440.00	73.40	—	1,513.40
6"	P1@2.50	Ea	1,440.00	91.80	—	1,531.80

Class 125 bronze body globe valve, brazed joints

Description	Craft@Hrs	Unit	Material $	Labor $	Equipment $	Total $
½"	P1@.240	Ea	34.10	8.81	—	42.91
¾"	P1@.300	Ea	45.40	11.00	—	56.40
1"	P1@.360	Ea	78.20	13.20	—	91.40
1¼"	P1@.480	Ea	110.00	17.60	—	127.60
1½"	P1@.540	Ea	149.00	19.80	—	168.80
2"	P1@.600	Ea	241.00	22.00	—	263.00

Class 125 iron body globe valve, flanged ends

Description	Craft@Hrs	Unit	Material $	Labor $	Equipment $	Total $
2½"	P1@.650	Ea	398.00	23.90	—	421.90
3"	P1@.750	Ea	477.00	27.50	—	504.50
4"	P1@1.35	Ea	644.00	49.50	—	693.50

200 PSIG iron body butterfly valve, lug-type, lever operated

Description	Craft@Hrs	Unit	Material $	Labor $	Equipment $	Total $
2"	P1@.450	Ea	156.00	16.50	—	172.50
2½"	P1@.450	Ea	162.00	16.50	—	178.50
3"	P1@.550	Ea	170.00	20.20	—	190.20
4"	P1@.550	Ea	212.00	20.20	—	232.20

200 PSIG iron body butterfly valve, wafer-type, lever operated

Description	Craft@Hrs	Unit	Material $	Labor $	Equipment $	Total $
2"	P1@.450	Ea	142.00	16.50	—	158.50
2½"	P1@.450	Ea	145.00	16.50	—	161.50
3"	P1@.550	Ea	156.00	20.20	—	176.20
4"	P1@.550	Ea	189.00	20.20	—	209.20

Copper, Type M with Brazed Joints

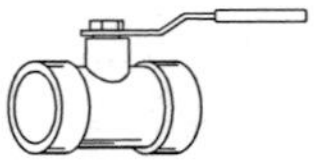

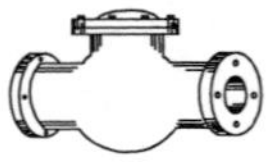

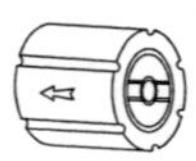

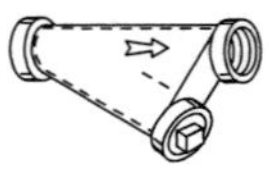

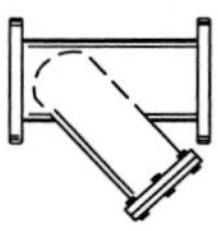

Description	Craft@Hrs	Unit	Material $	Labor $	Equipment $	Total $

Class 125 bronze body 2-piece ball valve, brazed joints

½"	P1@.240	Ea	10.10	8.81	—	18.91
¾"	P1@.300	Ea	13.40	11.00	—	24.40
1"	P1@.360	Ea	23.30	13.20	—	36.50
1¼"	P1@.480	Ea	38.80	17.60	—	56.40
1½"	P1@.540	Ea	52.40	19.80	—	72.20
2"	P1@.600	Ea	66.50	22.00	—	88.50
3"	P1@1.50	Ea	455.00	55.10	—	510.10
4"	P1@1.80	Ea	595.00	66.10	—	661.10

Class 125 bronze body swing check valve, brazed joints

½"	P1@.240	Ea	20.40	8.81	—	29.21
¾"	P1@.300	Ea	29.20	11.00	—	40.20
1"	P1@.360	Ea	38.10	13.20	—	51.30
1¼"	P1@.480	Ea	54.40	17.60	—	72.00
1½"	P1@.540	Ea	76.70	19.80	—	96.50
2"	P1@.600	Ea	127.00	22.00	—	149.00
2½"	P1@1.00	Ea	256.00	36.70	—	292.70
3"	P1@1.50	Ea	368.00	55.10	—	423.10

Class 125 iron body swing check valve, flanged ends

2"	P1@.500	Ea	170.00	18.40	—	188.40
2½"	P1@.600	Ea	226.00	22.00	—	248.00
3"	P1@.750	Ea	282.00	27.50	—	309.50
4"	P1@1.35	Ea	414.00	49.50	—	463.50

Class 125 iron body silent check valve, wafer-type

2"	P1@.500	Ea	131.00	18.40	—	149.40
2½"	P1@.600	Ea	145.00	22.00	—	167.00
3"	P1@.750	Ea	167.00	27.50	—	194.50
4"	P1@1.35	Ea	223.00	49.50	—	272.50

Class 125 bronze body strainer, threaded ends

½"	P1@.230	Ea	31.40	8.44	—	39.84
¾"	P1@.260	Ea	41.30	9.54	—	50.84
1"	P1@.330	Ea	50.60	12.10	—	62.70
1¼"	P1@.440	Ea	70.70	16.10	—	86.80
1½"	P1@.495	Ea	106.00	18.20	—	124.20
2"	P1@.550	Ea	185.00	20.20	—	205.20

Class 125 iron body strainer, flanged ends

2"	P1@.500	Ea	141.00	18.40	—	159.40
2½"	P1@.600	Ea	157.00	22.00	—	179.00
3"	P1@.750	Ea	183.00	27.50	—	210.50
4"	P1@1.35	Ea	308.00	49.50	—	357.50

Description	Craft@Hrs	Unit	Material $	Labor $	Equipment $	Total $

Installation of 2-way control valve, threaded joints

½"	P1@.210	Ea	—	7.71	—	7.71
¾"	P1@.275	Ea	—	10.10	—	10.10
1"	P1@.350	Ea	—	12.80	—	12.80
1¼"	P1@.430	Ea	—	15.80	—	15.80
1½"	P1@.505	Ea	—	18.50	—	18.50
2"	P1@.675	Ea	—	24.80	—	24.80
2½"	P1@.830	Ea	—	30.50	—	30.50
3"	P1@.990	Ea	—	36.30	—	36.30

Installation of 3-way control valve, threaded joints

½"	P1@.260	Ea	—	9.54	—	9.54
¾"	P1@.365	Ea	—	13.40	—	13.40
1"	P1@.475	Ea	—	17.40	—	17.40
1¼"	P1@.575	Ea	—	21.10	—	21.10
1½"	P1@.680	Ea	—	25.00	—	25.00
2"	P1@.910	Ea	—	33.40	—	33.40
2½"	P1@1.12	Ea	—	41.10	—	41.10
3"	P1@1.33	Ea	—	48.80	—	48.80

Companion flange, 150 pound cast brass

2"	P1@.290	Ea	305.00	10.60	—	315.60
2½"	P1@.380	Ea	376.00	13.90	—	389.90
3"	P1@.460	Ea	376.00	16.90	—	392.90
4"	P1@.600	Ea	599.00	22.00	—	621.00

Bolt and gasket sets

2"	P1@.500	Ea	3.52	18.40	—	21.92
2½"	P1@.650	Ea	4.13	23.90	—	28.03
3"	P1@.750	Ea	6.99	27.50	—	34.49
4"	P1@1.00	Ea	11.70	36.70	—	48.40

Thermometer with well

7"	P1@.250	Ea	151.00	9.18	—	160.18
9"	P1@.250	Ea	185.00	9.18	—	194.18

Dial-type pressure gauge

2½"	P1@.200	Ea	37.70	7.34	—	45.04
3½"	P1@.200	Ea	49.70	7.34	—	57.04

Pressure/temperature tap

Tap	P1@.150	Ea	17.30	5.51	—	22.81

Copper, Type M with Soft-Soldered Joints

Type M hard-drawn copper pipe with wrought copper fittings and soft-soldered joints is used in a wide variety of plumbing and HVAC systems such as potable water, heating hot water, chilled water, and A.C. condensate.

Soft-soldered joints are those made with solders having melting points in the 350 degree F. to 500 degree F. range. Maximum working pressure/temperature relationships for soft-soldered joints are approximately as follows:

Maximum Working Pressures (PSIG)*				
Soft-solder Type	Water Temperature (degrees F.)	Nominal Pipe Size (inches)		
		Up to 1	1¼ to 2	2½ x 4
50-50 tin-lead**	100	200	175	150
	150	150	125	100
	200	100	90	75
	250	85	75	50
95-5 tin-antimony	100	500	400	300
	150	400	350	275
	200	300	250	200
	250	200	175	150

*For copper pipe and solder-type fittings using soft-solders melting at approximately 350 degrees F. to 500 degrees F.
**The use of any solder containing lead is not allowed in potable water systems.

This section has been arranged to save the estimator's time by including all normally-used system components such as pipe, fittings, valves, hanger assemblies, riser clamps and miscellaneous items under one heading. Additional items can be found under "Plumbing and Piping Specialties." The cost estimates in this section are based on the conditions, limitations and wage rates described in the section "How to Use This Book" beginning on page 5.

Description	Craft@Hrs	Unit	Material $	Labor $	Equipment $	Total $

Type M copper pipe with soft-soldered joints installed horizontally.
Complete installation including six to twelve wrought copper tees and six to twelve wrought copper elbows every 100', and hangers spaced to meet plumbing code *(a tee & elbow every 8.3' for ½" dia., a tee and elbow every 16.6' for 4" dia.).*

Description	Craft@Hrs	Unit	Material $	Labor $	Equipment $	Total $
½"	P1@.110	LF	1.90	4.04	—	5.94
¾"	P1@.120	LF	2.93	4.40	—	7.33
1"	P1@.140	LF	6.08	5.14	—	11.22
1¼"	P1@.160	LF	10.30	5.87	—	16.17
1½"	P1@.180	LF	12.10	6.61	—	18.71
2"	P1@.210	LF	25.50	7.71	—	33.21
2½"	P1@.240	LF	37.30	8.81	—	46.11
3"	P1@.270	LF	46.80	9.91	—	56.71
4"	P1@.312	LF	68.90	11.50	—	80.40

Description	Craft@Hrs	Unit	Material $	Labor $	Equipment $	Total $

Type M copper pipe with soft-soldered joints, installed risers.

Complete installation including a wrought copper reducing tee every floor and a riser clamp every other floor.

Description	Craft@Hrs	Unit	Material $	Labor $	Equipment $	Total $
½"	P1@.060	LF	1.61	2.20	—	3.81
¾"	P1@.070	LF	2.60	2.57	—	5.17
1"	P1@.075	LF	5.09	2.75	—	7.84
1¼"	P1@.080	LF	7.51	2.94	—	10.45
1½"	P1@.100	LF	9.87	3.67	—	13.54
2"	P1@.130	LF	13.10	4.77	—	17.87
2½"	P1@.150	LF	28.50	5.51	—	34.01
3"	P1@.170	LF	35.00	6.24	—	41.24
4"	P1@.190	LF	51.90	6.97	—	58.87

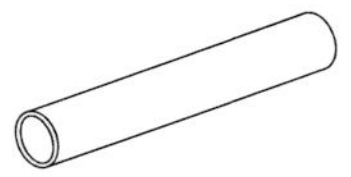

Type M copper pipe with soft-soldered joints, pipe only

Description	Craft@Hrs	Unit	Material $	Labor $	Equipment $	Total $
½"	P1@.032	LF	1.16	1.17	—	2.33
¾"	P1@.035	LF	1.90	1.28	—	3.18
1"	P1@.038	LF	2.97	1.39	—	4.36
1¼"	P1@.042	LF	4.45	1.54	—	5.99
1½"	P1@.046	LF	6.24	1.69	—	7.93
2"	P1@.053	LF	11.50	1.95	—	13.45
2½"	P1@.060	LF	15.90	2.20	—	18.10
3"	P1@.066	LF	19.90	2.42	—	22.32
4"	P1@.080	LF	32.00	2.94	—	34.94

Type M copper 45-degree ell C x C with soft-soldered joints

Description	Craft@Hrs	Unit	Material $	Labor $	Equipment $	Total $
½"	P1@.107	Ea	.82	3.93	—	4.75
¾"	P1@.150	Ea	1.41	5.51	—	6.92
1"	P1@.193	Ea	4.60	7.08	—	11.68
1¼"	P1@.236	Ea	5.18	8.66	—	13.84
1½"	P1@.278	Ea	9.28	10.20	—	19.48
2"	P1@.371	Ea	14.70	13.60	—	28.30
2½"	P1@.457	Ea	37.50	16.80	—	54.30
3"	P1@.543	Ea	49.60	19.90	—	69.50
4"	P1@.714	Ea	95.30	26.20	—	121.50

Type M copper 90-degree ell C x C with soft-soldered joints

Description	Craft@Hrs	Unit	Material $	Labor $	Equipment $	Total $
½"	P1@.107	Ea	.44	3.93	—	4.37
¾"	P1@.150	Ea	.92	5.51	—	6.43
1"	P1@.193	Ea	3.51	7.08	—	10.59
1¼"	P1@.236	Ea	5.67	8.66	—	14.33
1½"	P1@.278	Ea	8.55	10.20	—	18.75
2"	P1@.371	Ea	14.90	13.60	—	28.50
2½"	P1@.457	Ea	31.00	16.80	—	47.80
3"	P1@.543	Ea	39.30	19.90	—	59.20
4"	P1@.714	Ea	83.10	26.20	—	109.30

Copper, Type M with Soft-Soldered Joints

Description	Craft@Hrs	Unit	Material $	Labor $	Equipment $	Total $

Type M copper 90-degree ell ftg. x C with soft-soldered joint

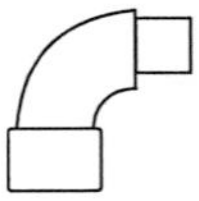

Description	Craft@Hrs	Unit	Material $	Labor $	Equipment $	Total $
½"	P1@.107	Ea	.77	3.93	—	4.70
¾"	P1@.150	Ea	1.66	5.51	—	7.17
1"	P1@.193	Ea	4.97	7.08	—	12.05
1¼"	P1@.236	Ea	6.93	8.66	—	15.59
1½"	P1@.278	Ea	10.30	10.20	—	20.50
2"	P1@.371	Ea	20.20	13.60	—	33.80
2½"	P1@.457	Ea	42.20	16.80	—	59.00
3"	P1@.543	Ea	47.50	19.90	—	67.40
4"	P1@.714	Ea	95.30	26.20	—	121.50

Type M copper tee C x C x C with soft-soldered joints

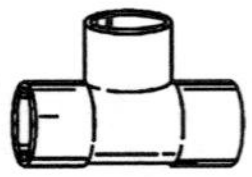

Description	Craft@Hrs	Unit	Material $	Labor $	Equipment $	Total $
½"	P1@.129	Ea	.74	4.73	—	5.47
¾"	P1@.181	Ea	1.99	6.64	—	8.63
1"	P1@.233	Ea	8.23	8.55	—	16.78
1¼"	P1@.285	Ea	13.10	10.50	—	23.60
1½"	P1@.337	Ea	19.20	12.40	—	31.60
2"	P1@.449	Ea	29.60	16.50	—	46.10
2½"	P1@.552	Ea	60.00	20.30	—	80.30
3"	P1@.656	Ea	92.80	24.10	—	116.90
4"	P1@.863	Ea	116.00	31.70	—	147.70

Type M copper branch reducing tee C x C x C with soft-soldered joints

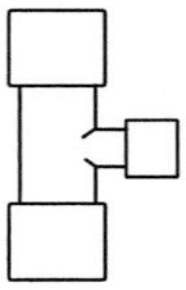

Description	Craft@Hrs	Unit	Material $	Labor $	Equipment $	Total $
½" x 3/8"	P1@.129	Ea	1.65	4.73	—	6.38
¾" x ½"	P1@.182	Ea	3.50	6.68	—	10.18
1" x ½"	P1@.195	Ea	6.85	7.16	—	14.01
1" x ¾"	P1@.219	Ea	6.85	8.04	—	14.89
1¼" x ½"	P1@.225	Ea	10.30	8.26	—	18.56
1¼" x ¾"	P1@.247	Ea	10.30	9.06	—	19.36
1¼" x 1"	P1@.268	Ea	10.30	9.84	—	20.14
1½" x ½"	P1@.272	Ea	12.70	9.98	—	22.68
1½" x ¾"	P1@.287	Ea	12.70	10.50	—	23.20
1½" x 1"	P1@.302	Ea	12.70	11.10	—	23.80
1½" x 1¼"	P1@.317	Ea	12.70	11.60	—	24.30
2" x ½"	P1@.325	Ea	16.60	11.90	—	28.50
2" x ¾"	P1@.348	Ea	16.60	12.80	—	29.40
2" x 1"	P1@.373	Ea	16.60	13.70	—	30.30
2" x 1¼"	P1@.398	Ea	16.60	14.60	—	31.20
2" x 1½"	P1@.422	Ea	16.60	15.50	—	32.10
2½" x ½"	P1@.437	Ea	50.00	16.00	—	66.00
2½" x ¾"	P1@.453	Ea	50.00	16.60	—	66.60
2½" x 1"	P1@.469	Ea	50.00	17.20	—	67.20
2½" x 1½"	P1@.502	Ea	50.00	18.40	—	68.40
2½" x 2"	P1@.519	Ea	50.00	19.00	—	69.00
3" x 1¼"	P1@.525	Ea	56.50	19.30	—	75.80
3" x 2"	P1@.585	Ea	58.50	21.50	—	80.00
3" x 2½"	P1@.617	Ea	58.50	22.60	—	81.10
4" x 1¼"	P1@.625	Ea	102.00	22.90	—	124.90
4" x 1½"	P1@.671	Ea	102.00	24.60	—	126.60
4" x 2"	P1@.718	Ea	102.00	26.40	—	128.40
4" x 2½"	P1@.765	Ea	102.00	28.10	—	130.10
4" x 3"	P1@.811	Ea	102.00	29.80	—	131.80

Description	Craft@Hrs	Unit	Material $	Labor $	Equipment $	Total $

Type M copper reducer with soft-soldered joints

Description	Craft@Hrs	Unit	Material $	Labor $	Equipment $	Total $
½" x 3/8"	P1@.124	Ea	.59	4.55	—	5.14
¾" x ½"	P1@.129	Ea	1.11	4.73	—	5.84
1" x ¾"	P1@.172	Ea	2.00	6.31	—	8.31
1¼" x ½"	P1@.172	Ea	3.36	6.31	—	9.67
1¼" x ¾"	P1@.193	Ea	3.36	7.08	—	10.44
1¼" x 1"	P1@.215	Ea	3.36	7.89	—	11.25
1½" x ½"	P1@.193	Ea	5.14	7.08	—	12.22
1½" x ¾"	P1@.214	Ea	5.14	7.85	—	12.99
1½" x 1"	P1@.236	Ea	5.14	8.66	—	13.80
1½" x 1¼"	P1@.257	Ea	5.14	9.43	—	14.57
2" x ½"	P1@.239	Ea	7.74	8.77	—	16.51
2" x ¾"	P1@.261	Ea	7.74	9.58	—	17.32
2" x 1"	P1@.282	Ea	7.74	10.30	—	18.04
2" x 1¼"	P1@.304	Ea	7.74	11.20	—	18.94
2" x 1½"	P1@.325	Ea	7.74	11.90	—	19.64
2½" x 1"	P1@.325	Ea	21.10	11.90	—	33.00
2½" x 1¼"	P1@.347	Ea	21.10	12.70	—	33.80
2½" x 1½"	P1@.368	Ea	21.10	13.50	—	34.60
2½" x 2"	P1@.414	Ea	21.10	15.20	—	36.30
3" x 2"	P1@.457	Ea	24.60	16.80	—	41.40
3" x 2½"	P1@.500	Ea	24.60	18.40	—	43.00
4" x 2"	P1@.543	Ea	47.50	19.90	—	67.40
4" x 2½"	P1@.586	Ea	47.50	21.50	—	69.00
4" x 3"	P1@.629	Ea	47.50	23.10	—	70.60

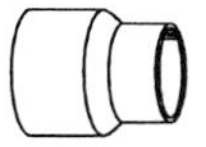

Type M copper adapter C x MPT with soft-soldered joint

Description	Craft@Hrs	Unit	Material $	Labor $	Equipment $	Total $
½"	P1@.075	Ea	1.46	2.75	—	4.21
¾"	P1@.105	Ea	2.29	3.85	—	6.14
1"	P1@.134	Ea	5.67	4.92	—	10.59
1¼"	P1@.164	Ea	10.60	6.02	—	16.62
1½"	P1@.194	Ea	11.90	7.12	—	19.02
2"	P1@.259	Ea	14.90	9.51	—	24.41
2½"	P1@.319	Ea	48.30	11.70	—	60.00
3"	P1@.378	Ea	80.10	13.90	—	94.00

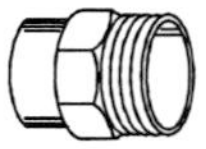

Type M copper adapter C x FPT with soft-soldered joint

Description	Craft@Hrs	Unit	Material $	Labor $	Equipment $	Total $
½"	P1@.075	Ea	2.43	2.75	—	5.18
¾"	P1@.105	Ea	4.41	3.85	—	8.26
1"	P1@.134	Ea	6.66	4.92	—	11.58
1¼"	P1@.164	Ea	14.00	6.02	—	20.02
1½"	P1@.194	Ea	16.90	7.12	—	24.02
2"	P1@.259	Ea	25.20	9.51	—	34.71
2½"	P1@.319	Ea	63.40	11.70	—	75.10
3"	P1@.378	Ea	87.20	13.90	—	101.10

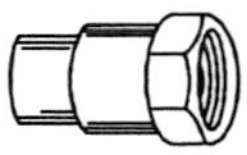

Copper, Type M with Soft-Soldered Joints

Description	Craft@Hrs	Unit	Material $	Labor $	Equipment $	Total $

Type M copper flush bushing with soft-soldered joints

Description	Craft@Hrs	Unit	Material $	Labor $	Equipment $	Total $
½" x 3/8"	P1@.124	Ea	1.48	4.55	—	6.03
¾" x ½"	P1@.129	Ea	2.53	4.73	—	7.26
1" x ¾"	P1@.172	Ea	6.54	6.31	—	12.85
1" x ½"	P1@.172	Ea	7.46	6.31	—	13.77
1¼" x 1"	P1@.215	Ea	7.76	7.89	—	15.65
1½" x 1¼"	P1@.257	Ea	9.87	9.43	—	19.30
2" x 1½"	P1@.325	Ea	18.00	11.90	—	29.90

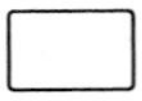

Type M copper union with soft-soldered joint

Description	Craft@Hrs	Unit	Material $	Labor $	Equipment $	Total $
½"	P1@.121	Ea	5.94	4.44	—	10.38
¾"	P1@.170	Ea	7.09	6.24	—	13.33
1"	P1@.218	Ea	15.90	8.00	—	23.90
1¼"	P1@.267	Ea	28.90	9.80	—	38.70
1½"	P1@.315	Ea	35.50	11.60	—	47.10
2"	P1@.421	Ea	51.30	15.50	—	66.80

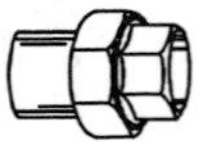

Type M copper dielectric union with soft-soldered joint

Description	Craft@Hrs	Unit	Material $	Labor $	Equipment $	Total $
½"	P1@.121	Ea	5.57	4.44	—	10.01
¾"	P1@.170	Ea	5.57	6.24	—	11.81
1"	P1@.218	Ea	11.50	8.00	—	19.50
1¼"	P1@.267	Ea	19.60	9.80	—	29.40
1½"	P1@.315	Ea	27.00	11.60	—	38.60
2"	P1@.421	Ea	36.10	15.50	—	51.60

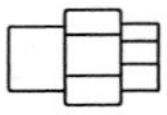

Type M copper cap with soft-soldered joint

Description	Craft@Hrs	Unit	Material $	Labor $	Equipment $	Total $
½"	P1@.069	Ea	.49	2.53	—	3.02
¾"	P1@.096	Ea	.99	3.52	—	4.51
1"	P1@.124	Ea	2.23	4.55	—	6.78
1¼"	P1@.151	Ea	3.21	5.54	—	8.75
1½"	P1@.178	Ea	4.68	6.53	—	11.21
2"	P1@.237	Ea	8.14	8.70	—	16.84
2½"	P1@.292	Ea	20.70	10.70	—	31.40
3"	P1@.347	Ea	32.30	12.70	—	45.00
4"	P1@.457	Ea	49.70	16.80	—	66.50

Type M copper coupling with soft-soldered joints

Description	Craft@Hrs	Unit	Material $	Labor $	Equipment $	Total $
½"	P1@.107	Ea	.31	3.93	—	4.24
¾"	P1@.150	Ea	.79	5.51	—	6.30
1"	P1@.193	Ea	2.12	7.08	—	9.20
1¼"	P1@.236	Ea	3.70	8.66	—	12.36
1½"	P1@.278	Ea	5.35	10.20	—	15.55
2"	P1@.371	Ea	7.86	13.60	—	21.46
2½"	P1@.457	Ea	18.60	16.80	—	35.40
3"	P1@.543	Ea	28.30	19.90	—	48.20
4"	P1@.714	Ea	55.40	26.20	—	81.60

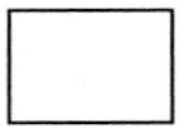

Copper, Type M with Soft-Soldered Joints

Description	Craft@Hrs	Unit	Material $	Labor $	Equipment $	Total $

Class 125 bronze body gate valve with soft-soldered joints

½"	P1@.200	Ea	18.40	7.34	—	25.74
¾"	P1@.249	Ea	23.00	9.14	—	32.14
1"	P1@.299	Ea	32.40	11.00	—	43.40
1¼"	P1@.398	Ea	40.70	14.60	—	55.30
1½"	P1@.448	Ea	54.90	16.40	—	71.30
2"	P1@.498	Ea	91.80	18.30	—	110.10
2½"	P1@.830	Ea	152.00	30.50	—	182.50
3"	P1@1.24	Ea	218.00	45.50	—	263.50

Class 125 iron body gate valve with flanged ends

2"	P1@.500	Ea	351.00	18.40	—	369.40
2½"	P1@.600	Ea	474.00	22.00	—	496.00
3"	P1@.750	Ea	514.00	27.50	—	541.50
4"	P1@1.35	Ea	754.00	49.50	—	803.50
5"	P1@2.00	Ea	1,440.00	73.40	—	1,513.40
6"	P1@2.50	Ea	1,440.00	91.80	—	1,531.80

Class 125 bronze body globe valve with soft-soldered joints

½"	P1@.200	Ea	34.10	7.34	—	41.44
¾"	P1@.249	Ea	45.40	9.14	—	54.54
1"	P1@.299	Ea	78.20	11.00	—	89.20
1¼"	P1@.398	Ea	110.00	14.60	—	124.60
1½"	P1@.448	Ea	149.00	16.40	—	165.40
2"	P1@.498	Ea	241.00	18.30	—	259.30

Class 125 iron body globe valve with flanged ends

2½"	P1@.600	Ea	398.00	22.00	—	420.00
3"	P1@.750	Ea	477.00	27.50	—	504.50
4"	P1@1.35	Ea	640.00	49.50	—	689.50

200 PSIG iron body butterfly valve, lug-type, lever operated

2"	P1@.450	Ea	156.00	16.50	—	172.50
2½"	P1@.450	Ea	162.00	16.50	—	178.50
3"	P1@.550	Ea	170.00	20.20	—	190.20
4"	P1@.550	Ea	212.00	20.20	—	232.20

200 PSIG iron body butterfly valve, wafer-type, lever operated

2"	P1@.450	Ea	142.00	16.50	—	158.50
2½"	P1@.450	Ea	145.00	16.50	—	161.50
3"	P1@.550	Ea	156.00	20.20	—	176.20
4"	P1@.550	Ea	189.00	20.20	—	209.20

Copper, Type M with Soft-Soldered Joints

Description	Craft@Hrs	Unit	Material $	Labor $	Equipment $	Total $

Class 125 bronze body 2-piece ball valve with soft-soldered joints

Description	Craft@Hrs	Unit	Material $	Labor $	Equipment $	Total $
½"	P1@.200	Ea	10.10	7.34	—	17.44
¾"	P1@.249	Ea	13.40	9.14	—	22.54
1"	P1@.299	Ea	23.30	11.00	—	34.30
1¼"	P1@.398	Ea	38.80	14.60	—	53.40
1½"	P1@.448	Ea	52.40	16.40	—	68.80
2"	P1@.498	Ea	66.50	18.30	—	84.80
3"	P1@1.24	Ea	455.00	45.50	—	500.50
4"	P1@1.45	Ea	595.00	53.20	—	648.20

Class 125 bronze body swing check valve with soft-soldered joints

Description	Craft@Hrs	Unit	Material $	Labor $	Equipment $	Total $
½"	P1@.200	Ea	20.40	7.34	—	27.74
¾"	P1@.249	Ea	29.20	9.14	—	38.34
1"	P1@.299	Ea	38.10	11.00	—	49.10
1¼"	P1@.398	Ea	54.40	14.60	—	69.00
1½"	P1@.448	Ea	76.70	16.40	—	93.10
2"	P1@.498	Ea	127.00	18.30	—	145.30
2½"	P1@.830	Ea	256.00	30.50	—	286.50
3"	P1@1.24	Ea	368.00	45.50	—	413.50

Class 125 iron body swing check valve, flanged ends

Description	Craft@Hrs	Unit	Material $	Labor $	Equipment $	Total $
2"	P1@.500	Ea	179.00	18.40	—	197.40
2½"	P1@.600	Ea	226.00	22.00	—	248.00
3"	P1@.750	Ea	282.00	27.50	—	309.50
4"	P1@1.35	Ea	414.00	49.50	—	463.50

Class 125 iron body silent check valve, wafer-type

Description	Craft@Hrs	Unit	Material $	Labor $	Equipment $	Total $
2"	P1@.500	Ea	131.00	18.40	—	149.40
2½"	P1@.600	Ea	145.00	22.00	—	167.00
3"	P1@.750	Ea	167.00	27.50	—	194.50
4"	P1@1.35	Ea	223.00	49.50	—	272.50

Class 125 bronze body strainer, threaded ends

Description	Craft@Hrs	Unit	Material $	Labor $	Equipment $	Total $
½"	P1@.230	Ea	31.40	8.44	—	39.84
¾"	P1@.260	Ea	41.30	9.54	—	50.84
1"	P1@.330	Ea	50.60	12.10	—	62.70
1¼"	P1@.440	Ea	70.70	16.10	—	86.80
1½"	P1@.495	Ea	106.00	18.20	—	124.20
2"	P1@.550	Ea	185.00	20.20	—	205.20

Class 125 iron body strainer, flanged ends

Description	Craft@Hrs	Unit	Material $	Labor $	Equipment $	Total $
2"	P1@.500	Ea	141.00	18.40	—	159.40
2½"	P1@.600	Ea	157.00	22.00	—	179.00
3"	P1@.750	Ea	183.00	27.50	—	210.50
4"	P1@1.35	Ea	308.00	49.50	—	357.50

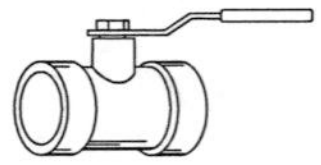

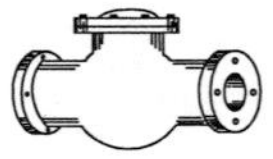

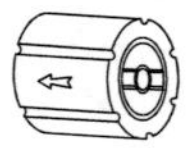

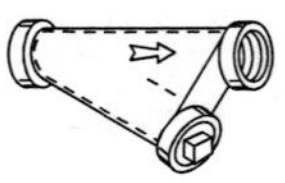

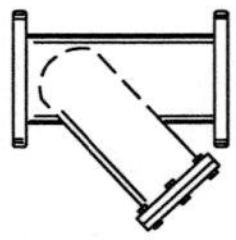

Description	Craft@Hrs	Unit	Material $	Labor $	Equipment $	Total $

Installation of 2-way control valve, threaded joints

Description	Craft@Hrs	Unit	Material $	Labor $	Equipment $	Total $
½"	P1@.210	Ea	—	7.71	—	7.71
¾"	P1@.275	Ea	—	10.10	—	10.10
1"	P1@.350	Ea	—	12.80	—	12.80
1¼"	P1@.430	Ea	—	15.80	—	15.80
1½"	P1@.505	Ea	—	18.50	—	18.50
2"	P1@.675	Ea	—	24.80	—	24.80
2½"	P1@.830	Ea	—	30.50	—	30.50
3"	P1@.990	Ea	—	36.30	—	36.30

Installation of 3-way control valve, threaded joints

Description	Craft@Hrs	Unit	Material $	Labor $	Equipment $	Total $
½"	P1@.260	Ea	—	9.54	—	9.54
¾"	P1@.365	Ea	—	13.40	—	13.40
1"	P1@.475	Ea	—	17.40	—	17.40
1¼"	P1@.575	Ea	—	21.10	—	21.10
1½"	P1@.680	Ea	—	25.00	—	25.00
2"	P1@.910	Ea	—	33.40	—	33.40
2½"	P1@1.12	Ea	—	41.10	—	41.10
3"	P1@1.33	Ea	—	48.80	—	48.80

Companion flange, 150 pound cast brass

Description	Craft@Hrs	Unit	Material $	Labor $	Equipment $	Total $
2"	P1@.290	Ea	315.00	10.60	—	325.60
2½"	P1@.380	Ea	386.00	13.90	—	399.90
3"	P1@.460	Ea	386.00	16.90	—	402.90
4"	P1@.600	Ea	616.00	22.00	—	638.00

Bolt and gasket sets

Description	Craft@Hrs	Unit	Material $	Labor $	Equipment $	Total $
2"	P1@.500	Ea	4.40	18.40	—	22.80
2½"	P1@.650	Ea	4.89	23.90	—	28.79
3"	P1@.750	Ea	8.31	27.50	—	35.81
4"	P1@1.00	Ea	13.80	36.70	—	50.50

Thermometer with well

Description	Craft@Hrs	Unit	Material $	Labor $	Equipment $	Total $
7"	P1@.250	Ea	180.00	9.18	—	189.18
9"	P1@.250	Ea	185.00	9.18	—	194.18

Dial-type pressure gauge

Description	Craft@Hrs	Unit	Material $	Labor $	Equipment $	Total $
2½"	P1@.200	Ea	37.70	7.34	—	45.04
3½"	P1@.200	Ea	49.70	7.34	—	57.04

Pressure/temperature tap

Description	Craft@Hrs	Unit	Material $	Labor $	Equipment $	Total $
Tap	P1@.150	Ea	17.30	5.51	—	22.81

Copper, Press Fit Fittings

Press Fit (Pressfast) Copper Pipe Fittings. For joining ½" to 4" type K, L or M hard drawn or annealed copper pipe and tube (ASTM 888). No brazing required. Includes a square cut and de-burr of the pipe ends and joining with a hand-held electric press fit tool. Equipment cost, when shown, is $1.90 per hour for pressfit tool. Fittings include an O-ring for each joint.

Description	Craft@Hrs	Unit	Material $	Labor $	Equipment $	Total $
45-degree Copper press fit ell (P x P)						
½"	P1@.170	Ea	2.20	6.24	.16	8.60
¾"	P1@.170	Ea	2.86	6.24	.16	9.26
1"	P1@.180	Ea	9.14	6.61	.17	15.92
1¼"	P1@.190	Ea	13.10	6.97	.18	20.25
1½"	P1@.200	Ea	21.20	7.34	.19	28.73
2"	P1@.200	Ea	29.70	7.34	.19	37.23
2½"	P1@.225	Ea	65.20	8.26	.21	73.67
3"	P1@.240	Ea	90.90	8.81	.23	99.94
4"	P1@.260	Ea	130.00	9.54	.25	139.79
90-degree Copper press fit ell (P x P)						
½"	P1@.170	Ea	1.82	6.24	.16	8.22
¾"	P1@.170	Ea	3.06	6.24	.16	9.46
1"	P1@.180	Ea	6.76	6.61	.17	13.54
1¼"	P1@.190	Ea	13.60	6.97	.18	20.75
1½"	P1@.200	Ea	25.70	7.34	.19	33.23
2"	P1@.200	Ea	35.80	7.34	.19	43.33
2½"	P1@.225	Ea	76.20	8.26	.21	84.67
3"	P1@.240	Ea	141.00	8.81	.23	150.04
4"	P1@.260	Ea	173.00	9.54	.25	182.79
Coupling, Copper, press fit (P x P)						
½"	P1@.170	Ea	1.63	6.24	.16	8.03
¾"	P1@.170	Ea	2.48	6.24	.16	8.88
1"	P1@.180	Ea	4.99	6.61	.17	11.77
1¼"	P1@.190	Ea	6.18	6.97	.18	13.33
1½"	P1@.200	Ea	11.50	7.34	.19	19.03
2"	P1@.200	Ea	14.70	7.34	.19	22.23
2½"	P1@.225	Ea	48.80	8.26	.21	57.27
3"	P1@.240	Ea	63.00	8.81	.23	72.04
4"	P1@.260	Ea	86.50	9.54	.25	96.29

Description	Craft@Hrs	Unit	Material $	Labor $	Equipment $	Total $

Tee, Copper, press fit (P x P)

Description	Craft@Hrs	Unit	Material $	Labor $	Equipment $	Total $
½"	P1@.170	Ea	2.20	6.24	.16	8.60
¾"	P1@.170	Ea	2.86	6.24	.16	9.26
1"	P1@.180	Ea	9.14	6.61	.17	15.92
1¼"	P1@.190	Ea	13.10	6.97	.18	20.25
1½"	P1@.200	Ea	21.20	7.34	.19	28.73
2"	P1@.200	Ea	29.70	7.34	.19	37.23
2½"	P1@.225	Ea	65.20	8.26	.21	73.67
3"	P1@.240	Ea	90.90	8.81	.23	99.94
4"	P1@.260	Ea	130.00	9.54	.25	139.79

Reducing tee, Copper press fit (P x P x P)

Description	Craft@Hrs	Unit	Material $	Labor $	Equipment $	Total $
¾ x ½"	P1@.200	Ea	6.68	7.34	.19	14.21
1 x ¾"	P1@.200	Ea	8.47	7.34	.19	16.00
1¼" x ¾"	P1@.240	Ea	17.20	8.81	.23	26.24
1¼" x 1"	P1@.240	Ea	12.30	8.81	.23	21.34
1½" x ¾"	P1@.255	Ea	26.20	9.36	.24	35.80
1½ x 1"	P1@.255	Ea	27.30	9.36	.24	36.90
2" x ¾"	P1@.270	Ea	42.50	9.91	.26	52.67
2" x 1"	P1@.270	Ea	43.90	9.91	.26	54.07
2" x 1½"	P1@.280	Ea	33.20	10.30	.27	43.77
3" x 1½"	P1@.380	Ea	158.00	13.80	.36	171.80
3" x 2"	P1@.400	Ea	147.00	14.70	.38	162.08
4" x 1½"	P1@.400	Ea	197.00	14.70	.38	212.08
4" x 2"	P1@.420	Ea	178.00	15.30	.40	193.30
4" x 3"	P1@.450	Ea	173.00	16.50	.43	189.93

Union, Copper press fit (P x M)

Description	Craft@Hrs	Unit	Material $	Labor $	Equipment $	Total $
½"	P1@.220	Ea	14.90	8.07	.21	23.18
¾"	P1@.220	Ea	19.50	8.07	.21	27.78
1"	P1@.260	Ea	31.20	9.54	.25	40.99
1¼"	P1@.280	Ea	44.60	10.30	.27	55.17
1½"	P1@.280	Ea	60.80	10.30	.27	71.37
2"	P1@.300	Ea	98.90	11.00	.29	110.19
2½"	P1@.320	Ea	151.00	11.70	.30	163.00
3"	P1@.380	Ea	163.00	13.90	.36	177.26
4"	P1@.420	Ea	189.00	15.40	.40	204.80

Copper, Press Fit Fittings

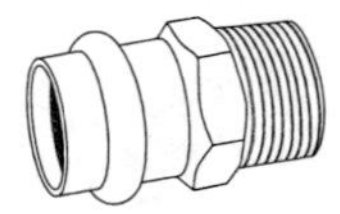

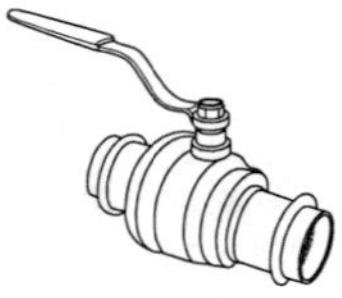

Description	Craft@Hrs	Unit	Material $	Labor $	Equipment $	Total $
Male threaded Copper press fit adapter (P x M)						
½"	P1@.220	Ea	2.35	8.07	.21	10.63
¾" x ½"	P1@.220	Ea	4.79	8.07	.21	13.07
¾"	P1@.220	Ea	4.23	8.07	.21	12.51
¾" x 1"	P1@.240	Ea	23.20	8.81	.23	32.24
1 x ¾"	P1@.250	Ea	12.30	9.18	.24	21.72
1"	P1@.260	Ea	7.75	9.54	.25	17.54
1¼"	P1@.280	Ea	16.90	10.30	.27	27.47
1½"	P1@.280	Ea	23.26	10.20	.27	33.46
2"	P1@.300	Ea	45.70	11.00	.29	56.99
2½"	P1@.320	Ea	96.41	11.60	.30	108.00
3"	P1@.380	Ea	121.00	13.90	.36	135.26
4"	P1@.420	Ea	153.00	15.40	.40	168.80
Female threaded Copper press fit adapter (P x F)						
½"	P1@.200	Ea	2.76	7.34	.19	10.29
¾"	P1@.220	Ea	4.49	8.07	.21	12.77
¾" x ½"	P1@.220	Ea	5.00	8.07	.21	13.28
¾"	P1@.220	Ea	4.23	8.07	.21	12.51
1" x ½"	P1@.240	Ea	14.80	8.81	.23	23.84
1 x ¾"	P1@.240	Ea	13.90	8.81	.23	22.94
1"	P1@.260	Ea	8.13	9.54	.25	17.92
1¼"	P1@.280	Ea	18.70	10.30	.27	29.27
1½"	P1@.280	Ea	27.20	10.30	.27	37.77
2"	P1@.300	Ea	45.21	10.90	.29	56.11
3"	P1@.380	Ea	127.00	13.90	.36	141.26
4"	P1@.420	Ea	154.00	15.40	.40	169.80
Ball Valve, Brass press fit (P x P)						
½"	P1@.220	Ea	19.80	8.07	.21	28.08
¾"	P1@.225	Ea	26.30	8.26	.21	34.77
1"	P1@.275	Ea	34.40	10.10	.26	44.76
1¼"	P1@.295	Ea	67.30	10.80	.28	78.38
1½"	P1@.295	Ea	83.60	10.80	.28	94.68
2"	P1@.330	Ea	137.00	12.10	.31	149.41
3"	P1@.395	Ea	294.00	14.50	.38	308.88
4"	P1@.435	Ea	372.00	16.00	.41	388.41

Hard drawn rigid copper roll-grooved pipe with factory grooved copper fittings is commonly used for domestic or potable water distribution systems.

Maximum operating pressure for fittings is 300 psi (2065kPa). Operating temperature for fittings ranges from −30 F. to 250 F. (−34 C. – 121 C.).

Consult manufacturer for pipe pressure and temperature ratings and for the various combinations of pipe, fittings and system applications.

The cost estimates in this section are based on the conditions, limitations and wage rates described in the section "How to Use This Book" beginning on page 5.

Description	Craft@Hrs	Unit	Material $	Labor $	Equipment $	Total $

Type K Copper pipe only, roll-grooved, no hangers or fittings

2"	P1@.080	LF	15.10	2.94	—	18.04
2½"	P1@.090	LF	22.00	3.30	—	25.30
3"	P1@.105	LF	29.90	3.85	—	33.75
4"	P1@.145	LF	46.70	5.32	—	52.02
6"	P1@.195	LF	71.40	7.16	—	78.56

Type L Copper pipe only, roll-grooved, no hangers or fittings

2"	P1@.070	LF	13.70	2.57	—	16.27
2½"	P1@.070	LF	19.80	2.57	—	22.37
3"	P1@.090	LF	26.80	3.30	—	30.10
4"	P1@.130	LF	41.90	4.77	—	46.67
6"	P1@.180	LF	64.20	6.61	—	70.81

90-degree copper elbow, roll-grooved, style #610, (Victaulic)

2"	P1@.410	Ea	106.00	15.00	—	121.00
2½"	P1@.430	Ea	117.00	15.80	—	132.80
3"	P1@.500	Ea	165.00	18.40	—	183.40
4"	P1@.690	Ea	352.00	25.30	—	377.30
6"	P1@1.19	Ea	2,080.00	43.70	—	2,123.70

45-degree copper elbow, roll-grooved, style #611, (Victaulic)

2"	P1@.410	Ea	93.60	15.00	—	108.60
2½"	P1@.430	Ea	102.00	15.80	—	117.80
3"	P1@.500	Ea	140.00	18.40	—	158.40
4"	P1@.690	Ea	323.00	25.30	—	348.30
6"	P1@1.19	Ea	1,770.00	43.70	—	1,813.70

Tee, copper, roll-grooved, style #620, (Victaulic)

2"	P1@.510	Ea	181.00	18.70	—	199.70
2½"	P1@.540	Ea	192.00	19.80	—	211.80
3"	P1@.630	Ea	289.00	23.10	—	312.10
4"	P1@.860	Ea	638.00	31.60	—	669.60
6"	P1@1.49	Ea	2,370.00	54.70	—	2,424.70

Copper, Type K & L with Roll-Grooved Joints

Description	Craft@Hrs	Unit	Material $	Labor $	Equipment $	Total $

Reducing tee, copper, roll-grooved, style #625, (Victaulic)

Description	Craft@Hrs	Unit	Material $	Labor $	Equipment $	Total $
2½" x 2"	P1@.540	Ea	295.00	19.80	—	314.80
3" x 2½"	P1@.630	Ea	317.00	23.10	—	340.10
4" x 3"	P1@.860	Ea	743.00	31.60	—	774.60

Reducing tee, copper, roll-grooved, style #626, (Victaulic)

Description	Craft@Hrs	Unit	Material $	Labor $	Equipment $	Total $
2½" x 1"	P1@.500	Ea	375.00	18.40	—	393.40
2½" x 1½"	P1@.520	Ea	375.00	19.10	—	394.10
3" x 1½"	P1@.600	Ea	396.00	22.00	—	418.00

Reducer, copper, roll-grooved, style #650, (Victaulic)

Description	Craft@Hrs	Unit	Material $	Labor $	Equipment $	Total $
2½" x 2"	P1@.430	Ea	158.00	15.80	—	173.80
3" x 2"	P1@.500	Ea	158.00	18.40	—	176.40
3" x 2½"	P1@.520	Ea	158.00	19.10	—	177.10
4" x 2"	P1@.690	Ea	311.00	25.30	—	336.30
4" x 3"	P1@.720	Ea	311.00	26.40	—	337.40
6" x 3"	P1@1.20	Ea	1,080.00	44.00	—	1,124.00

Roll-grooved copper coupling with gasket, style #650, (Victaulic)

Description	Craft@Hrs	Unit	Material $	Labor $	Equipment $	Total $
2"	P1@.300	Ea	52.30	11.00	—	63.30
2½"	P1@.350	Ea	59.10	12.80	—	71.90
3"	P1@.400	Ea	63.80	14.70	—	78.50
4"	P1@.500	Ea	82.10	18.40	—	100.50
5"	P1@.600	Ea	166.00	22.00	—	188.00
6"	P1@.700	Ea	190.00	25.70	—	215.70
8"	P1@.900	Ea	293.00	33.00	—	326.00
10"	P1@1.10	Ea	434.00	40.40	—	474.40
12"	P1@1.35	Ea	650.00	49.50	—	699.50

Flange adapter, copper, roll-grooved, style #641, (Victaulic)

Description	Craft@Hrs	Unit	Material $	Labor $	Equipment $	Total $
2½"	P1@.300	Ea	314.00	11.00	—	325.00
3"	P1@.400	Ea	335.00	14.70	—	349.70
4"	P1@.550	Ea	460.00	20.20	—	480.20

Butterfly valve, brass, roll-grooved, lever handle, style #608, (Victaulic)

Description	Craft@Hrs	Unit	Material $	Labor $	Equipment $	Total $
2½"	P1@.450	Ea	933.00	16.50	—	949.50
3"	P1@.450	Ea	1,200.00	16.50	—	1,216.50
4"	P1@.550	Ea	1,610.00	20.20	—	1,630.20

PVC (Polyvinyl Chloride) Schedule 40 pipe with Schedule 40 socket-type fittings and solvent-welded joints is widely used in process, chemical, A.C. condensate, potable water and irrigation piping systems. PVC is available in Type I (normal impact) and Type II (high impact). Type I has a maximum temperature rating of 150 degrees F., and Type II is rated at 140 degrees F.

Consult the manufacturers for maximum pressure ratings and recommended joint solvents for specific applications.

Because of the current unreliability of many plastic valves, most engineers specify standard bronze-body and iron-body valves for use in PVC water piping systems. Plastic valves, however, have to be used in systems conveying liquids that may be injurious to metallic valves.

This section has been arranged to save the estimator's time by including all normally-used system components such as pipe, fittings, valves, hanger assemblies, riser clamps and miscellaneous items under one heading. Additional items can be found under "Plumbing and Piping Specialties." The cost estimates in this section are based on the conditions, limitations and wage rates described in the section "How to Use This Book" beginning on page 5.

Equipment cost, where shown, is $4.33 per hour for a 2-ton chain hoist.

Description	Craft@Hrs	Unit	Material $	Labor $	Equipment $	Total $

Schedule 40 PVC pipe assembly with solvent-weld joints installed horizontally.
Complete installation including three to eight tees and three to eight elbows every 100' and hangers spaced to meet plumbing code *(a tee and elbow every 12' for 1/2" dia., a tee and elbow every 40' for 8" dia.)*.

Description	Craft@Hrs	Unit	Material $	Labor $	Equipment $	Total $
½"	P1@.084	LF	1.57	3.08	—	4.65
¾"	P1@.091	LF	1.88	3.34	—	5.22
1"	P1@.099	LF	1.97	3.63	—	5.60
1¼"	P1@.114	LF	2.06	4.18	—	6.24
1½"	P1@.121	LF	2.34	4.44	—	6.78
2"	P1@.130	LF	2.88	4.77	—	7.65
2½"	P1@.134	LF	4.66	4.92	—	9.58
3"	P1@.146	LF	7.30	5.36	—	12.66
4"	P1@.177	LF	9.26	6.50	—	15.76
6"	P1@.231	LF	15.50	8.48	—	23.98
8"	P1@.236	LF	27.50	8.66	—	36.16

Schedule 40 PVC pipe assembly with solvent-weld joints, installed risers.
Complete installation including a reducing tee every floor and a riser clamp every other.

Description	Craft@Hrs	Unit	Material $	Labor $	Equipment $	Total $
½"	P1@.040	LF	.76	1.47	—	2.23
¾"	P1@.047	LF	1.00	1.72	—	2.72
1"	P1@.052	LF	1.27	1.91	—	3.18
1¼"	P1@.061	LF	1.60	2.24	—	3.84
1½"	P1@.069	LF	1.97	2.53	—	4.50
2"	P1@.077	LF	2.58	2.83	—	5.41
2½"	P1@.088	LF	4.10	3.23	—	7.33
3"	P1@.099	LF	6.23	3.63	—	9.86
4"	P1@.125	LF	8.01	4.59	—	12.60
6"	P1@.202	LF	16.80	7.41	—	24.21
8"	P1@.266	LF	31.00	9.76	—	40.76

PVC, Schedule 40, with Solvent-Weld Joints

Description	Craft@Hrs	Unit	Material $	Labor $	Equipment $	Total $

Schedule 40 PVC pipe with solvent-weld joints

Description	Craft@Hrs	Unit	Material $	Labor $	Equipment $	Total $
½"	P1@.020	LF	.42	.73	—	1.15
¾"	P1@.025	LF	.64	.92	—	1.56
1"	P1@.030	LF	.86	1.10	—	1.96
1¼"	P1@.035	LF	1.05	1.28	—	2.33
1½"	P1@.040	LF	1.24	1.47	—	2.71
2"	P1@.045	LF	1.67	1.65	—	3.32
2½"	P1@.050	LF	2.63	1.84	—	4.47
3"	P1@.055	LF	4.22	2.02	—	6.24
4"	P1@.070	LF	4.92	2.57	—	7.49
5"	P1@.095	LF	8.42	3.49	—	11.91
6"	P1@.120	LF	8.42	4.40	—	12.82
8"	P1@.160	LF	13.10	5.87	—	18.97

Schedule 40 PVC 45-degree ell S x S with solvent-weld joints

Description	Craft@Hrs	Unit	Material $	Labor $	Equipment $	Total $
½"	P1@.100	Ea	.61	3.67	—	4.28
¾"	P1@.115	Ea	.94	4.22	—	5.16
1"	P1@.120	Ea	1.13	4.40	—	5.53
1¼"	P1@.160	Ea	1.59	5.87	—	7.46
1½"	P1@.180	Ea	2.00	6.61	—	8.61
2"	P1@.200	Ea	2.71	7.34	—	10.05
2½"	P1@.250	Ea	6.75	9.18	—	15.93
3"	P1@.300	Ea	10.50	11.00	—	21.50
4"	P1@.500	Ea	18.70	18.40	—	37.10
5"	P1@.550	Ea	37.20	20.20	—	57.40
6"	P1@.600	Ea	46.50	22.00	—	68.50
8"	P1@.800	Ea	111.00	29.40	—	140.40

Schedule 40 PVC 90-degree ell S x S with solvent-weld joints

Description	Craft@Hrs	Unit	Material $	Labor $	Equipment $	Total $
½"	P1@.100	Ea	.38	3.67	—	4.05
¾"	P1@.115	Ea	.42	4.22	—	4.64
1"	P1@.120	Ea	.73	4.40	—	5.13
1¼"	P1@.160	Ea	1.33	5.87	—	7.20
1½"	P1@.180	Ea	1.40	6.61	—	8.01
2"	P1@.200	Ea	2.21	7.34	—	9.55
2½"	P1@.250	Ea	6.73	9.18	—	15.91
3"	P1@.300	Ea	8.04	11.00	—	19.04
4"	P1@.500	Ea	14.50	18.40	—	32.90
5"	P1@.550	Ea	37.20	20.20	—	57.40
6"	P1@.600	Ea	46.00	22.00	—	68.00
8"	P1@.800	Ea	118.00	29.40	—	147.40

Description	Craft@Hrs	Unit	Material $	Labor $	Equipment $	Total $

Schedule 40 PVC tee S x S x S with solvent-weld joints

Description	Craft@Hrs	Unit	Material $	Labor $	Equipment $	Total $
½"	P1@.130	Ea	.49	4.77	—	5.26
¾"	P1@.140	Ea	.53	5.14	—	5.67
1"	P1@.170	Ea	.98	6.24	—	7.22
1¼"	P1@.220	Ea	1.54	8.07	—	9.61
1½"	P1@.250	Ea	1.71	9.18	—	10.89
2"	P1@.280	Ea	2.21	10.30	—	12.51
2½"	P1@.350	Ea	9.09	12.80	—	21.89
3"	P1@.420	Ea	11.60	15.40	—	27.00
4"	P1@.560	Ea	21.60	20.60	—	42.20
5"	P1@.710	Ea	51.90	26.10	—	78.00
6"	P1@.840	Ea	75.30	30.80	—	106.10
8"	P1@1.12	Ea	166.00	41.10	—	207.10

Schedule 40 PVC reducing tee S x S x S with solvent-weld joints

Description	Craft@Hrs	Unit	Material $	Labor $	Equipment $	Total $
¾" x ½"	P1@.140	Ea	.61	5.14	—	5.75
1" x ½"	P1@.170	Ea	1.13	6.24	—	7.37
1" x ¾"	P1@.170	Ea	1.69	6.24	—	7.93
1¼" x ½"	P1@.220	Ea	1.69	8.07	—	9.76
1¼" x ¾"	P1@.220	Ea	1.69	8.07	—	9.76
1¼" x 1"	P1@.220	Ea	1.69	8.07	—	9.76
1½" x ½"	P1@.250	Ea	2.92	9.18	—	12.10
1½" x ¾"	P1@.250	Ea	2.92	9.18	—	12.10
1½" x 1"	P1@.250	Ea	2.92	9.18	—	12.10
1½" x 1¼"	P1@.250	Ea	2.92	9.18	—	12.10
2" x ½"	P1@.280	Ea	4.16	10.30	—	14.46
2" x ¾"	P1@.280	Ea	4.16	10.30	—	14.46
2" x 1"	P1@.280	Ea	4.16	10.30	—	14.46
2" x 1¼"	P1@.280	Ea	4.16	10.30	—	14.46
2" x 1½"	P1@.280	Ea	4.16	10.30	—	14.46
2½" x 1"	P1@.350	Ea	8.66	12.80	—	21.46
2½" x 1¼"	P1@.350	Ea	8.66	12.80	—	21.46
2½" x 1½"	P1@.350	Ea	8.66	12.80	—	21.46
2½" x 2"	P1@.350	Ea	8.66	12.80	—	21.46
3" x 2"	P1@.420	Ea	12.80	15.40	—	28.20
3" x 2½"	P1@.420	Ea	12.80	15.40	—	28.20
4" x 2"	P1@.560	Ea	21.60	20.60	—	42.20
4" x 2½"	P1@.560	Ea	21.60	20.60	—	42.20
4" x 3"	P1@.560	Ea	21.60	20.60	—	42.20
5" x 3"	P1@.710	Ea	38.30	26.10	—	64.40
5" x 4"	P1@.710	Ea	38.30	26.10	—	64.40
6" x 3"	P1@.840	Ea	62.80	30.80	—	93.60
6" x 4"	P1@.840	Ea	62.80	30.80	—	93.60
6" x 5"	P1@.840	Ea	62.80	30.80	—	93.60
8" x 4"	P1@1.12	Ea	139.00	41.10	—	180.10
8" x 5"	P1@1.12	Ea	139.00	41.10	—	180.10
8" x 6"	P1@1.12	Ea	139.00	41.10	—	180.10

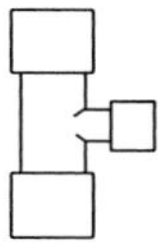

PVC, Schedule 40, with Solvent-Weld Joints

Description	Craft@Hrs	Unit	Material $	Labor $	Equipment $	Total $

Schedule 40 PVC reducing bushing SPIG x S with solvent-weld joints

Description	Craft@Hrs	Unit	Material $	Labor $	Equipment $	Total $
¾" x ½"	P1@.050	Ea	.40	1.84	—	2.24
1" x ½"	P1@.060	Ea	.71	2.20	—	2.91
1" x ¾"	P1@.060	Ea	.71	2.20	—	2.91
1¼" x ½"	P1@.080	Ea	.94	2.94	—	3.88
1¼" x ¾"	P1@.080	Ea	.94	2.94	—	3.88
1¼" x 1"	P1@.080	Ea	.94	2.94	—	3.88
1½" x ½"	P1@.100	Ea	.98	3.67	—	4.65
1½" x ¾"	P1@.100	Ea	.98	3.67	—	4.65
1½" x 1"	P1@.100	Ea	.98	3.67	—	4.65
1½" x 1¼"	P1@.100	Ea	.98	3.67	—	4.65
2" x 1"	P1@.110	Ea	1.64	4.04	—	5.68
2" x 1¼"	P1@.110	Ea	1.64	4.04	—	5.68
2" x 1½"	P1@.110	Ea	1.64	4.04	—	5.68
2½" x 1¼"	P1@.130	Ea	2.62	4.77	—	7.39
2½" x 1½"	P1@.130	Ea	2.62	4.77	—	7.39
2½" x 2"	P1@.130	Ea	2.62	4.77	—	7.39
3" x 2"	P1@.150	Ea	3.91	5.51	—	9.42
3" x 2½"	P1@.150	Ea	3.91	5.51	—	9.42
4" x 3"	P1@.200	Ea	8.74	7.34	—	16.08
5" x 3"	P1@.250	Ea	17.20	9.18	—	26.38
5" x 4"	P1@.250	Ea	17.20	9.18	—	26.38
6" x 4"	P1@.300	Ea	25.20	11.00	—	36.20
6" x 5"	P1@.300	Ea	25.20	11.00	—	36.20
8" x 4"	P1@.400	Ea	75.60	14.70	—	90.30
8" x 5"	P1@.400	Ea	75.60	14.70	—	90.30

Schedule 40 PVC adapter MPT x S with solvent-weld joints

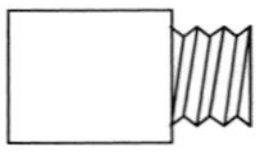

Description	Craft@Hrs	Unit	Material $	Labor $	Equipment $	Total $
½"	P1@.060	Ea	.34	2.20	—	2.54
¾"	P1@.065	Ea	.38	2.39	—	2.77
1"	P1@.070	Ea	.67	2.57	—	3.24
1¼"	P1@.100	Ea	.82	3.67	—	4.49
1½"	P1@.110	Ea	1.10	4.04	—	5.14
2"	P1@.120	Ea	3.60	4.40	—	8.00
2½"	P1@.150	Ea	4.21	5.51	—	9.72
3"	P1@.180	Ea	6.13	6.61	—	12.74
4"	P1@.240	Ea	7.81	8.81	—	16.62
5"	P1@.370	Ea	13.30	13.60	—	26.90
6"	P1@.420	Ea	20.40	15.40	—	35.80

Schedule 40 PVC adapter FPT x S with solvent-weld joints

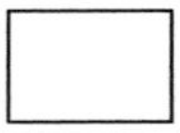

Description	Craft@Hrs	Unit	Material $	Labor $	Equipment $	Total $
½"	P1@.110	Ea	.42	4.04	—	4.46
¾"	P1@.120	Ea	.53	4.40	—	4.93
1"	P1@.130	Ea	.61	4.77	—	5.38
1¼"	P1@.170	Ea	.95	6.24	—	7.19
1½"	P1@.210	Ea	1.10	7.71	—	8.81
2"	P1@.290	Ea	1.48	10.60	—	12.08
2½"	P1@.330	Ea	3.70	12.10	—	15.80
3"	P1@.470	Ea	7.37	17.20	—	24.57
4"	P1@.630	Ea	13.60	23.10	—	36.70
5"	P1@.890	Ea	21.60	32.70	—	54.30
6"	P1@.970	Ea	30.30	35.60	—	65.90

Description	Craft@Hrs	Unit	Material $	Labor $	Equipment $	Total $

Schedule 40 PVC cap S with solvent-weld joints

½"	P1@.060	Ea	.34	2.20	—	2.54
¾"	P1@.065	Ea	.40	2.39	—	2.79
1"	P1@.070	Ea	.61	2.57	—	3.18
1¼"	P1@.100	Ea	.94	3.67	—	4.61
1½"	P1@.110	Ea	.95	4.04	—	4.99
2"	P1@.120	Ea	1.14	4.40	—	5.54
2½"	P1@.150	Ea	3.60	5.51	—	9.11
3"	P1@.180	Ea	3.97	6.61	—	10.58
4"	P1@.240	Ea	9.09	8.81	—	17.90
5"	P1@.370	Ea	15.10	13.60	—	28.70
6"	P1@.420	Ea	21.70	15.40	—	37.10

Schedule 40 PVC plug MPT with solvent-weld joints

½"	P1@.150	Ea	1.00	5.51	—	6.51
¾"	P1@.160	Ea	1.09	5.87	—	6.96
1"	P1@.180	Ea	1.77	6.61	—	8.38
1¼"	P1@.210	Ea	1.83	7.71	—	9.54
1½"	P1@.230	Ea	1.98	8.44	—	10.42
2"	P1@.240	Ea	2.58	8.81	—	11.39
2½"	P1@.260	Ea	3.55	9.54	—	13.09
3"	P1@.300	Ea	5.37	11.00	—	16.37
4"	P1@.400	Ea	12.30	14.70	—	27.00

Schedule 40 PVC coupling S x S with solvent-weld joints

½"	P1@.100	Ea	.23	3.67	—	3.90
¾"	P1@.115	Ea	.34	4.22	—	4.56
1"	P1@.120	Ea	.59	4.40	—	4.99
1¼"	P1@.160	Ea	.82	5.87	—	6.69
1½"	P1@.180	Ea	.86	6.61	—	7.47
2"	P1@.200	Ea	1.34	7.34	—	8.68
2½"	P1@.250	Ea	2.92	9.18	—	12.10
3"	P1@.300	Ea	4.61	11.00	—	15.61
4"	P1@.500	Ea	6.61	18.40	—	25.01
5"	P1@.550	Ea	12.30	20.20	—	32.50
6"	P1@.600	Ea	20.90	22.00	—	42.90
8"	P1@.800	Ea	39.10	29.40	—	68.50

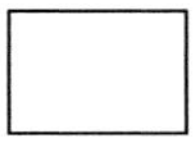

Schedule 40 PVC union S x S with solvent-weld joints

1/2"	P1@.125	Ea	3.07	4.59	—	7.66
3/4"	P1@.140	Ea	3.82	5.14	—	8.96
1"	P1@.160	Ea	3.97	5.87	—	9.84
1¼"	P1@.200	Ea	8.19	7.34	—	15.53
1½"	P1@.225	Ea	9.32	8.26	—	17.58
2"	P1@.250	Ea	14.50	9.18	—	23.68
2½"	P1@.275	Ea	21.60	10.10	—	31.70
3"	P1@.300	Ea	26.30	11.00	—	37.30

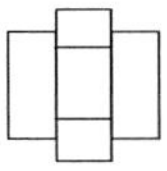

PVC, Schedule 40, with Solvent-Weld Joints

Description	Craft@Hrs	Unit	Material $	Labor $	Equipment $	Total $

Class 125 bronze body gate valve, threaded ends

½"	P1@.210	Ea	18.40	7.71	—	26.11
¾"	P1@.250	Ea	23.00	9.18	—	32.18
1"	P1@.300	Ea	32.40	11.00	—	43.40
1¼"	P1@.400	Ea	40.70	14.70	—	55.40
1½"	P1@.450	Ea	54.90	16.50	—	71.40
2"	P1@.500	Ea	91.80	18.40	—	110.20
2½"	P1@.750	Ea	152.00	27.50	—	179.50
3"	P1@.950	Ea	218.00	34.90	—	252.90

Class 125 iron body gate valve, flanged ends

2"	P1@.500	Ea	351.00	18.40	—	369.40
2½"	P1@.600	Ea	474.00	22.00	—	496.00
3"	P1@.750	Ea	514.00	27.50	—	541.50
4"	P1@1.35	Ea	754.00	49.50	—	803.50
5"	ER@2.00	Ea	1,440.00	78.80	1.24	1,520.04
6"	ER@2.50	Ea	1,440.00	98.50	1.55	1,540.05
8"	ER@3.00	Ea	2,370.00	118.00	1.86	2,489.86

Class 125 bronze body globe valve, threaded ends

½"	P1@.210	Ea	34.10	7.71	—	41.81
¾"	P1@.250	Ea	45.40	9.18	—	54.58
1"	P1@.300	Ea	65.10	11.00	—	76.10
1¼"	P1@.400	Ea	91.60	14.70	—	106.30
1½"	P1@.450	Ea	122.00	16.50	—	138.50
2"	P1@.500	Ea	200.00	18.40	—	218.40

200 PSIG iron body butterfly valve, lug-type, lever operated

2"	P1@.450	Ea	156.00	16.50	—	172.50
2½"	P1@.450	Ea	162.00	16.50	—	178.50
3"	P1@.550	Ea	170.00	20.20	—	190.20
4"	P1@.550	Ea	212.00	20.20	—	232.20
5"	ER@.800	Ea	279.00	31.50	.49	310.99
6"	ER@.800	Ea	341.00	31.50	.49	372.99
8"	ER@.900	Ea	466.00	35.50	.56	502.06

200 PSIG iron body butterfly valve, wafer-type, lever operated

2"	P1@.450	Ea	142.00	16.50	—	158.50
2½"	P1@.450	Ea	145.00	16.50	—	161.50
3"	P1@.550	Ea	156.00	20.20	—	176.20
4"	P1@.550	Ea	189.00	20.20	—	209.20
5"	ER@.800	Ea	252.00	31.50	.49	283.99
6"	ER@.800	Ea	316.00	31.50	.49	347.99
8"	ER@.900	Ea	438.00	35.50	.56	474.06

Description	Craft@Hrs	Unit	Material $	Labor $	Equipment $	Total $

Class 125 bronze body 2-piece ball valve, threaded ends

Description	Craft@Hrs	Unit	Material $	Labor $	Equipment $	Total $
½"	P1@.210	Ea	10.10	7.71	—	17.81
¾"	P1@.250	Ea	13.40	9.18	—	22.58
1"	P1@.300	Ea	23.30	11.00	—	34.30
1¼"	P1@.400	Ea	38.80	14.70	—	53.50
1½"	P1@.450	Ea	52.40	16.50	—	68.90
2"	P1@.500	Ea	66.50	18.40	—	84.90
3"	P1@.625	Ea	455.00	22.90	—	477.90
4"	P1@.690	Ea	595.00	25.30	—	620.30

PVC ball valve, solid body, solvent weld joints, EDPM (female socket)

Description	Craft@Hrs	Unit	Material $	Labor $	Equipment $	Total $
½"	P1@.170	Ea	5.68	6.24	—	11.92
¾"	P1@.200	Ea	6.43	7.34	—	13.77
1"	P1@.220	Ea	9.35	8.07	—	17.42
1¼"	P1@.280	Ea	10.80	10.30	—	21.10
1½"	P1@.315	Ea	18.10	11.60	—	29.70
2"	P1@.385	Ea	22.60	14.10	—	36.70

PVC ball valve, solid body, threaded joints, EDPM (female pipe thread)

Description	Craft@Hrs	Unit	Material $	Labor $	Equipment $	Total $
½"	P1@.200	Ea	5.68	7.34	—	13.02
¾"	P1@.240	Ea	6.43	8.81	—	15.24
1"	P1@.290	Ea	9.35	10.60	—	19.95
1¼"	P1@.390	Ea	10.80	14.30	—	25.10
1½"	P1@.420	Ea	18.10	15.40	—	33.50
2"	P1@.490	Ea	22.60	18.00	—	40.60

PVC ball valve, tru-union, threaded joints, EDPM (female pipe thread)

Description	Craft@Hrs	Unit	Material $	Labor $	Equipment $	Total $
2"	P1@.490	Ea	80.60	18.00	—	98.60
2½"	P1@.520	Ea	93.10	19.10	—	112.20
3"	P1@.590	Ea	215.00	21.70	—	236.70
4"	P1@.630	Ea	397.00	23.10	—	420.10

PVC, Schedule 40, with Solvent-Weld Joints

Description	Craft@Hrs	Unit	Material $	Labor $	Equipment $	Total $

PVC ball valve, union type body, solvent-weld joints (female socket)

Description	Craft@Hrs	Unit	Material $	Labor $	Equipment $	Total $
½"	P1@.185	Ea	5.68	6.79	—	12.47
¾"	P1@.225	Ea	6.43	8.26	—	14.69
1"	P1@.270	Ea	9.35	9.91	—	19.26
1¼"	P1@.360	Ea	10.80	13.20	—	24.00
1½"	P1@.400	Ea	18.10	14.70	—	32.80
2"	P1@.465	Ea	22.60	17.10	—	39.70
2½"	P1@.490	Ea	93.10	18.00	—	111.10
3"	P1@.540	Ea	215.00	19.80	—	234.80
4"	P1@.600	Ea	397.00	22.00	—	419.00

Class 125 bronze body swing check valve, threaded

Description	Craft@Hrs	Unit	Material $	Labor $	Equipment $	Total $
½"	P1@.210	Ea	20.40	7.71	—	28.11
¾"	P1@.250	Ea	29.20	9.18	—	38.38
1"	P1@.300	Ea	38.10	11.00	—	49.10
1¼"	P1@.400	Ea	54.40	14.70	—	69.10
1½"	P1@.450	Ea	76.70	16.50	—	93.20
2"	P1@.500	Ea	127.00	18.40	—	145.40

Class 125 iron body swing check valve, flanged ends

Description	Craft@Hrs	Unit	Material $	Labor $	Equipment $	Total $
2"	P1@.500	Ea	179.00	18.40	—	197.40
2½"	P1@.600	Ea	226.00	22.00	—	248.00
3"	P1@.750	Ea	282.00	27.50	—	309.50
4"	P1@1.35	Ea	414.00	49.50	—	463.50
5"	ER@2.00	Ea	797.00	78.80	1.24	877.04
6"	ER@2.50	Ea	797.00	98.50	1.55	897.05
8"	ER@3.00	Ea	1,440.00	118.00	1.86	1,559.86

Class 125 iron body silent check valve, wafer-type

Description	Craft@Hrs	Unit	Material $	Labor $	Equipment $	Total $
2"	P1@.500	Ea	131.00	18.40	—	149.40
2½"	P1@.600	Ea	145.00	22.00	—	167.00
3"	P1@.750	Ea	167.00	27.50	—	194.50
4"	P1@1.35	Ea	223.00	49.50	—	272.50
5"	ER@2.00	Ea	342.00	78.80	1.24	422.04
6"	ER@2.50	Ea	466.00	98.50	1.55	566.05
8"	ER@3.00	Ea	797.00	118.00	1.86	916.86

Class 125 bronze body strainer, threaded ends

Description	Craft@Hrs	Unit	Material $	Labor $	Equipment $	Total $
½"	P1@.210	Ea	31.40	7.71	—	39.11
¾"	P1@.250	Ea	41.30	9.18	—	50.48
1"	P1@.300	Ea	50.60	11.00	—	61.60
1¼"	P1@.400	Ea	70.70	14.70	—	85.40
1½"	P1@.450	Ea	106.00	16.50	—	122.50
2"	P1@.500	Ea	185.00	18.40	—	203.40

Description	Craft@Hrs	Unit	Material $	Labor $	Equipment $	Total $

Class 125 iron body strainer, flanged ends

Description	Craft@Hrs	Unit	Material $	Labor $	Equipment $	Total $
2"	P1@.500	Ea	141.00	18.40	—	159.40
2½"	P1@.600	Ea	157.00	22.00	—	179.00
3"	P1@.750	Ea	183.00	27.50	—	210.50
4"	P1@1.35	Ea	308.00	49.50	—	357.50
5"	ER@2.00	Ea	629.00	78.80	1.24	709.04
6"	ER@2.50	Ea	629.00	98.50	1.55	729.05
8"	ER@3.00	Ea	1,060.00	118.00	1.86	1,179.86

Installation of 2-way control valve, threaded joints

Description	Craft@Hrs	Unit	Material $	Labor $	Equipment $	Total $
½"	P1@.210	Ea	—	7.71	—	7.71
¾"	P1@.275	Ea	—	10.10	—	10.10
1"	P1@.350	Ea	—	12.80	—	12.80
1¼"	P1@.430	Ea	—	15.80	—	15.80
1½"	P1@.505	Ea	—	18.50	—	18.50
2"	P1@.675	Ea	—	24.80	—	24.80
2½"	P1@.830	Ea	—	30.50	—	30.50
3"	P1@.990	Ea	—	36.30	—	36.30

Installation of 3-way control valve, threaded joints

Description	Craft@Hrs	Unit	Material $	Labor $	Equipment $	Total $
½"	P1@.260	Ea	—	9.54	—	9.54
¾"	P1@.365	Ea	—	13.40	—	13.40
1"	P1@.475	Ea	—	17.40	—	17.40
1¼"	P1@.575	Ea	—	21.10	—	21.10
1½"	P1@.680	Ea	—	25.00	—	25.00
2"	P1@.910	Ea	—	33.40	—	33.40
2½"	P1@1.12	Ea	—	41.10	—	41.10
3"	P1@1.33	Ea	—	48.80	—	48.80

PVC companion flange

Description	Craft@Hrs	Unit	Material $	Labor $	Equipment $	Total $
2"	P1@.140	Ea	6.08	5.14	—	11.22
2½"	P1@.180	Ea	18.30	6.61	—	24.91
3"	P1@.210	Ea	18.90	7.71	—	26.61
4"	P1@.280	Ea	19.40	10.30	—	29.70
5"	P1@.310	Ea	43.70	11.40	—	55.10
6"	P1@.420	Ea	43.70	15.40	—	59.10
8"	P1@.560	Ea	65.20	20.60	—	85.80

Bolt and gasket sets

Description	Craft@Hrs	Unit	Material $	Labor $	Equipment $	Total $
2"	P1@.500	Ea	4.22	18.40	—	22.62
2½"	P1@.650	Ea	4.95	23.90	—	28.85
3"	P1@.750	Ea	8.43	27.50	—	35.93
4"	P1@1.00	Ea	14.10	36.70	—	50.80
5"	P1@1.10	Ea	23.30	40.40	—	63.70
6"	P1@1.20	Ea	23.30	44.00	—	67.30
8"	P1@1.25	Ea	25.90	45.90	—	71.80

PVC, Schedule 40, with Solvent-Weld Joints

Description	Craft@Hrs	Unit	Material $	Labor $	Equipment $	Total $

Thermometer with well

Description	Craft@Hrs	Unit	Material $	Labor $	Equipment $	Total $
7"	P1@.250	Ea	180.00	9.18	—	189.18
9"	P1@.250	Ea	185.00	9.18	—	194.18

Dial-type pressure gauge

Description	Craft@Hrs	Unit	Material $	Labor $	Equipment $	Total $
2½"	P1@.200	Ea	37.70	7.34	—	45.04
3½"	P1@.200	Ea	49.70	7.34	—	57.04

Pressure/temperature tap

Description	Craft@Hrs	Unit	Material $	Labor $	Equipment $	Total $
Tap	P1@.150	Ea	17.30	5.51	—	22.81

Hanger with swivel assembly

Description	Craft@Hrs	Unit	Material $	Labor $	Equipment $	Total $
½"	P1@.250	Ea	4.22	9.18	—	13.40
¾"	P1@.250	Ea	4.68	9.18	—	13.86
1"	P1@.250	Ea	4.89	9.18	—	14.07
1¼"	P1@.300	Ea	5.02	11.00	—	16.02
1½"	P1@.300	Ea	5.40	11.00	—	16.40
2"	P1@.300	Ea	5.63	11.00	—	16.63
2½"	P1@.350	Ea	7.65	12.80	—	20.45
3"	P1@.350	Ea	9.47	12.80	—	22.27
4"	P1@.350	Ea	10.40	12.80	—	23.20
5"	P1@.450	Ea	15.10	16.50	—	31.60
6"	P1@.450	Ea	17.10	16.50	—	33.60
8"	P1@.450	Ea	21.40	16.50	—	37.90

Riser clamp

Description	Craft@Hrs	Unit	Material $	Labor $	Equipment $	Total $
½"	P1@.100	Ea	2.72	3.67	—	6.39
¾"	P1@.100	Ea	4.02	3.67	—	7.69
1"	P1@.100	Ea	4.06	3.67	—	7.73
1¼"	P1@.105	Ea	4.89	3.85	—	8.74
1½"	P1@.110	Ea	5.17	4.04	—	9.21
2"	P1@.115	Ea	5.47	4.22	—	9.69
2½"	P1@.120	Ea	5.76	4.40	—	10.16
3"	P1@.120	Ea	6.23	4.40	—	10.63
4"	P1@.125	Ea	7.94	4.59	—	12.53
5"	P1@.180	Ea	11.40	6.61	—	18.01
6"	P1@.200	Ea	13.70	7.34	—	21.04
8"	P1@.200	Ea	22.40	7.34	—	29.74

PVC (Polyvinyl Chloride) Schedule 80 pipe with Schedule 80 socket-type fittings and solvent-welded joints is widely used in process, chemical, A.C. condensate, potable water and irrigation piping systems. PVC is available in Type I (normal impact) and Type II (high impact). Type I has a maximum temperature rating of 150 degrees F., and Type II is rated at 140 degrees F.

Schedule 80 PVC can be threaded, but it is more economical to use solvent-welded joints.

Manufacturers should be consulted for maximum pressure ratings and recommended joints solvents for specific applications.

Because of the current unreliability of many plastic valves, most engineers specify standard bronze-body and iron-body valves for use in PVC water piping systems. Plastic valves, however, have to be used in systems conveying liquids that may be injurious to metallic valves.

This section has been arranged to save the estimator's time by including all normally-used system components such as pipe, fittings, valves, hanger assemblies, riser clamps and miscellaneous items under one heading. Additional items can be found under "Plumbing and Piping Specialties." The cost estimates in this section are based on the conditions, limitations and wage rates described in the section "How to Use This Book" beginning on page 5.

Equipment cost, where shown, is $4.33 per hour for a 2-ton chain hoist.

Description	Craft@Hrs	Unit	Material $	Labor $	Equipment $	Total $

Schedule 80 PVC pipe assembly with solvent-weld joints installed horizontally.
Complete installation including three to eight tees and three to eight elbows every 100' and hangers spaced to meet plumbing code (a tee and elbow every 12' for 1/2" dia., a tee and elbow every 40' for 8"dia.).

Description	Craft@Hrs	Unit	Material $	Labor $	Equipment $	Total $
½"	P1@.085	LF	2.13	3.12	—	5.25
¾"	P1@.093	LF	2.66	3.41	—	6.07
1"	P1@.099	LF	2.93	3.63	—	6.56
1¼"	P1@.115	LF	3.43	4.22	—	7.65
1½"	P1@.117	LF	3.67	4.29	—	7.96
2"	P1@.120	LF	4.65	4.40	—	9.05
2½"	P1@.125	LF	7.66	4.59	—	12.25
3"	P1@.129	LF	10.40	4.73	—	15.13
4"	P1@.161	LF	14.90	5.91	—	20.81
6"	P1@.253	LF	22.60	9.29	—	31.89
8"	P1@.299	LF	37.80	11.00	—	48.80

PVC, Schedule 80, with Solvent-Weld Joints

Description	Craft@Hrs	Unit	Material $	Labor $	Equipment $	Total $

Schedule 80 PVC pipe assembly with solvent-weld joints, installed risers. Complete installation including a reducing tee every floor and a riser clamp every other.

Description	Craft@Hrs	Unit	Material $	Labor $	Equipment $	Total $
½"	P1@.041	LF	1.03	1.50	—	2.53
¾"	P1@.048	LF	1.49	1.76	—	3.25
1"	P1@.054	LF	1.80	1.98	—	3.78
1¼"	P1@.063	LF	2.12	2.31	—	4.43
1½"	P1@.071	LF	2.77	2.61	—	5.38
2"	P1@.079	LF	3.71	2.90	—	6.61
2½"	P1@.090	LF	6.07	3.30	—	9.37
3"	P1@.101	LF	7.86	3.71	—	11.57
4"	P1@.129	LF	11.20	4.73	—	15.93
6"	P1@.242	LF	20.10	8.88	—	28.98
8"	P1@.296	LF	33.90	10.90	—	44.80

Schedule 80 PVC pipe with solvent-weld joints

Description	Craft@Hrs	Unit	Material $	Labor $	Equipment $	Total $
½"	P1@.021	LF	.71	.77	—	1.48
¾"	P1@.026	LF	.95	.95	—	1.90
1"	P1@.032	LF	1.24	1.17	—	2.41
1¼"	P1@.037	LF	1.46	1.36	—	2.82
1½"	P1@.042	LF	1.97	1.54	—	3.51
2"	P1@.047	LF	2.65	1.72	—	4.37
2½"	P1@.052	LF	4.22	1.91	—	6.13
3"	P1@.057	LF	5.65	2.09	—	7.74
4"	P1@.074	LF	8.33	2.72	—	11.05
6"	P1@.160	LF	15.80	5.87	—	21.67
8"	P1@.190	LF	24.10	6.97	—	31.07

Schedule 80 PVC 45-degree ell S x S with solvent-weld joints

Description	Craft@Hrs	Unit	Material $	Labor $	Equipment $	Total $
½"	P1@.105	Ea	2.97	3.85	—	6.82
¾"	P1@.120	Ea	4.49	4.40	—	8.89
1"	P1@.125	Ea	6.76	4.59	—	11.35
1¼"	P1@.170	Ea	8.66	6.24	—	14.90
1½"	P1@.190	Ea	10.20	6.97	—	17.17
2"	P1@.210	Ea	13.20	7.71	—	20.91
2½"	P1@.262	Ea	27.70	9.62	—	37.32
3"	P1@.315	Ea	33.70	11.60	—	45.30
4"	P1@.520	Ea	60.70	19.10	—	79.80
6"	P1@.630	Ea	76.30	23.10	—	99.40
8"	P1@.840	Ea	165.00	30.80	—	195.80

Description	Craft@Hrs	Unit	Material $	Labor $	Equipment $	Total $

Schedule 80 PVC 90-degree ell S x S with solvent-weld joints

Description	Craft@Hrs	Unit	Material $	Labor $	Equipment $	Total $
½"	P1@.105	Ea	1.61	3.85	—	5.46
¾"	P1@.120	Ea	2.04	4.40	—	6.44
1"	P1@.125	Ea	3.21	4.59	—	7.80
1¼"	P1@.170	Ea	4.37	6.24	—	10.61
1½"	P1@.190	Ea	4.66	6.97	—	11.63
2"	P1@.210	Ea	5.58	7.71	—	13.29
2½"	P1@.262	Ea	13.00	9.62	—	22.62
3"	P1@.315	Ea	14.70	11.60	—	26.30
4"	P1@.520	Ea	22.40	19.10	—	41.50
6"	P1@.630	Ea	63.80	23.10	—	86.90
8"	P1@.840	Ea	174.00	30.80	—	204.80

Schedule 80 PVC tee S x S x S with solvent-weld joints

Description	Craft@Hrs	Unit	Material $	Labor $	Equipment $	Total $
½"	P1@.135	Ea	4.45	4.95	—	9.40
¾"	P1@.145	Ea	4.66	5.32	—	9.98
1"	P1@.180	Ea	5.77	6.61	—	12.38
1¼"	P1@.230	Ea	15.90	8.44	—	24.34
1½"	P1@.265	Ea	15.90	9.73	—	25.63
2"	P1@.295	Ea	19.90	10.80	—	30.70
2½"	P1@.368	Ea	21.70	13.50	—	35.20
3"	P1@.440	Ea	27.20	16.10	—	43.30
4"	P1@.585	Ea	31.50	21.50	—	53.00
6"	P1@.880	Ea	106.00	32.30	—	138.30
8"	P1@1.18	Ea	248.00	43.30	—	291.30

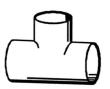

Schedule 80 PVC reducing tee S x S x S with solvent-weld joints

Description	Craft@Hrs	Unit	Material $	Labor $	Equipment $	Total $
¾" x ½"	P1@.145	Ea	5.03	5.32	—	10.35
1" x ½"	P1@.175	Ea	5.42	6.42	—	11.84
1" x ¾"	P1@.175	Ea	5.42	6.42	—	11.84
1¼" x ½"	P1@.230	Ea	17.80	8.44	—	26.24
1¼" x ¾"	P1@.230	Ea	17.80	8.44	—	26.24
1¼" x 1"	P1@.230	Ea	17.80	8.44	—	26.24
1½" x ¾"	P1@.260	Ea	19.10	9.54	—	28.64
1½" x 1"	P1@.260	Ea	19.10	9.54	—	28.64
1½" x 1¼"	P1@.260	Ea	19.10	9.54	—	28.64
2" x ½"	P1@.290	Ea	23.40	10.60	—	34.00
2" x ¾"	P1@.290	Ea	23.40	10.60	—	34.00
2" x 1"	P1@.290	Ea	23.40	10.60	—	34.00
2" x 1½"	P1@.290	Ea	23.40	10.60	—	34.00
3" x 2"	P1@.430	Ea	33.20	15.80	—	49.00
4" x 2"	P1@.570	Ea	37.70	20.90	—	58.60
4" x 3"	P1@.570	Ea	37.70	20.90	—	58.60
6" x 4"	P1@.850	Ea	127.00	31.20	—	158.20
8" x 4"	P1@1.16	Ea	285.00	42.60	—	327.60
8" x 5"	P1@1.16	Ea	285.00	42.60	—	327.60
8" x 6"	P1@1.16	Ea	285.00	42.60	—	327.60

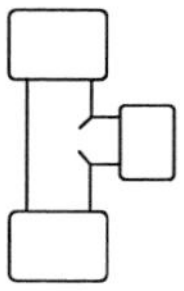

PVC, Schedule 80, with Solvent-Weld Joints

Description	Craft@Hrs	Unit	Material $	Labor $	Equipment $	Total $

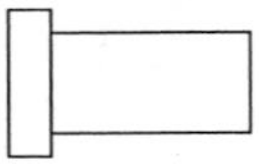

Schedule 80 PVC reducing bushing SPIG x S with solvent-weld joints

Description	Craft@Hrs	Unit	Material $	Labor $	Equipment $	Total $
¾" x ½"	P1@.053	Ea	.92	1.95	—	2.87
1" x ½"	P1@.063	Ea	2.62	2.31	—	4.93
1" x ¾"	P1@.063	Ea	2.62	2.31	—	4.93
1¼" x ½"	P1@.084	Ea	4.20	3.08	—	7.28
1¼" x ¾"	P1@.084	Ea	4.20	3.08	—	7.28
1¼" x 1"	P1@.084	Ea	4.20	3.08	—	7.28
1½" x ½"	P1@.105	Ea	5.58	3.85	—	9.43
1½" x ¾"	P1@.105	Ea	5.58	3.85	—	9.43
1½" x 1"	P1@.105	Ea	5.58	3.85	—	9.43
1½" x 1¼"	P1@.105	Ea	5.58	3.85	—	9.43
2" x 1"	P1@.116	Ea	7.99	4.26	—	12.25
2" x 1¼"	P1@.116	Ea	7.99	4.26	—	12.25
2" x 1½"	P1@.116	Ea	7.99	4.26	—	12.25
3" x 2"	P1@.158	Ea	22.10	5.80	—	27.90
4" x 2½"	P1@.210	Ea	30.40	7.71	—	38.11
4" x 3"	P1@.210	Ea	30.40	7.71	—	38.11
6" x 3"	P1@.315	Ea	42.30	11.60	—	53.90
6" x 4"	P1@.315	Ea	42.30	11.60	—	53.90
8" x 4"	P1@.420	Ea	102.00	15.40	—	117.40
8" x 6"	P1@.420	Ea	102.00	15.40	—	117.40

Schedule 80 PVC adapter MPT x S with solvent-weld joints

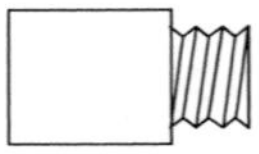

Description	Craft@Hrs	Unit	Material $	Labor $	Equipment $	Total $
½"	P1@.063	Ea	3.34	2.31	—	5.65
¾"	P1@.068	Ea	3.65	2.50	—	6.15
1"	P1@.074	Ea	6.27	2.72	—	8.99
1¼"	P1@.105	Ea	7.42	3.85	—	11.27
1½"	P1@.116	Ea	10.60	4.26	—	14.86
2"	P1@.126	Ea	15.30	4.62	—	19.92
2½"	P1@.158	Ea	17.70	5.80	—	23.50
3"	P1@.189	Ea	19.40	6.94	—	26.34
4"	P1@.252	Ea	34.30	9.25	—	43.55
5"	P1@.385	Ea	47.90	14.10	—	62.00
6"	P1@.435	Ea	110.00	16.00	—	126.00

Schedule 80 PVC adapter FPT x S with solvent-weld joints

Description	Craft@Hrs	Unit	Material $	Labor $	Equipment $	Total $
½"	P1@.116	Ea	2.68	4.26	—	6.94
¾"	P1@.126	Ea	3.97	4.62	—	8.59
1"	P1@.137	Ea	5.80	5.03	—	10.83
1¼"	P1@.179	Ea	9.41	6.57	—	15.98
1½"	P1@.221	Ea	11.40	8.11	—	19.51
2"	P1@.305	Ea	20.10	11.20	—	31.30
2½"	P1@.345	Ea	31.70	12.70	—	44.40
3"	P1@.490	Ea	35.60	18.00	—	53.60
4"	P1@.655	Ea	61.30	24.00	—	85.30
5"	P1@.920	Ea	99.70	33.80	—	133.50
6"	P1@1.10	Ea	122.00	40.40	—	162.40

Description	Craft@Hrs	Unit	Material $	Labor $	Equipment $	Total $

Schedule 80 PVC cap S with solvent-weld joints

Description	Craft@Hrs	Unit	Material $	Labor $	Equipment $	Total $
½"	P1@.063	Ea	2.80	2.31	—	5.11
¾"	P1@.070	Ea	2.90	2.57	—	5.47
1"	P1@.075	Ea	5.17	2.75	—	7.92
1¼"	P1@.105	Ea	6.27	3.85	—	10.12
1½"	P1@.115	Ea	6.27	4.22	—	10.49
2"	P1@.125	Ea	12.30	4.59	—	16.89
2½"	P1@.158	Ea	25.00	5.80	—	30.80
3"	P1@.189	Ea	29.70	6.94	—	36.64
4"	P1@.250	Ea	50.10	9.18	—	59.28
5"	P1@.390	Ea	103.00	14.30	—	117.30
6"	P1@.450	Ea	124.00	16.50	—	140.50

Schedule 80 PVC plug MPT with solvent-weld joints

Description	Craft@Hrs	Unit	Material $	Labor $	Equipment $	Total $
½"	P1@.155	Ea	2.71	5.69	—	8.40
¾"	P1@.165	Ea	2.80	6.06	—	8.86
1"	P1@.190	Ea	3.42	6.97	—	10.39
1¼"	P1@.220	Ea	5.03	8.07	—	13.10
1½"	P1@.240	Ea	6.08	8.81	—	14.89
2"	P1@.255	Ea	6.26	9.36	—	15.62
2½"	P1@.275	Ea	15.70	10.10	—	25.80
3"	P1@.320	Ea	20.40	11.70	—	32.10
4"	P1@.425	Ea	39.70	15.60	—	55.30

Schedule 80 PVC coupling S x S with solvent-weld joints

Description	Craft@Hrs	Unit	Material $	Labor $	Equipment $	Total $
½"	P1@.105	Ea	2.85	3.85	—	6.70
¾"	P1@.120	Ea	3.82	4.40	—	8.22
1"	P1@.125	Ea	3.96	4.59	—	8.55
1¼"	P1@.170	Ea	6.05	6.24	—	12.29
1½"	P1@.190	Ea	6.48	6.97	—	13.45
2"	P1@.210	Ea	6.94	7.71	—	14.65
2½"	P1@.262	Ea	17.00	9.62	—	26.62
3"	P1@.315	Ea	19.70	11.60	—	31.30
4"	P1@.520	Ea	24.80	19.10	—	43.90
6"	P1@.630	Ea	52.90	23.10	—	76.00
8"	P1@.840	Ea	72.80	30.80	—	103.60

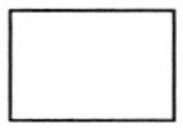

Schedule 80 PVC union S x S with solvent-weld joints

Description	Craft@Hrs	Unit	Material $	Labor $	Equipment $	Total $
½"	P1@.125	Ea	5.72	4.59	—	10.31
¾"	P1@.140	Ea	7.87	5.14	—	13.01
1"	P1@.160	Ea	8.30	5.87	—	14.17
1¼"	P1@.200	Ea	16.50	7.34	—	23.84
1½"	P1@.225	Ea	18.70	8.26	—	26.96
2"	P1@.250	Ea	25.30	9.18	—	34.48
2½"	P1@.275	Ea	40.00	10.10	—	50.10
3"	P1@.300	Ea	47.20	11.00	—	58.20

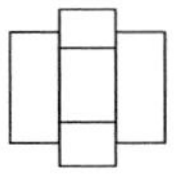

PVC, Schedule 80, with Solvent-Weld Joints

Description	Craft@Hrs	Unit	Material $	Labor $	Equipment $	Total $
Class 125 bronze body gate valve, threaded ends						
½"	P1@.210	Ea	18.40	7.71	—	26.11
¾"	P1@.250	Ea	23.00	9.18	—	32.18
1"	P1@.300	Ea	32.40	11.00	—	43.40
1¼"	P1@.400	Ea	40.70	14.70	—	55.40
1½"	P1@.450	Ea	54.90	16.50	—	71.40
2"	P1@.500	Ea	91.80	18.40	—	110.20
2½"	P1@.750	Ea	152.00	27.50	—	179.50
3"	P1@.950	Ea	218.00	34.90	—	252.90
Class 125 iron body gate valve, flanged ends						
2"	P1@.500	Ea	351.00	18.40	—	369.40
2½"	P1@.600	Ea	474.00	22.00	—	496.00
3"	P1@.750	Ea	514.00	27.50	—	541.50
4"	P1@1.35	Ea	754.00	49.50	—	803.50
5"	ER@2.00	Ea	1,440.00	78.80	1.24	1,520.04
6"	ER@2.50	Ea	1,440.00	98.50	1.55	1,540.05
8"	ER@3.00	Ea	2,370.00	118.00	1.86	2,489.86
Class 125 bronze body globe valve, threaded ends						
½"	P1@.210	Ea	34.10	7.71	—	41.81
¾"	P1@.250	Ea	45.40	9.18	—	54.58
1"	P1@.300	Ea	65.10	11.00	—	76.10
1¼"	P1@.400	Ea	91.60	14.70	—	106.30
1½"	P1@.450	Ea	122.00	16.50	—	138.50
2"	P1@.500	Ea	200.00	18.40	—	218.40
200 PSIG iron body butterfly valve, lug-type, lever operated						
2"	P1@.450	Ea	156.00	16.50	—	172.50
2½"	P1@.450	Ea	162.00	16.50	—	178.50
3"	P1@.550	Ea	170.00	20.20	—	190.20
4"	P1@.550	Ea	212.00	20.20	—	232.20
5"	ER@.800	Ea	279.00	31.50	.49	310.99
6"	ER@.800	Ea	341.00	31.50	.49	372.99
8"	ER@.900	Ea	466.00	35.50	.56	502.06
200 PSIG iron body butterfly valve, wafer-type, lever operated						
2"	P1@.450	Ea	142.00	16.50	—	158.50
2½"	P1@.450	Ea	145.00	16.50	—	161.50
3"	P1@.550	Ea	156.00	20.20	—	176.20
4"	P1@.550	Ea	189.00	20.20	—	209.20
5"	ER@.800	Ea	252.00	31.50	.49	283.99
6"	ER@.800	Ea	316.00	31.50	.49	347.99
8"	ER@.900	Ea	438.00	35.50	.56	474.06

Description	Craft@Hrs	Unit	Material $	Labor $	Equipment $	Total $

Class 125 bronze body 2-piece ball valve, threaded ends

Description	Craft@Hrs	Unit	Material $	Labor $	Equipment $	Total $
½"	P1@.210	Ea	10.10	7.71	—	17.81
¾"	P1@.250	Ea	13.40	9.18	—	22.58
1"	P1@.300	Ea	23.40	11.00	—	34.40
1¼"	P1@.400	Ea	39.00	14.70	—	53.70
1½"	P1@.450	Ea	52.60	16.50	—	69.10
2"	P1@.500	Ea	66.70	18.40	—	85.10
3"	P1@.625	Ea	456.00	22.90	—	478.90
4"	P1@.690	Ea	597.00	25.30	—	622.30

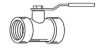

PVC ball valve, solid body, solvent-weld joints, EDPM (female socket)

Description	Craft@Hrs	Unit	Material $	Labor $	Equipment $	Total $
½"	P1@.170	Ea	6.05	6.24	—	12.29
¾"	P1@.200	Ea	6.84	7.34	—	14.18
1"	P1@.220	Ea	9.98	8.07	—	18.05
1¼"	P1@.280	Ea	11.40	10.30	—	21.70
1½"	P1@.315	Ea	19.20	11.60	—	30.80
2"	P1@.385	Ea	23.90	14.10	—	38.00

PVC ball valve, solid body, threaded joints, EDPM (female pipe thread)

Description	Craft@Hrs	Unit	Material $	Labor $	Equipment $	Total $
½"	P1@.200	Ea	6.05	7.34	—	13.39
¾"	P1@.240	Ea	6.84	8.81	—	15.65
1"	P1@.290	Ea	9.98	10.60	—	20.58
1¼"	P1@.390	Ea	11.40	14.30	—	25.70
1½"	P1@.420	Ea	19.20	15.40	—	34.60
2"	P1@.490	Ea	23.90	18.00	—	41.90

PVC ball valve, tru-union, threaded joints, EDPM (female pipe thread)

Description	Craft@Hrs	Unit	Material $	Labor $	Equipment $	Total $
2"	P1@.490	Ea	80.60	18.00	—	98.60
2½"	P1@.520	Ea	93.10	19.10	—	112.20
3"	P1@.590	Ea	215.00	21.70	—	236.70
4"	P1@.630	Ea	397.00	23.10	—	420.10

PVC ball valve, union-type body, solvent-weld joints (female socket)

Description	Craft@Hrs	Unit	Material $	Labor $	Equipment $	Total $
½"	P1@.185	Ea	6.05	6.79	—	12.84
¾"	P1@.225	Ea	6.84	8.26	—	15.10
1"	P1@.270	Ea	9.98	9.91	—	19.89
1¼"	P1@.360	Ea	11.40	13.20	—	24.60
1½"	P1@.400	Ea	19.20	14.70	—	33.90
2"	P1@.465	Ea	23.90	17.10	—	41.00
2½"	P1@.490	Ea	93.10	18.00	—	111.10
3"	P1@.540	Ea	215.00	19.80	—	234.80
4"	P1@.600	Ea	397.00	22.00	—	419.00

PVC, Schedule 80, with Solvent-Weld Joints

Description	Craft@Hrs	Unit	Material $	Labor $	Equipment $	Total $

Class 125 bronze body swing check valve, threaded ends

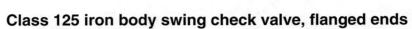

Description	Craft@Hrs	Unit	Material $	Labor $	Equipment $	Total $
½"	P1@.210	Ea	20.40	7.71	—	28.11
¾"	P1@.250	Ea	29.20	9.18	—	38.38
1"	P1@.300	Ea	38.10	11.00	—	49.10
1¼"	P1@.400	Ea	54.40	14.70	—	69.10
1½"	P1@.450	Ea	76.70	16.50	—	93.20
2"	P1@.500	Ea	127.00	18.40	—	145.40

Class 125 iron body swing check valve, flanged ends

Description	Craft@Hrs	Unit	Material $	Labor $	Equipment $	Total $
2"	P1@.500	Ea	179.00	18.40	—	197.40
2½"	P1@.600	Ea	226.00	22.00	—	248.00
3"	P1@.750	Ea	282.00	27.50	—	309.50
4"	P1@1.35	Ea	414.00	49.50	—	463.50
5"	ER@2.00	Ea	797.00	78.80	1.24	877.04
6"	ER@2.50	Ea	797.00	98.50	1.55	897.05
8"	ER@3.00	Ea	1,440.00	118.00	1.86	1,559.86

Class 125 iron body silent check valve, wafer-type

Description	Craft@Hrs	Unit	Material $	Labor $	Equipment $	Total $
2"	P1@.500	Ea	131.00	18.40	—	149.40
2½"	P1@.600	Ea	145.00	22.00	—	167.00
3"	P1@.750	Ea	167.00	27.50	—	194.50
4"	P1@1.35	Ea	223.00	49.50	—	272.50
5"	ER@2.00	Ea	342.00	78.80	1.24	422.04
6"	ER@2.50	Ea	466.00	98.50	1.55	566.05
8"	ER@3.00	Ea	797.00	118.00	1.86	916.86

Class 125 bronze body strainer, threaded ends

Description	Craft@Hrs	Unit	Material $	Labor $	Equipment $	Total $
½"	P1@.210	Ea	31.40	7.71	—	39.11
¾"	P1@.250	Ea	41.30	9.18	—	50.48
1"	P1@.300	Ea	50.60	11.00	—	61.60
1¼"	P1@.400	Ea	70.70	14.70	—	85.40
1½"	P1@.450	Ea	106.00	16.50	—	122.50
2"	P1@.500	Ea	185.00	18.40	—	203.40

Class 125 iron body strainer, flanged ends

Description	Craft@Hrs	Unit	Material $	Labor $	Equipment $	Total $
2"	P1@.500	Ea	188.00	18.40	—	206.40
2½"	P1@.600	Ea	209.00	22.00	—	231.00
3"	P1@.750	Ea	242.00	27.50	—	269.50
4"	P1@1.35	Ea	413.00	49.50	—	462.50
5"	ER@2.00	Ea	838.00	78.80	1.24	918.04
6"	ER@2.50	Ea	838.00	98.50	1.55	938.05
8"	ER@3.00	Ea	1,420.00	118.00	1.86	1,539.86

Description	Craft@Hrs	Unit	Material $	Labor $	Equipment $	Total $

Installation of 2-way control valve, threaded joints

½"	P1@.210	Ea	—	7.71	—	7.71
¾"	P1@.275	Ea	—	10.10	—	10.10
1"	P1@.350	Ea	—	12.80	—	12.80
1¼"	P1@.430	Ea	—	15.80	—	15.80
1½"	P1@.505	Ea	—	18.50	—	18.50
2"	P1@.675	Ea	—	24.80	—	24.80
2½"	P1@.830	Ea	—	30.50	—	30.50
3"	P1@.990	Ea	—	36.30	—	36.30

Installation of 3-way control valve, threaded joints

½"	P1@.260	Ea	—	9.54	—	9.54
¾"	P1@.365	Ea	—	13.40	—	13.40
1"	P1@.475	Ea	—	17.40	—	17.40
1¼"	P1@.575	Ea	—	21.10	—	21.10
1½"	P1@.680	Ea	—	25.00	—	25.00
2"	P1@.910	Ea	—	33.40	—	33.40
2½"	P1@1.12	Ea	—	41.10	—	41.10
3"	P1@1.33	Ea	—	48.80	—	48.80

PVC companion flange

2"	P1@.140	Ea	6.47	5.14	—	11.61
2½"	P1@.180	Ea	19.20	6.61	—	25.81
3"	P1@.210	Ea	20.10	7.71	—	27.81
4"	P1@.280	Ea	20.80	10.30	—	31.10
5"	P1@.310	Ea	46.10	11.40	—	57.50
6"	P1@.420	Ea	46.10	15.40	—	61.50
8"	P1@.560	Ea	68.90	20.60	—	89.50

Bolt and gasket set

2"	P1@.500	Ea	4.16	18.40	—	22.56
2½"	P1@.650	Ea	4.86	23.90	—	28.76
3"	P1@.750	Ea	8.22	27.50	—	35.72
4"	P1@1.00	Ea	13.70	36.70	—	50.40
5"	P1@1.10	Ea	22.90	40.40	—	63.30
6"	P1@1.20	Ea	22.90	44.00	—	66.90
8"	P1@1.25	Ea	25.60	45.90	—	71.50

Thermometer with well

| 7" | P1@.250 | Ea | 180.00 | 9.18 | — | 189.18 |
| 9" | P1@.250 | Ea | 185.00 | 9.18 | — | 194.18 |

Dial-type pressure gauge

| 2½" | P1@.200 | Ea | 37.70 | 7.34 | — | 45.04 |
| 3½" | P1@.200 | Ea | 49.70 | 7.34 | — | 57.04 |

PVC, Schedule 80, with Solvent-Weld Joints

Description	Craft@Hrs	Unit	Material $	Labor $	Equipment $	Total $
Pressure/temperature tap						
Tap	P1@.150	Ea	17.30	5.51	—	22.81
Hanger with swivel assembly						
½"	P1@.250	Ea	4.22	9.18	—	13.40
¾"	P1@.250	Ea	4.68	9.18	—	13.86
1"	P1@.250	Ea	4.89	9.18	—	14.07
1¼"	P1@.300	Ea	5.02	11.00	—	16.02
1½"	P1@.300	Ea	5.40	11.00	—	16.40
2"	P1@.300	Ea	5.63	11.00	—	16.63
2½"	P1@.350	Ea	7.65	12.80	—	20.45
3"	P1@.350	Ea	9.47	12.80	—	22.27
4"	P1@.350	Ea	10.40	12.80	—	23.20
5"	P1@.450	Ea	15.10	16.50	—	31.60
6"	P1@.450	Ea	17.10	16.50	—	33.60
8"	P1@.450	Ea	21.40	16.50	—	37.90
Riser clamp						
½"	P1@.100	Ea	2.72	3.67	—	6.39
¾"	P1@.100	Ea	4.02	3.67	—	7.69
1"	P1@.100	Ea	4.06	3.67	—	7.73
1¼"	P1@.105	Ea	4.89	3.85	—	8.74
1½"	P1@.110	Ea	5.17	4.04	—	9.21
2"	P1@.115	Ea	5.47	4.22	—	9.69
2½"	P1@.120	Ea	5.76	4.40	—	10.16
3"	P1@.120	Ea	6.23	4.40	—	10.63
4"	P1@.125	Ea	7.94	4.59	—	12.53
5"	P1@.180	Ea	11.40	6.61	—	18.01
6"	P1@.200	Ea	13.70	7.34	—	21.04
8"	P1@.200	Ea	22.40	7.34	—	29.74

HDPE – High Density Polyethylene pipe, or PEX pipe, is cross-linked at the molecular level. Composite pressure pipe is aluminum tube laminated between 2 layers of plastic pipe. An additional external layer of an EVOH polymer creates an oxygen barrier and restricts oxygen permeability. It is manufactured and sold in various coiled lengths (200' to 1,000' rolls). The jointing method can be either crimped or compression connections. It will not rust or corrode and, because of its flexibility, installations require approximately 40% less fittings. Expansion rates are similar to copper, but its thermal coefficient (heat loss) is over 800 times less than copper. It is also chemically resistant to most acids, salt solutions, alkalis, fats and oils.

Common applications include ice or snow melting systems, radiant floor heating systems, water service tubing, hot & cold domestic water service, chilled water systems, compressed air systems, solar and process piping applications. Ice or snow melting systems require an oxygen barrier type PEX pipe selection.

Maximum temperature	Maximum pressure	Pipe selection
73 degrees F	160 psi	PEX
140 degrees F	120 psi	PEX
180 degrees F	100 psi	PEX
210 degrees F	115 psi	PEX-AL

PEX is the designation for HDPE pipe
PEX-AL is the designation for cross linked polyethylene-aluminum composite pipe

Description	Craft@Hrs	Unit	Material $	Labor $	Equipment $	Total $

Cross linked HD Polyethylene-PEX pipe with crimped joints (PEX), pipe only

Description	Craft@Hrs	Unit	Material $	Labor $	Equipment $	Total $
3/8"	P1@.028	LF	.42	1.03	—	1.45
1/2"	P1@.028	LF	.61	1.03	—	1.64
5/8"	P1@.030	LF	.86	1.10	—	1.96
3/4"	P1@.030	LF	1.22	1.10	—	2.32
1"	P1@.034	LF	1.96	1.25	—	3.21
Add for oxygen barrier type		%	25.0	—	—	—
Add for PEX-AL		%	50.0	—	—	—

90-degree brass ell with crimped joints (PEX)

Description	Craft@Hrs	Unit	Material $	Labor $	Equipment $	Total $
1/2"	P1@.080	Ea	1.33	2.94	—	4.27
5/8"	P1@.085	Ea	1.60	3.12	—	4.72
3/4"	P1@.090	Ea	3.08	3.30	—	6.38
1"	P1@.105	Ea	3.36	3.85	—	7.21

90-degree brass ell with crimped joints x male pipe thread (PEX x MPT)

Description	Craft@Hrs	Unit	Material $	Labor $	Equipment $	Total $
1/2" x 1/2"	P1@.085	Ea	1.45	3.12	—	4.57
1/2" x 3/8"	P1@.095	Ea	1.83	3.49	—	5.32

Polyethylene-Aluminum Pipe with Crimped Joints

Description	Craft@Hrs	Unit	Material $	Labor $	Equipment $	Total $

90-degree brass ell with crimped joints x copper socket (PEX x C)

Description	Craft@Hrs	Unit	Material $	Labor $	Equipment $	Total $
1/2"	P1@.085	Ea	1.45	3.12	—	4.57
1/2" x 3/4"	P1@.090	Ea	1.83	3.30	—	5.13
5/8" x 3/4"	P1@.095	Ea	1.98	3.49	—	5.47
3/4"	P1@.095	Ea	1.61	3.49	—	5.10

90-degree brass wingback ell with crimped joints x female pipe thread (PEX x FPT)

Description	Craft@Hrs	Unit	Material $	Labor $	Equipment $	Total $
1/2"	P1@.085	Ea	2.83	3.12	—	5.95
3/4"	P1@.095	Ea	3.36	3.49	—	6.85

Tee (brass) with crimped joints (PEX)

Description	Craft@Hrs	Unit	Material $	Labor $	Equipment $	Total $
1/2"	P1@.095	Ea	1.42	3.49	—	4.91
5/8"	P1@.105	Ea	1.66	3.85	—	5.51
3/4"	P1@.110	Ea	3.36	4.04	—	7.40
1"	P1@.120	Ea	4.11	4.40	—	8.51

Reducing tee (brass) with crimped joints (PEX)

Description	Craft@Hrs	Unit	Material $	Labor $	Equipment $	Total $
1/2" x 1/2" x 5/8"	P1@.110	Ea	1.66	4.04	—	5.70
1/2" x 1/2" x 3/4"	P1@.115	Ea	1.94	4.22	—	6.16
5/8" x 1/2" x 1/2"	P1@.115	Ea	1.66	4.22	—	5.88
5/8" x 1/2" x 5/8"	P1@.115	Ea	1.66	4.22	—	5.88
5/8" x 5/8" x 1/2"	P1@.115	Ea	1.66	4.22	—	5.88
3/4" x 1/2" x 1/2"	P1@.115	Ea	1.94	4.22	—	6.16
3/4" x 1/2" x 3/4"	P1@.115	Ea	2.07	4.22	—	6.29
3/4" x 3/4" x 1/2"	P1@.115	Ea	2.07	4.22	—	6.29
3/4" x 3/4" x 5/8"	P1@.115	Ea	2.07	4.22	—	6.29
3/4" x 3/4" x 1"	P1@.120	Ea	3.98	4.40	—	8.38
1" x 1/2" x 1/2"	P1@.120	Ea	3.87	4.40	—	8.27
1" x 1/2" x 3/4"	P1@.120	Ea	3.61	4.40	—	8.01
1" x 1/2" x 1"	P1@.120	Ea	3.98	4.40	—	8.38
1" x 3/4" x 1/2"	P1@.120	Ea	3.98	4.40	—	8.38
1" x 3/4" x 3/4"	P1@.120	Ea	3.98	4.40	—	8.38
1" x 3/4" x 1"	P1@.120	Ea	3.98	4.40	—	8.38
1" x 1" x 1/2"	P1@.120	Ea	3.59	4.40	—	7.99
1" x 1" x 3/4"	P1@.120	Ea	4.11	4.40	—	8.51

Coupling (brass) with crimped joints (PEX)

Description	Craft@Hrs	Unit	Material $	Labor $	Equipment $	Total $
1/2"	P1@.080	Ea	.65	2.94	—	3.59
5/8"	P1@.085	Ea	.82	3.12	—	3.94
3/4"	P1@.090	Ea	1.81	3.30	—	5.11
1"	P1@.105	Ea	1.85	3.85	—	5.70

Description	Craft@Hrs	Unit	Material $	Labor $	Equipment $	Total $

Reducing coupling (brass) with crimped joints (PEX)

Description	Craft@Hrs	Unit	Material $	Labor $	Equipment $	Total $
1/2"	P1@.080	Ea	.82	2.94	—	3.76
5/8"	P1@.085	Ea	1.16	3.12	—	4.28
3/4"	P1@.090	Ea	1.24	3.30	—	4.54
1"	P1@.105	Ea	1.51	3.85	—	5.36

Adapter (brass) with crimped joints x male pipe thread (PEX x MPT)

Description	Craft@Hrs	Unit	Material $	Labor $	Equipment $	Total $
1/2" x 1/2"	P1@.080	Ea	.97	2.94	—	3.91
1/2" x 3/4"	P1@.085	Ea	1.24	3.12	—	4.36
5/8" x 3/4"	P1@.085	Ea	1.23	3.12	—	4.35
3/4" x 1/2"	P1@.090	Ea	1.66	3.30	—	4.96
3/4" x 3/4"	P1@.090	Ea	1.76	3.30	—	5.06
3/4" x 1"	P1@.105	Ea	2.83	3.85	—	6.68
1" x 1"	P1@.105	Ea	3.08	3.85	—	6.93

Adapter (brass) with crimped joints x female pipe thread (PEX x FPT)

Description	Craft@Hrs	Unit	Material $	Labor $	Equipment $	Total $
1/2" x 1/2"	P1@.080	Ea	1.01	2.94	—	3.95
1/2" x 3/4"	P1@.085	Ea	1.42	3.12	—	4.54
5/8" x 3/4"	P1@.085	Ea	1.42	3.12	—	4.54
3/4" x 1/2"	P1@.090	Ea	1.66	3.30	—	4.96
3/4" x 3/4"	P1@.090	Ea	1.83	3.30	—	5.13
3/4" x 1"	P1@.105	Ea	3.06	3.85	—	6.91
1" x 1"	P1@.105	Ea	3.36	3.85	—	7.21

Adapter (brass) with crimped joints x copper fitting (spigot) (PEX x C ftg)

Description	Craft@Hrs	Unit	Material $	Labor $	Equipment $	Total $
1/2" x 1/2"	P1@.080	Ea	1.16	2.94	—	4.10
1/2" x 3/4"	P1@.085	Ea	1.43	3.12	—	4.55
5/8" x 3/4"	P1@.085	Ea	1.42	3.12	—	4.54
3/4" x 3/4"	P1@.090	Ea	2.83	3.30	—	6.13
1" x 1"	P1@.105	Ea	3.36	3.85	—	7.21

Adapter (brass) with crimped joints x copper socket (PEX x C)

Description	Craft@Hrs	Unit	Material $	Labor $	Equipment $	Total $
1/2" x 1/2"	P1@.080	Ea	.97	2.94	—	3.91
1/2" x 3/4"	P1@.085	Ea	1.44	3.12	—	4.56
5/8" x 3/4"	P1@.085	Ea	1.50	3.12	—	4.62
3/4" x 1/2"	P1@.090	Ea	1.60	3.30	—	4.90
3/4" x 3/4"	P1@.090	Ea	2.32	3.30	—	5.62
3/4" x 1"	P1@.105	Ea	3.36	3.85	—	7.21
1" x 1"	P1@.105	Ea	3.87	3.85	—	7.72

Polyethylene-Aluminum Pipe with Crimped Joints

Description	Craft@Hrs	Unit	Material $	Labor $	Equipment $	Total $

Cap (brass) with crimped joints (PEX)

Description	Craft@Hrs	Unit	Material $	Labor $	Equipment $	Total $
1/2"	P1@.060	Ea	.44	2.20	—	2.64
5/8"	P1@.065	Ea	.62	2.39	—	3.01
3/4"	P1@.070	Ea	.82	2.57	—	3.39
1"	P1@.085	Ea	1.01	3.12	—	4.13

Mini ball valve (brass) with crimped joints (PEX)

Description	Craft@Hrs	Unit	Material $	Labor $	Equipment $	Total $
1/2" x 3/8" straight	P1@.150	Ea	7.07	5.51	—	12.58
1/2" x 3/8" angle	P1@.150	Ea	6.51	5.51	—	12.02
1/2" x copper	P1@.150	Ea	7.46	5.51	—	12.97
1/2" x MPT	P1@.150	Ea	7.86	5.51	—	13.37
1/2"	P1@.150	Ea	8.30	5.51	—	13.81
1/2" x comp	P1@.150	Ea	8.42	5.51	—	13.93
5/8"	P1@.165	Ea	11.80	6.06	—	17.86
5/8" x 3/4" copper	P1@.170	Ea	9.43	6.24	—	15.67
5/8" x 3/4" MPT	P1@.170	Ea	10.30	6.24	—	16.54
3/4"	P1@.170	Ea	12.70	6.24	—	18.94
3/4" x copper	P1@.170	Ea	11.90	6.24	—	18.14

Balancing valve (brass) with crimped joints (PEX)

Description	Craft@Hrs	Unit	Material $	Labor $	Equipment $	Total $
1/2" x copper	P1@.150	Ea	43.60	5.51	—	49.11
1/2" x MPT	P1@.150	Ea	50.40	5.51	—	55.91

Manifolds (headers) copper with crimped joints (PEX)

Description	Craft@Hrs	Unit	Material $	Labor $	Equipment $	Total $
3 - 1/2" outlets	P1@.750	Ea	37.80	27.50	—	65.30
3 - 5/8" outlets	P1@.750	Ea	42.80	27.50	—	70.30
4 - 1/2" outlets	P1@.850	Ea	48.60	31.20	—	79.80
4 - 5/8" outlets	P1@.850	Ea	56.20	31.20	—	87.40
5 - 1/2" outlets	P1@1.00	Ea	62.20	36.70	—	98.90
6 - 1/2" outlets	P1@1.20	Ea	70.60	44.00	—	114.60
8 - 1/2" outlets	P1@1.45	Ea	101.00	53.20	—	154.20
10 - 1/2" outlets	P1@1.65	Ea	130.00	60.60	—	190.60
12 - 1/2" outlets	P1@1.85	Ea	157.00	67.90	—	224.90

Manifolds (headers) copper with mini ball valves and crimped joints (PEX)

Description	Craft@Hrs	Unit	Material $	Labor $	Equipment $	Total $
3 - 1/2" outlets	P1@.750	Ea	141.00	27.50	—	168.50
3 - 5/8" outlets	P1@.750	Ea	171.00	27.50	—	198.50
4 - 1/2" outlets	P1@.850	Ea	185.00	31.20	—	216.20
4 - 5/8" outlets	P1@.850	Ea	224.00	31.20	—	255.20
5 - 1/2" outlets	P1@1.00	Ea	235.00	36.70	—	271.70
6 - 1/2" outlets	P1@1.20	Ea	268.00	44.00	—	312.00

Description	Craft@Hrs	Unit	Material $	Labor $	Equipment $	Total $
Crimp rings, nickel-plated soft copper						
1/2"	—	Ea	.18	—	—	.18
5/8"	—	Ea	.19	—	—	.19
3/4"	—	Ea	.22	—	—	.22
1"	—	Ea	.33	—	—	.33
Replacement O-rings						
1/2"	—	Ea	.12	—	—	.12
5/8"	—	Ea	.12	—	—	.12
3/4"	—	Ea	.18	—	—	.18
1"	—	Ea	.18	—	—	.18
Pipe hangers (polypropylene)						
1/2"	P1@.050	Ea	.06	1.84	—	1.90
5/8"	P1@.050	Ea	.06	1.84	—	1.90
3/4"	P1@.050	Ea	.06	1.84	—	1.90
1"	P1@.050	Ea	.08	1.84	—	1.92
Pipe clips (nail clips)						
1/2"	P1@.030	Ea	.13	1.10	—	1.23
5/8"	P1@.030	Ea	.22	1.10	—	1.32
1"	P1@.030	Ea	.49	1.10	—	1.59
Miscellaneous tools						
Crimp tool	—	Ea	281.00	—	—	281.00
Beveling tool	—	Ea	5.70	—	—	5.70
Reaming tool	—	Ea	34.10	—	—	34.10
Pipe bender kit	—	Ea	503.00	—	—	503.00
Bending spring	—	Ea	42.10	—	—	42.10
Pipe cutter	—	Ea	20.50	—	—	20.50

Polyethylene-Aluminum Pipe with Compression Joints

Description	Craft@Hrs	Unit	Material $	Labor $	Equipment $	Total $

HDPE – High Density Polyethylene pipe, or PEX pipe, is cross-linked at the molecular level. Composite pressure pipe is aluminum tube laminated between 2 layers of plastic pipe. An additional external layer of an EVOH polymer creates an oxygen barrier and restricts oxygen permeability. It is manufactured and sold in various coiled lengths (200' to 1,000' rolls). The jointing method can be either crimped or compression connections. It will not rust or corrode and, because of its flexibility, installations require approximately 40% less fittings. Expansion rates are similar to copper, but its thermal coefficient (heat loss) is over 800 times less than copper. It is also chemically resistant to most acids, salt solutions, alkalis, fats and oils.

Common applications include ice or snow melting systems, radiant floor heating systems, water service tubing, hot & cold domestic water service, chilled water systems, compressed air systems, solar and process piping applications. Ice or snow melting systems require an oxygen barrier type PEX pipe selection.

Maximum temperature	Maximum pressure	Pipe selection
73 degrees F	160 psi	PEX
140 degrees F	120 psi	PEX
180 degrees F	100 psi	PEX
210 degrees F	115 psi	PEX-AL

PEX is the designation for HDPE pipe
PEX-AL is the designation for cross linked polyethylene-aluminum composite pipe

Description	Craft@Hrs	Unit	Material $	Labor $	Equipment $	Total $

Cross linked HD Polyurethane PEX pipe with compression joints (PEX), pipe only

3/8"	P1@.028	LF	.38	1.03	—	1.41
1/2"	P1@.028	LF	.57	1.03	—	1.60
5/8"	P1@.030	LF	.78	1.10	—	1.88
3/4"	P1@.030	LF	1.08	1.10	—	2.18
1"	P1@.034	LF	1.74	1.25	—	2.99

90-degree brass ell with compression joints (PEX)

1/2"	P1@.092	Ea	3.02	3.38	—	6.40
5/8"	P1@.098	Ea	4.19	3.60	—	7.79
3/4"	P1@.104	Ea	6.12	3.82	—	9.94
1"	P1@.121	Ea	10.70	4.44	—	15.14

90-degree brass ell with compression joints x male pipe thread (PEX x MPT)

½" x ½"	P1@.098	Ea	4.72	3.60	—	8.32
½" x ¾"	P1@.105	Ea	5.72	3.85	—	9.57
¾" x ¾"	P1@.109	Ea	5.27	4.00	—	9.27

Description	Craft@Hrs	Unit	Material $	Labor $	Equipment $	Total $

90-degree brass ell with compression joints x copper socket (PEX x C)

Description	Craft@Hrs	Unit	Material $	Labor $	Equipment $	Total $
1/2"	P1@.098	Ea	2.58	3.60	—	6.18
1/2" x 3/4"	P1@.109	Ea	2.74	4.00	—	6.74
5/8" x 3/4"	P1@.112	Ea	3.27	4.11	—	7.38
3/4"	P1@.115	Ea	4.95	4.22	—	9.17

90-degree brass wingback ell with compression joints x female pipe thread (PEX x FPT)

Description	Craft@Hrs	Unit	Material $	Labor $	Equipment $	Total $
1/2"	P1@.098	Ea	2.24	3.60	—	5.84
5/8"	P1@.105	Ea	3.84	3.85	—	7.69
3/4"	P1@.115	Ea	5.89	4.22	—	10.11

Tee (brass) with compression joints (PEX)

Description	Craft@Hrs	Unit	Material $	Labor $	Equipment $	Total $
1/2"	P1@.109	Ea	3.69	4.00	—	7.69
5/8"	P1@.121	Ea	5.93	4.44	—	10.37
3/4"	P1@.127	Ea	8.76	4.66	—	13.42
1"	P1@.141	Ea	14.50	5.17	—	19.67

Reducing tee (brass) with compression joints (PEX)

Description	Craft@Hrs	Unit	Material $	Labor $	Equipment $	Total $
5/8" x 5/8" x 1/2"	P1@.138	Ea	5.43	5.06	—	10.49
3/4" x 1/2" x 3/4"	P1@.138	Ea	7.71	5.06	—	12.77
3/4" x 3/4" x 1/2"	P1@.138	Ea	7.88	5.06	—	12.94
3/4" x 1/2" x 3/4"	P1@.138	Ea	6.98	5.06	—	12.04
3/4" x 3/4" x 5/8"	P1@.138	Ea	7.98	5.06	—	13.04
1" x 1/2" x 1/2"	P1@.142	Ea	11.90	5.21	—	17.11
1" x 1" x 1/2"	P1@.142	Ea	16.00	5.21	—	21.21
1" x 3/4" x 3/4"	P1@.142	Ea	8.98	5.21	—	14.19
1" x 1" x 3/4"	P1@.142	Ea	13.90	5.21	—	19.11

Coupling (brass) with compression joints (PEX)

Description	Craft@Hrs	Unit	Material $	Labor $	Equipment $	Total $
3/8"	P1@.090	Ea	2.55	3.30	—	5.85
1/2"	P1@.092	Ea	2.78	3.38	—	6.16
5/8"	P1@.098	Ea	3.79	3.60	—	7.39
3/4"	P1@.104	Ea	5.61	3.82	—	9.43
1"	P1@.121	Ea	10.00	4.44	—	14.44

Reducing coupling (brass) with compression joints (PEX)

Description	Craft@Hrs	Unit	Material $	Labor $	Equipment $	Total $
5/8" x 1/2"	P1@.092	Ea	3.33	3.38	—	6.71
3/4" x 1/2"	P1@.098	Ea	5.03	3.60	—	8.63
3/4" x 5/8"	P1@.104	Ea	4.87	3.82	—	8.69
1" x 3/4"	P1@.121	Ea	6.77	4.44	—	11.21

Cap (brass) with compression joints (PEX)

Description	Craft@Hrs	Unit	Material $	Labor $	Equipment $	Total $
1/2"	P1@.069	Ea	1.87	2.53	—	4.40
5/8"	P1@.075	Ea	2.58	2.75	—	5.33
3/4"	P1@.081	Ea	3.38	2.97	—	6.35
1"	P1@.098	Ea	5.03	3.60	—	8.63

Polyethylene-Aluminum Pipe with Compression Joints

Description	Craft@Hrs	Unit	Material $	Labor $	Equipment $	Total $

Mini ball valve (brass) with compression joints (PEX)

Description	Craft@Hrs	Unit	Material $	Labor $	Equipment $	Total $
1/2" x 3/8" straight	P1@.173	Ea	8.55	6.35	—	14.90
1/2" x 3/8" angle	P1@.173	Ea	8.06	6.35	—	14.41
1/2" x copper	P1@.173	Ea	8.02	6.35	—	14.37
1/2" x MPT	P1@.173	Ea	8.44	6.35	—	14.79
1/2"	P1@.173	Ea	9.43	6.35	—	15.78
5/8"	P1@.173	Ea	12.40	6.35	—	18.75
5/8" x 3/4" copper	P1@.196	Ea	9.72	7.19	—	16.91
5/8" x 3/4" MPT	P1@.196	Ea	10.40	7.19	—	17.59
3/4"	P1@.196	Ea	11.60	7.19	—	18.79
3/4" x copper	P1@.196	Ea	11.90	7.19	—	19.09

Balancing valve (brass) with compression joints (PEX)

Description	Craft@Hrs	Unit	Material $	Labor $	Equipment $	Total $
1/2" x copper	P1@.173	Ea	48.20	6.35	—	54.55
1/2" x MPT	P1@.173	Ea	53.60	6.35	—	59.95

Crimp rings, nickel plated soft copper

Description	Craft@Hrs	Unit	Material $	Labor $	Equipment $	Total $
1/2"	—	Ea	.18	—	—	.18
5/8"	—	Ea	.20	—	—	.20
3/4"	—	Ea	.23	—	—	.23
1"	—	Ea	.34	—	—	.34

Replacement O-rings

Description	Craft@Hrs	Unit	Material $	Labor $	Equipment $	Total $
1/2"	—	Ea	.12	—	—	.12
5/8"	—	Ea	.12	—	—	.12
3/4"	—	Ea	.18	—	—	.18
1"	—	Ea	.18	—	—	.18

Pipe hangers (polypropylene)

Description	Craft@Hrs	Unit	Material $	Labor $	Equipment $	Total $
1/2"	P1@.050	Ea	.06	1.84	—	1.90
5/8"	P1@.050	Ea	.06	1.84	—	1.90
3/4"	P1@.050	Ea	.06	1.84	—	1.90
1"	P1@.050	Ea	.08	1.84	—	1.92

Pipe clips (nail clips)

Description	Craft@Hrs	Unit	Material $	Labor $	Equipment $	Total $
1/2"	P1@.030	Ea	.13	1.10	—	1.23
5/8"	P1@.030	Ea	.22	1.10	—	1.32
1"	P1@.030	Ea	.49	1.10	—	1.59

Miscellaneous tools

Description	Craft@Hrs	Unit	Material $	Labor $	Equipment $	Total $
Crimp tool	—	Ea	281.00	—	—	281.00
Beveling tool	—	Ea	5.70	—	—	5.70
Reaming tool	—	Ea	34.10	—	—	34.10
Pipe bender kit	—	Ea	503.00	—	—	503.00
Bending spring	—	Ea	42.10	—	—	42.10
Pipe cutter	—	Ea	20.50	—	—	20.50

Description	Craft@Hrs	Unit	Material $	Labor $	Equipment $	Total $

Water meters, turbine-type. Including brass connection unions

Description	Craft@Hrs	Unit	Material $	Labor $	Equipment $	Total $
½"	P1@.500	Ea	149.00	18.40	—	167.40
¾"	P1@.500	Ea	242.00	18.40	—	260.40
1"	P1@.650	Ea	324.00	23.90	—	347.90
1½"	P1@.700	Ea	870.00	25.70	—	895.70
2"	P1@.800	Ea	1,300.00	29.40	—	1,329.40
3"	P1@.900	Ea	2,000.00	33.00	—	2,033.00
4"	P1@.975	Ea	3,730.00	35.80	—	3,765.80

Water meters, compound-type. Including brass connection unions

Description	Craft@Hrs	Unit	Material $	Labor $	Equipment $	Total $
2"	P1@.800	Ea	3,170.00	29.40	—	3,199.40
3"	P1@.900	Ea	4,120.00	33.00	—	4,153.00
4"	P1@.975	Ea	6,500.00	35.80	—	6,535.80

Water meter by-pass and connection assembly. Includes three isolation ball valves, two tees, two 90-degree elbows, 10' of Type L copper pipe and standard support devices. Make additional allowances for the cost of the water meter.

Description	Craft@Hrs	Unit	Material $	Labor $	Equipment $	Total $
½"	P1@1.00	Ea	87.30	36.70	—	124.00
¾"	P1@1.50	Ea	132.00	55.10	—	187.10
1"	P1@1.95	Ea	229.00	71.60	—	300.60
1½"	P1@2.45	Ea	420.00	89.90	—	509.90
2"	P1@3.25	Ea	618.00	119.00	—	737.00
3"	P1@5.50	Ea	1,840.00	202.00	—	2,042.00
4"	P1@6.75	Ea	3,130.00	248.00	—	3,378.00

Backflow preventers – reduced pressure. Including valves and test ports

Description	Craft@Hrs	Unit	Material $	Labor $	Equipment $	Total $
¾"	P1@1.00	Ea	338.00	36.70	—	374.70
1"	P1@1.25	Ea	418.00	45.90	—	463.90
1½"	P1@2.00	Ea	673.00	73.40	—	746.40
2"	P1@2.15	Ea	835.00	78.90	—	913.90
2½"	P1@4.00	Ea	3,030.00	147.00	—	3,177.00
3"	P1@4.75	Ea	4,000.00	174.00	—	4,174.00
4"	P1@6.00	Ea	4,550.00	220.00	—	4,770.00
6"	P1@7.95	Ea	4,970.00	292.00	—	5,262.00

Backflow preventers – reduced pressure. Including integral ball valves and test ports

Description	Craft@Hrs	Unit	Material $	Labor $	Equipment $	Total $
½"	P1@.900	Ea	206.00	33.00	—	239.00
¾"	P1@1.00	Ea	243.00	36.70	—	279.70
1"	P1@1.25	Ea	304.00	45.90	—	349.90
1¼"	P1@1.50	Ea	446.00	55.10	—	501.10
1½"	P1@2.00	Ea	485.00	73.40	—	558.40
2"	P1@2.15	Ea	546.00	78.90	—	624.90

Plumbing and Piping Specialties

Description	Craft@Hrs	Unit	Material $	Labor $	Equipment $	Total $

Backflow preventers, double-check valve assembly. Including integral ball valves and test ports

Description	Craft@Hrs	Unit	Material $	Labor $	Equipment $	Total $
½"	P1@1.00	Ea	154.00	36.70	—	190.70
¾"	P1@1.00	Ea	171.00	36.70	—	207.70
1"	P1@1.25	Ea	206.00	45.90	—	251.90
1¼"	P1@1.95	Ea	345.00	71.60	—	416.60
1½"	P1@2.00	Ea	358.00	73.40	—	431.40
2"	P1@2.15	Ea	434.00	78.90	—	512.90
2½"	P1@2.95	Ea	1,780.00	108.00	—	1,888.00
3"	P1@3.45	Ea	2,280.00	127.00	—	2,407.00
4"	P1@3.95	Ea	3,090.00	145.00	—	3,235.00
6"	P1@5.75	Ea	5,210.00	211.00	—	5,421.00

Vacuum breakers, atmospheric. Female pipe thread

Description	Craft@Hrs	Unit	Material $	Labor $	Equipment $	Total $
¼"	P1@.350	Ea	51.60	12.80	—	64.40
½"	P1@.350	Ea	71.90	12.80	—	84.70
¾"	P1@.350	Ea	87.90	12.80	—	100.70

Vacuum breakers, hose connection. Male/female hose thread, polished brass

Description	Craft@Hrs	Unit	Material $	Labor $	Equipment $	Total $
½"	P1@.150	Ea	21.90	5.51	—	27.41
¾"	P1@.150	Ea	26.80	5.51	—	32.31

Suction diffusers

Description	Craft@Hrs	Unit	Material $	Labor $	Equipment $	Total $
2"	P1@1.25	Ea	355.00	45.90	—	400.90
3"	P1@2.50	Ea	596.00	91.80	—	687.80
4"	P1@3.25	Ea	757.00	119.00	—	876.00
6"	P1@4.80	Ea	1,080.00	176.00	10.40	1,266.40
8"	P1@5.95	Ea	2,020.00	218.00	12.90	2,250.90
10"	P1@6.75	Ea	2,730.00	248.00	14.60	2,992.60

Triple-duty valves

Description	Craft@Hrs	Unit	Material $	Labor $	Equipment $	Total $
2"	P1@1.25	Ea	469.00	45.90	—	514.90
3"	P1@2.50	Ea	636.00	91.80	—	727.80
4"	P1@3.25	Ea	1,150.00	119.00	—	1,269.00
6"	P1@4.80	Ea	1,900.00	176.00	10.40	2,086.40
8"	P1@5.95	Ea	2,780.00	218.00	12.90	3,010.90
10"	P1@6.75	Ea	4,080.00	248.00	14.60	4,342.60

In-line circulating pump, all bronze. 115 volt, including flange kit

Description	Craft@Hrs	Unit	Material $	Labor $	Equipment $	Total $
1/25 HP	P1@1.50	Ea	425.00	55.10	—	480.10
1/16 HP	P1@1.55	Ea	536.00	56.90	—	592.90
1/12 HP	P1@1.60	Ea	738.00	58.70	—	796.70
1/6 HP	P1@1.75	Ea	1,250.00	64.20	—	1,314.20
1/4 HP	P1@1.95	Ea	1,880.00	71.60	—	1,951.60
1 HP	P1@2.25	Ea	2,180.00	82.60	—	2,262.60
1½ HP	P1@2.85	Ea	2,630.00	105.00	—	2,735.00

Description	Craft@Hrs	Unit	Material $	Labor $	Equipment $	Total $

In-line circulating pump, iron body. 115 volt, including flange kit

1/25 HP	P1@1.50	Ea	200.00	55.10	—	255.10
1/16 HP	P1@1.55	Ea	223.00	56.90	—	279.90
1/12 HP	P1@1.60	Ea	420.00	58.70	—	478.70
1/6 HP	P1@1.75	Ea	715.00	64.20	—	779.20
1/4 HP	P1@1.95	Ea	828.00	71.60	—	899.60
1 HP	P1@2.25	Ea	971.00	82.60	—	1,053.60
1½ HP	P1@2.85	Ea	1,270.00	105.00	—	1,375.00

Primed steel cam-lock non-fire-rated access doors

8" x 8"	P1@.400	Ea	67.90	14.70	—	82.60
12" x 12"	P1@.500	Ea	78.20	18.40	—	96.60
16" x 16"	P1@.600	Ea	96.20	22.00	—	118.20
18" x 18"	P1@.800	Ea	110.00	29.40	—	139.40
24" x 24"	P1@1.25	Ea	164.00	45.90	—	209.90
36" x 36"	P1@1.60	Ea	302.00	58.70	—	360.70
Add for Allen key lock		Ea	4.87	—	—	4.87
Add for cylinder lock		Ea	11.90	—	—	11.90

Primed steel cam-lock 1½ hour fire-rated access doors

8" x 8"	P1@.400	Ea	250.00	14.70	—	264.70
12" x 12"	P1@.500	Ea	263.00	18.40	—	281.40
16" x 16"	P1@.600	Ea	311.00	22.00	—	333.00
18" x 18"	P1@.800	Ea	316.00	29.40	—	345.40
24" x 24"	P1@1.25	Ea	443.00	45.90	—	488.90
24" x 36"	P1@1.50	Ea	572.00	55.10	—	627.10
48" x 48"	P1@2.00	Ea	885.00	73.40	—	958.40
Add for cylinder lock		Ea	11.90	—	—	11.90

Threaded automatic air vent

Float type	P1@.150	Ea	12.70	5.51	—	18.21
35 PSIG	P1@.150	Ea	80.90	5.51	—	86.41
150 PSIG	P1@.150	Ea	132.00	5.51	—	137.51

Threaded manual air vent

175 PSIG	P1@.150	Ea	7.67	5.51	—	13.18

150 pound forged steel slip-on companion flange, welding-type

2½"	P1@.610	Ea	41.40	22.40	—	63.80
3"	P1@.730	Ea	41.40	26.80	—	68.20
4"	P1@.980	Ea	57.00	36.00	—	93.00
6"	ER@1.47	Ea	75.30	57.90	.91	134.11
8"	ER@1.77	Ea	139.00	69.70	1.09	209.79
10"	ER@2.20	Ea	233.00	86.70	1.36	321.06
12"	ER@2.64	Ea	348.00	104.00	1.63	453.63

Plumbing and Piping Specialties

Description	Craft@Hrs	Unit	Material $	Labor $	Equipment $	Total $

150 pound forged steel companion flange, threaded

Description	Craft@Hrs	Unit	Material $	Labor $	Equipment $	Total $
2"	P1@.290	Ea	44.40	10.60	—	55.00
2½"	P1@.380	Ea	66.80	13.90	—	80.70
3"	P1@.460	Ea	74.20	16.90	—	91.10
4"	P1@.600	Ea	99.50	22.00	—	121.50
6"	ER@.680	Ea	147.00	26.80	.42	174.22
8"	ER@.760	Ea	202.00	29.90	.47	232.37

300 pound forged steel slip-on companion flange, welding-type

Description	Craft@Hrs	Unit	Material $	Labor $	Equipment $	Total $
2½"	P1@.810	Ea	47.60	29.70	—	77.30
3"	P1@.980	Ea	47.60	36.00	—	83.60
4"	P1@1.30	Ea	69.50	47.70	—	117.20
6"	ER@1.95	Ea	127.00	76.80	1.21	205.01
8"	ER@2.35	Ea	179.00	92.60	1.45	273.05
10"	ER@2.90	Ea	295.00	114.00	1.79	410.79
12"	ER@3.50	Ea	471.00	138.00	2.16	611.16

PVC companion flange

Description	Craft@Hrs	Unit	Material $	Labor $	Equipment $	Total $
2"	P1@.140	Ea	4.56	5.14	—	9.70
2½"	P1@.180	Ea	13.70	6.61	—	20.31
3"	P1@.210	Ea	14.30	7.71	—	22.01
4"	P1@.280	Ea	14.70	10.30	—	25.00
5"	P1@.310	Ea	33.00	11.40	—	44.40
6"	P1@.420	Ea	33.00	15.40	—	48.40
8"	P1@.560	Ea	48.90	20.60	—	69.50

Bolt and gasket set

Description	Craft@Hrs	Unit	Material $	Labor $	Equipment $	Total $
2"	P1@.500	Ea	4.40	18.40	—	22.80
2½"	P1@.650	Ea	5.10	23.90	—	29.00
3"	P1@.750	Ea	8.71	27.50	—	36.21
4"	P1@1.00	Ea	14.60	36.70	—	51.30
5"	P1@1.10	Ea	16.70	40.40	—	57.10
6"	P1@1.20	Ea	16.70	44.00	—	60.70
8"	P1@1.25	Ea	21.70	45.90	—	67.60
10"	P1@1.70	Ea	26.00	62.40	—	88.40
12"	P1@2.20	Ea	34.90	80.70	—	115.60

Dielectric union, brazed joints

Description	Craft@Hrs	Unit	Material $	Labor $	Equipment $	Total $
1/2"	P1@.146	Ea	16.20	5.36	—	21.56
3/4"	P1@.205	Ea	16.20	7.52	—	23.72
1"	P1@.263	Ea	33.60	9.65	—	43.25
1¼"	P1@.322	Ea	57.80	11.80	—	69.60
1½"	P1@.380	Ea	78.70	13.90	—	92.60
2"	P1@.507	Ea	108.00	18.60	—	126.60

Description	Craft@Hrs	Unit	Material $	Labor $	Equipment $	Total $

125/150# galvanized ASME expansion tank

Description	Craft@Hrs	Unit	Material $	Labor $	Equipment $	Total $
15 gallon	P1@1.75	Ea	332.00	64.20	—	396.20
30 gallon	P1@2.00	Ea	598.00	73.40	—	671.40
40 gallon	P1@2.35	Ea	663.00	86.20	—	749.20
60 gallon	P1@2.65	Ea	900.00	97.30	—	997.30
80 gallon	P1@2.95	Ea	1,660.00	108.00	—	1,768.00
100 gallon	P1@3.25	Ea	1,990.00	119.00	—	2,109.00
120 gallon	P1@3.50	Ea	2,140.00	128.00	—	2,268.00

Expansion tank fitting

Description	Craft@Hrs	Unit	Material $	Labor $	Equipment $	Total $
¾"	P1@.300	Ea	75.10	11.00	—	86.10

Lead roof pipe flashing

Description	Craft@Hrs	Unit	Material $	Labor $	Equipment $	Total $
1½"	P1@.250	Ea	31.60	9.18	—	40.78
2"	P1@.250	Ea	41.60	9.18	—	50.78
3"	P1@.250	Ea	48.20	9.18	—	57.38
4"	P1@.250	Ea	53.10	9.18	—	62.28
6"	P1@.350	Ea	68.00	12.80	—	80.80
8"	P1@.400	Ea	86.00	14.70	—	100.70

Roof flashing, 20" x 20"

Description	Craft@Hrs	Unit	Material $	Labor $	Equipment $	Total $
Pitched roof, neoprene						
all pipe	P1@.300	Ea	20.90	11.00	—	31.90
Pitched roof, aluminum/neoprene						
all pipe	P1@.300	Ea	22.10	11.00	—	33.10
Flat roof, aluminum/neoprene						
all pipe	P1@.300	Ea	42.00	11.00	—	53.00

Braided stainless steel pipe connector, solder ends, brazed joints

Description	Craft@Hrs	Unit	Material $	Labor $	Equipment $	Total $
½"	P1@.240	Ea	47.30	8.81	—	56.11
¾"	P1@.300	Ea	62.60	11.00	—	73.60
1"	P1@.360	Ea	93.80	13.20	—	107.00
1¼"	P1@.480	Ea	109.00	17.60	—	126.60
1½"	P1@.540	Ea	125.00	19.80	—	144.80
2"	P1@.600	Ea	182.00	22.00	—	204.00

Braided stainless steel pipe connector, threaded joints

Description	Craft@Hrs	Unit	Material $	Labor $	Equipment $	Total $
½"	P1@.210	Ea	53.30	7.71	—	61.01
¾"	P1@.250	Ea	57.60	9.18	—	66.78
1"	P1@.300	Ea	71.80	11.00	—	82.80
1¼"	P1@.400	Ea	73.20	14.70	—	87.90
1½"	P1@.450	Ea	99.40	16.50	—	115.90
2"	P1@.500	Ea	142.00	18.40	—	160.40

Plumbing and Piping Specialties

Description	Craft@Hrs	Unit	Material $	Labor $	Equipment $	Total $
Braided stainless steel pipe connector, flanged joints						
2"	P1@.500	Ea	126.00	18.40	—	144.40
2½"	P1@.630	Ea	181.00	23.10	—	204.10
3"	P1@.750	Ea	235.00	27.50	—	262.50
4"	P1@1.50	Ea	307.00	55.10	—	362.10
6"	P1@2.50	Ea	609.00	91.80	—	700.80
8"	P1@3.00	Ea	1,370.00	110.00	—	1,480.00
Galvanized steel band pipe hanger						
½"	P1@.250	Ea	2.13	9.18	—	11.31
¾"	P1@.250	Ea	2.13	9.18	—	11.31
1"	P1@.250	Ea	2.13	9.18	—	11.31
1¼"	P1@.300	Ea	2.26	11.00	—	13.26
1½"	P1@.300	Ea	2.40	11.00	—	13.40
2"	P1@.300	Ea	2.51	11.00	—	13.51
2½"	P1@.350	Ea	3.08	12.80	—	15.88
3"	P1@.350	Ea	3.54	12.80	—	16.34
4"	P1@.350	Ea	4.49	12.80	—	17.29
6"	P1@.450	Ea	7.07	16.50	—	23.57
8"	P1@.450	Ea	10.50	16.50	—	27.00
Trapeze pipe hanger						
24"	P1@1.00	Ea	47.70	36.70	—	84.40
36"	P1@1.25	Ea	60.70	45.90	—	106.60
48"	P1@1.50	Ea	92.40	55.10	—	147.50
Wall bracket support with anchors and bolts						
6" x 6"	P1@.600	Ea	22.20	22.00	—	44.20
8" x 8"	P1@.650	Ea	25.60	23.90	—	49.50
10" x 10"	P1@.700	Ea	28.90	25.70	—	54.60
12" x 12"	P1@.800	Ea	34.90	29.40	—	64.30
15" x 15"	P1@.950	Ea	40.20	34.90	—	75.10
18" x 18"	P1@1.10	Ea	56.80	40.40	—	97.20
24" x 24"	P1@1.60	Ea	82.10	58.70	—	140.80
Galvanized U-bolts with nuts						
½"	P1@.160	Ea	1.93	5.87	—	7.80
¾"	P1@.165	Ea	1.93	6.06	—	7.99
1"	P1@.170	Ea	2.10	6.24	—	8.34
1¼"	P1@.175	Ea	2.40	6.42	—	8.82
1½"	P1@.180	Ea	2.90	6.61	—	9.51
2"	P1@.190	Ea	3.10	6.97	—	10.07
2½"	P1@.210	Ea	4.79	7.71	—	12.50
3"	P1@.250	Ea	5.54	9.18	—	14.72
4"	P1@.300	Ea	6.26	11.00	—	17.26
6"	P1@.400	Ea	9.80	14.70	—	24.50

Description	Craft@Hrs	Unit	Material $	Labor $	Equipment $	Total $
Nail-on wire pipe hooks						
½"	P1@.100	Ea	.09	3.67	—	3.76
¾"	P1@.100	Ea	.11	3.67	—	3.78
1"	P1@.100	Ea	.12	3.67	—	3.79
Galvanized steel pipe sleeves						
1"	P1@.120	Ea	4.49	4.40	—	8.89
1¼"	P1@.125	Ea	4.97	4.59	—	9.56
1½"	P1@.125	Ea	5.63	4.59	—	10.22
2"	P1@.130	Ea	6.07	4.77	—	10.84
2½"	P1@.150	Ea	6.16	5.51	—	11.67
3"	P1@.180	Ea	6.30	6.61	—	12.91
4"	P1@.220	Ea	7.17	8.07	—	15.24
5"	P1@.250	Ea	8.94	9.18	—	18.12
6"	P1@.270	Ea	9.66	9.91	—	19.57
8"	P1@.270	Ea	11.10	9.91	—	21.01
10"	P1@.290	Ea	12.90	10.60	—	23.50
12"	P1@.310	Ea	17.60	11.40	—	29.00
14"	P1@.330	Ea	32.70	12.10	—	44.80
Dial-type pressure gauge						
2½"	P1@.200	Ea	37.20	7.34	—	44.54
3½"	P1@.200	Ea	49.30	7.34	—	56.64
Riser clamp						
½"	P1@.100	Ea	2.65	3.67	—	6.32
¾"	P1@.100	Ea	3.88	3.67	—	7.55
1"	P1@.100	Ea	3.92	3.67	—	7.59
1¼"	P1@.105	Ea	4.75	3.85	—	8.60
1½"	P1@.110	Ea	5.01	4.04	—	9.05
2"	P1@.115	Ea	5.30	4.22	—	9.52
2½"	P1@.120	Ea	5.58	4.40	—	9.98
3"	P1@.120	Ea	6.05	4.40	—	10.45
4"	P1@.125	Ea	7.68	4.59	—	12.27
5"	P1@.180	Ea	11.10	6.61	—	17.71
6"	P1@.200	Ea	13.30	7.34	—	20.64
8"	P1@.200	Ea	21.70	7.34	—	29.04
10"	P1@.250	Ea	32.20	9.18	—	41.38
12"	P1@.250	Ea	38.20	9.18	—	47.38
15 PSIG float and thermostatic steam trap, threaded						
¾"	P1@.250	Ea	326.00	9.18	—	335.18
1"	P1@.300	Ea	326.00	11.00	—	337.00
1¼"	P1@.400	Ea	393.00	14.70	—	407.70
1½"	P1@.450	Ea	737.00	16.50	—	753.50
2"	P1@.500	Ea	945.00	18.40	—	963.40

Plumbing and Piping Specialties

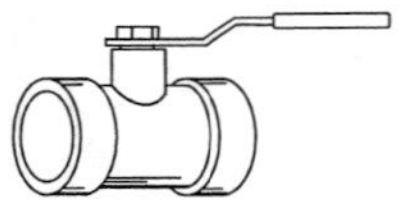

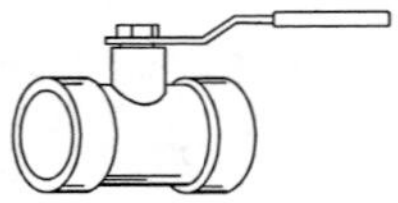

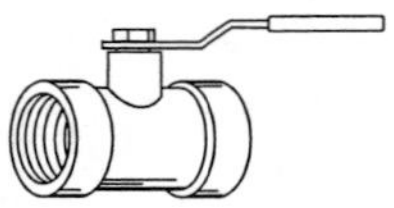

Description	Craft@Hrs	Unit	Material $	Labor $	Equipment $	Total $
15 PSIG inverted bucket steam trap, threaded joints						
½"	P1@.210	Ea	189.00	7.71	—	196.71
¾"	P1@.250	Ea	464.00	9.18	—	473.18
1"	P1@.300	Ea	664.00	11.00	—	675.00
1¼"	P1@.400	Ea	1,110.00	14.70	—	1,124.70
1½"	P1@.450	Ea	1,110.00	16.50	—	1,126.50
2"	P1@.500	Ea	1,660.00	18.40	—	1,678.40
Thermometer with well						
7"	P1@.250	Ea	180.00	9.18	—	189.18
9"	P1@.250	Ea	185.00	9.18	—	194.18
Class 125 bronze body 2-piece ball valve, brazed joints						
½"	P1@.240	Ea	10.10	8.81	—	18.91
¾"	P1@.300	Ea	13.40	11.00	—	24.40
1"	P1@.360	Ea	23.30	13.20	—	36.50
1¼"	P1@.480	Ea	38.80	17.60	—	56.40
1½"	P1@.540	Ea	52.40	19.80	—	72.20
2"	P1@.600	Ea	66.50	22.00	—	88.50
3"	P1@1.50	Ea	455.00	55.10	—	510.10
4"	P1@1.80	Ea	595.00	66.10	—	661.10
Class 125 bronze body 2-piece ball valve, soft-soldered joints						
½"	P1@.200	Ea	10.10	7.34	—	17.44
¾"	P1@.249	Ea	13.40	9.14	—	22.54
1"	P1@.299	Ea	23.30	11.00	—	34.30
1¼"	P1@.308	Ea	38.80	11.30	—	50.10
1½"	P1@.498	Ea	52.40	18.30	—	70.70
2"	P1@.498	Ea	66.50	18.30	—	84.80
3"	P1@1.24	Ea	455.00	45.50	—	500.50
4"	P1@1.45	Ea	595.00	53.20	—	648.20
Class 125 bronze body 2-piece ball valve, threaded joints						
½"	P1@.210	Ea	10.10	7.71	—	17.81
¾"	P1@.250	Ea	13.40	9.18	—	22.58
1"	P1@.300	Ea	23.30	11.00	—	34.30
1¼"	P1@.400	Ea	38.80	14.70	—	53.50
1½"	P1@.450	Ea	52.40	16.50	—	68.90
2"	P1@.500	Ea	66.50	18.40	—	84.90
3"	P1@.625	Ea	455.00	22.90	—	477.90
4"	P1@.690	Ea	595.00	25.30	—	620.30

Description	Craft@Hrs	Unit	Material $	Labor $	Equipment $	Total $

Class 150, 600 pound W.O.G. bronze body 2-piece ball valve, threaded joints

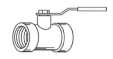

½"	P1@.210	Ea	16.50	7.71	—	24.21
¾"	P1@.250	Ea	27.00	9.18	—	36.18
1"	P1@.300	Ea	34.30	11.00	—	45.30
1¼"	P1@.400	Ea	42.70	14.70	—	57.40
1½"	P1@.450	Ea	57.50	16.50	—	74.00
2"	P1@.500	Ea	72.80	18.40	—	91.20

Bronze body circuit balance valve

½"	P1@.210	Ea	73.00	7.71	—	80.71
¾"	P1@.250	Ea	77.60	9.18	—	86.78
1"	P1@.300	Ea	90.20	11.00	—	101.20
1¼"	P1@.400	Ea	112.00	14.70	—	126.70
1½"	P1@.400	Ea	142.00	14.70	—	156.70
2"	P1@.500	Ea	198.00	18.40	—	216.40
2½"	P1@1.75	Ea	752.00	64.20	—	816.20
3"	P1@2.50	Ea	1,160.00	91.80	—	1,251.80
4"	P1@3.15	Ea	1,970.00	116.00	—	2,086.00

200 PSIG iron body butterfly valve, lug-type, lever operated

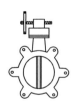

2"	P1@.450	Ea	156.00	16.50	—	172.50
2½"	P1@.450	Ea	162.00	16.50	—	178.50
3"	P1@.550	Ea	170.00	20.20	—	190.20
4"	P1@.505	Ea	212.00	18.50	—	230.50
6"	ER@.800	Ea	341.00	31.50	.49	372.99
8"	ER@.800	Ea	466.00	31.50	.49	497.99
10"	ER@.900	Ea	648.00	35.50	.56	684.06
12"	ER@1.00	Ea	847.00	39.40	.62	887.02

200 PSIG iron body butterfly valve, wafer-type, lever operated

2"	P1@.450	Ea	142.00	16.50	—	158.50
2½"	P1@.450	Ea	145.00	16.50	—	161.50
3"	P1@.550	Ea	156.00	20.20	—	176.20
4"	P1@.550	Ea	189.00	20.20	—	209.20
6"	ER@.800	Ea	316.00	31.50	.49	347.99
8"	ER@.800	Ea	438.00	31.50	.49	469.99
10"	ER@.900	Ea	612.00	35.50	.56	648.06
12"	ER@1.00	Ea	804.00	39.40	.62	844.02

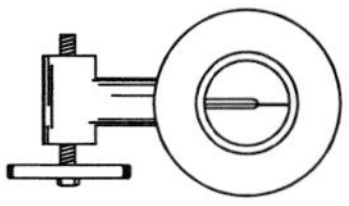

Class 125 bronze body swing check valve, with brazed joints

½"	P1@.240	Ea	20.40	8.81	—	29.21
¾"	P1@.300	Ea	29.20	11.00	—	40.20
1"	P1@.360	Ea	38.10	13.20	—	51.30
1¼"	P1@.480	Ea	54.40	17.60	—	72.00
1½"	P1@.540	Ea	76.70	19.80	—	96.50
2"	P1@.600	Ea	127.00	22.00	—	149.00
2½"	P1@1.00	Ea	256.00	36.70	—	292.70
3"	P1@1.50	Ea	368.00	55.10	—	423.10

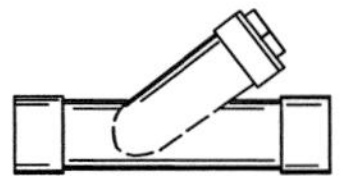

Plumbing and Piping Specialties

Description	Craft@Hrs	Unit	Material $	Labor $	Equipment $	Total $
Class 125 bronze body swing check valve with soft-soldered joints						
½"	P1@.200	Ea	20.40	7.34	—	27.74
¾"	P1@.249	Ea	29.20	9.14	—	38.34
1"	P1@.299	Ea	38.10	11.00	—	49.10
1¼"	P1@.398	Ea	54.40	14.60	—	69.00
1½"	P1@.448	Ea	76.70	16.40	—	93.10
2"	P1@.498	Ea	127.00	18.30	—	145.30
2½"	P1@.830	Ea	256.00	30.50	—	286.50
3"	P1@1.24	Ea	368.00	45.50	—	413.50

Description	Craft@Hrs	Unit	Material $	Labor $	Equipment $	Total $
Class 125 bronze body swing check valve, threaded joints						
½"	P1@.210	Ea	20.40	7.71	—	28.11
¾"	P1@.250	Ea	29.20	9.18	—	38.38
1"	P1@.300	Ea	38.10	11.00	—	49.10
1¼"	P1@.400	Ea	54.40	14.70	—	69.10
1½"	P1@.450	Ea	76.70	16.50	—	93.20
2"	P1@.500	Ea	127.00	18.40	—	145.40

Description	Craft@Hrs	Unit	Material $	Labor $	Equipment $	Total $
Class 300 bronze body swing check valve, threaded joints						
½"	P1@.210	Ea	68.90	7.71	—	76.61
¾"	P1@.250	Ea	81.40	9.18	—	90.58
1"	P1@.300	Ea	99.50	11.00	—	110.50
1¼"	P1@.400	Ea	144.00	14.70	—	158.70
1½"	P1@.450	Ea	207.00	16.50	—	223.50
2"	P1@.500	Ea	288.00	18.40	—	306.40

Description	Craft@Hrs	Unit	Material $	Labor $	Equipment $	Total $
Class 125 iron body swing check valve, flanged joints						
2"	P1@.500	Ea	179.00	18.40	—	197.40
2½"	P1@.600	Ea	226.00	22.00	—	248.00
3"	P1@.750	Ea	282.00	27.50	—	309.50
4"	P1@1.35	Ea	414.00	49.50	—	463.50
6"	ER@2.50	Ea	803.00	98.50	1.55	903.05
8"	ER@3.00	Ea	1,440.00	118.00	1.86	1,559.86
10"	ER@4.00	Ea	2,350.00	158.00	2.47	2,510.47
12"	ER@4.50	Ea	3,220.00	177.00	2.78	3,399.78

Description	Craft@Hrs	Unit	Material $	Labor $	Equipment $	Total $
Class 250 iron body swing check valve, flanged joints						
2"	P1@.500	Ea	461.00	18.40	—	479.40
2½"	P1@.600	Ea	537.00	22.00	—	559.00
3"	P1@.750	Ea	667.00	27.50	—	694.50
4"	P1@1.35	Ea	818.00	49.50	—	867.50
6"	ER@2.50	Ea	1,570.00	98.50	1.55	1,670.05
8"	ER@3.00	Ea	2,920.00	118.00	1.86	3,039.86

Description	Craft@Hrs	Unit	Material $	Labor $	Equipment $	Total $

Class 125 iron body silent check valve, wafer-type

Description	Craft@Hrs	Unit	Material $	Labor $	Equipment $	Total $
2"	P1@.500	Ea	131.00	18.40	—	149.40
2½"	P1@.600	Ea	145.00	22.00	—	167.00
3"	P1@.750	Ea	167.00	27.50	—	194.50
4"	P1@1.35	Ea	223.00	49.50	—	272.50
5"	P1@2.00	Ea	342.00	73.40	—	415.40
6"	P1@2.50	Ea	466.00	91.80	—	557.80
8"	P1@3.00	Ea	797.00	110.00	—	907.00

Class 125 iron body silent check valve, flanged joints

Description	Craft@Hrs	Unit	Material $	Labor $	Equipment $	Total $
2"	P1@.500	Ea	131.00	18.40	—	149.40
2½"	P1@.600	Ea	150.00	22.00	—	172.00
3"	P1@.750	Ea	170.00	27.50	—	197.50
4"	P1@1.35	Ea	222.00	49.50	—	271.50
6"	ER@2.50	Ea	399.00	98.50	1.55	499.05
8"	ER@3.00	Ea	719.00	118.00	1.86	838.86
10"	ER@4.00	Ea	1,110.00	158.00	2.47	1,270.47

Class 250 iron body silent check valve, flanged joints

Description	Craft@Hrs	Unit	Material $	Labor $	Equipment $	Total $
2"	P1@.500	Ea	193.00	18.40	—	211.40
2½"	P1@.600	Ea	222.00	22.00	—	244.00
3"	P1@.750	Ea	247.00	27.50	—	274.50
4"	P1@1.35	Ea	335.00	49.50	—	384.50
6"	ER@2.50	Ea	588.00	98.50	1.55	688.05
8"	ER@3.00	Ea	1,070.00	118.00	1.86	1,189.86
10"	ER@4.00	Ea	1,700.00	158.00	2.47	1,860.47

Class 125 bronze body gate valve, solder ends

Description	Craft@Hrs	Unit	Material $	Labor $	Equipment $	Total $
½"	P1@.240	Ea	18.90	8.81	—	27.71
¾"	P1@.300	Ea	23.60	11.00	—	34.60
1"	P1@.360	Ea	33.20	13.20	—	46.40
1¼"	P1@.480	Ea	41.80	17.60	—	59.40
1½"	P1@.540	Ea	56.30	19.80	—	76.10
2"	P1@.600	Ea	94.20	22.00	—	116.20
2½"	P1@1.00	Ea	156.00	36.70	—	192.70
3"	P1@1.50	Ea	224.00	55.10	—	279.10

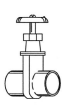

Class 125 bronze body gate valve, threaded joints

Description	Craft@Hrs	Unit	Material $	Labor $	Equipment $	Total $
½"	P1@.210	Ea	18.90	7.71	—	26.61
¾"	P1@.250	Ea	23.60	9.18	—	32.78
1"	P1@.300	Ea	33.20	11.00	—	44.20
1¼"	P1@.400	Ea	41.80	14.70	—	56.50
1½"	P1@.450	Ea	56.30	16.50	—	72.80
2"	P1@.500	Ea	94.20	18.40	—	112.60
2½"	P1@.750	Ea	156.00	27.50	—	183.50
3"	P1@.950	Ea	224.00	34.90	—	258.90

Plumbing and Piping Specialties

Description	Craft@Hrs	Unit	Material $	Labor $	Equipment $	Total $
Class 300 bronze body gate valve, threaded joints						
½"	P1@.210	Ea	38.40	7.71	—	46.11
¾"	P1@.250	Ea	50.20	9.18	—	59.38
1"	P1@.300	Ea	60.80	11.00	—	71.80
1¼"	P1@.400	Ea	82.80	14.70	—	97.50
1½"	P1@.450	Ea	105.00	16.50	—	121.50
2"	P1@.500	Ea	154.00	18.40	—	172.40
2½"	P1@.750	Ea	314.00	27.50	—	341.50
3"	P1@.950	Ea	446.00	34.90	—	480.90
Class 125 iron body gate valve, flanged joints						
2"	P1@.500	Ea	300.00	18.40	—	318.40
2½"	P1@.600	Ea	406.00	22.00	—	428.00
3"	P1@.750	Ea	440.00	27.50	—	467.50
4"	P1@1.35	Ea	645.00	49.50	—	694.50
6"	ER@2.50	Ea	1,220.00	98.50	1.55	1,320.05
8"	ER@3.00	Ea	2,020.00	118.00	1.86	2,139.86
10"	ER@4.00	Ea	3,320.00	158.00	2.47	3,480.47
12"	ER@4.50	Ea	4,590.00	177.00	2.78	4,769.78
Class 250 iron body gate valve, flanged joints						
2"	P1@.500	Ea	525.00	18.40	—	543.40
2½"	P1@.600	Ea	379.00	22.00	—	401.00
3"	P1@.750	Ea	479.00	27.50	—	506.50
4"	P1@1.35	Ea	813.00	49.50	—	862.50
6"	ER@2.50	Ea	1,520.00	98.50	1.55	1,620.05
8"	ER@3.00	Ea	2,340.00	118.00	1.86	2,459.86
10"	ER@4.00	Ea	4,170.00	158.00	2.47	4,330.47
12"	ER@4.50	Ea	6,790.00	177.00	2.78	6,969.78
Class 125 bronze body globe valve, brazed joints						
½"	P1@.240	Ea	35.00	8.81	—	43.81
¾"	P1@.300	Ea	46.60	11.00	—	57.60
1"	P1@.360	Ea	66.80	13.20	—	80.00
1¼"	P1@.480	Ea	94.00	17.60	—	111.60
1½"	P1@.540	Ea	125.00	19.80	—	144.80
2"	P1@.600	Ea	205.00	22.00	—	227.00
Class 125 bronze body globe valve, soft-soldered joints						
½"	P1@.200	Ea	35.00	7.34	—	42.34
¾"	P1@.249	Ea	46.60	9.14	—	55.74
1"	P1@.299	Ea	66.80	11.00	—	77.80
1¼"	P1@.398	Ea	94.00	14.60	—	108.60
1½"	P1@.448	Ea	125.00	16.40	—	141.40
2"	P1@.498	Ea	205.00	18.30	—	223.30

Description	Craft@Hrs	Unit	Material $	Labor $	Equipment $	Total $

Class 125 bronze body globe valve, threaded ends

½"	P1@.210	Ea	35.00	7.71	—	42.71
¾"	P1@.250	Ea	46.60	9.18	—	55.78
1"	P1@.300	Ea	66.80	11.00	—	77.80
1¼"	P1@.400	Ea	94.00	14.70	—	108.70
1½"	P1@.450	Ea	125.00	16.50	—	141.50
2"	P1@.500	Ea	205.00	18.40	—	223.40

Class 300 bronze body globe valve, threaded joints

½"	P1@.210	Ea	65.70	7.71	—	73.41
¾"	P1@.250	Ea	83.40	9.18	—	92.58
1"	P1@.300	Ea	118.00	11.00	—	129.00
1¼"	P1@.400	Ea	172.00	14.70	—	186.70
1½"	P1@.450	Ea	230.00	16.50	—	246.50
2"	P1@.500	Ea	317.00	18.40	—	335.40

Class 125 iron body globe valve, flanged joints

2"	P1@.500	Ea	241.00	18.40	—	259.40
2½"	P1@.600	Ea	372.00	22.00	—	394.00
3"	P1@.750	Ea	448.00	27.50	—	475.50
4"	P1@1.35	Ea	602.00	49.50	—	651.50
6"	ER@2.50	Ea	1,060.00	98.50	1.55	1,160.05
8"	ER@3.00	Ea	1,620.00	118.00	1.86	1,739.86
10"	ER@4.00	Ea	2,470.00	158.00	2.47	2,630.47

Class 250 iron body globe valve, flanged joints

2"	P1@.500	Ea	528.00	18.40	—	546.40
2½"	P1@.600	Ea	658.00	22.00	—	680.00
3"	P1@.750	Ea	782.00	27.50	—	809.50
4"	P1@1.35	Ea	1,080.00	49.50	—	1,129.50
6"	ER@2.50	Ea	2,130.00	98.50	1.55	2,230.05
8"	ER@3.00	Ea	4,730.00	118.00	1.86	4,849.86

Class 125 bronze body strainer, threaded joints

½"	P1@.230	Ea	31.40	8.44	—	39.84
¾"	P1@.260	Ea	41.30	9.54	—	50.84
1"	P1@.330	Ea	50.60	12.10	—	62.70
1¼"	P1@.440	Ea	70.70	16.10	—	86.80
1½"	P1@.495	Ea	106.00	18.20	—	124.20
2"	P1@.550	Ea	185.00	20.20	—	205.20

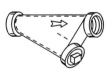
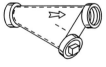

Class 125 iron body strainer, flanged joints

2"	P1@.500	Ea	130.00	18.40	—	148.40
2½"	P1@.600	Ea	144.00	22.00	—	166.00
3"	P1@.750	Ea	166.00	27.50	—	193.50
4"	P1@1.35	Ea	285.00	49.50	—	334.50
6"	ER@2.50	Ea	577.00	98.50	1.55	677.05
8"	ER@3.00	Ea	974.00	118.00	1.86	1,093.86

Plumbing and Piping Specialties

Description	Craft@Hrs	Unit	Material $	Labor $	Equipment $	Total $

Class 250 iron body strainer, flanged joints

Description	Craft@Hrs	Unit	Material $	Labor $	Equipment $	Total $
2"	P1@.500	Ea	220.00	18.40	—	238.40
2½"	P1@.600	Ea	241.00	22.00	—	263.00
3"	P1@.750	Ea	283.00	27.50	—	310.50
4"	P1@1.35	Ea	500.00	49.50	—	549.50
6"	ER@2.50	Ea	801.00	98.50	1.55	901.05
8"	ER@3.00	Ea	1,460.00	118.00	1.86	1,579.86

Installation of 2-way control valve, threaded joints

Description	Craft@Hrs	Unit	Material $	Labor $	Equipment $	Total $
½"	P1@.210	Ea	—	7.71	—	7.71
¾"	P1@.250	Ea	—	9.18	—	9.18
1"	P1@.300	Ea	—	11.00	—	11.00
1¼"	P1@.400	Ea	—	14.70	—	14.70
1½"	P1@.450	Ea	—	16.50	—	16.50
2"	P1@.500	Ea	—	18.40	—	18.40
2½"	P1@.830	Ea	—	30.50	—	30.50
3"	P1@.990	Ea	—	36.30	—	36.30

Installation of 2-way control valve, flanged joints

Description	Craft@Hrs	Unit	Material $	Labor $	Equipment $	Total $
2"	P1@.500	Ea	—	18.40	—	18.40
2½"	P1@.600	Ea	—	22.00	—	22.00
3"	P1@.750	Ea	—	27.50	—	27.50
4"	P1@1.35	Ea	—	49.50	—	49.50
6"	P1@2.50	Ea	—	91.80	5.41	97.21
8"	P1@3.00	Ea	—	110.00	6.49	116.49
10"	P1@4.00	Ea	—	147.00	8.66	155.66
12"	P1@4.50	Ea	—	165.00	9.74	174.74

Installation of 3-way control valve, threaded joints

Description	Craft@Hrs	Unit	Material $	Labor $	Equipment $	Total $
½"	P1@.260	Ea	—	9.54	—	9.54
¾"	P1@.365	Ea	—	13.40	—	13.40
1"	P1@.475	Ea	—	17.40	—	17.40
1¼"	P1@.575	Ea	—	21.10	—	21.10
1½"	P1@.680	Ea	—	25.00	—	25.00
2"	P1@.910	Ea	—	33.40	—	33.40
2½"	P1@1.12	Ea	—	41.10	—	41.10
3"	P1@1.33	Ea	—	48.80	—	48.80

Description	Craft@Hrs	Unit	Material $	Labor $	Equipment $	Total $

Installation of 3-way control valve, flanged joints

2"	P1@.910	Ea	—	33.40	—	33.40
2½"	P1@1.12	Ea	—	41.10	—	41.10
3"	P1@1.33	Ea	—	48.80	—	48.80
4"	P1@2.00	Ea	—	73.40	—	73.40
6"	P1@3.70	Ea	—	136.00	8.01	144.01
8"	P1@4.40	Ea	—	161.00	9.53	170.53
10"	P1@5.90	Ea	—	217.00	12.80	229.80
12"	P1@6.50	Ea	—	239.00	14.10	253.10

Lever-handled gas valve, threaded joints

½"	P1@.210	Ea	30.60	7.71	—	38.31
¾"	P1@.250	Ea	33.90	9.18	—	43.08
1"	P1@.300	Ea	42.10	11.00	—	53.10
1¼"	P1@.400	Ea	56.40	14.70	—	71.10
1½"	P1@.450	Ea	76.00	16.50	—	92.50
2"	P1@.500	Ea	110.00	18.40	—	128.40

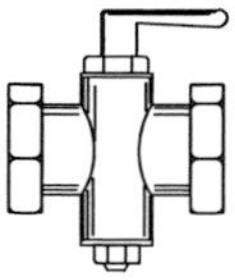

Plug-type gas valve, flanged joints, lubricated

2"	P1@.500	Ea	232.00	18.40	—	250.40
2½"	P1@.600	Ea	376.00	22.00	—	398.00
3"	P1@.750	Ea	528.00	27.50	—	555.50
4"	P1@1.35	Ea	723.00	49.50	—	772.50
6"	P1@2.50	Ea	1,290.00	91.80	—	1,381.80

¾" hose bibb

Low quality	P1@.350	Ea	12.20	12.80	—	25.00
Med. quality	P1@.350	Ea	17.60	12.80	—	30.40
High quality	P1@.350	Ea	23.20	12.80	—	36.00

Water pressure reducing valve, threaded joints, 25-75 PSI

½"	P1@.210	Ea	108.00	7.71	—	115.71
¾"	P1@.250	Ea	139.00	9.18	—	148.18
1"	P1@.300	Ea	227.00	11.00	—	238.00
1¼"	P1@.400	Ea	277.00	14.70	—	291.70
1½"	P1@.450	Ea	395.00	16.50	—	411.50
2"	P1@.500	Ea	625.00	18.40	—	643.40

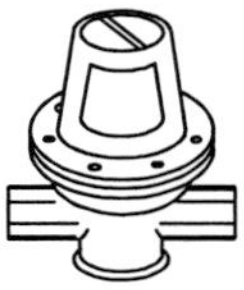

Water hammer arrester, threaded joints

½"	P1@.160	Ea	47.80	5.87	—	53.67
¾"	P1@.190	Ea	55.50	6.97	—	62.47
1"	P1@.230	Ea	140.00	8.44	—	148.44
1¼"	P1@.300	Ea	164.00	11.00	—	175.00
1½"	P1@.340	Ea	235.00	12.50	—	247.50
2"	P1@.380	Ea	358.00	13.90	—	371.90

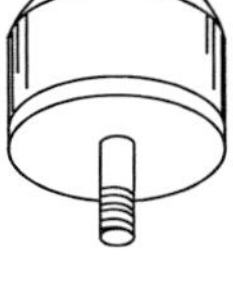

Plumbing and Piping Specialties

Description	Craft@Hrs	Unit	Material $	Labor $	Equipment $	Total $
Carbon steel weldolet, Schedule 40						
½"	P1@.330	Ea	16.50	12.10	1.07	29.67
¾"	P1@.500	Ea	17.10	18.40	1.62	37.12
1"	P1@.670	Ea	17.90	24.60	2.17	44.67
1¼"	P1@.830	Ea	21.60	30.50	2.68	54.78
1½"	P1@1.00	Ea	21.60	36.70	3.23	61.53
2"	P1@1.33	Ea	23.90	48.80	4.30	77.00
2½"	P1@1.67	Ea	52.50	61.30	5.40	119.20
3"	P1@2.00	Ea	59.70	73.40	6.47	139.57
4"	P1@2.67	Ea	77.00	98.00	8.64	183.64
6"	P1@4.00	Ea	318.00	147.00	12.90	477.90
8"	P1@4.80	Ea	467.00	176.00	15.50	658.50
10"	P1@6.00	Ea	646.00	220.00	19.40	885.40
12"	P1@7.20	Ea	794.00	264.00	23.30	1,081.30
Carbon steel weldolet, Schedule 80						
½"	P1@.440	Ea	18.30	16.10	1.42	35.82
¾"	P1@.670	Ea	18.70	24.60	2.17	45.47
1"	P1@.890	Ea	20.60	32.70	2.88	56.18
1¼"	P1@1.11	Ea	25.90	40.70	3.59	70.19
1½"	P1@1.33	Ea	25.90	48.80	4.30	79.00
2"	P1@1.77	Ea	27.90	65.00	5.73	98.63
2½"	P1@2.22	Ea	62.30	81.50	7.18	150.98
3"	P1@2.66	Ea	64.10	97.60	8.60	170.30
4"	P1@3.55	Ea	142.00	130.00	11.50	283.50
6"	P1@5.32	Ea	447.00	195.00	17.20	659.20
8"	P1@6.38	Ea	597.00	234.00	20.60	851.60
Carbon steel threadolet, Schedule 40						
¾"	P1@.330	Ea	10.30	12.10	—	22.40
1"	P1@.440	Ea	11.40	16.10	—	27.50
1¼"	P1@.560	Ea	15.90	20.60	—	36.50
1½"	P1@.670	Ea	17.90	24.60	—	42.50
2"	P1@.890	Ea	20.90	32.70	—	53.60
2½"	P1@1.11	Ea	64.30	40.70	—	105.00
Carbon steel threadolet, Schedule 80						
½"	P1@.300	Ea	15.40	11.00	—	26.40
¾"	P1@.440	Ea	17.00	16.10	—	33.10
1"	P1@.590	Ea	21.00	21.70	—	42.70
1¼"	P1@.740	Ea	94.40	27.20	—	121.60
1½"	P1@.890	Ea	94.40	32.70	—	127.10
2"	P1@1.18	Ea	123.00	43.30	—	166.30

Cast Iron, DWV, Service Weight, No-Hub with Coupled Joints

Service weight cast iron soil pipe with hub-less coupled joints is widely used for non-pressurized underground and in-building drain, waste and vent (DWV) systems.

This section has been arranged to save the estimator's time by including all normally-used system components such as pipe, fittings, hanger assemblies and riser clamps under one heading. Additional items can be found under "Plumbing and Piping Specialties." The cost estimates in this section are based on the conditions, limitations and wage rates described in the section "How to Use This Book" beginning on page 5.

Equipment cost, where shown, is $4.33 per hour for a 2-ton chain hoist.

Description	Craft@Hrs	Unit	Material $	Labor $	Equipment $	Total $

DWV cast iron no-hub (MJ) horizontal pipe assembly. Horizontally hung in a building. Assembly includes fittings, couplings, hangers and rod. Based on a reducing wye and a 45-degree elbow every 15 feet and a hanger every 6 feet for 2" pipe. A reducing wye and a 45-degree elbow every 40 feet and a hanger every 8 feet for 10" pipe. *(Distance between fittings in the pipe assembly increases as pipe diameter increases.)* Use these figures for preliminary estimates.

Description	Craft@Hrs	Unit	Material $	Labor $	Equipment $	Total $
2"	P1@.130	LF	10.90	4.77	—	15.67
3"	P1@.160	LF	12.00	5.87	—	17.87
4"	P1@.190	LF	14.10	6.97	—	21.07
6"	ER@.220	LF	27.30	8.67	.14	36.11
8"	ER@.260	LF	52.40	10.20	.16	62.76
10"	ER@.310	LF	81.30	12.20	.19	93.69

DWV cast iron no-hub (MJ), pipe only

Description	Craft@Hrs	Unit	Material $	Labor $	Equipment $	Total $
2"	P1@.050	LF	5.99	1.84	—	7.83
3"	P1@.060	LF	7.13	2.20	—	9.33
4"	P1@.080	LF	9.30	2.94	—	12.24
6"	ER@.110	LF	17.40	4.33	.14	21.87
8"	ER@.130	LF	32.20	5.12	.16	37.48
10"	ER@.140	LF	56.00	5.52	.19	61.71

DWV cast iron no-hub 1/8 bend

Description	Craft@Hrs	Unit	Material $	Labor $	Equipment $	Total $
2"	P1@.250	Ea	6.94	9.18	—	16.12
3"	P1@.350	Ea	9.04	12.80	—	21.84
4"	P1@.500	Ea	11.40	18.40	—	29.80
6"	ER@.650	Ea	36.70	25.60	.40	62.70
8"	ER@.710	Ea	98.40	28.00	.44	126.84
10"	ER@.800	Ea	147.00	31.50	.49	178.99

Cast Iron, DWV, Service Weight, No-Hub with Coupled Joints

Description	Craft@Hrs	Unit	Material $	Labor $	Equipment $	Total $
DWV cast iron no-hub ¼ bend						
2"	P1@.250	Ea	8.47	9.18	—	17.65
3"	P1@.350	Ea	10.50	12.80	—	23.30
4"	P1@.500	Ea	16.00	18.40	—	34.40
6"	ER@.650	Ea	38.90	25.60	.40	64.90
8"	ER@.710	Ea	151.00	28.00	.44	179.44
10"	ER@.710	Ea	165.00	28.00	.44	193.44
DWV cast iron no-hub long sweep ¼ bend						
2"	P1@.250	Ea	14.80	9.18	—	23.98
3"	P1@.350	Ea	17.80	12.80	—	30.60
4"	P1@.500	Ea	28.20	18.40	—	46.60
5"	ER@.580	Ea	51.70	22.90	.36	74.96
6"	ER@.650	Ea	62.90	25.60	.40	88.90
DWV cast iron no-hub low-heel outlet ¼ bend						
3"	P1@.470	Ea	15.00	17.20	—	32.20
4"	P1@.650	Ea	18.30	23.90	—	42.20
DWV cast iron no-hub closet bend						
3"	P1@.350	Ea	34.50	12.80	—	47.30
4"	P1@.500	Ea	35.40	18.40	—	53.80
DWV cast iron no-hub closet flange						
4" x 3"	P1@.500	Ea	50.50	18.40	—	68.90
DWV cast iron no-hub P-trap						
2"	P1@.250	Ea	13.70	9.18	—	22.88
3"	P1@.350	Ea	42.60	12.80	—	55.40
4"	P1@.500	Ea	52.60	18.40	—	71.00
6"	ER@.650	Ea	187.00	25.60	.40	213.00
DWV cast iron no-hub wye						
2"	P1@.380	Ea	10.80	13.90	—	24.70
3"	P1@.520	Ea	15.10	19.10	—	34.20
4"	P1@.700	Ea	23.70	25.70	—	49.40
6"	ER@.900	Ea	63.90	35.50	.56	99.96
8"	ER@1.15	Ea	162.00	45.30	.71	208.01
10"	ER@1.50	Ea	389.00	59.10	.93	449.03

Cast Iron, DWV, Service Weight, No-Hub with Coupled Joints

Description	Craft@Hrs	Unit	Material $	Labor $	Equipment $	Total $

DWV cast iron no-hub reducing wye

Description	Craft@Hrs	Unit	Material $	Labor $	Equipment $	Total $
3" x 2"	P1@.470	Ea	14.50	17.20	—	31.70
4" x 2"	P1@.600	Ea	19.30	22.00	—	41.30
4" x 3"	P1@.650	Ea	20.40	23.90	—	44.30
6" x 2"	ER@.770	Ea	51.80	30.30	.48	82.58
6" x 3"	ER@.770	Ea	56.30	30.30	.48	87.08
6" x 4"	ER@.800	Ea	55.30	31.50	.49	87.29
8" x 2"	ER@1.00	Ea	132.00	39.40	.62	172.02
8" x 3"	ER@1.00	Ea	134.00	39.40	.62	174.02
8" x 4"	ER@1.00	Ea	136.00	39.40	.62	176.02
8" x 6"	ER@1.10	Ea	141.00	43.30	.68	184.98

DWV cast iron no-hub combination upright wye and 1/8 bend

Description	Craft@Hrs	Unit	Material $	Labor $	Equipment $	Total $
2"	P1@.380	Ea	26.80	13.90	—	40.70
3"	P1@.520	Ea	43.20	19.10	—	62.30
4"	P1@.700	Ea	57.00	25.70	—	82.70

DWV cast iron no-hub combination upright reducing wye and 1/8 bend

Description	Craft@Hrs	Unit	Material $	Labor $	Equipment $	Total $
3" x 2"	P1@.470	Ea	46.80	17.20	—	64.00
4" x 2"	P1@.600	Ea	55.80	22.00	—	77.80
4" x 3"	P1@.650	Ea	58.40	23.90	—	82.30

DWV cast iron no-hub double wye

Description	Craft@Hrs	Unit	Material $	Labor $	Equipment $	Total $
2"	P1@.500	Ea	33.30	18.40	—	51.70
3"	P1@.700	Ea	33.70	25.70	—	59.40
4"	P1@.950	Ea	52.30	34.90	—	87.20
6"	ER@1.10	Ea	182.00	43.30	.68	225.98
8"	ER@1.45	Ea	332.00	57.10	.90	390.00

DWV cast iron no-hub double reducing wye

Description	Craft@Hrs	Unit	Material $	Labor $	Equipment $	Total $
3" x 2"	P1@.520	Ea	33.30	19.10	—	52.40
4" x 2"	P1@.740	Ea	46.30	27.20	—	73.50
4" x 3"	P1@1.00	Ea	47.70	36.70	—	84.40
6" x 4"	ER@1.75	Ea	179.00	69.00	1.08	249.08

DWV cast iron no-hub combination wye and 1/8 bend

Description	Craft@Hrs	Unit	Material $	Labor $	Equipment $	Total $
2"	P1@.380	Ea	20.80	13.90	—	34.70
3"	P1@.520	Ea	29.60	19.10	—	48.70
4"	P1@.700	Ea	47.30	25.70	—	73.00
6"	ER@.900	Ea	161.00	35.50	.56	197.06
8"	ER@1.15	Ea	324.00	45.30	.71	370.01

Cast Iron, DWV, Service Weight, No-Hub with Coupled Joints

Description	Craft@Hrs	Unit	Material $	Labor $	Equipment $	Total $

DWV cast iron no-hub combination reducing wye and 1/8 bend

Description	Craft@Hrs	Unit	Material $	Labor $	Equipment $	Total $
3" x 2"	P1@.470	Ea	25.20	17.20	—	42.40
4" x 2"	P1@.600	Ea	38.30	22.00	—	60.30
4" x 3"	P1@.650	Ea	41.00	23.90	—	64.90
6" x 2"	ER@.770	Ea	183.00	30.30	.48	213.78
6" x 3"	ER@.770	Ea	192.00	30.30	.48	222.78
6" x 4"	ER@.800	Ea	198.00	31.50	.49	229.99
8" x 4"	ER@1.20	Ea	359.00	47.30	.74	407.04

DWV cast iron no-hub double combination wye and 1/8 bend

Description	Craft@Hrs	Unit	Material $	Labor $	Equipment $	Total $
2"	P1@.500	Ea	36.00	18.40	—	54.40
3"	P1@.700	Ea	57.80	25.70	—	83.50
4"	P1@.950	Ea	71.20	34.90	—	106.10

DWV cast iron no-hub double combination reducing wye and 1/8 bend

Description	Craft@Hrs	Unit	Material $	Labor $	Equipment $	Total $
3" x 2"	P1@.600	Ea	40.80	22.00	—	62.80
4" x 2"	P1@1.25	Ea	64.40	45.90	—	110.30
4" x 3"	P1@1.35	Ea	67.30	49.50	—	116.80

DWV cast iron no-hub sanitary tee (TY)

Description	Craft@Hrs	Unit	Material $	Labor $	Equipment $	Total $
2"	P1@.380	Ea	12.60	13.90	—	26.50
3"	P1@.520	Ea	15.10	19.10	—	34.20
4"	P1@.700	Ea	24.90	25.70	—	50.60
6"	ER@.900	Ea	75.70	35.50	.56	111.76
8"	ER@1.15	Ea	200.00	45.30	.71	246.01

DWV cast iron no-hub reducing sanitary tee (TY)

Description	Craft@Hrs	Unit	Material $	Labor $	Equipment $	Total $
3" x 2"	P1@.470	Ea	14.30	17.20	—	31.50
4" x 2"	P1@.600	Ea	19.10	22.00	—	41.10
4" x 3"	P1@.650	Ea	22.00	23.90	—	45.90
6" x 2"	ER@.770	Ea	60.00	30.30	.48	90.78
6" x 3"	ER@.770	Ea	56.70	30.30	.48	87.48
6" x 4"	ER@.800	Ea	61.50	31.50	.49	93.49
8" x 3"	ER@1.00	Ea	158.00	39.40	.62	198.02
8" x 4"	ER@1.00	Ea	154.00	39.40	.62	194.02
8" x 6"	ER@1.10	Ea	162.00	43.30	.68	205.98

DWV cast iron no-hub tapped sanitary tee (tapped TY)

Description	Craft@Hrs	Unit	Material $	Labor $	Equipment $	Total $
2" x 1½"	P1@.200	Ea	35.20	7.34	—	42.54
2" x 2"	P1@.250	Ea	35.20	9.18	—	44.38
3" x 2"	P1@.350	Ea	44.90	12.80	—	57.70
4" x 2"	P1@.425	Ea	53.20	15.60	—	68.80
6" x 2"	ER@.700	Ea	137.00	27.60	.43	165.03

Cast Iron, DWV, Service Weight, No-Hub with Coupled Joints

Description	Craft@Hrs	Unit	Material $	Labor $	Equipment $	Total $
DWV cast iron no-hub test tee						
2"	P1@.370	Ea	44.60	13.60	—	58.20
3"	P1@.520	Ea	62.90	19.10	—	82.00
4"	P1@.750	Ea	80.20	27.50	—	107.70
6"	ER@1.12	Ea	287.00	44.10	.69	331.79
8"	ER@1.25	Ea	521.00	49.30	.77	571.07
DWV cast iron no-hub cross (double TY)						
2"	P1@.500	Ea	35.40	18.40	—	53.80
3"	P1@.700	Ea	37.70	25.70	—	63.40
4"	P1@.950	Ea	54.40	34.90	—	89.30
6"	ER@1.30	Ea	202.00	51.20	.80	254.00
DWV cast iron no-hub reducing cross (double TY)						
3" x 2"	P1@.600	Ea	33.40	22.00	—	55.40
4" x 2"	P1@.900	Ea	46.20	33.00	—	79.20
4" x 3"	P1@.900	Ea	47.30	33.00	—	80.30
6" x 4"	ER@1.20	Ea	185.00	47.30	.74	233.04
8" x 4"	ER@1.50	Ea	465.00	59.10	.93	525.03
DWV cast iron no-hub reducer						
2"	P1@.120	Ea	6.97	4.40	—	11.37
3"	P1@.130	Ea	7.50	4.77	—	12.27
4"	P1@.150	Ea	11.10	5.51	—	16.61
6"	ER@.200	Ea	29.70	7.88	.12	37.70
8"	ER@.400	Ea	76.40	15.80	.25	92.45
10"	ER@.650	Ea	118.00	25.60	.40	144.00
DWV cast iron no-hub cap						
2"	P1@.125	Ea	4.25	4.59	—	8.84
3"	P1@.175	Ea	7.29	6.42	—	13.71
4"	P1@.250	Ea	9.49	9.18	—	18.67
6"	ER@.375	Ea	18.00	14.80	.23	33.03
8"	ER@.500	Ea	33.30	19.70	.31	53.31
10"	ER@.750	Ea	51.40	29.60	.46	81.46
MJ (mechanical joint) coupling with gasket (labor included with fittings)						
2"	—	Ea	8.72	—	—	8.72
3"	—	Ea	9.53	—	—	9.53
4"	—	Ea	10.20	—	—	10.20
6"	—	Ea	27.80	—	—	27.80
8"	—	Ea	51.80	—	—	51.80
10"	—	Ea	71.90	—	—	71.90

Cast Iron, DWV, Service Weight, No-Hub with Coupled Joints

Description	Craft@Hrs	Unit	Material $	Labor $	Equipment $	Total $
Hanger with swivel assembly						
2"	P1@.400	Ea	7.18	14.70	—	21.88
3"	P1@.450	Ea	9.25	16.50	—	25.75
4"	P1@.450	Ea	11.50	16.50	—	28.00
5"	P1@.500	Ea	17.60	18.40	—	36.00
6"	P1@.500	Ea	19.90	18.40	—	38.30
8"	P1@.550	Ea	25.90	20.20	—	46.10
10"	P1@.650	Ea	34.20	23.90	—	58.10
Riser clamp						
2"	P1@.115	Ea	5.75	4.22	—	9.97
3"	P1@.120	Ea	6.56	4.40	—	10.96
4"	P1@.125	Ea	8.35	4.59	—	12.94
5"	P1@.180	Ea	11.90	6.61	—	18.51
6"	P1@.200	Ea	14.40	7.34	—	21.74
8"	P1@.200	Ea	23.60	7.34	—	30.94
10"	P1@.250	Ea	34.90	9.18	—	44.08

Cast Iron, DWV, Service Weight, Hub & Spigot with Gasketed Joints

Service weight hub and spigot cast iron soil pipe with gasketed joints is widely used for non-pressurized underground and in-building drain, waste and vent (DWV) systems.

This section has been arranged to save the estimator's time by including all normally-used system components such as pipe, fittings, valves, hanger assemblies and riser clamps under one heading. Additional items can be found under "Plumbing and Piping Specialties." The cost estimates in this section are based on the conditions, limitations and wage rates described in the section "How to Use This Book" beginning on page 5.

Equipment cost, where shown, is $4.33 per hour for a 2-ton chain hoist.

Description	Craft@Hrs	Unit	Material $	Labor $	Equipment $	Total $

DWV cast iron pipe, hub and spigot, gasketed joints

Description	Craft@Hrs	Unit	Material $	Labor $	Equipment $	Total $
2"	P1@.060	LF	10.30	2.20	—	12.50
3"	P1@.070	LF	11.30	2.57	—	13.87
4"	P1@.090	LF	14.90	3.30	—	18.20
6"	ER@.110	LF	28.30	4.33	.07	32.70
8"	ER@.140	LF	47.60	5.52	.09	53.21
10"	ER@.180	LF	76.00	7.09	.11	83.20
12"	ER@.230	LF	98.30	9.06	.14	107.50
15"	ER@.280	LF	148.00	11.00	.17	159.17

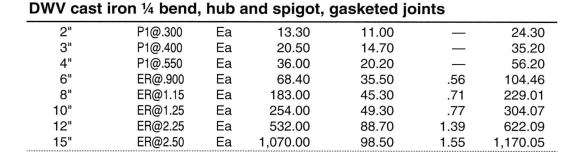

DWV cast iron 1/8 bend, hub and spigot, gasketed joints

Description	Craft@Hrs	Unit	Material $	Labor $	Equipment $	Total $
2"	P1@.300	Ea	12.00	11.00	—	23.00
3"	P1@.400	Ea	16.30	14.70	—	31.00
4"	P1@.550	Ea	22.90	20.20	—	43.10
6"	ER@.700	Ea	57.60	27.60	.43	85.63
8"	ER@1.15	Ea	136.00	45.30	.71	182.01
10"	ER@1.25	Ea	185.00	49.30	.77	235.07
12"	ER@2.25	Ea	426.00	88.70	1.39	516.09
15"	ER@2.50	Ea	714.00	98.50	1.55	814.05

DWV cast iron ¼ bend, hub and spigot, gasketed joints

Description	Craft@Hrs	Unit	Material $	Labor $	Equipment $	Total $
2"	P1@.300	Ea	13.30	11.00	—	24.30
3"	P1@.400	Ea	20.50	14.70	—	35.20
4"	P1@.550	Ea	36.00	20.20	—	56.20
6"	ER@.900	Ea	68.40	35.50	.56	104.46
8"	ER@1.15	Ea	183.00	45.30	.71	229.01
10"	ER@1.25	Ea	254.00	49.30	.77	304.07
12"	ER@2.25	Ea	532.00	88.70	1.39	622.09
15"	ER@2.50	Ea	1,070.00	98.50	1.55	1,170.05

DWV cast iron heel inlet ¼ bend, hub and spigot, gasketed joints

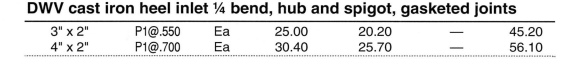

Description	Craft@Hrs	Unit	Material $	Labor $	Equipment $	Total $
3" x 2"	P1@.550	Ea	25.00	20.20	—	45.20
4" x 2"	P1@.700	Ea	30.40	25.70	—	56.10

Cast Iron, DWV, Service Weight, Hub & Spigot with Gasketed Joints

Description	Craft@Hrs	Unit	Material $	Labor $	Equipment $	Total $

DWV cast iron 1/16 bend, hub and spigot, gasketed joints

2"	P1@.300	Ea	7.55	11.00	—	18.55
3"	P1@.400	Ea	12.70	14.70	—	27.40
4"	P1@.550	Ea	17.60	20.20	—	37.80
6"	ER@.700	Ea	29.30	27.60	.43	57.33
8"	ER@1.15	Ea	104.00	45.30	.71	150.01

DWV cast iron closet bend, hub and spigot, gasketed joints

4" x 4"	P1@.400	Ea	43.80	14.70	—	58.50

DWV cast iron closet flange, hub and spigot, gasketed joints

4" x 2"	P1@.075	Ea	14.20	2.75	—	16.95
4" x 3"	P1@.100	Ea	19.70	3.67	—	23.37
4" x 4"	P1@.125	Ea	25.40	4.59	—	29.99

DWV cast iron regular P-trap, hub and spigot, gasketed joints

2"	P1@.300	Ea	50.70	11.00	—	61.70
3"	P1@.400	Ea	70.50	14.70	—	85.20
4"	P1@.550	Ea	109.00	20.20	—	129.20
6"	ER@.700	Ea	239.00	27.60	.43	267.03
8"	ER@1.15	Ea	705.00	45.30	.71	751.01
10"	ER@2.00	Ea	967.00	78.80	1.24	1,047.04

DWV cast iron sanitary tee (TY), hub and spigot, gasketed joints

2"	P1@.400	Ea	24.30	14.70	—	39.00
3"	P1@.600	Ea	42.60	22.00	—	64.60
4"	P1@.750	Ea	85.40	27.50	—	112.90
6"	ER@.900	Ea	197.00	35.50	.56	233.06
8"	ER@1.50	Ea	284.00	59.10	.93	344.03

DWV cast iron sanitary reducing tee (TY), hub and spigot, gasketed joints

3" x 2"	P1@.400	Ea	34.50	14.70	—	49.20
4" x 2"	P1@.650	Ea	90.30	23.90	—	114.20
4" x 3"	P1@.700	Ea	90.30	25.70	—	116.00
6" x 2"	ER@.750	Ea	122.00	29.60	.46	152.06
6" x 3"	ER@.800	Ea	122.00	31.50	.49	153.99
6" x 4"	ER@.850	Ea	122.00	33.50	.53	156.03

DWV cast iron sanitary tapped tee (tapped TY), hub and spigot, gasketed joints

2"	P1@.300	Ea	20.90	11.00	—	31.90
3"	P1@.400	Ea	30.80	14.70	—	45.50
4"	P1@.550	Ea	30.40	20.20	—	50.60

Cast Iron, DWV, Service Weight, Hub & Spigot with Gasketed Joints

Description	Craft@Hrs	Unit	Material $	Labor $	Equipment $	Total $

DWV cast iron combination wye and 1/8 bend, hub and spigot, gasketed joints

2"	P1@.400	Ea	21.60	14.70	—	36.30
3"	P1@.600	Ea	32.60	22.00	—	54.60
4"	P1@.750	Ea	45.20	27.50	—	72.70
5"	ER@.850	Ea	85.10	33.50	.53	119.13
6"	ER@.900	Ea	108.00	35.50	.56	144.06

DWV cast iron combination reducing wye and 1/8 bend, hub and spigot, gasketed joints

3" x 2"	P1@.550	Ea	24.30	20.20	—	44.50
4" x 2"	P1@.650	Ea	33.40	23.90	—	57.30
4" x 3"	P1@.700	Ea	38.80	25.70	—	64.50
5" x 2"	ER@.700	Ea	68.70	27.60	.43	96.73
5" x 4"	ER@.800	Ea	81.40	31.50	.49	113.39
6" x 2"	ER@.750	Ea	69.70	29.60	.46	99.76
6" x 3"	ER@.750	Ea	74.00	29.60	.46	104.06
6" x 4"	ER@.800	Ea	77.70	31.50	.49	109.69

DWV cast iron combination double wye and 1/8 bend, hub and spigot, gasketed joints

2"	P1@.500	Ea	41.30	18.40	—	59.70
3"	P1@.750	Ea	56.50	27.50	—	84.00
4"	P1@.950	Ea	80.30	34.90	—	115.20
6"	ER@1.10	Ea	276.00	43.30	.68	319.98

DWV cast iron combination double reducing wye and 1/8 bend, hub and spigot, gasketed joints

3" x 2"	P1@.550	Ea	45.40	20.20	—	65.60
4" x 2"	P1@.650	Ea	57.10	23.90	—	81.00
4" x 3"	P1@.700	Ea	64.80	25.70	—	90.50
6" x 4"	ER@.800	Ea	182.00	31.50	.49	213.99

DWV cast iron wye, hub and spigot, gasketed joints

2"	P1@.400	Ea	20.00	14.70	—	34.70
3"	P1@.600	Ea	34.30	22.00	—	56.30
4"	P1@.750	Ea	45.90	27.50	—	73.40
6"	ER@.900	Ea	110.00	35.50	.56	146.06
8"	ER@1.50	Ea	246.00	59.10	.93	306.03
10"	ER@2.00	Ea	413.00	78.80	1.24	493.04
12"	ER@2.50	Ea	930.00	98.50	1.55	1,030.05
15"	ER@3.00	Ea	1,850.00	118.00	1.86	1,969.86

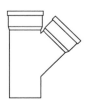

Cast Iron, DWV, Service Weight, Hub & Spigot with Gasketed Joints

Description	Craft@Hrs	Unit	Material $	Labor $	Equipment $	Total $
DWV cast iron reducing wye, hub and spigot, gasketed joints						
4" x 2"	P1@.650	Ea	34.50	23.90	—	58.40
4" x 3"	P1@.700	Ea	42.40	25.70	—	68.10
6" x 2"	ER@.750	Ea	88.20	29.60	.46	118.26
6" x 3"	ER@.750	Ea	88.20	29.60	.46	118.26
6" x 4"	ER@.800	Ea	89.10	31.50	.49	121.09
8" x 4"	ER@1.35	Ea	205.00	53.20	.84	259.04
8" x 6"	ER@1.45	Ea	218.00	57.10	.90	276.00
10" x 4"	ER@1.85	Ea	353.00	72.90	1.14	427.04
10" x 6"	ER@1.90	Ea	364.00	74.90	1.18	440.08
10" x 8"	ER@1.95	Ea	388.00	76.80	1.21	466.01
12" x 6"	ER@2.35	Ea	657.00	92.60	1.45	751.05
12" x 8"	ER@2.40	Ea	721.00	94.60	1.48	817.08
12" x 10"	ER@2.45	Ea	777.00	96.50	1.52	875.02
15" x 6"	ER@2.85	Ea	1,060.00	112.00	1.76	1,173.76
15" x 8"	ER@2.85	Ea	1,070.00	112.00	1.76	1,183.76
15" x 10"	ER@2.90	Ea	1,210.00	114.00	1.79	1,325.79
15" x 12"	ER@2.95	Ea	1,220.00	116.00	1.82	1,337.82
DWV cast iron double wye, hub and spigot, gasketed joints						
2"	P1@.500	Ea	25.90	18.40	—	44.30
3"	P1@.750	Ea	44.90	27.50	—	72.40
4"	P1@.950	Ea	62.30	34.90	—	97.20
6"	ER@1.10	Ea	145.00	43.30	.68	188.98
8"	ER@1.85	Ea	357.00	72.90	1.14	431.04
10"	ER@2.50	Ea	499.00	98.50	1.55	599.05
12"	ER@2.75	Ea	722.00	108.00	1.70	831.70
15"	ER@3.50	Ea	2,180.00	138.00	2.16	2,320.16
DWV cast iron reducing double wye, hub and spigot, gasketed joints						
3" x 2"	P1@.550	Ea	37.90	20.20	—	58.10
4" x 2"	P1@.650	Ea	46.30	23.90	—	70.20
4" x 3"	P1@.700	Ea	51.40	25.70	—	77.10
6" x 4"	ER@.800	Ea	109.00	31.50	.49	140.99
DWV cast iron reducer, hub and spigot, gasketed joints						
3"	P1@.150	Ea	13.30	5.51	—	18.81
4"	P1@.200	Ea	34.50	7.34	—	41.84
6"	ER@.320	Ea	54.00	12.60	.20	66.80
8"	ER@.350	Ea	114.00	13.80	.22	128.02
10"	ER@.570	Ea	241.00	22.50	.35	263.85
12"	ER@.620	Ea	320.00	24.40	.38	344.78
15"	ER@1.12	Ea	592.00	44.10	.69	636.79

Cast Iron, DWV, Service Weight, Hub & Spigot with Gasketed Joints

Description	Craft@Hrs	Unit	Material $	Labor $	Equipment $	Total $
Gaskets for hub and spigot joints						
2"	—	Ea	3.39	—	—	3.39
3"	—	Ea	4.49	—	—	4.49
4"	—	Ea	5.71	—	—	5.71
5"	—	Ea	8.76	—	—	8.76
6"	—	Ea	9.23	—	—	9.23
8"	—	Ea	20.00	—	—	20.00
10"	—	Ea	31.10	—	—	31.10
12"	—	Ea	39.60	—	—	39.60
15"	—	Ea	47.30	—	—	47.30
Hanger with swivel assembly						
2"	P1@.400	Ea	7.18	14.70	—	21.88
3"	P1@.450	Ea	9.25	16.50	—	25.75
4"	P1@.450	Ea	11.50	16.50	—	28.00
5"	P1@.500	Ea	17.60	18.40	—	36.00
6"	P1@.500	Ea	19.90	18.40	—	38.30
8"	P1@.550	Ea	25.90	20.20	—	46.10
10"	P1@.650	Ea	34.20	23.90	—	58.10
12"	P1@.750	Ea	47.10	27.50	—	74.60
15"	P1@.850	Ea	62.40	31.20	—	93.60
Riser clamp						
2"	P1@.115	Ea	5.75	4.22	—	9.97
3"	P1@.120	Ea	6.56	4.40	—	10.96
4"	P1@.125	Ea	8.35	4.59	—	12.94
5"	P1@.180	Ea	11.90	6.61	—	18.51
6"	P1@.200	Ea	14.40	7.34	—	21.74
8"	P1@.200	Ea	23.60	7.34	—	30.94
10"	P1@.250	Ea	34.90	9.18	—	44.08
12"	P1@.250	Ea	43.40	9.18	—	52.58

Copper, DWV with Soft-Soldered Joints

Copper drainage tubing is used for non-pressurized above-grade, interior building drain, waste and vent (DWV) systems.

Because of its thin walls, it must be handled carefully and protected from damage during building construction.

This section has been arranged to save the estimator's time by including all normally-used system components such as pipe, fittings, hanger assemblies and riser clamps under one heading. Additional items can be found under "Plumbing and Piping Specialties." The cost estimates in this section are based on the conditions, limitations and wage rates described in the section "How to Use This Book" beginning on page 5.

Description	Craft@Hrs	Unit	Material $	Labor $	Equipment $	Total $

DWV copper pipe assembly, installed horizontally.
Assembly includes fittings, couplings, hangers and rod. Based on a reducing wye and a 45-degree elbow every 12 feet for 1¼" pipe, a reducing wye and a 45-degree elbow every 25 feet for 4" pipe, and hangers spaced to meet plumbing code.

Description	Craft@Hrs	Unit	Material $	Labor $	Equipment $	Total $
1¼"	P1@.130	LF	10.90	4.77	—	15.67
1½"	P1@.132	LF	11.10	4.84	—	15.94
2"	P1@.137	LF	14.30	5.03	—	19.33
3"	P1@.179	LF	25.60	6.57	—	32.17
4"	P1@.202	LF	47.90	7.41	—	55.31

DWV copper pipe assembly, installed risers.
Complete installation, including a reducing tee every floor and a riser clamp every other.

Description	Craft@Hrs	Unit	Material $	Labor $	Equipment $	Total $
1¼"	P1@.069	LF	8.49	2.53	—	11.02
1½"	P1@.074	LF	9.57	2.72	—	12.29
2"	P1@.086	LF	13.10	3.16	—	16.26
3"	P1@.110	LF	25.50	4.04	—	29.54
4"	P1@.134	LF	42.60	4.92	—	47.52

DWV copper pipe, soft-soldered joints

Description	Craft@Hrs	Unit	Material $	Labor $	Equipment $	Total $
1¼"	P1@.040	LF	4.49	1.47	—	5.96
1½"	P1@.045	LF	5.69	1.65	—	7.34
2"	P1@.050	LF	7.48	1.84	—	9.32
3"	P1@.060	LF	11.70	2.20	—	13.90
4"	P1@.070	LF	15.10	2.57	—	17.67

DWV copper 1/8 bend, soft-soldered joints

Description	Craft@Hrs	Unit	Material $	Labor $	Equipment $	Total $
1¼"	P1@.170	Ea	5.03	6.24	—	11.27
1½"	P1@.200	Ea	4.62	7.34	—	11.96
2"	P1@.270	Ea	8.36	9.91	—	18.27
3"	P1@.400	Ea	20.70	14.70	—	35.40
4"	P1@.540	Ea	85.00	19.80	—	104.80

Description	Craft@Hrs	Unit	Material $	Labor $	Equipment $	Total $

DWV copper ¼ bend, soft-soldered joints

Description	Craft@Hrs	Unit	Material $	Labor $	Equipment $	Total $
1¼"	P1@.170	Ea	5.65	6.24	—	11.89
1½"	P1@.200	Ea	6.45	7.34	—	13.79
2"	P1@.270	Ea	10.40	9.91	—	20.31
3"	P1@.400	Ea	27.40	14.70	—	42.10
4"	P1@.540	Ea	132.00	19.80	—	151.80

DWV copper closet flange, soft-soldered joints

Description	Craft@Hrs	Unit	Material $	Labor $	Equipment $	Total $
4" x 3"	P1@.680	Ea	26.90	25.00	—	51.90
4" x 4"	P1@.980	Ea	33.10	36.00	—	69.10

DWV copper P-trap, soft-soldered joints

Description	Craft@Hrs	Unit	Material $	Labor $	Equipment $	Total $
1¼"	P1@.240	Ea	38.20	8.81	—	47.01
1½"	P1@.240	Ea	33.20	8.81	—	42.01
2"	P1@.240	Ea	50.60	8.81	—	59.41
3"	P1@.320	Ea	146.00	11.70	—	157.70

DWV copper sanitary tee (TY), soft-soldered joints

Description	Craft@Hrs	Unit	Material $	Labor $	Equipment $	Total $
1½"	P1@.240	Ea	11.70	8.81	—	20.51
2"	P1@.320	Ea	16.00	11.70	—	27.70
3"	P1@.490	Ea	55.30	18.00	—	73.30
4"	P1@.650	Ea	139.00	23.90	—	162.90

DWV copper sanitary cross, soft-soldered joints

Description	Craft@Hrs	Unit	Material $	Labor $	Equipment $	Total $
1½"	P1@.640	Ea	77.80	23.50	—	101.30
2"	P1@.900	Ea	112.00	33.00	—	145.00
3"	P1@1.36	Ea	318.00	49.90	—	367.90

DWV copper combination wye and 1/8 bend, soft-soldered joints

Description	Craft@Hrs	Unit	Material $	Labor $	Equipment $	Total $
1½"	P1@.240	Ea	24.70	8.81	—	33.51
2"	P1@.320	Ea	42.20	11.70	—	53.90
3"	P1@.490	Ea	139.00	18.00	—	157.00
4"	P1@.650	Ea	314.00	23.90	—	337.90

DWV copper wye, soft-soldered joints

Description	Craft@Hrs	Unit	Material $	Labor $	Equipment $	Total $
1½"	P1@.240	Ea	19.30	8.81	—	28.11
2"	P1@.320	Ea	27.60	11.70	—	39.30
3"	P1@.490	Ea	69.10	18.00	—	87.10
4"	P1@.650	Ea	139.00	23.90	—	162.90

DWV copper cleanout with threaded plug

Description	Craft@Hrs	Unit	Material $	Labor $	Equipment $	Total $
1¼"	P1@.100	Ea	9.78	3.67	—	13.45
1½"	P1@.160	Ea	14.60	5.87	—	20.47
2"	P1@.230	Ea	19.00	8.44	—	27.44

Copper, DWV with Soft-Soldered Joints

Description	Craft@Hrs	Unit	Material $	Labor $	Equipment $	Total $

DWV copper female adapter

Description	Craft@Hrs	Unit	Material $	Labor $	Equipment $	Total $
1½"	P1@.130	Ea	7.52	4.77	—	12.29
2"	P1@.170	Ea	14.40	6.24	—	20.64
3"	P1@.230	Ea	57.10	8.44	—	65.54
4"	P1@.300	Ea	97.80	11.00	—	108.80

DWV copper male adapter

Description	Craft@Hrs	Unit	Material $	Labor $	Equipment $	Total $
1½"	P1@.130	Ea	8.50	4.77	—	13.27
2"	P1@.170	Ea	10.20	6.24	—	16.44
3"	P1@.230	Ea	50.10	8.44	—	58.54
4"	P1@.300	Ea	83.60	11.00	—	94.60

DWV copper soil adapter (copper x MJ)

Description	Craft@Hrs	Unit	Material $	Labor $	Equipment $	Total $
2"	P1@.170	Ea	12.20	6.24	—	18.44
3"	P1@.250	Ea	25.00	9.18	—	34.18
4"	P1@.320	Ea	46.80	11.70	—	58.50

DWV copper test cap, soft-soldered joint

Description	Craft@Hrs	Unit	Material $	Labor $	Equipment $	Total $
1½"	P1@.100	Ea	.29	3.67	—	3.96
2"	P1@.130	Ea	.41	4.77	—	5.18
3"	P1@.200	Ea	.52	7.34	—	7.86
4"	P1@.250	Ea	1.09	9.18	—	10.27

DWV copper test tee, soft-soldered joints

Description	Craft@Hrs	Unit	Material $	Labor $	Equipment $	Total $
1½"	P1@.200	Ea	28.10	7.34	—	35.44
2"	P1@.270	Ea	36.60	9.91	—	46.51
3"	P1@.400	Ea	98.60	14.70	—	113.30

DWV copper fitting reducer, soft-soldered joints

Description	Craft@Hrs	Unit	Material $	Labor $	Equipment $	Total $
1½"	P1@.100	Ea	4.62	3.67	—	8.29
2"	P1@.140	Ea	5.93	5.14	—	11.07
3"	P1@.200	Ea	20.70	7.34	—	28.04
4"	P1@.270	Ea	32.90	9.91	—	42.81

DWV copper coupling, soft-soldered joints

Description	Craft@Hrs	Unit	Material $	Labor $	Equipment $	Total $
1¼"	P1@.170	Ea	2.47	6.24	—	8.71
1½"	P1@.200	Ea	2.84	7.34	—	10.18
2"	P1@.270	Ea	4.06	9.91	—	13.97
3"	P1@.400	Ea	7.52	14.70	—	22.22
4"	P1@.540	Ea	29.60	19.80	—	49.40

Description	Craft@Hrs	Unit	Material $	Labor $	Equipment $	Total $

Hanger with swivel assembly

Description	Craft@Hrs	Unit	Material $	Labor $	Equipment $	Total $
1¼"	P1@.300	Ea	5.02	11.00	—	16.02
1½"	P1@.300	Ea	5.40	11.00	—	16.40
2"	P1@.300	Ea	5.63	11.00	—	16.63
3"	P1@.350	Ea	9.47	12.80	—	22.27
4"	P1@.350	Ea	10.40	12.80	—	23.20

Riser clamp

Description	Craft@Hrs	Unit	Material $	Labor $	Equipment $	Total $
1¼"	P1@.105	Ea	4.89	3.85	—	8.74
1½"	P1@.110	Ea	5.17	4.04	—	9.21
2"	P1@.115	Ea	5.47	4.22	—	9.69
3"	P1@.120	Ea	6.23	4.40	—	10.63
4"	P1@.125	Ea	7.71	4.59	—	12.30

ABS, DWV with Solvent-Weld Joints

ABS (Acrylonitrile Butadiene Styrene) ASTM D 2661-73 is made from various grades of rubber-based resins and is primarily used for non-pressurized drain, waste and vent (DWV) systems.

ABS can also be used for low-temperature (100 degree F. maximum) water distribution systems, but consult the manufacturers for the proper ASTM type, maximum pressure limitations and recommended joint solvent.

The current Uniform Plumbing Code (UPC) states in Sections 401 and 503 that "ABS piping systems shall be limited to those structures where combustible construction is allowed." This does not preclude its use for underground DWV systems.

This section has been arranged to save the estimator's time by including all normally-used system components such as pipe, fittings, hanger assemblies and riser clamps under one heading. The cost estimates in this section are based on the conditions, limitations and wage rates described in the section "How to Use This Book" beginning on page 5.

Description	Craft@Hrs	Unit	Material $	Labor $	Equipment $	Total $
ABS DWV pipe with solvent-weld joints						
1½"	P1@.020	LF	.83	.73	—	1.56
2"	P1@.025	LF	1.30	.92	—	2.22
3"	P1@.030	LF	2.26	1.10	—	3.36
4"	P1@.040	LF	3.67	1.47	—	5.14
6"	P1@.055	LF	6.81	2.02	—	8.83
ABS DWV 1/8 bend with solvent-weld joints						
1½"	P1@.120	Ea	.88	4.40	—	5.28
2"	P1@.125	Ea	1.56	4.59	—	6.15
3"	P1@.175	Ea	3.77	6.42	—	10.19
4"	P1@.250	Ea	7.13	9.18	—	16.31
6"	P1@.325	Ea	56.50	11.90	—	68.40
ABS DWV 1/8 street bend with solvent-weld joints						
1½"	P1@.060	Ea	.95	2.20	—	3.15
2"	P1@.065	Ea	1.82	2.39	—	4.21
3"	P1@.088	Ea	4.01	3.23	—	7.24
4"	P1@.125	Ea	11.10	4.59	—	15.69
6"	P1@.163	Ea	55.40	5.98	—	61.38
ABS DWV ¼ bend with solvent-weld joints						
1½"	P1@.120	Ea	1.05	4.40	—	5.45
2"	P1@.125	Ea	1.79	4.59	—	6.38
3"	P1@.175	Ea	4.51	6.42	—	10.93
4"	P1@.250	Ea	10.30	9.18	—	19.48
6"	P1@.325	Ea	58.20	11.90	—	70.10
ABS DWV closet bend with solvent-weld joints						
3" x 4"	P1@.250	Ea	15.30	9.18	—	24.48
ABS DWV closet flange with solvent-weld joints						
4"	P1@.125	Ea	6.89	4.59	—	11.48

Description	Craft@Hrs	Unit	Material $	Labor $	Equipment $	Total $

ABS DWV P-trap with solvent-weld joints

1½"	P1@.150	Ea	3.17	5.51	—	8.68
2"	P1@.190	Ea	7.83	6.97	—	14.80
3"	P1@.260	Ea	18.60	9.54	—	28.14
4"	P1@.350	Ea	36.00	12.80	—	48.80

ABS DWV sanitary tee (TY) with solvent-weld joints

1½"	P1@.150	Ea	1.52	5.51	—	7.03
2"	P1@.190	Ea	3.51	6.97	—	10.48
3"	P1@.260	Ea	6.76	9.54	—	16.30
4"	P1@.350	Ea	14.30	12.80	—	27.10
6"	P1@.450	Ea	65.20	16.50	—	81.70

ABS DWV reducing sanitary tee (TY) with solvent-weld joints

2" x 1½"	P1@.160	Ea	2.54	5.87	—	8.41
3" x 1½"	P1@.210	Ea	4.34	7.71	—	12.05
3" x 2"	P1@.235	Ea	4.84	8.62	—	13.46
4" x 2"	P1@.300	Ea	14.00	11.00	—	25.00
4" x 3"	P1@.325	Ea	16.60	11.90	—	28.50

ABS DWV double sanitary tee (TY) with solvent-weld joints

1½"	P1@.210	Ea	4.65	7.71	—	12.36
2"	P1@.250	Ea	7.00	9.18	—	16.18
3"	P1@.350	Ea	11.10	12.80	—	23.90
4"	P1@.475	Ea	43.60	17.40	—	61.00

ABS DWV combination wye and 1/8 bend with solvent-weld joints

1½"	P1@.150	Ea	9.18	5.51	—	14.69
2"	P1@.190	Ea	11.90	6.97	—	18.87
3"	P1@.260	Ea	16.00	9.54	—	25.54
4"	P1@.350	Ea	28.00	12.80	—	40.80

ABS DWV combination reducing wye and 1/8 bend with solvent-weld joints

2" x 1½"	P1@.160	Ea	13.30	5.87	—	19.17
3" x 1½"	P1@.210	Ea	20.30	7.71	—	28.01
3" x 2"	P1@.235	Ea	14.20	8.62	—	22.82
4" x 2"	P1@.300	Ea	25.60	11.00	—	36.60
4" x 3"	P1@.325	Ea	30.30	11.90	—	42.20

ABS, DWV with Solvent-Weld Joints

Description	Craft@Hrs	Unit	Material $	Labor $	Equipment $	Total $

ABS DWV bushing with solvent-weld joints

Description	Craft@Hrs	Unit	Material $	Labor $	Equipment $	Total $
2" x 1½"	P1@.062	Ea	.80	2.28	—	3.08
3" x 1½"	P1@.088	Ea	3.94	3.23	—	7.17
3" x 2"	P1@.088	Ea	2.81	3.23	—	6.04
4" x 2"	P1@.125	Ea	5.69	4.59	—	10.28
4" x 3"	P1@.125	Ea	5.69	4.59	—	10.28
6" x 4"	P1@.163	Ea	28.30	5.98	—	34.28

ABS DWV FIP adapter

Description	Craft@Hrs	Unit	Material $	Labor $	Equipment $	Total $
1½"	P1@.080	Ea	1.19	2.94	—	4.13
2"	P1@.085	Ea	2.21	3.12	—	5.33
3"	P1@.155	Ea	5.23	5.69	—	10.92
4"	P1@.220	Ea	9.95	8.07	—	18.02
6"	P1@.285	Ea	38.30	10.50	—	48.80

ABS DWV test cap

Description	Craft@Hrs	Unit	Material $	Labor $	Equipment $	Total $
1½"	P1@.030	Ea	.60	1.10	—	1.70
2"	P1@.040	Ea	.97	1.47	—	2.44
3"	P1@.050	Ea	1.53	1.84	—	3.37
4"	P1@.090	Ea	2.41	3.30	—	5.71

ABS DWV cleanout plug

Description	Craft@Hrs	Unit	Material $	Labor $	Equipment $	Total $
1½"	P1@.090	Ea	2.79	3.30	—	6.09
2"	P1@.120	Ea	4.09	4.40	—	8.49
3"	P1@.240	Ea	7.28	8.81	—	16.09
4"	P1@.320	Ea	15.70	11.70	—	27.40
6"	P1@.450	Ea	50.30	16.50	—	66.80

ABS DWV cleanout tee with solvent-weld joints

Description	Craft@Hrs	Unit	Material $	Labor $	Equipment $	Total $
1½"	P1@.120	Ea	4.33	4.40	—	8.73
2"	P1@.125	Ea	7.00	4.59	—	11.59
3"	P1@.175	Ea	11.60	6.42	—	18.02
4"	P1@.250	Ea	23.70	9.18	—	32.88

ABS DWV wye with solvent-weld joints

Description	Craft@Hrs	Unit	Material $	Labor $	Equipment $	Total $
1½"	P1@.150	Ea	2.03	5.51	—	7.54
2"	P1@.190	Ea	3.77	6.97	—	10.74
3"	P1@.260	Ea	7.00	9.54	—	16.54
4"	P1@.350	Ea	16.20	12.80	—	29.00
6"	P1@.450	Ea	68.20	16.50	—	84.70

ABS DWV reducing wye with solvent-weld joints

Description	Craft@Hrs	Unit	Material $	Labor $	Equipment $	Total $
2" x 1½"	P1@.160	Ea	3.51	5.87	—	9.38
3" x 1½"	P1@.210	Ea	4.89	7.71	—	12.60
3" x 2"	P1@.235	Ea	5.47	8.62	—	14.09
4" x 2"	P1@.300	Ea	11.70	11.00	—	22.70
4" x 3"	P1@.325	Ea	14.70	11.90	—	26.60
6" x 4"	P1@.400	Ea	56.50	14.70	—	71.20

ABS, DWV with Solvent-Weld Joints

Description	Craft@Hrs	Unit	Material $	Labor $	Equipment $	Total $

ABS DWV reducer with solvent-weld joints

2" x 1½"	P1@.125	Ea	.80	4.59	—	5.39
3" x 1½"	P1@.150	Ea	3.94	5.51	—	9.45
3" x 2"	P1@.150	Ea	2.81	5.51	—	8.32
4" x 2"	P1@.185	Ea	5.69	6.79	—	12.48
4" x 3"	P1@.210	Ea	5.63	7.71	—	13.34

ABS DWV coupling with solvent-weld joints

1½"	P1@.120	Ea	.67	4.40	—	5.07
2"	P1@.125	Ea	.99	4.59	—	5.58
3"	P1@.175	Ea	2.31	6.42	—	8.73
4"	P1@.250	Ea	4.20	9.18	—	13.38
6"	P1@.325	Ea	24.80	11.90	—	36.70

Hanger with swivel assembly

1½"	P1@.300	Ea	5.40	11.00	—	16.40
2"	P1@.300	Ea	5.63	11.00	—	16.63
3"	P1@.350	Ea	9.47	12.80	—	22.27
4"	P1@.350	Ea	10.40	12.80	—	23.20
6"	P1@.450	Ea	17.10	16.50	—	33.60

Riser clamp

1½"	P1@.110	Ea	5.17	4.04	—	9.21
2"	P1@.115	Ea	5.47	4.22	—	9.69
3"	P1@.120	Ea	6.23	4.40	—	10.63
4"	P1@.125	Ea	7.94	4.59	—	12.53
6"	P1@.130	Ea	13.70	4.77	—	18.47

Galvanized steel pipe sleeves

2"	P1@.130	Ea	7.07	4.77	—	11.84
2½"	P1@.150	Ea	7.21	5.51	—	12.72
3"	P1@.180	Ea	7.36	6.61	—	13.97
4"	P1@.220	Ea	8.38	8.07	—	16.45
5"	P1@.250	Ea	10.40	9.18	—	19.58
6"	P1@.270	Ea	11.20	9.91	—	21.11

Lead roof pipe flashing

1½"	P1@.250	Ea	32.60	9.18	—	41.78
2"	P1@.250	Ea	42.90	9.18	—	52.08
3"	P1@.250	Ea	49.80	9.18	—	58.98
4"	P1@.250	Ea	54.60	9.18	—	63.78
6"	P1@.350	Ea	69.90	12.80	—	82.70

PVC, DWV with Solvent-Weld Joints

PVC (Polyvinyl Chloride) meets ASTM – D3034, D3212, F1336, F679 & CSA 182.2 (CSA – Canadian Standards Association – similar to UL – Underwriters Laboratories). The applications for PVC DWV have increased in number due to the unique flame-spread characteristics of their compounds, as well as the development of new fire stop devices for use with combustible piping systems.

PVC will not rust, pit or degrade when exposed to moisture and is extremely resistant to a broad range of corrosive agents.

PVC DWV is suitable for sewer, vent and storm water drainage applications in both below- and above-grade applications.

When penetrating a vertical or horizontal fire separation (fire-rated wall or floor) with combustible piping, certified fire stop devices must be used in compliance with local and national codes.

PVC DWV is not permitted in a ceiling space used as a return air plenum and is not permitted in a vertical shaft. PVC DWV has a flame-spread rating of 15.

This section has been arranged to save the estimator's time by including all normally-used system components such as pipe, fittings, hanger assemblies and riser clamps under one heading. The cost estimates in this section are based on the conditions, limitations and wage rates described in the section "How to Use This Book" beginning on page 5.

Description	Craft@Hrs	Unit	Material $	Labor $	Equipment $	Total $

PVC DWV pipe with solvent-weld joints

Description	Craft@Hrs	Unit	Material $	Labor $	Equipment $	Total $
1½"	P1@.021	LF	.90	.77	—	1.67
2"	P1@.026	LF	1.44	.95	—	2.39
3"	P1@.032	LF	3.06	1.17	—	4.23
4"	P1@.042	LF	3.76	1.54	—	5.30
6"	P1@.058	LF	6.76	2.13	—	8.89

PVC DWV 90-degree elbow with solvent-weld joints

Description	Craft@Hrs	Unit	Material $	Labor $	Equipment $	Total $
1½"	P1@.126	Ea	1.50	4.62	—	6.12
2"	P1@.131	Ea	2.26	4.81	—	7.07
3"	P1@.184	Ea	5.94	6.75	—	12.69
4"	P1@.263	Ea	11.70	9.65	—	21.35
6"	P1@.341	Ea	49.40	12.50	—	61.90

PVC DWV 90-degree street elbow with solvent-weld joints

Description	Craft@Hrs	Unit	Material $	Labor $	Equipment $	Total $
1½"	P1@.063	Ea	1.58	2.31	—	3.89
2"	P1@.068	Ea	2.70	2.50	—	5.20
3"	P1@.092	Ea	7.89	3.38	—	11.27
4"	P1@.131	Ea	19.00	4.81	—	23.81
6"	P1@.171	Ea	69.80	6.28	—	76.08

PVC DWV 45-degree elbow with solvent-weld joints

Description	Craft@Hrs	Unit	Material $	Labor $	Equipment $	Total $
1½"	P1@.126	Ea	1.26	4.62	—	5.88
2"	P1@.131	Ea	1.39	4.81	—	6.20
3"	P1@.184	Ea	3.48	6.75	—	10.23
4"	P1@.263	Ea	8.80	9.65	—	18.45
6"	P1@.341	Ea	53.40	12.50	—	65.90

Description	Craft@Hrs	Unit	Material $	Labor $	Equipment $	Total $

PVC DWV wye with solvent-weld joints

Description	Craft@Hrs	Unit	Material $	Labor $	Equipment $	Total $
1½"	P1@.158	Ea	2.34	5.80	—	8.14
2"	P1@.200	Ea	4.78	7.34	—	12.12
3"	P1@.273	Ea	7.88	10.00	—	17.88
4"	P1@.368	Ea	12.90	13.50	—	26.40
6"	P1@.473	Ea	50.60	17.40	—	68.00

PVC DWV reducing wye with solvent-weld joints

Description	Craft@Hrs	Unit	Material $	Labor $	Equipment $	Total $
2" x 1½"	P1@.168	Ea	4.14	6.17	—	10.31
3" x 1½"	P1@.221	Ea	4.75	8.11	—	12.86
3" x 2"	P1@.247	Ea	6.82	9.06	—	15.88
4" x 2"	P1@.315	Ea	13.20	11.60	—	24.80
4" x 3"	P1@.341	Ea	15.60	12.50	—	28.10
6" x 3"	P1@.473	Ea	57.70	17.40	—	75.10
6" x 4"	P1@.473	Ea	61.40	17.40	—	78.80

PVC DWV sanitary tee (TY) with solvent-weld joints

Description	Craft@Hrs	Unit	Material $	Labor $	Equipment $	Total $
1½"	P1@.158	Ea	1.66	5.80	—	7.46
2"	P1@.200	Ea	2.82	7.34	—	10.16
3"	P1@.273	Ea	6.07	10.00	—	16.07
4"	P1@.368	Ea	11.60	13.50	—	25.10
6"	P1@.473	Ea	63.60	17.40	—	81.00

PVC DWV reducing sanitary tee (TY) with solvent-weld joints

Description	Craft@Hrs	Unit	Material $	Labor $	Equipment $	Total $
2" x 1½"	P1@.168	Ea	2.20	6.17	—	8.37
3" x 1½"	P1@.221	Ea	3.78	8.11	—	11.89
3" x 2"	P1@.247	Ea	5.08	9.06	—	14.14
4" x 2"	P1@.315	Ea	11.10	11.60	—	22.70
4" x 3"	P1@.341	Ea	12.80	12.50	—	25.30
6" x 3"	P1@.473	Ea	48.80	17.40	—	66.20
6" x 4"	P1@.473	Ea	48.80	17.40	—	66.20

PVC DWV double wye with solvent-weld joints

Description	Craft@Hrs	Unit	Material $	Labor $	Equipment $	Total $
2" x 1½"	P1@.263	Ea	6.22	9.65	—	15.87
3" x 2"	P1@.368	Ea	12.60	13.50	—	26.10
4" x 3"	P1@.473	Ea	41.20	17.40	—	58.60
6" x 4"	P1@.578	Ea	49.80	21.20	—	71.00

PVC DWV closet flange with solvent-weld joints, adjustable with test plate

Description	Craft@Hrs	Unit	Material $	Labor $	Equipment $	Total $
4" x 3"	P1@.131	Ea	10.70	4.81	—	15.51

PVC DWV closet flange with solvent-weld joints, non-adjustable with test plate

Description	Craft@Hrs	Unit	Material $	Labor $	Equipment $	Total $
4" x 3"	P1@.131	Ea	8.16	4.81	—	12.97

PVC, DWV with Solvent-Weld Joints

Description	Craft@Hrs	Unit	Material $	Labor $	Equipment $	Total $

PVC DWV closet flange with solvent-weld joints, non-adjustable without test plate

Description	Craft@Hrs	Unit	Material $	Labor $	Equipment $	Total $
4" x 3"	P1@.131	Ea	7.40	4.81	—	12.21

PVC DWV closet flange with solvent-weld joints, adjustable, offset without test plate

Description	Craft@Hrs	Unit	Material $	Labor $	Equipment $	Total $
4" x 3"	P1@.170	Ea	15.90	6.24	—	22.14

PVC DWV P-trap with solvent-weld joints

Description	Craft@Hrs	Unit	Material $	Labor $	Equipment $	Total $
1½"	P1@.160	Ea	3.78	5.87	—	9.65
2"	P1@.200	Ea	8.48	7.34	—	15.82
3"	P1@.285	Ea	19.90	10.50	—	30.40
4"	P1@.370	Ea	39.10	13.60	—	52.70
6"	P1@.580	Ea	150.00	21.30	—	171.30

PVC DWV bushing with solvent-weld joints

Description	Craft@Hrs	Unit	Material $	Labor $	Equipment $	Total $
2" x 1½"	P1@.062	Ea	.71	2.28	—	2.99
3" x 2"	P1@.088	Ea	2.10	3.23	—	5.33
4" x 3"	P1@.125	Ea	4.40	4.59	—	8.99
6" x 4"	P1@.163	Ea	22.70	5.98	—	28.68

PVC DWV FIP adapter

Description	Craft@Hrs	Unit	Material $	Labor $	Equipment $	Total $
1½"	P1@.080	Ea	1.04	2.94	—	3.98
2"	P1@.085	Ea	1.69	3.12	—	4.81
3"	P1@.155	Ea	4.09	5.69	—	9.78
4"	P1@.220	Ea	8.00	8.07	—	16.07
6"	P1@.285	Ea	35.90	10.50	—	46.40

PVC DWV MIP adapter

Description	Craft@Hrs	Unit	Material $	Labor $	Equipment $	Total $
1½"	P1@.080	Ea	1.15	2.94	—	4.09
2"	P1@.085	Ea	1.76	3.12	—	4.88
3"	P1@.155	Ea	3.64	5.69	—	9.33
4"	P1@.220	Ea	9.76	8.07	—	17.83
6"	P1@.285	Ea	17.80	10.50	—	28.30

PVC DWV coupling with solvent-weld joints

Description	Craft@Hrs	Unit	Material $	Labor $	Equipment $	Total $
1½"	P1@.120	Ea	.55	4.40	—	4.95
2"	P1@.125	Ea	.80	4.59	—	5.39
3"	P1@.175	Ea	1.86	6.42	—	8.28
4"	P1@.250	Ea	3.91	9.18	—	13.09
6"	P1@.325	Ea	19.00	11.90	—	30.90

Description	Craft@Hrs	Unit	Material $	Labor $	Equipment $	Total $

PVC DWV reducer with solvent-weld joints

Description	Craft@Hrs	Unit	Material $	Labor $	Equipment $	Total $
2" x 1½"	P1@.125	Ea	1.18	4.59	—	5.77
3" x 2"	P1@.150	Ea	3.56	5.51	—	9.07
4" x 3"	P1@.185	Ea	4.65	6.79	—	11.44
6" x 4"	P1@.210	Ea	47.00	7.71	—	54.71

PVC DWV in-line cleanout with solvent-weld joints

Description	Craft@Hrs	Unit	Material $	Labor $	Equipment $	Total $
1½"	P1@.120	Ea	4.82	4.40	—	9.22
2"	P1@.125	Ea	7.58	4.59	—	12.17
3"	P1@.175	Ea	11.10	6.42	—	17.52
4"	P1@.250	Ea	23.20	9.18	—	32.38
6" x 4"	P1@.250	Ea	78.80	9.18	—	87.98
6"	P1@.250	Ea	78.80	9.18	—	87.98

PVC DWV end cleanout

Description	Craft@Hrs	Unit	Material $	Labor $	Equipment $	Total $
1½"	P1@.090	Ea	2.12	3.30	—	5.42
2"	P1@.120	Ea	3.17	4.40	—	7.57
3"	P1@.240	Ea	6.42	8.81	—	15.23
4"	P1@.320	Ea	11.90	11.70	—	23.60
6"	P1@.450	Ea	41.70	16.50	—	58.20

PVC DWV fitting end cleanout

Description	Craft@Hrs	Unit	Material $	Labor $	Equipment $	Total $
1½"	P1@.090	Ea	2.02	3.30	—	5.32
2"	P1@.120	Ea	3.67	4.40	—	8.07
3"	P1@.240	Ea	4.28	8.81	—	13.09
4"	P1@.320	Ea	9.12	11.70	—	20.82
6"	P1@.450	Ea	47.50	16.50	—	64.00

Hanger with swivel assembly

Description	Craft@Hrs	Unit	Material $	Labor $	Equipment $	Total $
1½"	P1@.300	Ea	6.87	11.00	—	17.87
2"	P1@.300	Ea	7.20	11.00	—	18.20
3"	P1@.350	Ea	12.20	12.80	—	25.00
4"	P1@.350	Ea	13.20	12.80	—	26.00
6"	P1@.450	Ea	21.60	16.50	—	38.10

Riser clamp

Description	Craft@Hrs	Unit	Material $	Labor $	Equipment $	Total $
1½"	P1@.110	Ea	5.17	4.04	—	9.21
2"	P1@.115	Ea	5.47	4.22	—	9.69
3"	P1@.120	Ea	6.23	4.40	—	10.63
4"	P1@.125	Ea	7.94	4.59	—	12.53
6"	P1@.130	Ea	13.70	4.77	—	18.47

PVC, DWV with Solvent-Weld Joints

Description	Craft@Hrs	Unit	Material $	Labor $	Equipment $	Total $
Galvanized steel pipe sleeves						
2"	P1@.130	Ea	7.07	4.77	—	11.84
2½"	P1@.150	Ea	7.21	5.51	—	12.72
3"	P1@.180	Ea	7.36	6.61	—	13.97
4"	P1@.220	Ea	8.38	8.07	—	16.45
5"	P1@.250	Ea	10.40	9.18	—	19.58
6"	P1@.270	Ea	11.20	9.91	—	21.11
Lead roof pipe flashing						
1½"	P1@.400	Ea	34.10	14.70	—	48.80
2"	P1@.450	Ea	45.00	16.50	—	61.50
3"	P1@.500	Ea	52.00	18.40	—	70.40
4"	P1@.550	Ea	57.50	20.20	—	77.70

PVC, DWV with Gasketed Bell and Spigot Joints

PVC (Polyvinyl Chloride) is polymerized vinyl chloride produced from acetylene and anhydrous hydrochloric acid.

PVC sewer pipe (ASTM D-3034) 10' lay length, is for underground installation and is used primarily to convey sewage and rain water. It is of the bell & spigot type with joints made water-tight by means of an elastomeric gasket (O-ring) factory-installed in the hub end of the pipe and fittings.

This section includes pipe and all commonly-used fittings. Additional items can be found under "Plumbing and Piping Specialties." The cost estimates in this section are based on the conditions, limitations, and wage rates described in the section "How to Use This Book" beginning on page 5.

Description	Craft@Hrs	Unit	Material $	Labor $	Equipment $	Total $

PVC SDR35 sewer pipe with bell and spigot gasketed joints

4"	P1@.050	LF	3.90	1.84	—	5.74
6"	P1@.070	LF	8.57	2.57	—	11.14
8"	P1@.090	LF	13.60	3.30	—	16.90
10"	P1@.110	LF	20.20	4.04	—	24.24
12"	P1@.140	LF	25.50	5.14	—	30.64
15"	P1@.180	LF	37.70	6.61	—	44.31
18"	P1@.230	LF	58.70	8.44	—	67.14

PVC sewer pipe 1/16 bend B x B with bell and spigot gasketed joints

4"	P1@.320	Ea	18.10	11.70	—	29.80
6"	P1@.400	Ea	35.40	14.70	—	50.10
8"	P1@.480	Ea	104.00	17.60	—	121.60
10"	P1@.800	Ea	282.00	29.40	—	311.40
12"	P1@1.06	Ea	371.00	38.90	—	409.90
15"	P1@1.34	Ea	981.00	49.20	—	1,030.20
18"	P1@1.60	Ea	1,420.00	58.70	—	1,478.70

PVC sewer pipe 1/16 bend B x S with bell and spigot gasketed joints

4"	P1@.320	Ea	17.10	11.70	—	28.80
6"	P1@.400	Ea	33.80	14.70	—	48.50
8"	P1@.480	Ea	108.00	17.60	—	125.60
10"	P1@.800	Ea	278.00	29.40	—	307.40
12"	P1@1.06	Ea	360.00	38.90	—	398.90
15"	P1@1.34	Ea	784.00	49.20	—	833.20
18"	P1@1.60	Ea	1,200.00	58.70	—	1,258.70

PVC sewer pipe 1/8 bend B x B with bell and spigot gasketed joints

4"	P1@.320	Ea	18.20	11.70	—	29.90
6"	P1@.400	Ea	36.60	14.70	—	51.30
8"	P1@.480	Ea	98.10	17.60	—	115.70
10"	P1@.800	Ea	265.00	29.40	—	294.40
12"	P1@1.06	Ea	391.00	38.90	—	429.90
15"	P1@1.34	Ea	864.00	49.20	—	913.20
18"	P1@1.60	Ea	1,380.00	58.70	—	1,438.70

PVC, DWV with Gasketed Bell and Spigot Joints

Description	Craft@Hrs	Unit	Material $	Labor $	Equipment $	Total $

PVC sewer pipe 1/8 bend B x S with bell and spigot gasketed joints

Description	Craft@Hrs	Unit	Material $	Labor $	Equipment $	Total $
4"	P1@.320	Ea	16.30	11.70	—	28.00
6"	P1@.400	Ea	32.60	14.70	—	47.30
8"	P1@.480	Ea	98.10	17.60	—	115.70
10"	P1@.800	Ea	261.00	29.40	—	290.40
12"	P1@1.06	Ea	378.00	38.90	—	416.90
15"	P1@1.34	Ea	677.00	49.20	—	726.20
18"	P1@1.60	Ea	1,140.00	58.70	—	1,198.70

PVC sewer pipe ¼ bend B x B with bell and spigot gasketed joints

Description	Craft@Hrs	Unit	Material $	Labor $	Equipment $	Total $
4"	P1@.320	Ea	26.00	11.70	—	37.70
6"	P1@.400	Ea	48.20	14.70	—	62.90
8"	P1@.480	Ea	130.00	17.60	—	147.60
10"	P1@.800	Ea	448.00	29.40	—	477.40
12"	P1@1.06	Ea	574.00	38.90	—	612.90
15"	P1@1.34	Ea	1,220.00	49.20	—	1,269.20
18"	P1@1.60	Ea	2,050.00	58.70	—	2,108.70

PVC sewer pipe ¼ bend B x S with bell and spigot gasketed joints

Description	Craft@Hrs	Unit	Material $	Labor $	Equipment $	Total $
4"	P1@.320	Ea	23.80	11.70	—	35.50
6"	P1@.400	Ea	50.60	14.70	—	65.30
8"	P1@.480	Ea	138.00	17.60	—	155.60
10"	P1@.800	Ea	443.00	29.40	—	472.40
12"	P1@1.06	Ea	559.00	38.90	—	597.90
15"	P1@1.34	Ea	1,090.00	49.20	—	1,139.20
18"	P1@1.60	Ea	1,910.00	58.70	—	1,968.70

PVC sewer pipe coupling B x B with bell and spigot gasketed joints

Description	Craft@Hrs	Unit	Material $	Labor $	Equipment $	Total $
4"	P1@.320	Ea	22.30	11.70	—	34.00
6"	P1@.400	Ea	44.80	14.70	—	59.50
8"	P1@.480	Ea	76.20	17.60	—	93.80
10"	P1@.800	Ea	167.00	29.40	—	196.40
12"	P1@1.06	Ea	246.00	38.90	—	284.90
15"	P1@1.34	Ea	506.00	49.20	—	555.20
18"	P1@1.60	Ea	990.00	58.70	—	1,048.70

PVC sewer pipe wye B x B x B with bell and spigot gasketed joints

Description	Craft@Hrs	Unit	Material $	Labor $	Equipment $	Total $
4"	P1@.480	Ea	30.30	17.60	—	47.90
6"	P1@.600	Ea	69.40	22.00	—	91.40
8"	P1@.720	Ea	192.00	26.40	—	218.40
10"	P1@1.20	Ea	559.00	44.00	—	603.00
12"	P1@1.59	Ea	792.00	58.40	—	850.40
15"	P1@2.01	Ea	1,380.00	73.80	—	1,453.80
18"	P1@2.40	Ea	2,420.00	88.10	—	2,508.10

Description	Craft@Hrs	Unit	Material $	Labor $	Equipment $	Total $

PVC sewer pipe wye B x S x B with bell and spigot gasketed joints

Description	Craft@Hrs	Unit	Material $	Labor $	Equipment $	Total $
10"	P1@1.20	Ea	551.00	44.00	—	595.00
12"	P1@1.59	Ea	784.00	58.40	—	842.40
15"	P1@2.01	Ea	1,260.00	73.80	—	1,333.80
18"	P1@2.40	Ea	2,230.00	88.10	—	2,318.10

PVC sewer pipe reducing wye B x B x B with bell and spigot gasketed joints

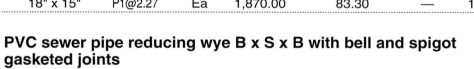

Description	Craft@Hrs	Unit	Material $	Labor $	Equipment $	Total $
6" x 4"	P1@.560	Ea	61.70	20.60	—	82.30
8" x 4"	P1@.640	Ea	92.10	23.50	—	115.60
8" x 6"	P1@.680	Ea	109.00	25.00	—	134.00
10" x 4"	P1@.960	Ea	278.00	35.20	—	313.20
10" x 6"	P1@1.00	Ea	282.00	36.70	—	318.70
10" x 8"	P1@1.04	Ea	448.00	38.20	—	486.20
12" x 4"	P1@1.22	Ea	403.00	44.80	—	447.80
12" x 6"	P1@1.26	Ea	414.00	46.20	—	460.20
12" x 8"	P1@1.30	Ea	617.00	47.70	—	664.70
12" x 10"	P1@1.46	Ea	753.00	53.60	—	806.60
15" x 4"	P1@1.50	Ea	607.00	55.10	—	662.10
15" x 6"	P1@1.54	Ea	718.00	56.50	—	774.50
15" x 8"	P1@1.58	Ea	812.00	58.00	—	870.00
15" x 10"	P1@1.74	Ea	971.00	63.90	—	1,034.90
15" x 12"	P1@1.87	Ea	1,080.00	68.60	—	1,148.60
18" x 4"	P1@1.76	Ea	1,380.00	64.60	—	1,444.60
18" x 6"	P1@1.80	Ea	1,400.00	66.10	—	1,466.10
18" x 8"	P1@1.84	Ea	1,480.00	67.50	—	1,547.50
18" x 10"	P1@2.00	Ea	1,630.00	73.40	—	1,703.40
18" x 12"	P1@2.13	Ea	1,760.00	78.20	—	1,838.20
18" x 15"	P1@2.27	Ea	1,870.00	83.30	—	1,953.30

PVC sewer pipe reducing wye B x S x B with bell and spigot gasketed joints

Description	Craft@Hrs	Unit	Material $	Labor $	Equipment $	Total $
10" x 4"	P1@.960	Ea	271.00	35.20	—	306.20
10" x 6"	P1@1.00	Ea	285.00	36.70	—	321.70
10" x 8"	P1@1.04	Ea	460.00	38.20	—	498.20
12" x 4"	P1@1.22	Ea	408.00	44.80	—	452.80
12" x 6"	P1@1.26	Ea	414.00	46.20	—	460.20
12" x 8"	P1@1.30	Ea	609.00	47.70	—	656.70
12" x 10"	P1@1.46	Ea	740.00	53.60	—	793.60
15" x 4"	P1@1.50	Ea	620.00	55.10	—	675.10
15" x 6"	P1@1.54	Ea	701.00	56.50	—	757.50
15" x 8"	P1@1.58	Ea	824.00	58.00	—	882.00
15" x 10"	P1@1.74	Ea	938.00	63.90	—	1,001.90
15" x 12"	P1@1.87	Ea	1,090.00	68.60	—	1,158.60
18" x 4"	P1@1.76	Ea	1,290.00	64.60	—	1,354.60
18" x 6"	P1@1.80	Ea	1,400.00	66.10	—	1,466.10
18" x 8"	P1@1.84	Ea	1,480.00	67.50	—	1,547.50
18" x 10"	P1@2.00	Ea	1,630.00	73.40	—	1,703.40
18" x 12"	P1@2.13	Ea	1,760.00	78.20	—	1,838.20
18" x 15"	P1@2.27	Ea	1,870.00	83.30	—	1,953.30

PVC, DWV with Gasketed Bell and Spigot Joints

Description	Craft@Hrs	Unit	Material $	Labor $	Equipment $	Total $

PVC sewer pipe tee-wye B x B x B with bell and spigot gasketed joints

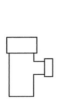

Description	Craft@Hrs	Unit	Material $	Labor $	Equipment $	Total $
4"	P1@.480	Ea	33.80	17.60	—	51.40
6"	P1@.600	Ea	85.50	22.00	—	107.50
6" x 4"	P1@.560	Ea	72.90	20.60	—	93.50
8" x 4"	P1@.640	Ea	92.10	23.50	—	115.60
8" x 6"	P1@.680	Ea	106.00	25.00	—	131.00

PVC sewer pipe tee B x S x B with bell and spigot gasketed joints

Description	Craft@Hrs	Unit	Material $	Labor $	Equipment $	Total $
10"	P1@1.20	Ea	466.00	44.00	—	510.00
12"	P1@1.59	Ea	645.00	58.40	—	703.40
15"	P1@2.01	Ea	1,030.00	73.80	—	1,103.80
18"	P1@2.40	Ea	1,930.00	88.10	—	2,018.10

PVC sewer pipe reducing tee B x S x B with bell and spigot gasketed joints

Description	Craft@Hrs	Unit	Material $	Labor $	Equipment $	Total $
10" x 4"	P1@.960	Ea	296.00	35.20	—	331.20
10" x 6"	P1@1.00	Ea	308.00	36.70	—	344.70
10" x 8"	P1@1.04	Ea	482.00	38.20	—	520.20
12" x 4"	P1@1.22	Ea	363.00	44.80	—	407.80
12" x 6"	P1@1.26	Ea	371.00	46.20	—	417.20
12" x 8"	P1@1.30	Ea	509.00	47.70	—	556.70
12" x 10"	P1@1.46	Ea	609.00	53.60	—	662.60
15" x 4"	P1@1.50	Ea	588.00	55.10	—	643.10
15" x 6"	P1@1.54	Ea	617.00	56.50	—	673.50
15" x 8"	P1@1.58	Ea	632.00	58.00	—	690.00
15" x 10"	P1@1.74	Ea	740.00	63.90	—	803.90
15" x 12"	P1@1.87	Ea	812.00	68.60	—	880.60
18" x 4"	P1@1.76	Ea	1,490.00	64.60	—	1,554.60
18" x 6"	P1@1.80	Ea	1,520.00	66.10	—	1,586.10
18" x 8"	P1@1.84	Ea	1,610.00	67.50	—	1,677.50
18" x 10"	P1@2.00	Ea	1,650.00	73.40	—	1,723.40
18" x 12"	P1@2.13	Ea	1,710.00	78.20	—	1,788.20
18" x 15"	P1@2.27	Ea	1,820.00	83.30	—	1,903.30

PVC sewer pipe reducer S x B with bell and spigot gasketed joints

6" x 4"	P1@.360	Ea	32.60	13.20	—	45.80
8" x 4"	P1@.400	Ea	89.30	14.70	—	104.00
8" x 6"	P1@.440	Ea	100.00	16.10	—	116.10
10" x 4"	P1@.560	Ea	234.00	20.60	—	254.60
10" x 6"	P1@.600	Ea	238.00	22.00	—	260.00
10" x 8"	P1@.640	Ea	287.00	23.50	—	310.50
12" x 4"	P1@.690	Ea	308.00	25.30	—	333.30
12" x 6"	P1@.730	Ea	315.00	26.80	—	341.80
12" x 8"	P1@.770	Ea	363.00	28.30	—	391.30
12" x 10"	P1@.930	Ea	403.00	34.10	—	437.10
15" x 4"	P1@.830	Ea	518.00	30.50	—	548.50
15" x 6"	P1@.870	Ea	537.00	31.90	—	568.90
15" x 8"	P1@.910	Ea	609.00	33.40	—	642.40
15" x 10"	P1@1.07	Ea	609.00	39.30	—	648.30
15" x 12"	P1@1.20	Ea	701.00	44.00	—	745.00
18" x 8"	P1@1.04	Ea	753.00	38.20	—	791.20
18" x 10"	P1@1.20	Ea	762.00	44.00	—	806.00
18" x 12"	P1@1.33	Ea	766.00	48.80	—	814.80
18" x 15"	P1@1.47	Ea	784.00	53.90	—	837.90

PVC sewer pipe adapter B x S with bell and spigot gasketed joints

4"	P1@.320	Ea	19.10	11.70	—	30.80
6"	P1@.400	Ea	35.40	14.70	—	50.10
8"	P1@.480	Ea	91.10	17.60	—	108.70
10"	P1@.800	Ea	112.00	29.40	—	141.40
12"	P1@1.06	Ea	171.00	38.90	—	209.90
15"	P1@1.34	Ea	260.00	49.20	—	309.20
18"	P1@1.60	Ea	569.00	58.70	—	627.70

PVC sewer pipe cap with bell and spigot gasketed joints

4"	P1@.160	Ea	11.60	5.87	—	17.47
6"	P1@.200	Ea	21.70	7.34	—	29.04
8"	P1@.240	Ea	58.80	8.81	—	67.61
10"	P1@.400	Ea	183.00	14.70	—	197.70
12"	P1@.530	Ea	278.00	19.50	—	297.50
15"	P1@.670	Ea	448.00	24.60	—	472.60
18"	P1@.800	Ea	645.00	29.40	—	674.40

PVC sewer pipe test plug with bell and spigot gasketed joints

4"	P1@.160	Ea	8.79	5.87	—	14.66
6"	P1@.200	Ea	13.40	7.34	—	20.74
8"	P1@.240	Ea	48.20	8.81	—	57.01
10"	P1@.400	Ea	160.00	14.70	—	174.70
12"	P1@.530	Ea	191.00	19.50	—	210.50
15"	P1@.670	Ea	378.00	24.60	—	402.60
18"	P1@.800	Ea	524.00	29.40	—	553.40

Polypropylene, Schedule 40, with Heat-Fusioned Joints

Polypropylene is used almost exclusively for acid and laboratory drain, waste and vent (DWV) systems.

Joints are made by applying low-voltage current to electrical resistance coils imbedded in propylene collars, which are slipped over the ends of the pipe, then inserted into the hubs of the fittings. Compression clamps are placed around the assemblies and tightened to compress the joints prior to applying electrical current from the power unit to fuse the joint.

This section has been arranged to save the estimator's time by including all normally-used system components such as pipe, fittings, hanger assemblies, and riser clamps under one heading. The cost estimates in this section are based on the conditions, limitations and wage rates described in the section "How to Use This Book" beginning on page 5.

Description	Craft@Hrs	Unit	Material $	Labor $	Equipment $	Total $

Schedule 40 DWV polypropylene pipe with heat-fusioned joints

Description	Craft@Hrs	Unit	Material $	Labor $	Equipment $	Total $
1½"	P1@.050	LF	5.59	1.84	—	7.43
2"	P1@.055	LF	7.59	2.02	—	9.61
3"	P1@.060	LF	13.50	2.20	—	15.70
4"	P1@.080	LF	19.40	2.94	—	22.34
6"	P1@.110	LF	35.00	4.04	—	39.04

Schedule 40 DWV polypropylene 1/8 bend with heat-fusioned joints

Description	Craft@Hrs	Unit	Material $	Labor $	Equipment $	Total $
1½"	P1@.350	Ea	22.80	12.80	—	35.60
2"	P1@.400	Ea	27.80	14.70	—	42.50
3"	P1@.450	Ea	50.80	16.50	—	67.30
4"	P1@.500	Ea	57.50	18.40	—	75.90
6"	P1@.750	Ea	158.00	27.50	—	185.50

Schedule 40 DWV polypropylene ¼ bend with heat-fusioned joints

Description	Craft@Hrs	Unit	Material $	Labor $	Equipment $	Total $
1½"	P1@.350	Ea	23.00	12.80	—	35.80
2"	P1@.400	Ea	28.40	14.70	—	43.10
3"	P1@.450	Ea	48.80	16.50	—	65.30
4"	P1@.500	Ea	77.30	18.40	—	95.70
6"	P1@.750	Ea	190.00	27.50	—	217.50

Schedule 40 DWV polypropylene long sweep ¼ bend with heat-fusioned joints

Description	Craft@Hrs	Unit	Material $	Labor $	Equipment $	Total $
1½"	P1@.350	Ea	23.40	12.80	—	36.20
2"	P1@.400	Ea	32.80	14.70	—	47.50
3"	P1@.450	Ea	55.80	16.50	—	72.30
4"	P1@.500	Ea	79.80	18.40	—	98.20

Schedule 40 DWV polypropylene P-trap with heat-fusioned joints

Description	Craft@Hrs	Unit	Material $	Labor $	Equipment $	Total $
1½"	P1@.500	Ea	45.10	18.40	—	63.50
2"	P1@.560	Ea	62.80	20.60	—	83.40
3"	P1@.630	Ea	113.00	23.10	—	136.10
4"	P1@.700	Ea	204.00	25.70	—	229.70

Polypropylene, Schedule 40, with Heat-Fusioned Joints

Description	Craft@Hrs	Unit	Material $	Labor $	Equipment $	Total $

Schedule 40 DWV polypropylene sanitary tee (TY) with heat-fusioned joints

1½"	P1@.500	Ea	28.40	18.40	—	46.80
2"	P1@.560	Ea	34.50	20.60	—	55.10
3"	P1@.630	Ea	57.20	23.10	—	80.30
4"	P1@.700	Ea	102.00	25.70	—	127.70

Schedule 40 DWV polypropylene reducing sanitary tee (TY) with heat-fusioned joints

2" x 1½"	P1@.510	Ea	37.80	18.70	—	56.50
3" x 1½"	P1@.530	Ea	65.60	19.50	—	85.10
3" x 2"	P1@.580	Ea	70.00	21.30	—	91.30
4" x 2"	P1@.650	Ea	101.00	23.90	—	124.90
4" x 3"	P1@.700	Ea	105.00	25.70	—	130.70
6" x 4"	P1@.850	Ea	190.00	31.20	—	221.20

Schedule 40 DWV polypropylene combination wye and 1/8 bend with heat-fusioned joints

1½"	P1@.500	Ea	41.90	18.40	—	60.30
2"	P1@.560	Ea	56.10	20.60	—	76.70
3"	P1@.630	Ea	89.20	23.10	—	112.30
4"	P1@.700	Ea	125.00	25.70	—	150.70

Schedule 40 DWV polypropylene combination reducing wye and 1/8 bend with heat-fusioned joints

2" x 1½"	P1@.500	Ea	50.20	18.40	—	68.60
3" x 1½"	P1@.520	Ea	74.10	19.10	—	93.20
3" x 2"	P1@.570	Ea	80.30	20.90	—	101.20
4" x 2"	P1@.580	Ea	110.00	21.30	—	131.30
4" x 3"	P1@.630	Ea	117.00	23.10	—	140.10

Schedule 40 DWV polypropylene wye with heat-fusioned joints

1½"	P1@.500	Ea	32.20	18.40	—	50.60
2"	P1@.560	Ea	45.70	20.60	—	66.30
3"	P1@.630	Ea	80.30	23.10	—	103.40
4"	P1@.700	Ea	117.00	25.70	—	142.70
6"	P1@1.05	Ea	296.00	38.50	—	334.50

Polypropylene, Schedule 40, with Heat-Fusioned Joints

Description	Craft@Hrs	Unit	Material $	Labor $	Equipment $	Total $

Schedule 40 DWV polypropylene reducing wye with heat-fusioned joints

Description	Craft@Hrs	Unit	Material $	Labor $	Equipment $	Total $
2" x 1½"	P1@.510	Ea	44.70	18.70	—	63.40
3" x 2"	P1@.580	Ea	74.60	21.30	—	95.90
4" x 2"	P1@.600	Ea	107.00	22.00	—	129.00
4" x 3"	P1@.650	Ea	111.00	23.90	—	134.90
6" x 2"	P1@.900	Ea	188.00	33.00	—	221.00
6" x 3"	P1@.900	Ea	190.00	33.00	—	223.00
6" x 4"	P1@.950	Ea	193.00	34.90	—	227.90

Schedule 40 DWV polypropylene female adapter with heat-fusioned joints

Description	Craft@Hrs	Unit	Material $	Labor $	Equipment $	Total $
1½"	P1@.280	Ea	15.60	10.30	—	25.90
2"	P1@.330	Ea	20.90	12.10	—	33.00
3"	P1@.350	Ea	35.20	12.80	—	48.00
4"	P1@.400	Ea	67.20	14.70	—	81.90

Schedule 40 DWV polypropylene male adapter with heat-fusioned joints

Description	Craft@Hrs	Unit	Material $	Labor $	Equipment $	Total $
1½"	P1@.280	Ea	15.30	10.30	—	25.60
2"	P1@.330	Ea	18.60	12.10	—	30.70
3"	P1@.350	Ea	30.80	12.80	—	43.60
4"	P1@.400	Ea	63.70	14.70	—	78.40

Schedule 40 DWV polypropylene cleanout adapter with heat-fusioned joints

Description	Craft@Hrs	Unit	Material $	Labor $	Equipment $	Total $
1½"	P1@.210	Ea	16.30	7.71	—	24.01
2"	P1@.280	Ea	19.30	10.30	—	29.60
3"	P1@.430	Ea	35.80	15.80	—	51.60
4"	P1@.640	Ea	61.40	23.50	—	84.90
6"	P1@.850	Ea	106.00	31.20	—	137.20

Schedule 40 DWV polypropylene plug with heat-fusioned joints

Description	Craft@Hrs	Unit	Material $	Labor $	Equipment $	Total $
1½"	P1@.060	Ea	5.65	2.20	—	7.85
2"	P1@.065	Ea	6.17	2.39	—	8.56
3"	P1@.070	Ea	10.80	2.57	—	13.37
4"	P1@.090	Ea	12.10	3.30	—	15.40

Schedule 40 DWV polypropylene coupling with heat-fusioned joints

Description	Craft@Hrs	Unit	Material $	Labor $	Equipment $	Total $
1½"	P1@.350	Ea	19.90	12.80	—	32.70
2"	P1@.400	Ea	24.50	14.70	—	39.20
3"	P1@.450	Ea	31.40	16.50	—	47.90
4"	P1@.500	Ea	44.70	18.40	—	63.10
6"	P1@.750	Ea	71.40	27.50	—	98.90

Polypropylene, Schedule 40, with Heat-Fusioned Joints

Description	Craft@Hrs	Unit	Material $	Labor $	Equipment $	Total $

Schedule 40 DWV polypropylene reducer with heat-fusioned joints

Description	Craft@Hrs	Unit	Material $	Labor $	Equipment $	Total $
2" x 1½"	P1@.360	Ea	20.90	13.20	—	34.10
3" x 2"	P1@.410	Ea	24.00	15.00	—	39.00
4" x 2"	P1@.430	Ea	53.70	15.80	—	69.50
4" x 3"	P1@.450	Ea	55.20	16.50	—	71.70

Schedule 40 DWV polypropylene flange with heat-fusioned joints

Description	Craft@Hrs	Unit	Material $	Labor $	Equipment $	Total $
2"	P1@.350	Ea	69.10	12.80	—	81.90
3"	P1@.380	Ea	143.00	13.90	—	156.90
4"	P1@.430	Ea	186.00	15.80	—	201.80
6"	P1@.630	Ea	244.00	23.10	—	267.10

Schedule 40 DWV polypropylene bolt and gasket set

Description	Craft@Hrs	Unit	Material $	Labor $	Equipment $	Total $
2"	P1@.500	Ea	6.80	18.40	—	25.20
2½"	P1@.650	Ea	7.20	23.90	—	31.10
3"	P1@.750	Ea	7.57	27.50	—	35.07
4"	P1@1.00	Ea	13.60	36.70	—	50.30

Hanger with swivel assembly

Description	Craft@Hrs	Unit	Material $	Labor $	Equipment $	Total $
1½"	P1@.300	Ea	5.40	11.00	—	16.40
2"	P1@.300	Ea	5.63	11.00	—	16.63
3"	P1@.350	Ea	9.47	12.80	—	22.27
4"	P1@.350	Ea	10.40	12.80	—	23.20
6"	P1@.450	Ea	17.10	16.50	—	33.60

Riser clamp

Description	Craft@Hrs	Unit	Material $	Labor $	Equipment $	Total $
1½"	P1@.110	Ea	5.17	4.04	—	9.21
2"	P1@.115	Ea	5.47	4.22	—	9.69
3"	P1@.120	Ea	6.23	4.40	—	10.63
4"	P1@.125	Ea	7.94	4.59	—	12.53
6"	P1@.200	Ea	13.70	7.34	—	21.04

Pipe sleeve

Description	Craft@Hrs	Unit	Material $	Labor $	Equipment $	Total $
2"	P1@.130	Ea	7.95	4.77	—	12.72
2½"	P1@.150	Ea	8.06	5.51	—	13.57
3"	P1@.180	Ea	8.27	6.61	—	14.88
4"	P1@.220	Ea	9.39	8.07	—	17.46
5"	P1@.250	Ea	11.60	9.18	—	20.78
6"	P1@.270	Ea	12.70	9.91	—	22.61
8"	P1@.270	Ea	14.50	9.91	—	24.41

Lead roof pipe flashing

Description	Craft@Hrs	Unit	Material $	Labor $	Equipment $	Total $
1½"	P1@.250	Ea	34.30	9.18	—	43.48
2"	P1@.250	Ea	45.50	9.18	—	54.68
3"	P1@.250	Ea	52.30	9.18	—	61.48
4"	P1@.250	Ea	57.90	9.18	—	67.08

Floor, Area, Roof and Planter Drains

Description	Craft@Hrs	Unit	Material $	Labor $	Equipment $	Total $

Floor drain, cast iron, nickel-bronze grate. Add for rough-in (P-trap, fittings, hangers, etc.).

Description	Craft@Hrs	Unit	Material $	Labor $	Equipment $	Total $
3" x 2"	P1@.500	Ea	154.00	18.40	—	172.40
5" x 3"	P1@.500	Ea	190.00	18.40	—	208.40
6" x 4"	P1@.500	Ea	245.00	18.40	—	263.40
8" x 6"	P1@.500	Ea	324.00	18.40	—	342.40

Area drain, cast iron grate. Add for rough-in (P-trap, fittings, hangers, etc.).

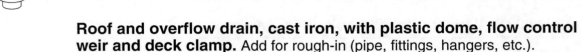

Description	Craft@Hrs	Unit	Material $	Labor $	Equipment $	Total $
10" x 3"	P1@.650	Ea	324.00	23.90	—	347.90
12" x 4"	P1@.650	Ea	355.00	23.90	—	378.90
14" x 6"	P1@.700	Ea	382.00	25.70	—	407.70

Roof and overflow drain, cast iron, with plastic dome, flow control weir and deck clamp. Add for rough-in (pipe, fittings, hangers, etc.).

Description	Craft@Hrs	Unit	Material $	Labor $	Equipment $	Total $
8" x 3"	P1@1.50	Ea	437.00	55.10	—	492.10
10" x 4"	P1@1.50	Ea	462.00	55.10	—	517.10
12" x 6"	P1@1.65	Ea	462.00	60.60	—	522.60

Planter drain, cast iron. Add for rough-in (pipe, fittings, hangers, etc.).

Description	Craft@Hrs	Unit	Material $	Labor $	Equipment $	Total $
2"	P1@1.00	Ea	83.90	36.70	—	120.60
3"	P1@1.10	Ea	96.70	40.40	—	137.10
4"	P1@1.15	Ea	117.00	42.20	—	159.20
6"	P1@1.25	Ea	171.00	45.90	—	216.90

Planter drain, plastic. Add for rough-in (pipe, fittings, hangers, etc.).

Description	Craft@Hrs	Unit	Material $	Labor $	Equipment $	Total $
4"	P1@.800	Ea	25.60	29.40	—	55.00
5"	P1@1.00	Ea	27.20	36.70	—	63.90
6"	P1@1.05	Ea	44.50	38.50	—	83.00
8"	P1@1.10	Ea	67.90	40.40	—	108.30

Description	Craft@Hrs	Unit	Material $	Labor $	Equipment $	Total $

Cleanout, in-line, (Barrett) cast iron, (MJ) brass plug. Add for MJ couplings.

2"	P1@.350	Ea	72.80	12.80	—	85.60
3"	P1@.350	Ea	93.20	12.80	—	106.00
4"	P1@.400	Ea	132.00	14.70	—	146.70
6"	P1@.400	Ea	463.00	14.70	—	477.70

Cleanout, in-line (Barrett), ABS/PVC

2"	P1@.300	Ea	21.30	11.00	—	32.30
3"	P1@.300	Ea	34.30	11.00	—	45.30
4"	P1@.350	Ea	62.20	12.80	—	75.00
6"	P1@.350	Ea	242.00	12.80	—	254.80
8"	P1@.400	Ea	470.00	14.70	—	484.70
10"	P1@.550	Ea	580.00	20.20	—	600.20

Cleanout, end of line (Malcolm), cast iron body, brass top. Add for MJ coupling.

2" raised	P1@.250	Ea	26.20	9.18	—	35.38
2" recessed	P1@.250	Ea	65.40	9.18	—	74.58
3" raised	P1@.250	Ea	30.50	9.18	—	39.68
3" recessed	P1@.300	Ea	96.00	11.00	—	107.00
4" raised	P1@.300	Ea	40.80	11.00	—	51.80
4" recessed	P1@.350	Ea	125.00	12.80	—	137.80
6" raised	P1@.350	Ea	129.00	12.80	—	141.80
6" recessed	P1@.400	Ea	239.00	14.70	—	253.70
8" raised	P1@.450	Ea	278.00	16.50	—	294.50

Cleanout, end-of-line (Malcolm), ABS/PVC

2"	P1@.250	Ea	11.30	9.18	—	20.48
3"	P1@.250	Ea	19.00	9.18	—	28.18
4"	P1@.300	Ea	34.30	11.00	—	45.30
6"	P1@.300	Ea	144.00	11.00	—	155.00
8"	P1@.400	Ea	261.00	14.70	—	275.70
10"	P1@.450	Ea	297.00	16.50	—	313.50
12"	P1@.550	Ea	347.00	20.20	—	367.20

Fire Protection Sprinklers

Description	Unit	Material $	Labor $	Equipment $	Total $

Complete wet sprinkler system (square foot costs). System includes subcontractors' overhead and profit. Cost includes design drawings, zone and supervision valves, sprinkler heads, pipe, and connection to water supply service. Make additional allowances if a fire pump is required. Costs are based on ordinary hazard coverage (110 sf per head). Coverage density can vary depending on room sizes and/or hazard rating. Use these costs for preliminary estimates.

Description	Unit	Material $	Labor $	Equipment $	Total $
Exposed piping to 5,000	SF	—	—	—	4.82
Exposed piping 5,000-15,000	SF	—	—	—	4.05
Exposed piping over 15,000	SF	—	—	—	3.63
Concealed piping to 5,000	SF	—	—	—	4.49
Concealed piping 5,000-15,000	SF	—	—	—	3.87
Concealed piping over 15,000	SF	—	—	—	3.03
Add for dry system	%	—	—	—	20.0

Complete wet sprinkler system (per head costs). Cost includes sprinkler heads, sprinkler mains, branch piping, and supports. Make additional allowances for zone and supervision valves, alarms, connection to water service and fire pump when required. Costs are based on ordinary hazard 155 degree heads. Use these costs for preliminary estimates.

Description	Unit	Material $	Labor $	Equipment $	Total $
Upright heads to 5,000 SF	Ea	—	—	—	337.00
Upright heads 5,000 to 15,000 SF	Ea	—	—	—	321.00
Upright heads over 15,000 SF	Ea	—	—	—	305.00
Pendent heads to 5,000 SF	Ea	—	—	—	509.00
Pendent heads 5,000 to 15,000 SF	Ea	—	—	—	424.00
Pendent heads over 15,000 SF	Ea	—	—	—	359.00

Description	Craft@Hrs	Unit	Material $	Labor $	Equipment $	Total $

Supervision (zone) valves, flanged. Add for alarm trim if required.

Description	Craft@Hrs	Unit	Material $	Labor $	Equipment $	Total $
2½" OS&Y gate valve	SL@1.70	Ea	375.00	62.00	—	437.00
3" OS&Y gate vlv	SL@1.80	Ea	397.00	65.60	—	462.60
4" OS&Y gate vlv	SL@2.20	Ea	458.00	80.20	—	538.20
6" OS&Y gate vlv	SL@2.80	Ea	703.00	102.00	—	805.00
8" OS&Y gate vlv	SL@3.50	Ea	1,120.00	128.00	—	1,248.00
2½" butterfly valve (lug)	SL@.600	Ea	570.00	21.90	—	591.90
3" butterfly vlv	SL@.700	Ea	575.00	25.50	—	600.50
4" butterfly vlv	SL@.850	Ea	590.00	31.00	—	621.00
6" butterfly vlv	SL@1.15	Ea	818.00	41.90	—	859.90
8" butterfly vlv	SL@1.85	Ea	1,230.00	67.40	—	1,297.40

Description	Craft@Hrs	Unit	Material $	Labor $	Equipment $	Total $

Supervision (zone) valves, grooved. Add for alarm trim if required

Description	Craft@Hrs	Unit	Material $	Labor $	Equipment $	Total $
4" butterfly valve	SL@.550	Ea	440.00	20.00	—	460.00
6" butterfly valve	SL@.850	Ea	572.00	31.00	—	603.00
8" butterfly valve	SL@1.25	Ea	866.00	45.60	—	911.60

Alarm valves (flanged or grooved)

Description	Craft@Hrs	Unit	Material $	Labor $	Equipment $	Total $
4" alarm valve	SL@3.50	Ea	602.00	128.00	—	730.00
6" alarm valve	SL@4.00	Ea	741.00	146.00	—	887.00
8" alarm valve	SL@4.75	Ea	1,080.00	173.00	—	1,253.00
4" alarm vlv trim	SL@1.25	Ea	403.00	45.60	—	448.60
6" alarm vlv trim	SL@1.25	Ea	403.00	45.60	—	448.60
8" alarm vlv trim	SL@1.25	Ea	403.00	45.60	—	448.60
4" alarm vlv pkg	SL@4.75	Ea	1,890.00	173.00	—	2,063.00
6" alarm vlv pkg	SL@5.50	Ea	2,090.00	200.00	—	2,290.00
8" alarm vlv pkg	SL@6.00	Ea	2,540.00	219.00	—	2,759.00

Dry valves and trim (deluge/pre-action system)

Description	Craft@Hrs	Unit	Material $	Labor $	Equipment $	Total $
2" dry valve	SL@2.25	Ea	1,130.00	82.00	—	1,212.00
3" dry valve	SL@2.75	Ea	1,170.00	100.00	—	1,270.00
4" dry valve	SL@3.05	Ea	1,310.00	111.00	—	1,421.00
6" dry valve	SL@3.65	Ea	1,690.00	133.00	—	1,823.00
2" dry valve trim	SL@1.25	Ea	551.00	45.60	—	596.60
3" dry valve trim	SL@1.25	Ea	551.00	45.60	—	596.60
4" dry valve trim	SL@1.25	Ea	551.00	45.60	—	596.60
6" dry valve trim	SL@1.25	Ea	551.00	45.60	—	596.60

Check valves, grooved

Description	Craft@Hrs	Unit	Material $	Labor $	Equipment $	Total $
2½"	SL@.450	Ea	215.00	16.40	—	231.40
3"	SL@.500	Ea	237.00	18.20	—	255.20
4"	SL@.550	Ea	212.00	20.00	—	232.00
6"	SL@.850	Ea	418.00	31.00	—	449.00

Check valves, wafer-type, flanged

Description	Craft@Hrs	Unit	Material $	Labor $	Equipment $	Total $
2½"	SL@1.65	Ea	268.00	60.10	—	328.10
3"	SL@1.70	Ea	278.00	62.00	—	340.00
4"	SL@2.10	Ea	301.00	76.50	—	377.50
6"	SL@2.60	Ea	491.00	94.80	—	585.80
8"	SL@3.20	Ea	699.00	117.00	—	816.00

Double-check detector valve assembly (flanged)

Description	Craft@Hrs	Unit	Material $	Labor $	Equipment $	Total $
2½"	SL@4.00	Ea	2,790.00	146.00	—	2,936.00
3"	SL@4.35	Ea	3,000.00	159.00	—	3,159.00
4"	SL@4.95	Ea	3,260.00	180.00	—	3,440.00
6"	SL@6.25	Ea	5,110.00	228.00	—	5,338.00
8"	SL@6.95	Ea	9,200.00	253.00	—	9,453.00

Fire Protection Sprinklers

Description	Craft@Hrs	Unit	Material $	Labor $	Equipment $	Total $

Switches: flow, pressure, monitor, supervisory

Description	Craft@Hrs	Unit	Material $	Labor $	Equipment $	Total $
Flow switch	SL@1.00	Ea	187.00	36.50	—	223.50
Pressure switch	SL@1.00	Ea	140.00	36.50	—	176.50
Monitor switch	SL@1.00	Ea	136.00	36.50	—	172.50
Supervisory sw.	SL@1.00	Ea	125.00	36.50	—	161.50

Sprinkler heads, 155 degree

Description	Craft@Hrs	Unit	Material $	Labor $	Equipment $	Total $
Pendent, brass	SL@.350	Ea	8.16	12.80	—	20.96
Pendent, chrome	SL@.350	Ea	8.74	12.80	—	21.54
Upright, brass	SL@.350	Ea	8.16	12.80	—	20.96
Upright, chrome	SL@.350	Ea	20.50	12.80	—	33.30
Sidewall, brass	SL@.350	Ea	10.60	12.80	—	23.40
Sidewall, chrome	SL@.350	Ea	11.30	12.80	—	24.10

Sprinkler heads, dry, 165 degree

Description	Craft@Hrs	Unit	Material $	Labor $	Equipment $	Total $
Dry pendent	SL@.450	Ea	81.40	16.40	—	97.80
Dry sidewall	SL@.450	Ea	113.00	16.40	—	129.40

Sprinkler heads, 200 degree

Description	Craft@Hrs	Unit	Material $	Labor $	Equipment $	Total $
Pendent, brass	SL@.350	Ea	8.16	12.80	—	20.96
Pendent, chrome	SL@.350	Ea	8.74	12.80	—	21.54
Upright, brass	SL@.350	Ea	8.16	12.80	—	20.96
Upright, chrome	SL@.350	Ea	20.50	12.80	—	33.30
Sidewall, brass	SL@.350	Ea	10.60	12.80	—	23.40
Sidewall, chrome	SL@.350	Ea	11.30	12.80	—	24.10

Sprinkler heads, 286 degree

Description	Craft@Hrs	Unit	Material $	Labor $	Equipment $	Total $
Pendent, brass	SL@.350	Ea	8.16	12.80	—	20.96
Pendent, chrome	SL@.350	Ea	8.74	12.80	—	21.54
Upright, brass	SL@.350	Ea	8.16	12.80	—	20.96
Upright, chrome	SL@.350	Ea	20.50	12.80	—	33.30
Sidewall, brass	SL@.350	Ea	10.60	12.80	—	23.40
Sidewall, chrome	SL@.350	Ea	11.30	12.80	—	24.10

Sprinkler heads, 360 degree

Description	Craft@Hrs	Unit	Material $	Labor $	Equipment $	Total $
Pendent, brass	SL@.350	Ea	5.50	12.80	—	18.30
Upright, brass	SL@.350	Ea	5.50	12.80	—	18.30

Sprinkler heads, 400 degree

Description	Craft@Hrs	Unit	Material $	Labor $	Equipment $	Total $
Pendent, brass	SL@.350	Ea	24.30	12.80	—	37.10
Upright, brass	SL@.350	Ea	24.30	12.80	—	37.10

Description	Craft@Hrs	Unit	Material $	Labor $	Equipment $	Total $

Fire hose cabinet, complete assembly. Includes recessed steel cabinet, 2½" and 1½" angle valves, 100' fire hose, hose rack and fog nozzle.

Description	Craft@Hrs	Unit	Material $	Labor $	Equipment $	Total $
FHC, chrome trim	SL@1.95	Ea	393.00	71.10	—	464.10
FHC, brass trim	SL@1.95	Ea	384.00	71.10	—	455.10
FHC, plastic trim	SL@1.95	Ea	373.00	71.10	—	444.10

Fire hose cabinet components

Description	Craft@Hrs	Unit	Material $	Labor $	Equipment $	Total $
Recessed cabinet	SL@1.00	Ea	177.00	36.50	—	213.50
Surface cabinet	SL@.850	Ea	205.00	31.00	—	236.00
Cabinet glass	SL@.350	Ea	13.00	12.80	—	25.80
100' fire hose	SL@.250	Ea	166.00	9.11	—	175.11
75' hose rack	SL@.350	Ea	51.30	12.80	—	64.10
100' hose rack	SL@.350	Ea	69.50	12.80	—	82.30
1½" brass fire hose angle valve	SL@.350	Ea	83.30	12.80	—	96.10
1½" chrome fire hose angle valve	SL@.350	Ea	139.00	12.80	—	151.80
2½" brass fire hose angle valve	SL@.450	Ea	160.00	16.40	—	176.40
2½" chrome fire hose angle valve	SL@.450	Ea	63.20	16.40	—	79.60
Brass fog nozzle	SL@.150	Ea	62.20	5.47	—	67.67
Plastic fog nozzle	SL@.150	Ea	17.20	5.47	—	22.67
Chrome fog nozzle	SL@.150	Ea	63.80	5.47	—	69.27

Fire Protection Equipment

Description	Craft@Hrs	Unit	Material $	Labor $	Equipment $	Total $
Electric fire pump, skid-mounted with controller and jockey pump						
15 KW, 500 gpm	—	Ea	—	—	—	48,400.00
22 KW, 500 gpm	—	Ea	—	—	—	51,900.00
30 KW, 750 gpm	—	Ea	—	—	—	54,800.00
38 KW, 1,000 gpm	—	Ea	—	—	—	60,700.00
56 KW, 1,000 gpm	—	Ea	—	—	—	61,600.00
Diesel fire pump, skid-mounted with controller and jockey pump						
250 gpm	—	Ea	—	—	—	72,300.00
500 gpm	—	Ea	—	—	—	75,400.00
750 gpm	—	Ea	—	—	—	76,800.00
1,000 gpm	—	Ea	—	—	—	78,200.00
Excess pressure pump						
1/4 hp	SL@1.35	Ea	539.00	49.20	—	588.20
1/3 hp	SL@1.35	Ea	541.00	49.20	—	590.20
1/2 hp	SL@1.50	Ea	562.00	54.70	—	616.70
Siamese connection, flush, fire dept. connection						
4" brass	SL@4.00	Ea	759.00	146.00	—	905.00
4" chrome	SL@4.00	Ea	819.00	146.00	—	965.00
6" brass	SL@4.50	Ea	881.00	164.00	—	1,045.00
6" chrome	SL@4.50	Ea	897.00	164.00	—	1,061.00
Siamese connection, surface, fire dept. connection						
4" brass	SL@2.85	Ea	180.00	104.00	—	284.00
6" brass	SL@3.25	Ea	340.00	118.00	—	458.00
Siamese connection, sidewalk, fire dept. connection						
4" brass	SL@3.25	Ea	686.00	118.00	—	804.00
6" brass	SL@3.75	Ea	970.00	137.00	—	1,107.00

Description	Craft@Hrs	Unit	Material $	Labor $	Equipment $	Total $
Water motor gong						
Gong	SL@1.50	Ea	326.00	54.70	—	380.70
Multi-purpose fire extinguisher						
2 lb.	SL@.500	Ea	38.60	18.20	—	56.80
5 lb.	SL@.500	Ea	54.20	18.20	—	72.40
10 lb.	SL@.500	Ea	77.70	18.20	—	95.90
20 lb.	SL@.500	Ea	163.00	18.20	—	181.20
Fire hydrant with spool						
6"	P1@5.00	Ea	1,500.00	184.00	—	1,684.00
Indicator post						
6"	P1@6.50	Ea	864.00	239.00	—	1,103.00

Fire Protection Sprinkler Pipe and Fittings (Roll-Grooved)

Description	Craft@Hrs	Unit	Material $	Labor $	Equipment $	Total $

Black steel pipe (A53), installed horizontal, roll-grooved, Schedule 40. Standard weight, including hangers every 10' and a fitting every 33'. Use these figures for preliminary estimates.

Description	Craft@Hrs	Unit	Material $	Labor $	Equipment $	Total $
2" (50mm)	SL@.160	LF	7.69	5.83	—	13.52
2½" (65mm)	SL@.180	LF	12.30	6.56	—	18.86
3" (75mm)	SL@.210	LF	15.80	7.65	—	23.45
4" (10cm)	SL@.300	LF	21.40	10.90	—	32.30
6" (15cm)	SK@.450	LF	42.60	17.60	.28	60.48
8" (20cm)	SK@.630	LF	63.40	24.60	.39	88.39
10" (25cm)	SK@.750	LF	123.00	29.30	.46	152.76
12" (30cm)	SK@.800	LF	161.00	31.30	.49	192.79

Black steel pipe (A53), installed riser, roll-grooved, Schedule 40. Standard weight, including a riser clamp every other floor and a tee at every floor. Make additional allowances for sleeving or coring as required. Use these figures for preliminary estimates.

Description	Craft@Hrs	Unit	Material $	Labor $	Equipment $	Total $
2" (50mm)	SL@.110	LF	7.95	4.01	—	11.96
2½" (65mm)	SL@.120	LF	12.80	4.37	—	17.17
3" (75mm)	SL@.150	LF	15.20	5.47	—	20.67
4" (10cm)	SL@.190	LF	21.70	6.93	—	28.63
6" (15cm)	SK@.300	LF	41.30	11.70	.19	53.19
8" (20cm)	SK@.400	LF	64.20	15.60	.25	80.05
10" (25cm)	SK@.510	LF	100.00	19.90	.32	120.22
12" (30cm)	SK@.560	LF	145.00	21.90	.35	167.25

Black steel pipe (A53), installed horizontal, roll-grooved, thin wall, Schedule 10. Light weight, including hangers every 10' and a fitting every 33'. Use these figures for preliminary estimates.

Description	Craft@Hrs	Unit	Material $	Labor $	Equipment $	Total $
2" (50mm)	SL@.160	LF	6.39	5.83	—	12.22
2½" (65mm)	SL@.180	LF	9.32	6.56	—	15.88
3" (75mm)	SL@.210	LF	10.90	7.65	—	18.55
4" (10cm)	SL@.300	LF	15.70	10.90	—	26.60
6" (15cm)	SK@.450	LF	30.30	17.60	.28	48.18
8" (20cm)	SK@.630	LF	56.20	24.60	.39	81.19

Black steel pipe (A53), installed riser, roll-grooved, thin wall, Schedule 10. Light weight, including a riser clamp every other floor and a tee at every floor. Make additional allowances for sleeving or coring as required. Use these figures for preliminary estimates.

Description	Craft@Hrs	Unit	Material $	Labor $	Equipment $	Total $
2" (50mm)	SL@.110	LF	7.18	4.01	—	11.19
2½" (65mm)	SL@.120	LF	10.40	4.37	—	14.77
3" (75mm)	SL@.150	LF	12.10	5.47	—	17.57
4" (10cm)	SL@.190	LF	17.00	6.93	—	23.93
6" (15cm)	SK@.300	LF	32.60	11.70	.19	44.49
8" (20cm)	SK@.400	LF	63.40	15.60	.25	79.25

Description	Craft@Hrs	Unit	Material $	Labor $	Equipment $	Total $

Black steel pipe only (A53), roll-grooved, Schedule 40. Standard weight, no hangers or fittings.

2" (50mm)	SL@.070	LF	3.40	2.55	—	5.95
2½" (65mm)	SL@.070	LF	7.26	2.55	—	9.81
3" (75mm)	SL@.090	LF	8.08	3.28	—	11.36
4" (10cm)	SL@.130	LF	11.40	4.74	—	16.14
6" (15cm)	SK@.160	LF	16.70	6.26	.10	23.06
8" (20cm)	SK@.210	LF	25.30	8.21	.13	33.64
10" (25cm)	SK@.280	LF	36.60	11.00	.17	47.77
12" (30cm)	SK@.340	LF	61.80	13.30	.21	75.31
14" (36cm)	SK@.400	LF	71.30	15.60	.25	87.15
16" (41cm)	SK@.550	LF	82.80	21.50	.34	104.64

Black steel pipe only, (A53) roll-grooved, thin wall, Schedule 10. Light weight, no hangers or fittings.

2" (50mm)	SL@.070	LF	2.54	2.55	—	5.09
2½" (65mm)	SL@.070	LF	5.38	2.55	—	7.93
3" (75mm)	SL@.090	LF	5.96	3.28	—	9.24
4" (10cm)	SL@.130	LF	8.55	4.74	—	13.29
6" (15cm)	SK@.160	LF	12.30	6.26	.10	18.66
8" (20cm)	SK@.210	LF	21.20	8.21	.13	29.54
10" (25cm)	SK@.280	LF	30.70	11.00	.17	41.87
12" (30cm)	SK@.340	LF	52.00	13.30	.21	65.51
14" (36cm)	SK@.400	LF	60.20	15.60	.25	76.05
16" (41cm)	SK@.550	LF	70.00	21.50	.34	91.84

90-degree elbow, roll-grooved, Style #10 (Victaulic)

2"	SL@.410	Ea	29.20	14.90	—	44.10
2½"	SL@.430	Ea	29.20	15.70	—	44.90
3"	SL@.500	Ea	38.60	18.20	—	56.80
4"	SL@.690	Ea	60.30	25.20	—	85.50
6"	SL@1.19	Ea	131.00	43.40	—	174.40
8"	SL@1.69	Ea	224.00	61.60	—	285.60
10"	SL@2.23	Ea	431.00	81.30	—	512.30
12"	SL@2.48	Ea	680.00	90.40	—	770.40

45-degree elbow, roll-grooved, Style #11 (Victaulic)

2"	SL@.410	Ea	29.20	14.90	—	44.10
2½"	SL@.430	Ea	29.20	15.70	—	44.90
3"	SL@.500	Ea	38.60	18.20	—	56.80
4"	SL@.690	Ea	60.30	25.20	—	85.50
6"	SL@1.19	Ea	131.00	43.40	—	174.40
8"	SL@1.69	Ea	224.00	61.60	—	285.60
10"	SL@2.23	Ea	431.00	81.30	—	512.30
12"	SL@2.48	Ea	680.00	90.40	—	770.40

Fire Protection Sprinkler Pipe and Fittings (Roll-Grooved)

Description	Craft@Hrs	Unit	Material $	Labor $	Equipment $	Total $

22½-degree elbow, roll-grooved, Style #12 (Victaulic)

Description	Craft@Hrs	Unit	Material $	Labor $	Equipment $	Total $
2"	SL@.410	Ea	34.90	14.90	—	49.80
2½"	SL@.430	Ea	34.90	15.70	—	50.60
3"	SL@.500	Ea	36.40	18.20	—	54.60
4"	SL@.690	Ea	68.70	25.20	—	93.90
6"	SL@1.19	Ea	149.00	43.40	—	192.40
8"	SL@1.69	Ea	306.00	61.60	—	367.60
10"	SL@2.23	Ea	433.00	81.30	—	514.30
12"	SL@2.48	Ea	463.00	90.40	—	553.40

Tee, roll-grooved, Style #20 (Victaulic)

Description	Craft@Hrs	Unit	Material $	Labor $	Equipment $	Total $
2"	SL@.510	Ea	56.50	18.60	—	75.10
2½"	SL@.540	Ea	56.50	19.70	—	76.20
3"	SL@.630	Ea	68.10	23.00	—	91.10
4"	SL@.860	Ea	95.00	31.30	—	126.30
6"	SL@1.49	Ea	219.00	54.30	—	273.30
8"	SL@2.11	Ea	371.00	76.90	—	447.90
10"	SL@2.79	Ea	655.00	102.00	—	757.00
12"	SL@3.10	Ea	768.00	113.00	—	881.00

Reducing tee, roll-grooved, Style #25 (Victaulic)

Description	Craft@Hrs	Unit	Material $	Labor $	Equipment $	Total $
2"	SL@.510	Ea	56.50	18.60	—	75.10
2½"	SL@.540	Ea	74.20	19.70	—	93.90
3"	SL@.630	Ea	67.10	23.00	—	90.10
4"	SL@.860	Ea	114.00	31.30	—	145.30
6"	SL@1.49	Ea	212.00	54.30	—	266.30
8"	SL@2.11	Ea	385.00	76.90	—	461.90
10"	SL@2.79	Ea	509.00	102.00	—	611.00
12"	SL@3.10	Ea	562.00	113.00	—	675.00

Reducer, roll-grooved, Style #50 (Victaulic)

Description	Craft@Hrs	Unit	Material $	Labor $	Equipment $	Total $
2"	SL@.410	Ea	18.70	14.90	—	33.60
2½"	SL@.430	Ea	18.70	15.70	—	34.40
3"	SL@.500	Ea	26.10	18.20	—	44.30
4"	SL@.690	Ea	43.80	25.20	—	69.00
6"	SL@1.19	Ea	63.60	43.40	—	107.00
8"	SL@1.69	Ea	112.00	61.60	—	173.60
10"	SL@2.23	Ea	233.00	81.30	—	314.30

Cap, roll-grooved, Style #60 (Victaulic)

Description	Craft@Hrs	Unit	Material $	Labor $	Equipment $	Total $
2"	SL@.270	Ea	16.50	9.84	—	26.34
2½"	SL@.280	Ea	16.50	10.20	—	26.70
3"	SL@.330	Ea	20.80	12.00	—	32.80
4"	SL@.450	Ea	24.00	16.40	—	40.40
6"	SL@.770	Ea	46.80	28.10	—	74.90
8"	SL@1.10	Ea	75.60	40.10	—	115.70
10"	SL@1.61	Ea	323.00	58.70	—	381.70

Description	Craft@Hrs	Unit	Material $	Labor $	Equipment $	Total $

Mechanical tee (saddle tee), roll-grooved, Style #920 (Victaulic)

Description	Craft@Hrs	Unit	Material $	Labor $	Equipment $	Total $
3"	SL@.750	Ea	57.00	27.30	—	84.30
4"	SL@1.00	Ea	64.70	36.50	—	101.20
6"	SL@1.30	Ea	76.90	47.40	—	124.30
8"	SL@1.50	Ea	121.00	54.70	—	175.70

Coupling, roll-grooved, 800 PSI, Style #77 (Victaulic). Labor included with fittings.

Description	Craft@Hrs	Unit	Material $	Labor $	Equipment $	Total $
3"	—	Ea	26.60	—	—	26.60
4"	—	Ea	44.10	—	—	44.10
6"	—	Ea	72.30	—	—	72.30
8"	—	Ea	98.60	—	—	98.60
10"	—	Ea	133.00	—	—	133.00
12"	—	Ea	177.00	—	—	177.00

Coupling, roll-grooved, light weight, Style #75 (Victaulic). Labor included with fittings.

Description	Craft@Hrs	Unit	Material $	Labor $	Equipment $	Total $
3"	—	Ea	21.50	—	—	21.50
4"	—	Ea	23.20	—	—	23.20
6"	—	Ea	41.40	—	—	41.40
8"	—	Ea	63.50	—	—	63.50

Coupling, roll-grooved, zeroflex, Style #07 (Victaulic). Labor included with fittings.

Description	Craft@Hrs	Unit	Material $	Labor $	Equipment $	Total $
3"	—	Ea	24.80	—	—	24.80
4"	—	Ea	32.60	—	—	32.60
6"	—	Ea	47.90	—	—	47.90
8"	—	Ea	72.90	—	—	72.90
10"	—	Ea	107.00	—	—	107.00
12"	—	Ea	148.00	—	—	148.00

Flange, roll-grooved, Style #741 (Victaulic)

Description	Craft@Hrs	Unit	Material $	Labor $	Equipment $	Total $
3"	SL@.500	Ea	90.80	18.20	—	109.00
4"	SL@.690	Ea	123.00	25.20	—	148.20
6"	SL@1.19	Ea	150.00	43.40	—	193.40
8"	SL@1.69	Ea	174.00	61.60	—	235.60
10"	SL@2.23	Ea	259.00	81.30	—	340.30
12"	SL@2.48	Ea	353.00	90.40	—	443.40

Fire Protection Branch Pipe & Fittings

Description	Craft@Hrs	Unit	Material $	Labor $	Equipment $	Total $

Schedule 40 carbon steel pipe, threaded and coupled, pipe only. Add for hangers, fittings etc.

Description	Craft@Hrs	Unit	Material $	Labor $	Equipment $	Total $
1"	P1@.090	LF	2.68	3.30	—	5.98
1¼"	P1@.100	LF	3.51	3.67	—	7.18
1½"	P1@.110	LF	4.17	4.04	—	8.21
2"	P1@.120	LF	5.63	4.40	—	10.03

125# cast iron 90-degree ell, threaded joints

Description	Craft@Hrs	Unit	Material $	Labor $	Equipment $	Total $
1" x ¾"	P1@.130	Ea	4.75	4.77	—	9.52
1"	P1@.180	Ea	4.41	6.61	—	11.02
1¼"	P1@.240	Ea	6.23	8.81	—	15.04
1½"	P1@.300	Ea	8.68	11.00	—	19.68
2"	P1@.380	Ea	13.20	13.90	—	27.10

125# cast iron 45-degree ell, threaded joints

Description	Craft@Hrs	Unit	Material $	Labor $	Equipment $	Total $
1"	P1@.180	Ea	5.78	6.61	—	12.39
1¼"	P1@.240	Ea	8.23	8.81	—	17.04
1½"	P1@.300	Ea	10.60	11.00	—	21.60
2"	P1@.380	Ea	14.30	13.90	—	28.20

125# cast iron tee, threaded joints

Description	Craft@Hrs	Unit	Material $	Labor $	Equipment $	Total $
1"	P1@.230	Ea	6.07	8.44	—	14.51
1¼"	P1@.310	Ea	11.00	11.40	—	22.40
1½"	P1@.390	Ea	14.10	14.30	—	28.40
2"	P1@.490	Ea	19.30	18.00	—	37.30

125# cast iron reducing tee, threaded joints

Description	Craft@Hrs	Unit	Material $	Labor $	Equipment $	Total $
1" x ¾"	P1@.220	Ea	7.74	8.07	—	15.81
1¼" x ¾"	P1@.290	Ea	9.87	10.60	—	20.47
1¼" x 1"	P1@.290	Ea	10.80	10.60	—	21.40
1½" x ¾"	P1@.370	Ea	13.70	13.60	—	27.30
1½" x 1"	P1@.370	Ea	15.00	13.60	—	28.60
1½" x 1¼"	P1@.370	Ea	16.50	13.60	—	30.10
2" x 1"	P1@.460	Ea	21.50	16.90	—	38.40
2" x 1¼"	P1@.460	Ea	23.00	16.90	—	39.90
2" x 1½"	P1@.460	Ea	24.80	16.90	—	41.70

125# cast iron reducer, threaded joints

Description	Craft@Hrs	Unit	Material $	Labor $	Equipment $	Total $
1" x ¾"	P1@.200	Ea	6.07	7.34	—	13.41
1¼" x 1"	P1@.210	Ea	7.30	7.71	—	15.01
1½" x 1"	P1@.270	Ea	8.96	9.91	—	18.87
1½" x 1¼"	P1@.270	Ea	9.47	9.91	—	19.38
2" x 1¼"	P1@.340	Ea	14.20	12.50	—	26.70
2" x 1½"	P1@.340	Ea	14.90	12.50	—	27.40

Description	Craft@Hrs	Unit	Material $	Labor $	Equipment $	Total $

125# cast iron cross, threaded joints

Description	Craft@Hrs	Unit	Material $	Labor $	Equipment $	Total $
1"	P1@.360	Ea	14.20	13.20	—	27.40
1¼"	P1@.480	Ea	23.30	17.60	—	40.90
1½"	P1@.600	Ea	28.40	22.00	—	50.40
2"	P1@.700	Ea	46.80	25.70	—	72.50

125# cast iron cap, threaded joint

Description	Craft@Hrs	Unit	Material $	Labor $	Equipment $	Total $
¾"	P1@.100	Ea	2.98	3.67	—	6.65
1"	P1@.140	Ea	3.61	5.14	—	8.75
1¼"	P1@.180	Ea	4.61	6.61	—	11.22
1½"	P1@.230	Ea	5.85	8.44	—	14.29
2"	P1@.290	Ea	9.20	10.60	—	19.80

125# cast iron plug, threaded joint

Description	Craft@Hrs	Unit	Material $	Labor $	Equipment $	Total $
¾"	P1@.100	Ea	1.94	3.67	—	5.61
1"	P1@.140	Ea	2.57	5.14	—	7.71
1¼"	P1@.180	Ea	4.00	6.61	—	10.61
1½"	P1@.230	Ea	5.08	8.44	—	13.52
2"	P1@.290	Ea	6.29	10.60	—	16.89

125# cast iron coupling, threaded joints

Description	Craft@Hrs	Unit	Material $	Labor $	Equipment $	Total $
1"	P1@.180	Ea	5.26	6.61	—	11.87
1¼"	P1@.240	Ea	6.49	8.81	—	15.30
1½"	P1@.300	Ea	8.16	11.00	—	19.16
2"	P1@.380	Ea	11.00	13.90	—	24.90

Fire Protection Sprinkler Branch Fittings

Description	Craft@Hrs	Unit	Material $	Labor $	Equipment $	Total $

Steel pipe nipples — 3" long, standard right-hand thread

Description	Craft@Hrs	Unit	Material $	Labor $	Equipment $	Total $
1"	P1@.025	Ea	2.04	.92	—	2.96
1¼"	P1@.025	Ea	2.61	.92	—	3.53
1½"	P1@.025	Ea	3.11	.92	—	4.03
2"	P1@.030	Ea	3.87	1.10	—	4.97

Steel pipe nipples — 4" long, standard right-hand thread

Description	Craft@Hrs	Unit	Material $	Labor $	Equipment $	Total $
1"	P1@.025	Ea	2.43	.92	—	3.35
1¼"	P1@.030	Ea	3.03	1.10	—	4.13
1½"	P1@.030	Ea	3.74	1.10	—	4.84
2"	P1@.030	Ea	4.82	1.10	—	5.92

Steel pipe nipples — 6" long, standard right-hand thread

Description	Craft@Hrs	Unit	Material $	Labor $	Equipment $	Total $
1"	P1@.030	Ea	3.11	1.10	—	4.21
1¼"	P1@.030	Ea	3.90	1.10	—	5.00
1½"	P1@.030	Ea	4.75	1.10	—	5.85
2"	P1@.035	Ea	6.38	1.28	—	7.66

Description	Craft@Hrs	Unit	Material $	Labor $	Equipment $	Total $

Chlorinated Polyvinyl Chloride pipe (CPVC), CPVC pipe only, solvent weld. No hangers or fittings.

Description	Craft@Hrs	Unit	Material $	Labor $	Equipment $	Total $
¾"	SL@.055	LF	1.10	2.00	—	3.10
1"	SL@.055	LF	1.60	2.00	—	3.60
1¼"	SL@.065	LF	2.51	2.37	—	4.88
1½"	SL@.075	LF	3.53	2.73	—	6.26
2"	SL@.085	LF	5.21	3.10	—	8.31
2½"	SL@.110	LF	8.08	4.01	—	12.09
3"	SL@.125	LF	12.30	4.56	—	16.86

CPVC 90-degree elbow, solvent weld

Description	Craft@Hrs	Unit	Material $	Labor $	Equipment $	Total $
¾"	SL@.100	Ea	3.18	3.65	—	6.83
1"	SL@.120	Ea	4.79	4.37	—	9.16
1¼"	SL@.170	Ea	6.13	6.20	—	12.33
1½"	SL@.180	Ea	9.24	6.56	—	15.80
2"	SL@.200	Ea	10.70	7.29	—	17.99

CPVC 45-degree elbow, solvent weld

Description	Craft@Hrs	Unit	Material $	Labor $	Equipment $	Total $
¾"	SL@.100	Ea	3.29	3.65	—	6.94
1"	SL@.120	Ea	4.30	4.37	—	8.67
1¼"	SL@.170	Ea	5.65	6.20	—	11.85
1½"	SL@.180	Ea	7.60	6.56	—	14.16
2"	SL@.200	Ea	9.48	7.29	—	16.77

CPVC tee, solvent weld

Description	Craft@Hrs	Unit	Material $	Labor $	Equipment $	Total $
¾"	SL@.150	Ea	3.51	5.47	—	8.98
1"	SL@.170	Ea	5.98	6.20	—	12.18
1¼"	SL@.230	Ea	9.42	8.38	—	17.80
1½"	SL@.250	Ea	12.40	9.11	—	21.51
2"	SL@.280	Ea	19.60	10.20	—	29.80

CPVC reducing tee, solvent weld

Description	Craft@Hrs	Unit	Material $	Labor $	Equipment $	Total $
1"	SL@.170	Ea	5.15	6.20	—	11.35
1¼"	SL@.230	Ea	9.35	8.38	—	17.73
1½"	SL@.250	Ea	12.50	9.11	—	21.61
2"	SL@.280	Ea	20.00	10.20	—	30.20
3"	SL@.350	Ea	31.50	12.80	—	44.30

CPVC coupling, solvent weld

Description	Craft@Hrs	Unit	Material $	Labor $	Equipment $	Total $
¾"	SL@.100	Ea	2.31	3.65	—	5.96
1"	SL@.120	Ea	3.13	4.37	—	7.50
1¼"	SL@.170	Ea	4.95	6.20	—	11.15
1½"	SL@.180	Ea	6.24	6.56	—	12.80
2"	SL@.200	Ea	7.86	7.29	—	15.15
3"	SL@.280	Ea	16.10	10.20	—	26.30

Fire Protection Sprinkler Fittings (CPVC)

Description	Craft@Hrs	Unit	Material $	Labor $	Equipment $	Total $
CPVC head adapter, solvent weld						
¾" x ½"	SL@.150	Ea	5.24	5.47	—	10.71
1" x ½"	SL@.165	Ea	5.60	6.01	—	11.61
CPVC cap, solvent weld						
¾"	SL@.060	Ea	1.72	2.19	—	3.91
1"	SL@.065	Ea	2.40	2.37	—	4.77
1¼"	SL@.075	Ea	4.11	2.73	—	6.84
1½"	SL@.085	Ea	5.27	3.10	—	8.37
2"	SL@.095	Ea	6.14	3.46	—	9.60
CPVC flange, solvent weld						
1"	SL@.165	Ea	6.14	6.01	—	12.15
1¼"	SL@.175	Ea	7.90	6.38	—	14.28
1½"	SL@.185	Ea	12.30	6.74	—	19.04
2"	SL@.200	Ea	18.50	7.29	—	25.79
3"	SL@.225	Ea	30.80	8.20	—	39.00
CPVC female adapter, solvent weld						
1"	SL@.065	Ea	15.20	2.37	—	17.57
1¼"	SL@.075	Ea	25.30	2.73	—	28.03
1½"	SL@.085	Ea	32.40	3.10	—	35.50
2"	SL@.095	Ea	41.90	3.46	—	45.36

Boilers for industrial, commercial and institutional applications. ASME "H" or "U" stamped, LFUE (Levelized Fuel Utilization Efficiency) certified 90% or better. U.S. Green Building council certified. Cleaver-Brooks or equal. Shipped factory-assembled and skid-mounted. Includes the boiler shell and structure, I-beam skids, combustion chamber, fire tubes, crown sheet, tube sheets (for fire tube), water tubes, mud drum(s), steam drum (for water tube boiler), finned water tubes and check valves (for hot water generators), refractory brick lining, explosion doors, combustion air inlets and stack outlet. Add the cost of the fuel train piping, electrical wiring, refractory, stack, feedwater system, combustion controls, water treatment, expansion tanks (if hot water) or condensate return (if steam) and system start-up. See the sections following packaged boilers. Costs are based on boiler horsepower rating (BHP). Add the cost of commissioning, insurance inspection, permit, permit inspection, hydraulic test, calibration and lockout sealing of emissions monitor, and final run test. Equipment cost assumes a 17-ton hydraulic truck-mounted crane (at $132 per hour). LEED (Leadership in Energy & Environmental Design) certification requires 88% boiler efficiency, recording controls, compliance with SCAQMD 1146.2 emission standards and zone recording thermostats. A LEED-certified boiler with a U.S. Green Building Council rating will add 4% to 15% to the cost of the boiler, as identified in the cost tables that follow.

Description	Craft@Hrs	Unit	Material $	Labor $	Equipment $	Total $

Hot water or steam cast iron gas boiler, 30 PSIG.

Costs shown are based on boiler horsepower rating (BHP). On-site assembly. Includes allowances for associated appurtenances and pipe connections.

Description	Craft@Hrs	Unit	Material $	Labor $	Equipment $	Total $
4.5 BHP	SN@8.50	Ea	5,130.00	329.00	160.00	5,619.00
5.9 BHP	SN@9.50	Ea	6,400.00	367.00	179.00	6,946.00
9.0 BHP	SN@10.0	Ea	9,540.00	387.00	189.00	10,116.00
15.0 BHP	SN@12.0	Ea	14,900.00	464.00	227.00	15,591.00
20.0 BHP	SN@14.0	Ea	19,700.00	541.00	264.00	20,505.00
30.0 BHP	SN@16.0	Ea	25,400.00	618.00	302.00	26,320.00
45.0 BHP	SN@18.0	Ea	36,800.00	696.00	340.00	37,836.00
60.0 BHP	SN@20.0	Ea	47,800.00	773.00	378.00	48,951.00
Add for LEED-certified boiler	—	%	6.0	—	—	—
Add for LEED inspection	—	Ea	—	—	—	2,060.00
Add for central monitor/ recorder	—	Ea	1,740.00	—	—	1,740.00
Add for recording thermostat	—	Ea	406.00	—	—	406.00

Commercial Boilers

Description	Craft@Hrs	Unit	Material $	Labor $	Equipment $	Total $

Hot water Scotch marine firetube gas boiler, 30 PSIG. Costs shown are based on boiler horsepower rating (BHP). Set in place only. Make additional allowances for associated appurtenances and pipe connections.

Description	Craft@Hrs	Unit	Material $	Labor $	Equipment $	Total $
30 BHP	SN@10.0	Ea	36,600.00	387.00	189.00	37,176.00
40 BHP	SN@10.0	Ea	38,800.00	387.00	189.00	39,376.00
50 BHP	SN@10.0	Ea	40,400.00	387.00	189.00	40,976.00
60 BHP	SN@12.0	Ea	44,900.00	464.00	227.00	45,591.00
70 BHP	SN@12.0	Ea	48,500.00	464.00	227.00	49,191.00
80 BHP	SN@14.0	Ea	51,600.00	541.00	264.00	52,405.00
100 BHP	SN@14.0	Ea	61,500.00	541.00	264.00	62,305.00
125 BHP	SN@16.0	Ea	66,000.00	618.00	302.00	66,920.00
150 BHP	SN@18.0	Ea	70,800.00	696.00	340.00	71,836.00
200 BHP	SN@18.0	Ea	83,500.00	696.00	340.00	84,536.00
250 BHP	SN@26.0	Ea	91,900.00	1,000.00	491.00	93,391.00
300 BHP	SN@26.0	Ea	98,700.00	1,000.00	491.00	100,191.00
350 BHP	SN@26.0	Ea	113,000.00	1,000.00	490.00	114,490.00
400 BHP	SN@32.0	Ea	116,000.00	1,240.00	604.00	117,844.00
500 BHP	SN@32.0	Ea	137,000.00	1,240.00	604.00	138,844.00
600 BHP	SN@32.0	Ea	155,000.00	1,240.00	604.00	156,844.00
750 BHP	SN@40.0	Ea	161,000.00	1,550.00	755.00	163,305.00
Add for LEED-certified boiler	—	%	4.0	—	—	—
Add for LEED inspection	—	Ea	—	—	—	2,060.00
Add for central monitor/ recorder	—	Ea	4,060.00	—	—	4,060.00
Add for recording thermostat	—	Ea	406.00	—	—	406.00

Description	Craft@Hrs	Unit	Material $	Labor $	Equipment $	Total $

Packaged Scotch marine firetube oil or gas fired boiler, 150 PSIG.
Costs shown are based on boiler horsepower rating (BHP). Set in place only.
Make additional allowances for associated appurtenances and pipe connections.

Description	Craft@Hrs	Unit	Material $	Labor $	Equipment $	Total $
60 BHP	SN@12.0	Ea	51,600.00	464.00	227.00	52,291.00
70 BHP	SN@12.0	Ea	55,500.00	464.00	227.00	56,191.00
80 BHP	SN@14.0	Ea	59,500.00	541.00	264.00	60,305.00
100 BHP	SN@14.0	Ea	71,300.00	541.00	264.00	72,105.00
125 BHP	SN@16.0	Ea	75,600.00	618.00	302.00	76,520.00
150 BHP	SN@18.0	Ea	80,800.00	696.00	340.00	81,836.00
200 BHP	SN@18.0	Ea	107,000.00	696.00	340.00	108,036.00
250 BHP	SN@26.0	Ea	114,000.00	1,000.00	491.00	115,491.00
300 BHP	SN@26.0	Ea	128,000.00	1,000.00	491.00	129,491.00
350 BHP	SN@26.0	Ea	132,000.00	1,000.00	491.00	133,491.00
400 BHP	SN@32.0	Ea	158,000.00	1,240.00	604.00	159,844.00
500 BHP	SN@32.0	Ea	175,000.00	1,240.00	604.00	176,844.00
600 BHP	SN@32.0	Ea	180,000.00	1,240.00	604.00	181,844.00
700 BHP	SN@40.0	Ea	186,000.00	1,550.00	755.00	188,305.00
Add for LEED-certified boiler	—	%	4.0	—	—	—
Add for LEED inspection	—	Ea	—	—	—	2,060.00
Add for central monitor/ recorder	—	Ea	4,060.00	—	—	4,060.00
Add for recording thermostat	—	Ea	406.00	—	—	406.00

Packaged hot water firebox oil or gas fired boiler.
Costs shown are based on boiler horsepower rating (BHP). Set in place only. Make additional allowances for associated appurtenances and pipe connections.

Description	Craft@Hrs	Unit	Material $	Labor $	Equipment $	Total $
13 BHP	SN@8.00	Ea	19,400.00	309.00	151.00	19,860.00
16 BHP	SN@8.00	Ea	19,700.00	309.00	151.00	20,160.00
19 BHP	SN@8.00	Ea	20,300.00	309.00	151.00	20,760.00
22 BHP	SN@8.00	Ea	20,900.00	309.00	151.00	21,360.00
28 BHP	SN@8.00	Ea	21,900.00	309.00	151.00	22,360.00
34 BHP	SN@8.00	Ea	25,100.00	309.00	151.00	25,560.00
40 BHP	SN@8.00	Ea	27,700.00	309.00	151.00	28,160.00
46 BHP	SN@8.00	Ea	31,600.00	309.00	151.00	32,060.00
52 BHP	SN@8.00	Ea	36,900.00	309.00	151.00	37,360.00
61 BHP	SN@8.00	Ea	40,700.00	309.00	151.00	41,160.00
70 BHP	SN@8.00	Ea	43,500.00	309.00	151.00	43,960.00
80 BHP	SN@8.00	Ea	44,900.00	309.00	151.00	45,360.00
100 BHP	SN@8.00	Ea	50,700.00	309.00	151.00	51,160.00
125 BHP	SN@16.0	Ea	60,300.00	618.00	302.00	61,220.00
150 BHP	SN@16.0	Ea	64,700.00	618.00	302.00	65,620.00
Add for LEED-certified boiler	—	%	4.0	—	—	—
Add for LEED inspection	—	Ea	—	—	—	2,060.00
Add for central monitor/ recorder	—	Ea	2,900.00	—	—	2,900.00
Add for recording thermostat	—	Ea	406.00	—	—	406.00

Commercial Boilers

Description	Craft@Hrs	Unit	Material $	Labor $	Equipment $	Total $

Packaged watertube industrial high pressure boilers. Costs shown are based on boiler horsepower rating (BHP). Set in place only. Make additional allowances for associated appurtenances and pipe connections.

Description	Craft@Hrs	Unit	Material $	Labor $	Equipment $	Total $
480 BHP	SN@32.0	Ea	366,000.00	1,240.00	604.00	367,844.00
526 BHP	SN@32.0	Ea	374,000.00	1,240.00	604.00	375,844.00
572 BHP	SN@32.0	Ea	396,000.00	1,240.00	604.00	397,844.00
618 BHP	SN@32.0	Ea	403,000.00	1,240.00	604.00	404,844.00
664 BHP	SN@32.0	Ea	421,000.00	1,240.00	604.00	422,844.00
710 BHP	SN@64.0	Ea	427,000.00	2,470.00	1,210.00	430,680.00
747 BHP	SN@64.0	Ea	433,000.00	2,470.00	1,210.00	436,680.00
792 BHP	SN@64.0	Ea	447,000.00	2,470.00	1,210.00	450,680.00
838 BHP	SN@64.0	Ea	461,000.00	2,470.00	1,210.00	464,680.00
884 BHP	SN@64.0	Ea	474,000.00	2,470.00	1,210.00	477,680.00
930 BHP	SN@64.0	Ea	480,000.00	2,470.00	1,210.00	483,680.00
1,000 BHP	SN@64.0	Ea	489,000.00	2,470.00	1,210.00	492,680.00
Add for LEED-certified boiler	—	%	4.0	—	—	—
Add for LEED inspection	—	Ea	—	—	—	2,060.00
Add for central monitor/ recorder	—	Ea	4,990.00	—	—	4,990.00
Add for recording thermostat	—	Ea	406.00	—	—	406.00

Gas fired hot water steel boilers, 12 PSIG Mid-efficiency (82%) atmospheric draft, standing pilot light, c/w primary loop circulating pump. Labor includes setting in place and connecting pump and controls. Make additional allowances for distribution piping connections, safety devices, flue, gas connections, wiring, etc.

Description	Craft@Hrs	Unit	Material $	Labor $	Equipment $	Total $
50,000 Btu	P1@4.00	Ea	1,390.00	147.00	75.50	1,612.50
70,000 Btu	P1@4.25	Ea	1,590.00	156.00	79.90	1,825.90
105,000 Btu	P1@4.50	Ea	1,750.00	165.00	85.00	2,000.00
140,000 Btu	P1@4.75	Ea	2,010.00	174.00	89.70	2,273.70
175,000 Btu	P1@5.00	Ea	2,240.00	184.00	94.40	2,518.40
210,000 Btu	P1@5.50	Ea	2,840.00	202.00	104.00	3,146.00
245,000 Btu	P1@7.00	Ea	3,160.00	257.00	132.00	3,549.00
Electronic spark ignition Add	—	Ea	302.00	—	—	302.00
Add for LEED-certified boiler	—	%	15.0	—	—	—
Add for LEED inspection	—	Ea	—	—	—	2,060.00
Add for central monitor/ recorder	—	Ea	1,270.00	—	—	1,270.00
Add for recording thermostat	—	Ea	174.00	—	—	174.00

Description	Craft@Hrs	Unit	Material $	Labor $	Equipment $	Total $

Gas fired hot water cast iron boilers, 12 PSIG Mid-efficiency (82%) direct vent, forced draft electronic spark ignition, c/w primary loop circulating pump. Labor includes setting in place and connecting pump and controls. Make additional allowances for distribution piping connections, safety devices, flue, gas connections, wiring, etc.

Description	Craft@Hrs	Unit	Material $	Labor $	Equipment $	Total $
50,000 Btu	P1@4.00	Ea	1,740.00	147.00	75.50	1,962.50
65,000 Btu	P1@4.25	Ea	1,950.00	156.00	80.20	2,186.20
100,000 Btu	P1@4.50	Ea	2,130.00	165.00	85.00	2,380.00
130,000 Btu	P1@4.75	Ea	2,360.00	174.00	89.70	2,623.70
170,000 Btu	P1@5.00	Ea	2,730.00	184.00	94.40	3,008.40
200,000 Btu	P1@5.50	Ea	3,000.00	202.00	104.00	3,306.00
235,000 Btu	P1@7.00	Ea	3,310.00	257.00	132.00	3,699.00
Add for LEED-certified boiler	—	%	15.0	—	—	—
Add for LEED inspection	—	Ea	—	—	—	2,060.00
Add for central monitor/ recorder	—	Ea	1,270.00	—	—	1,270.00
Add for recording thermostat	—	Ea	174.00	—	—	174.00

Description	Craft@Hrs	Unit	Material $	Labor $	Total $

Steam heating boilers for residential and light commercial applications. Parker or equal 15 PSIG Spiral watertube type, with ASTM SA-53 grade tubing, full American Society of Mechanical Engineers "M", "H" and "S" ratings and Underwriters Laboratories certifications. Specifically designed for "twinning" and multiple unit additions and retrofits. Packaged type units with ASME rated pressure relief valve, low water cutoff valve, main and check valves, pressure gauge, and pre-wired Fireye/McDonnell Miller combustion and steam regulator controls. Also includes pre-assembled FM/IRI approved natural gas combustion train w/Fisher or equal diaphragm regulator valve, check trim and vent line fittings or pre-assembled FM/IRI fuel oil combustion train with oil preheater, valved recirculation loop, filter check trim and fuel oil pump. Does not include sheet metal stack or vents, installation permits and approvals, or inspection costs. (BHP = boiler horse power)

Description	Craft@Hrs	Unit	Material $	Labor $	Total $
37,500 Btu (1 BHP)	P1@3.00	Ea	1,900.00	110.00	2,010.00
50,000 Btu (1.5 BHP)	P1@3.00	Ea	2,310.00	110.00	2,420.00
100,000 Btu (2.0 BHP)	P1@4.00	Ea	2,830.00	147.00	2,977.00
150,000 Btu (2.5 BHP)	P1@4.25	Ea	3,570.00	156.00	3,726.00
200,000 Btu (3.0 BHP)	P1@5.00	Ea	3,780.00	184.00	3,964.00
Add for LEED-certified boiler	—	%	15.0	—	—
Add for LEED inspection	—	Ea	—	—	2,060.00
Add for central monitor/ recorder	—	Ea	1,330.00	—	1,330.00
Add for recording thermostat	—	Ea	174.00	—	174.00

Commercial Boiler Connections

Description	Craft@Hrs	Unit	Material $	Labor $	Equipment $	Total $

Commercial hot water boiler connection. Includes pipe, fittings, water feed and pressure relief (drains), pipe insulation, valves, gauges, vents and thermometers to connect boiler to distribution system.

Description	Craft@Hrs	Unit	Material $	Labor $	Equipment $	Total $
2½" supply	SN@18.0	Ea	2,240.00	696.00	11.10	2,947.10
3" supply	SN@20.5	Ea	2,810.00	792.00	12.70	3,614.70
4" supply	SN@26.0	Ea	4,060.00	1,000.00	16.10	5,076.10
6" supply	SN@32.0	Ea	5,960.00	1,240.00	19.80	7,219.80
8" supply	SN@39.5	Ea	7,950.00	1,530.00	24.40	9,504.40
10" supply	SN@48.0	Ea	11,300.00	1,860.00	29.70	13,189.70
12" supply	SN@54.0	Ea	14,100.00	2,090.00	33.40	16,223.40

Commercial steam boiler connection. Includes pipe, fittings, water feed and pressure relief (drains), pipe insulation, valves, gauges, vents and thermometers to connect boiler to distribution system.

Description	Craft@Hrs	Unit	Material $	Labor $	Equipment $	Total $
2½" supply	SN@34.0	Ea	4,290.00	1,310.00	21.00	5,621.00
3" supply	SN@38.0	Ea	4,510.00	1,470.00	23.50	6,003.50
4" supply	SN@43.0	Ea	5,250.00	1,660.00	26.60	6,936.60
6" supply	SN@53.0	Ea	7,200.00	2,050.00	32.80	9,282.80
8" supply	SN@62.0	Ea	9,270.00	2,400.00	38.40	11,708.40
10" supply	SN@70.0	Ea	12,700.00	2,710.00	43.30	15,453.30
12" supply	SN@82.0	Ea	15,800.00	3,170.00	50.70	19,020.70

Additional costs for industrial, commercial and institutional packaged boilers. See more detailed installation costs and boiler trim costs in the sections that follow.

Description	Craft@Hrs	Unit	Material $	Labor $	Equipment $	Total $
Form and pour an interior slab	CF@.156	SF	9.36	5.51	—	14.87
Place vibration pad and anchors	CF@.100	SF	5.69	3.53	—	9.22
Boltdown of boiler	P1@3.00	Ea	—	110.00	—	110.00
Fuel train and feedwater lines	P1@.200	LF	11.40	7.34	—	18.74
Mount interior boiler drain	P1@.500	LF	8.84	18.40	—	27.24
Bore pipe hole in concrete wall	P1@.250	Ea	—	9.18	—	9.18
Bore stack vent in concrete wall	P1@.250	Ea	—	9.18	—	9.18
Mount and edge-seal stack	SN@.100	LF	9.56	3.87	—	13.43
Mount circulating pump	P1@.450	Ea	11.40	16.50	—	27.90
Expansion tank or condensate return lines	P1@.250	LF	3.66	9.18	—	12.84
Electrical wiring	BE@.150	LF	2.17	6.06	—	8.23
Combustion controls	BE@.200	LF	4.00	8.08	—	12.08
Install flue gas monitoring system	BE@.200	LF	3.99	8.08	—	12.07

Description	Craft@Hrs	Unit	Material $	Labor $	Equipment $	Total $

Additional costs for steam heating boilers for residential and light commercial applications. See more detailed installation costs and boiler trim costs in the sections that follow.

Description	Craft@Hrs	Unit	Material $	Labor $	Equipment $	Total $
Form and pour a 4' x 4' interior slab	CF@2.50	Ea	148.00	88.40	—	236.40
Place 4' x 4' x 1/2" vibration pads	CF@.750	Ea	43.40	26.50	—	69.90
Boltdown of boiler	P1@1.00	Ea	—	36.70	—	36.70
Fuel train and feedwater lines	P1@2.50	Ea	—	91.80	—	91.80
Mount interior boiler drain	P1@.500	LF	8.73	18.40	—	27.13
Bore piping hole in basement wall	P1@.250	Ea	—	9.18	—	9.18
Bore stack vent in concrete wall	P1@.250	Ea	—	9.18	—	9.18
Mount and edge-seal stack	P1@.500	Ea	45.30	18.40	—	63.70
Mount circulating pump	P1@.450	Ea	—	16.50	—	16.50
Install expansion tank						
2.1 gallon	P1@.300	Ea	39.80	11.00	—	50.80
4.5 gallon	P1@.300	Ea	66.80	11.00	—	77.80

Commercial Boiler Components and Accessories

Costs for Combustion Trains for Boilers. Factory Mutual, Industrial Risk Insurers, and Underwriters' Laboratories certified, U.S. Green Building Council rated, LFUE (Levelized Fuel Utilization Efficiency) rated at 90% or better. Includes the burner, burner head, burner mount flange, associated refractory and gasketing, main fuel valve, draft damper and fan (for a power burner), or burner face plate (for an atmospheric or induced-draft burner). Note: BHP multiplied by 33.5 is the approximate required 1,000 Btu per hour input (MBH input). Cost of fuel line relief vent piping and trip valve to outside of the building are not required in all jurisdictions and are not assumed in these costs.

Description	Craft@Hrs	Unit	Material $	Labor $	Equipment $	Total $

Combustion train for Scotch marine firetube boilers. Costs shown are based on boiler horsepower rating (BHP).

Description	Craft@Hrs	Unit	Material $	Labor $	Equipment $	Total $
20 BHP	SN@8.00	Ea	3,000.00	309.00	4.95	3,313.95
30 BHP	SN@8.00	Ea	4,050.00	309.00	4.95	4,363.95
40 BHP	SN@8.00	Ea	5,560.00	309.00	4.95	5,873.95
50 BHP	SN@8.00	Ea	6,880.00	309.00	4.95	7,193.95
60 BHP	SN@8.00	Ea	7,660.00	309.00	4.95	7,973.95
70 BHP	SN@8.00	Ea	8,380.00	309.00	4.95	8,693.95
80 BHP	SN@8.00	Ea	8,540.00	309.00	4.95	8,853.95
100 BHP	SN@8.00	Ea	9,330.00	309.00	4.95	9,643.95
125 BHP	SN@8.00	Ea	9,680.00	309.00	4.95	9,993.95
150 BHP	SN@8.00	Ea	10,700.00	309.00	4.95	11,013.95
200 BHP	SN@8.00	Ea	11,100.00	309.00	4.95	11,413.95
250 BHP	SN@8.00	Ea	12,400.00	309.00	4.95	12,713.95
300 BHP	SN@8.00	Ea	13,900.00	309.00	151.00	14,360.00
350 BHP	SN@8.00	Ea	14,100.00	309.00	151.00	14,560.00
400 BHP	SN@8.00	Ea	14,400.00	309.00	151.00	14,860.00
500 BHP	SN@16.0	Ea	15,600.00	618.00	302.00	16,520.00
600 BHP	SN@16.0	Ea	17,000.00	618.00	302.00	17,920.00
750 BHP	SN@32.0	Ea	18,000.00	1,240.00	604.00	19,844.00
800 BHP	SN@32.0	Ea	18,700.00	1,240.00	604.00	20,544.00

Combustion train for package firebox boilers. Costs shown are based on boiler horsepower rating (BHP).

Description	Craft@Hrs	Unit	Material $	Labor $	Equipment $	Total $
13.4 BHP	SN@8.00	Ea	1,530.00	309.00	151.00	1,990.00
16.4 BHP	SN@8.00	Ea	2,050.00	309.00	151.00	2,510.00
17.4 BHP	SN@8.00	Ea	2,420.00	309.00	151.00	2,880.00
22.4 BHP	SN@8.00	Ea	2,920.00	309.00	151.00	3,380.00
28.4 BHP	SN@8.00	Ea	3,080.00	309.00	151.00	3,540.00
34.4 BHP	SN@8.00	Ea	3,400.00	309.00	151.00	3,860.00
40.3 BHP	SN@8.00	Ea	3,700.00	309.00	151.00	4,160.00
46.3 BHP	SN@8.00	Ea	4,070.00	309.00	151.00	4,530.00
52.3 BHP	SN@8.00	Ea	4,540.00	309.00	151.00	5,000.00
61.1 BHP	SN@8.00	Ea	4,940.00	309.00	151.00	5,400.00
70.0 BHP	SN@8.00	Ea	5,040.00	309.00	151.00	5,500.00
80.0 BHP	SN@8.00	Ea	5,250.00	309.00	151.00	5,710.00
100.0 BHP	SN@8.00	Ea	5,560.00	309.00	151.00	6,020.00
125.0 BHP	SN@8.00	Ea	5,940.00	309.00	151.00	6,400.00
150.0 BHP	SN@8.00	Ea	6,440.00	309.00	151.00	6,900.00

Description	Craft@Hrs	Unit	Material $	Labor $	Equipment $	Total $

Natural gas fuel train piping for firetube and watertube boilers. Meets requirements of Underwriters' Laboratories. Shipped pre-assembled. Add the cost of venting and piping to a gas meter. Costs shown are based on boiler horsepower rating (BHP).

Description	Craft@Hrs	Unit	Material $	Labor $	Equipment $	Total $
20 BHP	SN@8.00	Ea	782.00	309.00	151.00	1,242.00
30 BHP	SN@8.00	Ea	1,070.00	309.00	151.00	1,530.00
40 BHP	SN@8.00	Ea	1,640.00	309.00	151.00	2,100.00
50 BHP	SN@8.00	Ea	1,940.00	309.00	151.00	2,400.00
60 BHP	SN@8.00	Ea	2,130.00	309.00	151.00	2,590.00
70 BHP	SN@8.00	Ea	2,260.00	309.00	151.00	2,720.00
80 BHP	SN@8.00	Ea	2,450.00	309.00	151.00	2,910.00
100 BHP	SN@8.00	Ea	2,840.00	309.00	151.00	3,300.00
125 BHP	SN@8.00	Ea	3,380.00	309.00	151.00	3,840.00
150 BHP	SN@8.00	Ea	4,010.00	309.00	151.00	4,470.00
200 BHP	SN@8.00	Ea	4,810.00	309.00	151.00	5,270.00
250 BHP	SN@8.00	Ea	5,940.00	309.00	151.00	6,400.00
300 BHP	SN@8.00	Ea	6,260.00	309.00	151.00	6,720.00
350 BHP	SN@8.00	Ea	6,710.00	309.00	151.00	7,170.00
400 BHP	SN@8.00	Ea	7,210.00	309.00	151.00	7,670.00
500 BHP	SN@16.0	Ea	8,100.00	618.00	302.00	9,020.00
600 BHP	SN@16.0	Ea	9,330.00	618.00	302.00	10,250.00
750 BHP	SN@16.0	Ea	10,100.00	618.00	302.00	11,020.00
800 BHP	SN@16.0	Ea	10,500.00	618.00	302.00	11,420.00

Natural gas fuel train piping for package firebox boilers.
Costs shown are based on boiler horsepower rating (BHP).

Description	Craft@Hrs	Unit	Material $	Labor $	Equipment $	Total $
13 BHP	SN@8.00	Ea	1,040.00	309.00	151.00	1,500.00
16 BHP	SN@8.00	Ea	1,210.00	309.00	151.00	1,670.00
19 BHP	SN@8.00	Ea	1,290.00	309.00	151.00	1,750.00
22 BHP	SN@8.00	Ea	1,410.00	309.00	151.00	1,870.00
28 BHP	SN@8.00	Ea	1,480.00	309.00	151.00	1,940.00
34 BHP	SN@8.00	Ea	1,640.00	309.00	151.00	2,100.00
40 BHP	SN@8.00	Ea	1,940.00	309.00	151.00	2,400.00
45 BHP	SN@8.00	Ea	2,050.00	309.00	151.00	2,510.00
50 BHP	SN@8.00	Ea	2,450.00	309.00	151.00	2,910.00
60 BHP	SN@8.00	Ea	2,790.00	309.00	151.00	3,250.00
70 BHP	SN@8.00	Ea	3,130.00	309.00	151.00	3,590.00
80 BHP	SN@8.00	Ea	3,280.00	309.00	151.00	3,740.00
100 BHP	SN@8.00	Ea	3,930.00	309.00	151.00	4,390.00
125 BHP	SN@8.00	Ea	4,060.00	309.00	151.00	4,520.00
150 BHP	SN@8.00	Ea	4,450.00	309.00	151.00	4,910.00
Add for code approval:						
Factory Mutual	—	%	7.0	—	—	—
Industrial Risk	—	%	10.0	—	—	—

Commercial Boiler Components and Accessories

Oil Fuel Train Piping. Costs shown include recirculating pump, electric oil preheater, regulating valve, check valves, isolating pump stop cocks, oil line pressure relief valve and pressure gauges. Add the cost of concrete pump pad, electrical connections, relief line piping back to the oil tank and tank connecting piping. Add for number 4-6 fuel oil as shown.

Description	Craft@Hrs	Unit	Material $	Labor $	Equipment $	Total $

Oil fuel train piping for boilers. Costs shown are based on boiler horsepower rating (BHP) using number 2 fuel oil.

Description	Craft@Hrs	Unit	Material $	Labor $	Equipment $	Total $
20 to 30 BHP	SN@4.00	Ea	1,660.00	155.00	75.50	1,890.50
40 to 60 BHP	SN@4.00	Ea	1,870.00	155.00	75.50	2,100.50
70 to 80 BHP	SN@4.00	Ea	2,040.00	155.00	75.50	2,270.50
100 BHP	SN@8.00	Ea	2,610.00	309.00	151.00	3,070.00
125 to 150 BHP	SN@8.00	Ea	3,380.00	309.00	151.00	3,840.00
200 BHP	SN@8.00	Ea	4,460.00	309.00	151.00	4,920.00
250 BHP	SN@8.00	Ea	5,280.00	309.00	151.00	5,740.00
300 BHP	SN@16.0	Ea	5,650.00	618.00	302.00	6,570.00
350 BHP	SN@16.0	Ea	6,310.00	618.00	302.00	7,230.00
400 BHP	SN@16.0	Ea	7,580.00	618.00	302.00	8,500.00
500 BHP	SN@32.0	Ea	7,670.00	1,240.00	604.00	9,514.00
600 BHP	SN@32.0	Ea	8,040.00	1,240.00	604.00	9,884.00
750 to 800 BHP	SN@32.0	Ea	8,780.00	1,240.00	604.00	10,624.00
900 to 1,000 BHP	SN@32.0	Ea	10,300.00	1,240.00	604.00	12,144.00
Add for #4-6 oil	—	%	10.0	—	—	—
Add for tie-in with:						
dual-fuel burner	—	%	5.0	—	—	—

Dual-fuel boiler burners. For a new installation or retrofit for Scotch marine firetube, firebox or package watertube boilers. Rated at the best available control technology (BACT) for natural gas. Complies with SCAQMD rules 1146.2, 222 and 219 and the Clean Air Act section 307(d). Includes control of CO_2 emissions as well as NOx, SOx, VMs. Cleaver Brooks or equal. Factory Mutual, Industrial Risk Insurers, and Underwriters' Laboratories certified, U.S. Green Building Council rated, LFUE (Levelized Fuel Utilization Efficiency) rated at 90% or better. Includes the burner, burner head, burner mount flange, associated refractory and gasketing, main natural gas and oil fuel valves, draft damper and fan (for a power burner), or burner face plate (for an atmospheric or induced-draft burner) and the natural gas fuel train. Add the cost of an oil fuel train to complete dual-fuel bid. The control panel includes the control circuit, burner mounted control panel, motor starter(s), panel signal lights and full modulation with manual potentiometer, flame safeguard controls, flue gas recirculator air inlet butterfly valve and relay actuator, combustion air proving switch and high fire air interlock switch. Note: BHP multiplied by 33.5 is the approximate required 1,000 Btu per hour input (MBH input). Add the installation costs below and the cost of natural gas fuel line relief vent piping and a trip valve to outside of the building if required.

Description	Craft@Hrs	Unit	Material $	Labor $	Equipment $	Total $

Dual-fuel boiler burners. Costs shown are based on boiler horsepower rating (BHP).

Description	Craft@Hrs	Unit	Material $	Labor $	Equipment $	Total $
20 BHP	SN@8.00	Ea	5,330.00	309.00	—	5,639.00
30 BHP	SN@8.00	Ea	6,190.00	309.00	—	6,499.00
40 BHP	SN@8.00	Ea	7,650.00	309.00	—	7,959.00
50 BHP	SN@8.00	Ea	8,730.00	309.00	—	9,039.00
60 BHP	SN@8.00	Ea	9,930.00	309.00	—	10,239.00
70 BHP	SN@8.00	Ea	10,900.00	309.00	—	11,209.00
80 BHP	SN@8.00	Ea	12,500.00	309.00	—	12,809.00
100 BHP	SN@8.00	Ea	14,600.00	309.00	—	14,909.00
125 BHP	SN@8.00	Ea	16,500.00	309.00	—	16,809.00
150 BHP	SN@8.00	Ea	17,300.00	309.00	—	17,609.00
200 BHP	SN@8.00	Ea	22,000.00	309.00	—	22,309.00
250 BHP	SN@8.00	Ea	28,100.00	309.00	—	28,409.00
300 BHP	SN@8.00	Ea	33,200.00	309.00	—	33,509.00
350 BHP	SN@8.00	Ea	36,900.00	309.00	—	37,209.00
400 BHP	SN@8.00	Ea	39,200.00	309.00	—	39,509.00
500 BHP	SN@16.0	Ea	40,200.00	618.00	—	40,818.00
600 BHP	SN@16.0	Ea	45,100.00	618.00	—	45,718.00
750 BHP	SN@32.0	Ea	48,500.00	1,240.00	—	49,740.00
800 BHP	SN@32.0	Ea	49,200.00	1,240.00	—	50,440.00

Commercial Boiler Components and Accessories

Description	Craft@Hrs	Unit	Material $	Labor $	Equipment $	Total $
Additional costs for installing dual-fuel boiler burners						
Form and pour 4' x 4' x 2" interior slab	CF@2.50	Ea	148.00	88.40	—	236.40
Bore fuel line hole through wall or pad	P1@.250	Ea	—	9.18	—	9.18
Bore flue gas hole through wall or pad	P1@.250	Ea	—	9.18	—	9.18
Place burner and control panel enclosure	ER@8.00	Ea	—	315.00	—	315.00
Mount, level and grout burner, duct and control panel	P1@.450	Ea	468.00	16.50	—	484.50
Piping for fuel lines and flue gas recirculator	P1@.500	LF	8.84	18.40	—	27.24
Electrical wiring	BE@.300	LF	3.85	12.10	—	15.95
Wire control panel	BE@.300	LF	1.79	12.10	—	13.89
Ground fault protection wiring	BE@.300	LF	5.64	12.10	—	17.74
Commission and test	BE@8.00	Ea	—	323.00	—	323.00

Costs for Electrical Service for Boilers, Subcontract. Costs shown include typical main power tie-in and fusing, circuit breaker panel, and main bus. Assumes controls, fan, blower switchgear and control console are pre-wired electrical service. These costs include the electrical subcontractor's overhead and profit.

Description	Craft@Hrs	Unit	Material $	Labor $	Total $
Electrical service for firetube and watertube boilers. Costs shown are based on boiler horsepower rating (BHP).					
20 to 100 BHP	—	Ea	—	—	1,460.00
125 to 200 BHP	—	Ea	—	—	1,770.00
250 to 350 BHP	—	Ea	—	—	2,010.00
400 to 500 BHP	—	Ea	—	—	2,300.00
500 to 600 BHP	—	Ea	—	—	3,320.00
750 to 800 BHP	—	Ea	—	—	5,990.00
Electrical service for package firebox boilers. Costs shown are based on boiler horsepower rating (BHP).					
13 to 70 BHP	—	Ea	—	—	1,420.00
80 to 125 BHP	—	Ea	—	—	1,450.00
150 BHP	—	Ea	—	—	1,790.00

Commercial Boiler Components and Accessories

Costs for Refractory for Boilers, Subcontract. For package Scotch marine firetube, firebox and watertube boilers. Most new boilers are shipped from the factory with the refractory brick installed. If refractory brick is installed on the job site to repair shipping damage or to create firedoor seals, use the figures below. These costs include the refractory subcontractor's overhead and profit. Costs shown are per 100 pounds of refractory used.

Description	Craft@Hrs	Unit	Material $	Labor $	Equipment $	Total $
Repairs and seals	—	100 lb	—	—	—	245.00

Costs for Boiler Feedwater Pumps. Costs shown are for the pump only. Add the cost of piping, pump actuating valves and concrete pump base.

Feedwater pumps for firetube and watertube boilers. Costs shown are based on boiler horsepower rating (BHP).

20 to 40 BHP	SN@8.00	Ea	11,100.00	309.00	151.00	11,560.00
50 to 80 BHP	SN@8.00	Ea	13,400.00	309.00	151.00	13,860.00
100 BHP	SN@8.00	Ea	14,800.00	309.00	151.00	15,260.00
125 to 200 BHP	SN@16.0	Ea	15,200.00	618.00	302.00	16,120.00
250 to 400 BHP	SN@32.0	Ea	17,000.00	1,240.00	604.00	18,844.00
500 BHP	SN@32.0	Ea	18,000.00	1,240.00	604.00	19,844.00
600 BHP	SN@32.0	Ea	18,500.00	1,240.00	604.00	20,344.00
750 BHP	SN@32.0	Ea	19,000.00	1,240.00	604.00	20,844.00
800 BHP	SN@32.0	Ea	25,700.00	1,240.00	604.00	27,544.00

Feedwater pumps for package firebox boilers. Costs shown are based on boiler horsepower rating (BHP).

13 to 35 BHP	SN@8.00	Ea	5,960.00	309.00	151.00	6,420.00
40 to 45 BHP	SN@8.00	Ea	7,010.00	309.00	151.00	7,470.00
50 to 60 BHP	SN@8.00	Ea	8,780.00	309.00	151.00	9,240.00
70 to 80 BHP	SN@8.00	Ea	10,400.00	309.00	151.00	10,860.00
100 BHP	SN@8.00	Ea	13,400.00	309.00	151.00	13,860.00
125 BHP	SN@16.0	Ea	13,400.00	618.00	302.00	14,320.00
150 BHP	SN@16.0	Ea	14,800.00	618.00	302.00	15,720.00

Costs for Natural Gas Combustion Controls for Boilers. Instrumentation Society of America, Underwriters' Laboratories, Factory Mutual, and Industrial Risk Insurers certified. Fireye or equal. Includes automatic recycling or continuous pilot, spark igniter and coil (for electric ignition), timer to control the purge cycle, pilot, main burner (hi-lo-off) or main burner (proportional control), flame safeguard and sensor and combustion control panel.

Natural gas combustion controls for firetube and watertube boilers. Costs shown are based on boiler horsepower rating (BHP).

20 to 80 BHP	SN@4.00	Ea	4,540.00	155.00	75.50	4,770.50
100 BHP	SN@4.00	Ea	5,040.00	155.00	75.50	5,270.50
125 to 200 BHP	SN@8.00	Ea	5,040.00	309.00	151.00	5,500.00
250 BHP	SN@16.0	Ea	5,040.00	618.00	302.00	5,960.00
300 BHP	SN@16.0	Ea	6,050.00	618.00	302.00	6,970.00
350 or 400 BHP	SN@16.0	Ea	7,090.00	618.00	302.00	8,010.00
500 BHP	SN@32.0	Ea	7,090.00	1,240.00	604.00	8,934.00
600 BHP	SN@32.0	Ea	8,470.00	1,240.00	604.00	10,314.00
750 BHP	SN@64.0	Ea	9,870.00	2,470.00	1,210.00	13,550.00
800 BHP	SN@64.0	Ea	11,400.00	2,470.00	1,210.00	15,080.00

Commercial Boiler Components and Accessories

Description	Craft@Hrs	Unit	Material $	Labor $	Equipment $	Total $

Natural gas combustion controls for firebox boilers. Costs shown are based on boiler horsepower rating (BHP).

Description	Craft@Hrs	Unit	Material $	Labor $	Equipment $	Total $
13.4-80 BHP	SN@4.00	Ea	4,360.00	155.00	75.50	4,590.50
100.0 BHP	SN@4.00	Ea	5,040.00	155.00	51.50	5,246.50
126.9 BHP	SN@8.00	Ea	5,040.00	309.00	151.00	5,500.00
150.0 BHP	SN@8.00	Ea	9,410.00	309.00	151.00	9,870.00
Add if burner controls are for:						
#2 oil-fired	—	%	—	—	—	10.0
Heavy oil-fired	—	%	—	—	—	20.0
Dual fuel burners	—	%	—	—	—	70.0

Costs for Water Softening Systems for Use with Boilers. Costs shown include the metering valve, tank, pump, and zeolite media. Add the cost of piping to connect the water softening system to the main feedwater tank and pump.

Water softening system for firetube and watertube boilers. Costs shown are based on boiler horsepower rating (BHP).

Description	Craft@Hrs	Unit	Material $	Labor $	Equipment $	Total $
20 to 80 BHP	SN@4.00	Ea	7,270.00	155.00	75.50	7,500.50
100 BHP	SN@4.00	Ea	8,530.00	155.00	75.50	8,760.50
125 BHP	SN@8.00	Ea	9,890.00	309.00	151.00	10,350.00
150 or 200 BHP	SN@8.00	Ea	12,700.00	309.00	151.00	13,160.00
250 BHP	SN@8.00	Ea	14,800.00	309.00	151.00	15,260.00
300 or 350 BHP	SN@16.0	Ea	19,000.00	618.00	302.00	19,920.00
400 or 500 BHP	SN@16.0	Ea	23,800.00	618.00	302.00	24,720.00
600 BHP	SN@32.0	Ea	28,800.00	1,240.00	604.00	30,644.00
750 or 800 BHP	SN@64.0	Ea	33,500.00	2,470.00	1,210.00	37,180.00

Water softening system for package firebox boilers. Costs shown are based on boiler horsepower rating (BHP).

Description	Craft@Hrs	Unit	Material $	Labor $	Equipment $	Total $
13.4 to 80 BHP	SN@4.00	Ea	8,550.00	155.00	75.50	8,780.50
100.0 BHP	SN@4.00	Ea	8,550.00	155.00	75.50	8,780.50
126.9 BHP	SN@8.00	Ea	10,100.00	309.00	151.00	10,560.00
150.0 BHP	SN@8.00	Ea	11,600.00	309.00	151.00	12,060.00

Costs for Start-up, Shakedown, and Calibration of Boilers. Costs shown include minor changes and adjustments to meet requirements of the boiler inspector, adjustment of combustion efficiency at the burner by a control technician, pump and metering calibration testing and steam or hot water metering on start-up.

Start-up, shakedown & calibration of firetube and watertube boilers. Costs shown are based on boiler horsepower rating (BHP.) If a factory technician performs the start-up, add this cost at a minimum of $800 (8 hours at $100.00 per hour).

Description	Craft@Hrs	Unit	Material $	Labor $	Equipment $	Total $
13.4 to 100 BHP	S2@4.00	Ea	—	144.00	—	144.00
125 to 200 BHP	S2@8.00	Ea	—	288.00	—	288.00
250 to 400 BHP	S2@16.0	Ea	—	577.00	—	577.00
500 to 600 BHP	S2@24.0	Ea	—	865.00	—	865.00
750 to 800 BHP	S2@30.0	Ea	—	1,080.00	—	1,080.00

Costs for Stack Economizer Units for Boilers. Costs shown include piping connection to existing boiler, economizer fin tube coils, feedwater pump, check valves, thermostatic regulating valve, stack dampers, pressure relief valve, stack mount flanges and insulating enclosure, but no architectural modifications.

Description	Craft@Hrs	Unit	Material $	Labor $	Equipment $	Total $

Stack waste heat recovery system for boilers. Costs shown are based on boiler horsepower rating (BHP).

Description	Craft@Hrs	Unit	Material $	Labor $	Equipment $	Total $
500 BHP	SN@40.0	Ea	42,500.00	1,550.00	755.00	44,805.00
750 BHP	SN@48.0	Ea	52,900.00	1,860.00	906.00	55,666.00
800 BHP	SN@48.0	Ea	57,100.00	1,860.00	906.00	59,866.00
1,000 BHP	SN@56.0	Ea	71,200.00	2,160.00	1,060.00	74,420.00

Pollution control modules for gas-fired industrial package watertube boilers. N0x, S0x, VM (volatile materials) and particulates emissions reduction module. Free-standing unit. Verantis, Croll-Reynolds or equal. Designed to meet or exceed 2010 California Code, Tier IV, EPA and European ISO standards. Either vertical counter-flow or packed-bed type. For multi-dwelling, process and industrial applications. By boiler rating in millions of Btu's per hour (MMBtu/hr). See installation and efficiency optimizing control costs below. Oil-fired boilers will add about 15% to the installed cost, including a second stage reduction module, second stage pump, sulfur extraction, venturi particulate separator and sludge tank. To calculate the annual lease rate for a complete installation, divide the total installed cost by five. Then add 2% per year to cover vendor-supplied maintenance.

Description	Craft@Hrs	Unit	Material $	Labor $	Equipment $	Total $
1.0 MMBtu/Hr	—	Ea	39,500.00	—	—	39,500.00
1.7 MMBtu/Hr	—	Ea	70,200.00	—	—	70,200.00
3.4 MMBtu/Hr	—	Ea	125,000.00	—	—	125,000.00
4.1 MMBtu/Hr	—	Ea	130,000.00	—	—	130,000.00
5.0 MMBtu/Hr	—	Ea	140,000.00	—	—	140,000.00
6.7 MMBtu/Hr	—	Ea	150,000.00	—	—	150,000.00
13.4 MMBtu/Hr	—	Ea	156,000.00	—	—	156,000.00
20.1 MMBtu/Hr	—	Ea	174,000.00	—	—	174,000.00
26.8 MMBtu/Hr	—	Ea	184,000.00	—	—	184,000.00
33.5 MMBtu/Hr	—	Ea	201,000.00	—	—	201,000.00

Installation costs for pollution control modules for gas-fired industrial package watertube boilers.

Description	Craft@Hrs	Unit	Material $	Labor $	Equipment $	Total $
Form and pour foundation	ER@.040	SF	2.39	1.58	—	3.97
Position main housing, fans and pumps	ER@4.00	Ea	—	158.00	—	158.00
Tie-down and leveling	ER@4.00	Ea	575.00	158.00	—	733.00
Mount piping system frame supports	ER@6.00	Ea	1.74	236.00	—	237.74
Rig and mount plenum and ducts	SN@5.00	Ea	—	193.00	—	193.00
Rig and mount tank for emission reduction, fluid and housing inlet metering valve	ER@4.00	Ea	—	158.00	—	158.00

Commercial Boiler Components and Accessories

Description	Craft@Hrs	Unit	Material $	Labor $	Equipment $	Total $

Installation costs for pollution control modules for gas-fired industrial package watertube boilers – (continued).

Description	Craft@Hrs	Unit	Material $	Labor $	Equipment $	Total $
Weld systems piping	ER@8.00	Ea	2.42	315.00	—	317.42
Wire electrical system	BE@8.00	Ea	2.07	323.00	—	325.07
Wire controls system	BE@16.0	Ea	2.19	646.00	—	648.19
Level and balance rotating equipment	P1@8.00	Ea	231.00	294.00	—	525.00
Calibration and test	ER@2.00	Ea	—	78.80	—	78.80
Witnessed run trials (add the permit fee)	P1@4.00	Ea	—	147.00	—	147.00

Efficiency-optimizing centralized boiler burner controls. Retrofit or new construction. Preferred Instruments or equal. For use with emissions monitoring and digital reporting pollution control equipment installed for a gas-fired industrial boiler. Complies with the Clean Air Act section 307(d), California and Massachusetts boiler codes and NFPA 85. Includes a boiler-mounted control panel with chassis, servo motor controls for burner fuel line valves, LCD touchpad control panel, ultra-violet, infra-red or self-checking flame scanner, operator interface panel and interface to a remote PC (not included). Controls variable speed drives for burner forced-draft fans, induced draft fans and feedwater pumps.

Description	Craft@Hrs	Unit	Material $	Labor $	Equipment $	Total $
Mount automation module, less wiring	BE@.250	Ea	1,280.00	10.10	—	1,290.10
Mount flame safety sensors less wiring	BE@.250	Ea	203.00	10.10	—	213.10
Servomotor mounting less wiring	BE@.250	Ea	540.00	10.10	—	550.10
Mount operator interface terminal with power supply less wiring	BE@.250	Ea	92.40	10.10	—	102.50
Operator interface terminal bridge, less wiring	BE@.250	Ea	4,390.00	10.10	—	4,400.10
Bore control wiring hole through basement wall	CF@.250	Ea	—	8.84	—	8.84
Twisted pair Cat 6, 1 GBPS wire 4 pair	BE@.008	LF	.55	.32	—	.87
Panel wiring to emissions monitoring array	BE@.250	LF	2.91	10.10	—	13.01
Sensor wiring to controller	BE@.300	LF	2.65	12.10	—	14.75
Electric servomotor wiring to controller	BE@.400	LF	2.42	16.20	—	18.62
Mount and wire PC module	BE@2.00	LF	23.20	80.80	—	104.00
Ground fault protection wiring	BE@1.25	LF	144.00	50.50	—	194.50
Startup, programming and system test	BE@3.00	Ea	—	121.00	—	121.00
Commissioning and final test	BE@8.00	Ea	—	323.00	—	323.00

Costs for Condensate Receiver and Pumping Units for Boilers. Costs shown include a stainless steel factory assembled tank and pump with motor, float switch, condensate inlet, feedwater tank, connection, level gauge, thermometer, solenoid valve for direct boiler feed. No concrete mounting pad required.

Description	Craft@Hrs	Unit	Material $	Labor $	Equipment $	Total $

Condensate receiver and pumping unit for boilers.
Unit sizing and costs are based on storage for one minute at the designed steam return rate. Set in place only. Make additional allowances for pipe and electrical connections.

Description	Craft@Hrs	Unit	Material $	Labor $	Equipment $	Total $
750 lbs per hour	SN@4.00	Ea	5,540.00	155.00	2.47	5,697.47
1,500 lbs per hour	SN@4.00	Ea	5,540.00	155.00	2.47	5,697.47
3,000 lbs per hour	SN@4.00	Ea	7,040.00	155.00	2.47	7,197.47
5,000 lbs per hour	SN@6.00	Ea	7,760.00	232.00	3.71	7,995.71
6,250 lbs per hour	SN@6.00	Ea	8,650.00	232.00	3.71	8,885.71

Costs for Deaerator/Condenser Units for Boilers. Pressurized jet spray type. Costs shown include an ASME storage receiver, stand, jet spray deaerator head, makeup control valve, makeup controller, pressure control valve, level controls, pressure and temperature controls, pumped condensate inlet, pumps and motor, steam inlet and system accessories. Add for a concrete mounting pad and piping to the condensate return line (inlet) and piping to the boiler feedwater system (outlet).

Deaerator/condenser unit for boilers.
Costs shown are based on boiler horsepower rating (BHP).

Description	Craft@Hrs	Unit	Material $	Labor $	Equipment $	Total $
100 BHP	SN@8.00	Ea	58,000.00	309.00	4.95	58,313.95
200 BHP	SN@12.0	Ea	60,200.00	464.00	7.42	60,671.42
300 BHP	SN@14.0	Ea	63,900.00	541.00	8.66	64,449.66
400 BHP	SN@16.0	Ea	72,400.00	618.00	9.90	73,027.90
600 BHP	SN@24.0	Ea	77,600.00	928.00	14.80	78,542.80
800 BHP	SN@36.0	Ea	84,900.00	1,390.00	22.30	86,312.30
1,000 BHP	SN@48.0	Ea	99,100.00	1,860.00	29.70	100,989.70

Rolairtrol type air separators.
Priced by capacity in gallons per minute (GPM) and by pipe connection size. Includes hanging separator but no connection, (2" is a screwed connection, 2½" and up are flanged connections) using small tools.

Description		Craft@Hrs	Unit	Material $	Labor $	Equipment $	Total $
56 GPM,	2"	S2@2.50	Ea	1,040.00	90.10	1.55	1,131.65
90 GPM,	2½"	S2@2.50	Ea	1,140.00	90.10	1.55	1,231.65
170 GPM,	3"	S2@3.00	Ea	1,480.00	108.00	1.86	1,589.86
300 GPM,	4"	S2@3.50	Ea	2,240.00	126.00	2.16	2,368.16
500 GPM,	6"	S2@4.75	Ea	2,920.00	171.00	2.94	3,093.94
700 GPM,	6"	S2@4.75	Ea	3,770.00	171.00	2.94	3,943.94
1,300 GPM,	8"	SN@6.00	Ea	5,060.00	232.00	3.71	5,295.71
2,000 GPM,	10"	SN@6.00	Ea	7,080.00	232.00	3.71	7,315.71
2,750 GPM,	12"	SN@12.0	Ea	9,190.00	464.00	7.42	9,661.42

Commercial Boiler Components and Accessories

Description	Craft@Hrs	Unit	Material $	Labor $	Equipment $	Total $

Costs for Chemical Feed Duplex Pump, Packaged Units for Boilers. Costs shown include the chemical tanks, electronic metering, valve and stand. Add the cost of piping to the feedwater tank.

Chemical feed duplex pump for boilers. Costs shown are based on boiler horsepower rating (BHP).

Description	Craft@Hrs	Unit	Material $	Labor $	Equipment $	Total $
20 to 100 BHP	SN@3.00	Ea	2,130.00	116.00	56.60	2,302.60
25 to 200 BHP	SN@3.00	Ea	2,790.00	116.00	56.60	2,962.60
200 to 400 BHP	SN@6.00	Ea	2,790.00	232.00	113.00	3,135.00
500 to 600 BHP	SN@6.00	Ea	6,330.00	232.00	113.00	6,675.00
700 to 800 BHP	SN@8.00	Ea	7,070.00	309.00	151.00	7,530.00
1,000 BHP	SN@10.0	Ea	9,560.00	387.00	189.00	10,136.00

Chemical feed simple in-line system for hot water heating system.
Costs shown include 3 gallon pot feeder, two 3/4" isolation valves, 10' of ¾" pipe and two ¾" threadolets. Chemicals are drawn in through suction created by hot water system circulation pumps.

Description	Craft@Hrs	Unit	Material $	Labor $	Equipment $	Total $
50 to 20 BHP	SN@3.00	Ea	494.00	116.00	56.60	666.60

Packaged Boiler Feedwater Systems for Boilers. Costs shown include duplex pumps with heater assembly, thermal lining, pressure gauges, makeup feeder valve, ASME feedwater tank and stand, single phase electric motor and level gauge. Add the costs for a concrete mounting pad and piping to the boilers.

Packaged boiler feedwater systems. Costs shown are based on boiler horsepower rating (BHP).

Description	Craft@Hrs	Unit	Material $	Labor $	Equipment $	Total $
15 to 50 BHP	SN@5.00	Ea	14,800.00	193.00	94.40	15,087.40
60 to 80 BHP	SN@5.00	Ea	16,400.00	193.00	94.40	16,687.40
100 BHP	SN@6.00	Ea	16,800.00	232.00	113.00	17,145.00
125 to 150 BHP	SN@6.00	Ea	18,000.00	232.00	113.00	18,345.00
200 BHP	SN@8.00	Ea	18,300.00	309.00	151.00	18,760.00
250 BHP	SN@8.00	Ea	19,100.00	309.00	151.00	19,560.00
300 BHP	SN@8.00	Ea	20,400.00	309.00	151.00	20,860.00
400 BHP	SN@8.00	Ea	21,000.00	309.00	151.00	21,460.00
500 BHP	SN@8.00	Ea	25,500.00	309.00	151.00	25,960.00
600 BHP	SN@12.0	Ea	26,600.00	464.00	227.00	27,291.00
750 BHP	SN@12.0	Ea	26,500.00	464.00	227.00	27,191.00

Continuous Blowdown Heat Recovery System. Includes a flash receiver (50 PSI ASME rated tank), plate and frame heat exchanger, pneumatic level controller and valve, air filter regulator, water gauge, pressure gauge, backflush piping, pressure relief valve connected to mud drum and drain. Add for concrete pad mount, piping and electrical wiring. Add the cost of automatic controls from the following section.

Continuous blowdown heat recovery systems. Costs shown by pounds per hour (lbs per hr) of rated makeup capacity.

Description	Craft@Hrs	Unit	Material $	Labor $	Equipment $	Total $
5,000 lbs per hr	SN@8.00	Ea	49,900.00	309.00	151.00	50,360.00
10,000 lbs per hr	SN@8.00	Ea	98,700.00	309.00	151.00	99,160.00
25,000 lbs per hr	SN@8.00	Ea	126,000.00	309.00	151.00	126,460.00
50,000 lbs per hr	SN@8.00	Ea	200,000.00	309.00	151.00	200,460.00

Commercial Boiler Components and Accessories

Boiler blowdown waste heat recovery system controls. Typical costs in rated pounds per hour for high-efficiency commercial or industrial controls for a manually-operated boiler blowdown waste heat recovery system. Wilson Equipment, Spirax Sarco, or equal. Meets EPA efficiency specifications and liquid waste control requirements. For use with package industrial watertube, firebox and firetube boilers up to 250 PSIG and up to 800 boiler horsepower. Includes timer, total dissolved solids sensor, digital control minipanel, valve, valve actuator and T-type security key for manual actuation. For new construction, add the appropriately sized blowdown waste heat recovery system shown above.

Description	Craft@Hrs	Unit	Material $	Labor $	Equipment $	Total $
5,000 lbs per hr	SN@8.00	Ea	12,800.00	309.00	—	13,109.00
10,000 lbs per hr	SN@8.00	Ea	17,500.00	309.00	—	17,809.00
25,000 lbs per hr	SN@8.00	Ea	30,600.00	309.00	—	30,909.00
50,000 lbs per hr	SN@8.00	Ea	37,000.00	309.00	—	37,309.00
Add for piping for blowdown valve	P1@.500	LF	8.95	18.40	—	27.35
Add for wiring the valve actuator	BE@.300	LF	3.67	12.10	—	15.77
Add for control panel wiring	BE@.300	LF	1.75	12.10	—	13.85
Add for ground fault protection wiring	BE@.300	LF	5.77	12.10	—	17.87
Add for commissioning and tests	BE@8.00	Ea	—	323.00	—	323.00

Boiler Stacks. Costs shown assume a 20-foot stack height and include rigging, breeching, connection of stack sections, guy wires where required, vendor-supplied stack supports and insulation. Add the cost of a concrete base.

Boiler stack. Costs shown assume a 20-foot stack height.

10" diameter	SN@12.0	Ea	3,550.00	464.00	227.00	4,241.00
12" diameter	SN@12.0	Ea	4,430.00	464.00	227.00	5,121.00
16" diameter	SN@16.0	Ea	5,290.00	618.00	302.00	6,210.00
20" diameter	SN@20.0	Ea	6,650.00	773.00	378.00	7,801.00
24" diameter	SN@24.0	Ea	8,780.00	928.00	453.00	10,161.00
30" diameter	SN@32.0	Ea	10,400.00	1,240.00	604.00	12,244.00
56" diameter	SN@40.0	Ea	20,600.00	1,550.00	755.00	22,905.00
Add for each 10' section (or portion thereof) beyond 20' high						
per 10' sect.	—	%	—	—	—	15.0

Costs for Tankless Indirect Water Heater. Provides domestic hot water from boiler coil insert. Costs shown include heat exchanger, coil, circulating pump, thermostatically controlled valve, bypass piping and air bleeder valve. Add the cost of an equipment base, electrical wiring, piping and an insulated hot water storage tank.

Tankless indirect water heater. Costs by gallons per hour (GPH).

240 GPH	SN@8.00	Ea	20,700.00	309.00	151.00	21,160.00
360 GPH	SN@8.00	Ea	25,900.00	309.00	151.00	26,360.00
480 GPH	SN@10.0	Ea	26,600.00	387.00	189.00	27,176.00
840 GPH	SN@12.0	Ea	60,300.00	464.00	227.00	60,991.00
920 GPH	SN@16.0	Ea	67,300.00	618.00	302.00	68,220.00
1,240 GPH	SN@20.0	Ea	75,400.00	773.00	378.00	76,551.00

Commercial Boiler Components and Accessories

Costs for Boiler Trim. Costs shown include the pressure relief valve, level gauge (for steam systems), level controller & low water cutoff, pressure and temperature gauge, feedwater valve & actuator, main valve, blowdown valve, cleanout manholes and gaskets.

Description	Craft@Hrs	Unit	Material $	Labor $	Equipment $	Total $

Boiler trim. Costs shown are per boiler horsepower rating (BHP).

Description	Craft@Hrs	Unit	Material $	Labor $	Equipment $	Total $
20 to 100 BHP	SN@4.00	Ea	3,300.00	155.00	2.47	3,457.47
100 to 200 BHP	SN@6.00	Ea	3,720.00	232.00	3.71	3,955.71
200 to 400 BHP	SN@8.00	Ea	4,580.00	309.00	4.95	4,893.95
400 to 600 BHP	SN@8.00	Ea	7,000.00	309.00	4.95	7,313.95
600 to 800 BHP	SN@8.00	Ea	8,260.00	309.00	4.95	8,573.95
800 to 1,000 BHP	SN@10.0	Ea	8,270.00	387.00	6.19	8,663.19

Heat reclaimer condensate return & feedwater system, stainless steel non-siphon, including flash tank with level control and vacuum breaker, overflow relief drain, simplex centrifugal pump, pre-assembled piping to feedwater tank, structural steel stand and frame, check trim and pre-wired controls. Also provides potable hot water as part of the condensed steam return cycle. Capacity rating is in GPH (gallons per hour of condensate)

Description	Craft@Hrs	Unit	Material $	Labor $	Total $
33 gal. ½ HP pump	P1@3.00	Ea	981.00	110.00	1,091.00
46 gal. ¾ HP pump	P1@4.00	Ea	1,210.00	147.00	1,357.00

Vacuum breakers, rough brass, threaded, to be mounted on finned radiation tube inlet valve fitting

Description	Craft@Hrs	Unit	Material $	Labor $	Total $
½" vacuum breaker	P1@.250	Ea	45.80	9.18	54.98

Steam traps, float valve Sarco type or equal w/drain fitting, brazed

Description	Craft@Hrs	Unit	Material $	Labor $	Total $
¾" with check valve	P1@.250	Ea	43.60	9.18	52.78
1.0" with check valve	P1@.450	Ea	69.80	16.50	86.30

Baseboard fin tube radiation, per linear foot, 15 PSIG test thinwall copper tube with aluminum fins, braze fitted wall mounted

Description	Craft@Hrs	Unit	Material $	Labor $	Total $
¾" element with 8" cover	P1@.280	LF	11.10	10.30	21.40
1" element with 8" cover	P1@.350	LF	18.10	12.80	30.90
Add for corners, fillers and caps, average per foot	—	%	—	15.0	—

Solenoid valves, zone thermostatically-controlled steam electric, 15 PSIG rating, braze fitting. Thermostatic Honeywell-type or equal solenoid relay valves

Description	Craft@Hrs	Unit	Material $	Labor $	Total $
¾", brazed	P1@.250	Ea	14.40	9.18	23.58
1", brazed	P1@.250	Ea	14.40	9.18	23.58

Description	Craft@Hrs	Unit	Material $	Labor $	Equipment $	Total $

Additional costs for boiler pumps, condenser units, feedwater systems, heat recovery and oil fuel train piping.

Description	Craft@Hrs	Unit	Material $	Labor $	Equipment $	Total $
Form and pour concrete pad	CF@1.56	SF	9.26	55.10	—	64.36
Place vibration pads	CF@.750	Ea	44.40	26.50	—	70.90
Mount interior drain	P1@.500	LF	8.73	18.40	—	27.13
Bore hole through wall or pad	P1@.250	Ea	—	9.18	—	9.18
Piping to load (typical)	P1@.500	LF	8.77	18.40	—	27.17
Bore burner stack vent in exterior wall	P1@.250	Ea	—	9.18	—	9.18
Mounting and edge-sealing stack	P1@.500	Ea	44.90	18.40	—	63.30

Emissions sensing and recording equipment. For stack gas combustion monitoring of boilers. Monitors and records NOx, SOx, CO, H_2S, volatile materials and particulates emissions to meet 2011 EPA greenhouse gas regulations. CO_2 monitoring and recording is required on boilers producing 25,000 tons or more a year of carbon emissions. Signal Group Limited or equal. Includes a stack-mounted pulse type infra-red single-beam light source with solid-state detector. Sensitivity range is variable between .1 ppm and 1,000 ppm. Includes a hot-wire radiation source, rotating shutter with reference target for continuous unmanned recalibration, 240 x 64 pixel LCD display with remote panel-mounted solid-state digital analyzer and lockout type inspector-only access recording circuitry. Cost includes calibration and testing by a factory technician. Add the cost of the permit, inspection, test and documentation.

Description	Craft@Hrs	Unit	Material $	Labor $	Equipment $	Total $
Analyzer and recorder installation to remote panel	BE@3.00	Ea	29,400.00	121.00	—	29,521.00
Infrared sensor unit installation to stack	BE@2.00	Ea	9,210.00	80.80	—	9,290.80
Wiring of sensor to analyzer and recorder	BE@8.00	Ea	575.00	323.00	—	898.00
Circuit testing and grounding	BE@2.00	Ea	—	80.80	—	80.80

Centrifugal Pumps and Pump Connections

Description	Craft@Hrs	Unit	Material $	Labor $	Equipment $	Total $

Base-mounted, bronze fitted centrifugal pump, set in place only. Make additional allowances for pipe connection assembly.

Description	Craft@Hrs	Unit	Material $	Labor $	Equipment $	Total $
½ HP	P1@1.50	Ea	1,410.00	55.10	—	1,465.10
¾ HP	P1@1.75	Ea	1,790.00	64.20	—	1,854.20
1 HP	P1@2.00	Ea	2,190.00	73.40	—	2,263.40
1½ HP	P1@2.40	Ea	2,660.00	88.10	—	2,748.10
2 HP	P1@2.70	Ea	3,270.00	99.10	—	3,369.10
3 HP	P1@3.00	Ea	3,900.00	110.00	—	4,010.00
5 HP	P1@3.50	Ea	4,450.00	128.00	—	4,578.00
7½ HP	P1@4.00	Ea	4,930.00	147.00	—	5,077.00
10 HP	P1@4.50	Ea	7,190.00	165.00	—	7,355.00
15 HP	ER@5.00	Ea	8,330.00	197.00	3.09	8,530.09
20 HP	ER@6.00	Ea	9,630.00	236.00	3.71	9,869.71
25 HP	ER@7.00	Ea	10,000.00	276.00	4.33	10,280.33
30 HP	ER@7.40	Ea	11,900.00	292.00	4.58	12,196.58
40 HP	ER@7.70	Ea	11,800.00	303.00	4.76	12,107.76
50 HP	ER@8.00	Ea	14,400.00	315.00	4.95	14,719.95

Base-mounted centrifugal pump connection assembly. Includes pipe, fittings, pipe insulation, valves, strainers, suction guides and gauges.

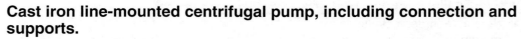

Description	Craft@Hrs	Unit	Material $	Labor $	Equipment $	Total $
2½" supply	P1@12.0	Ea	2,420.00	440.00	—	2,860.00
3" supply	P1@13.5	Ea	2,810.00	495.00	—	3,305.00
4" supply	P1@16.0	Ea	3,850.00	587.00	—	4,437.00
6" supply	P1@22.0	Ea	6,370.00	807.00	—	7,177.00
8" supply	P1@28.0	Ea	9,230.00	1,030.00	—	10,260.00
10" supply	P1@35.0	Ea	15,300.00	1,280.00	—	16,580.00
12" supply	P1@42.0	Ea	20,300.00	1,540.00	—	21,840.00

Cast iron line-mounted centrifugal pump, including connection and supports.

Description	Craft@Hrs	Unit	Material $	Labor $	Equipment $	Total $
1/6 HP	P1@1.25	Ea	364.00	45.90	—	409.90
1/4 HP	P1@1.40	Ea	466.00	51.40	—	517.40
1/3 HP	P1@1.50	Ea	575.00	55.10	—	630.10
1/2 HP	P1@1.70	Ea	735.00	62.40	—	797.40
3/4 HP	P1@1.80	Ea	990.00	66.10	—	1,056.10
1 HP	P1@2.00	Ea	1,180.00	73.40	—	1,253.40

Description	Craft@Hrs	Unit	Material $	Labor $	Equipment $	Total $

2-pass shell and tube type heat exchanger, set in place only. Make additional allowances for pipe connection assembly.

Description	Craft@Hrs	Unit	Material $	Labor $	Equipment $	Total $
4" dia. x 30"	SN@2.00	Ea	1,720.00	77.30	1.24	1,798.54
4" dia. x 60"	SN@2.50	Ea	1,950.00	96.60	1.55	2,048.15
6" dia. x 30"	SN@3.25	Ea	2,390.00	126.00	2.01	2,518.01
6" dia. x 60"	SN@3.50	Ea	2,970.00	135.00	2.16	3,107.16
8" dia. x 30"	SN@4.25	Ea	3,690.00	164.00	2.63	3,856.63
8" dia. x 60"	SN@4.50	Ea	4,760.00	174.00	2.78	4,936.78
10" dia. x 30"	SN@5.50	Ea	5,470.00	213.00	3.40	5,686.40
10" dia. x 60"	SN@5.75	Ea	7,230.00	222.00	3.56	7,455.56
10" dia. x 90"	SN@6.00	Ea	9,010.00	232.00	3.71	9,245.71
12" dia. x 60"	SN@6.50	Ea	9,480.00	251.00	4.02	9,735.02
12" dia. x 90"	SN@7.00	Ea	12,100.00	271.00	4.33	12,375.33
12" dia. x 120"	SN@7.25	Ea	14,800.00	280.00	4.48	15,084.48
14" dia. x 60"	SN@7.50	Ea	14,000.00	290.00	4.64	14,294.64
14" dia. x 90"	SN@8.00	Ea	17,400.00	309.00	4.95	17,713.95
14" dia. x 120"	SN@8.25	Ea	21,000.00	319.00	5.10	21,324.10
16" dia. x 60"	SN@8.50	Ea	17,300.00	329.00	5.26	17,634.26
16" dia. x 90"	SN@9.00	Ea	21,600.00	348.00	5.57	21,953.57
16" dia. x 120"	SN@9.25	Ea	24,400.00	358.00	5.72	24,763.72

Heat exchanger connection assembly, hot water to hot water. Includes pipe, fittings, pipe insulation, valves, thermometers gauges and vents.

Description	Craft@Hrs	Unit	Material $	Labor $	Equipment $	Total $
1½" supply	P1@32.0	Ea	3,800.00	1,170.00	—	4,970.00
2" supply	P1@38.0	Ea	4,680.00	1,390.00	—	6,070.00
2½" supply	P1@50.0	Ea	6,100.00	1,840.00		7,940.00
3" supply	P1@55.0	Ea	6,410.00	2,020.00	—	8,430.00
4" supply	P1@62.0	Ea	7,800.00	2,280.00	—	10,080.00
6" supply	P1@85.0	Ea	12,600.00	3,120.00	5,620.00	21,340.00
8" supply	P1@102.	Ea	17,900.00	3,740.00	6,740.00	28,380.00
10" supply	P1@130.	Ea	23,400.00	4,770.00	8,590.00	36,760.00

Heat exchanger connection assembly, steam to hot water. Includes pipe, fittings, pipe insulation, valves, thermometers gauges and vents.

Description	Craft@Hrs	Unit	Material $	Labor $	Equipment $	Total $
1½" supply	P1@22.0	Ea	4,350.00	807.00	—	5,157.00
2" supply	P1@26.0	Ea	5,080.00	954.00	—	6,034.00
2½" supply	P1@41.0	Ea	6,310.00	1,500.00	—	7,810.00
3" supply	P1@44.0	Ea	7,310.00	1,610.00	—	8,920.00
4" supply	P1@54.5	Ea	8,330.00	2,000.00	—	10,330.00
6" supply	P1@67.0	Ea	13,200.00	2,460.00	4,430.00	20,090.00
8" supply	P1@80.0	Ea	18,700.00	2,940.00	5,290.00	26,930.00

Fan Coil Units and Connections

Description	Craft@Hrs	Unit	Material $	Labor $	Equipment $	Total $

Ceiling-hung horizontal, 2 pipe hydronic fan coil unit with cabinet, set in place only. Make additional allowances for pipe connection assembly.

Description	Craft@Hrs	Unit	Material $	Labor $	Equipment $	Total $
200 CFM	P1@2.00	Ea	914.00	73.40	—	987.40
400 CFM	P1@2.50	Ea	1,080.00	91.80	—	1,171.80
600 CFM	P1@3.00	Ea	1,280.00	110.00	—	1,390.00
800 CFM	P1@3.25	Ea	1,610.00	119.00	—	1,729.00
1,000 CFM	P1@3.50	Ea	2,020.00	128.00	—	2,148.00
Deduct for "less cabinet"—		%	−20.0	—	—	—
Add for "floor-mounted" —		%	10.0	—	—	—

Vertical, 4 pipe hydronic fan coil unit, set in place only. Make additional allowances for pipe connection assembly.

Description	Craft@Hrs	Unit	Material $	Labor $	Equipment $	Total $
200 CFM	P1@2.15	Ea	1,030.00	78.90	—	1,108.90
400 CFM	P1@2.75	Ea	1,210.00	101.00	—	1,311.00
600 CFM	P1@3.25	Ea	1,420.00	119.00	—	1,539.00
800 CFM	P1@3.45	Ea	1,760.00	127.00	—	1,887.00
1,000 CFM	P1@3.60	Ea	2,210.00	132.00	—	2,342.00

2 pipe fan coil unit connection. Coil connection includes type L copper pipe and wrought fittings, pipe insulation, condensate drain, isolation valves and 2-way control valve.

Description	Craft@Hrs	Unit	Material $	Labor $	Equipment $	Total $
½" supply	P1@2.50	Ea	148.00	91.80	—	239.80
¾" supply	P1@2.80	Ea	180.00	103.00	—	283.00
1" supply	P1@3.40	Ea	272.00	125.00	—	397.00

4 pipe fan coil unit connection. Coil connection includes type L copper pipe and wrought fittings, pipe insulation, condensate drain, isolation valves and 3-way control valve.

Description	Craft@Hrs	Unit	Material $	Labor $	Equipment $	Total $
½" supply	P1@2.75	Ea	180.00	101.00	—	281.00
¾" supply	P1@2.95	Ea	208.00	108.00	—	316.00

Description	Craft@Hrs	Unit	Material $	Labor $	Equipment $	Total $

Duct-mounted flanged hot water reheat coil, set in place only. Make additional allowances for pipe connection assembly.

12" x 6"	SN@1.00	Ea	370.00	38.70	—	408.70
12" x 8"	SN@1.00	Ea	389.00	38.70	—	427.70
12" x 10"	SN@1.00	Ea	405.00	38.70	—	443.70
12" x 12"	SN@1.00	Ea	442.00	38.70	—	480.70
18" x 6"	SN@1.25	Ea	468.00	48.30	—	516.30
18" x 12"	SN@1.25	Ea	482.00	48.30	—	530.30
18" x 18"	SN@1.25	Ea	517.00	48.30	—	565.30
24" x 12"	SN@1.50	Ea	558.00	58.00	—	616.00
24" x 18"	SN@1.50	Ea	583.00	58.00	—	641.00
24" x 24"	SN@1.75	Ea	656.00	67.60	—	723.60

Duct-mounted hot water reheat coil connection assembly. Includes pipe, fittings, pipe insulation, valves and strainers.

½" steel	P1@2.25	Ea	189.00	82.60	—	271.60
¾" steel	P1@2.35	Ea	225.00	86.20	—	311.20
1" steel	P1@2.65	Ea	294.00	97.30	—	391.30
1¼" steel	P1@3.00	Ea	355.00	110.00	—	465.00
1½" steel	P1@3.25	Ea	379.00	119.00	—	498.00
2" steel	P1@3.75	Ea	422.00	138.00	—	560.00

Duct-mounted slip-in electric reheat coil, set in place only. Make additional allowances for electrical connection.

12" x 6"	SN@1.10	Ea	824.00	42.50	—	866.50
12" x 8"	SN@1.20	Ea	939.00	46.40	—	985.40
12" x 10"	SN@1.30	Ea	1,130.00	50.20	—	1,180.20
12" x 12"	SN@1.40	Ea	1,250.00	54.10	—	1,304.10
18" x 6"	SN@1.50	Ea	1,380.00	58.00	—	1,438.00
18" x 12"	SN@1.60	Ea	1,480.00	61.80	—	1,541.80
18" x 18"	SN@1.70	Ea	1,580.00	65.70	—	1,645.70
24" x 12"	SN@1.90	Ea	1,700.00	73.40	—	1,773.40
24" x 18"	SN@2.10	Ea	1,790.00	81.20	—	1,871.20
24" x 24"	SN@2.40	Ea	2,190.00	92.80	—	2,282.80

Unit Heaters and Connections

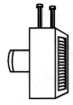

Description	Craft@Hrs	Unit	Material $	Labor $	Equipment $	Total $

Gas-fired unit heater, hang in place only. Make additional allowances for gas connection assembly.

Description	Craft@Hrs	Unit	Material $	Labor $	Equipment $	Total $
60mbh, 200cfm	P1@2.00	Ea	1,710.00	73.40	—	1,783.40
100mbh, 400cfm	P1@2.50	Ea	1,980.00	91.80	—	2,071.80
150mbh, 600cfm	P1@3.00	Ea	2,420.00	110.00	—	2,530.00
175mbh, 800cfm	P1@3.25	Ea	2,610.00	119.00	—	2,729.00
200mbh, 1,000cfm	P1@3.50	Ea	2,810.00	128.00	—	2,938.00
250mbh, 1,500cfm	P1@3.75	Ea	3,290.00	138.00	—	3,428.00
300mbh, 2,000cfm	P1@4.00	Ea	3,670.00	147.00	—	3,817.00

Hot water unit heater, hang in place only. Make additional allowances for pipe connection assembly.

Description	Craft@Hrs	Unit	Material $	Labor $	Equipment $	Total $
8mbh, 400cfm	P1@2.00	Ea	451.00	73.40	4.33	528.73
12mbh, 400cfm	P1@2.00	Ea	580.00	73.40	4.33	657.73
16mbh, 400cfm	P1@2.00	Ea	612.00	73.40	4.33	689.73
30mbh, 700cfm	P1@3.00	Ea	818.00	110.00	6.49	934.49
43mbh, 1,040cfm	P1@3.50	Ea	1,010.00	128.00	7.58	1,145.58
57mbh, 1,370cfm	P1@4.00	Ea	1,080.00	147.00	8.66	1,235.66
68mbh, 1,715cfm	P1@4.00	Ea	1,710.00	147.00	8.66	1,865.66
105mbh, 2,360cfm	P1@5.00	Ea	1,970.00	184.00	2,154.00	2,459.00
123mbh, 2,900cfm	P1@6.00	Ea	2,530.00	220.00	2,750.00	3,116.00
140mbh, 3,530cfm	P1@6.00	Ea	2,930.00	220.00	3,150.00	3,516.00
156mbh, 4,250cfm	P1@6.00	Ea	3,080.00	220.00	3,300.00	3,666.00
210mbh, 4,450cfm	P1@8.00	Ea	5,460.00	294.00	5,754.00	6,241.00
123mbh, 5,200cfm	P1@6.00	Ea	4,510.00	220.00	4,730.00	5,096.00
257mbh, 5,350cfm	P1@8.00	Ea	9,250.00	294.00	9,544.00	10,031.00

Hot water unit heater connection. Includes Schedule 40 threaded and coupled supply and return piping with malleable iron fittings, pipe insulation, valves and vents Make additional allowances for electrical connection.

Description	Craft@Hrs	Unit	Material $	Labor $	Equipment $	Total $
¾" supply	P1@2.50	Ea	261.00	91.80	—	352.80
1" supply	P1@3.00	Ea	306.00	110.00	—	416.00
1¼" supply	P1@3.30	Ea	364.00	121.00	—	485.00
1½" supply	P1@3.75	Ea	423.00	138.00	—	561.00
2" supply	P1@4.10	Ea	547.00	150.00	—	697.00

Gas-fired unit heater flue connection

Description	Craft@Hrs	Unit	Material $	Labor $	Equipment $	Total $
2" B vent	SN@.090	LF	5.68	3.48	—	9.16
3" B vent	SN@.100	LF	7.04	3.87	—	10.91
4" B vent	SN@.110	LF	9.41	4.25	—	13.66
6" B vent	SN@.130	LF	10.70	5.02	—	15.72

Gas-fired unit heater fuel connection. Assembly includes 8' of black steel pipe, malleable fittings, plug valve and regulator.

Description	Craft@Hrs	Unit	Material $	Labor $	Equipment $	Total $
¾" gas conn.	P1@1.30	Ea	94.90	47.70	—	142.60
1" gas conn.	P1@1.60	Ea	103.00	58.70	—	161.70
1¼" gas conn.	P1@2.10	Ea	184.00	77.10	—	261.10
1½" gas conn.	P1@2.50	Ea	212.00	91.80	—	303.80

Chillers and Chiller Connections

Description	Craft@Hrs	Unit	Material $	Labor $	Equipment $	Total $

Centrifugal water-cooled chiller, set in place only. Unit comes factory assembled complete. Make additional allowances for cooling tower pipe connections.

Description	Craft@Hrs	Unit	Material $	Labor $	Equipment $	Total $
100 tons	SN@20.0	Ea	98,200.00	773.00	378.00	99,351.00
150 tons	SN@20.0	Ea	149,000.00	773.00	378.00	150,151.00
200 tons	SN@20.0	Ea	160,000.00	773.00	378.00	161,151.00
250 tons	SN@20.0	Ea	188,000.00	773.00	378.00	189,151.00
300 tons	SN@25.0	Ea	202,000.00	966.00	472.00	203,438.00
350 tons	SN@25.0	Ea	208,000.00	966.00	472.00	209,438.00
400 tons	SN@25.0	Ea	246,000.00	966.00	472.00	247,438.00
500 tons	SN@25.0	Ea	307,000.00	966.00	472.00	308,438.00

Reciprocating water-cooled chiller, set in place only. Unit comes factory assembled complete. Make additional allowances for cooling tower pipe connections.

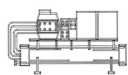

Description	Craft@Hrs	Unit	Material $	Labor $	Equipment $	Total $
20 tons	SN@12.0	Ea	30,100.00	464.00	227.00	30,791.00
25 tons	SN@12.0	Ea	34,500.00	464.00	227.00	35,191.00
30 tons	SN@12.0	Ea	37,600.00	464.00	227.00	38,291.00
40 tons	SN@12.0	Ea	45,200.00	464.00	227.00	45,891.00
50 tons	SN@16.0	Ea	63,800.00	618.00	302.00	64,720.00
60 tons	SN@16.0	Ea	82,000.00	618.00	302.00	82,920.00
75 tons	SN@16.0	Ea	88,100.00	618.00	302.00	89,020.00
100 tons	SN@20.0	Ea	93,600.00	773.00	378.00	94,751.00
125 tons	SN@20.0	Ea	114,000.00	773.00	378.00	115,151.00
150 tons	SN@20.0	Ea	130,000.00	773.00	378.00	131,151.00

Water-cooled chiller connection assembly. Includes condenser and chilled water pipe, fittings, pipe insulation, valves, thermometers and gauges.

Description	Craft@Hrs	Unit	Material $	Labor $	Equipment $	Total $
2½" chiller conn.	SN@36.0	Ea	2,990.00	1,390.00	—	4,380.00
3" chiller conn.	SN@38.0	Ea	3,290.00	1,470.00	—	4,760.00
4" chiller conn.	SN@51.0	Ea	4,380.00	1,970.00	—	6,350.00
6" chiller conn.	SN@65.0	Ea	7,350.00	2,510.00	1,230.00	11,090.00
8" chiller conn.	SN@77.0	Ea	11,200.00	2,980.00	1,450.00	15,630.00
10" chiller conn.	SN@100.	Ea	20,500.00	3,870.00	1,890.00	26,260.00
12" chiller conn.	SN@125.	Ea	23,700.00	4,830.00	2,360.00	30,890.00

Condensing Units and Cooling Towers

Description	Craft@Hrs	Unit	Material $	Labor $	Equipment $	Total $

Air-cooled condensing unit, set in place only. Unit comes factory assembled complete. Make additional allowances for pipe and electrical connections.

Description	Craft@Hrs	Unit	Material $	Labor $	Equipment $	Total $
3 tons	SN@2.00	Ea	3,960.00	77.30	37.80	4,075.10
5 tons	SN@3.00	Ea	6,740.00	116.00	56.60	6,912.60
7½ tons	SN@4.00	Ea	8,720.00	155.00	75.50	8,950.50
10 tons	SN@4.50	Ea	11,600.00	174.00	85.00	11,859.00
12½ tons	SN@5.00	Ea	13,700.00	193.00	94.40	13,987.40
15 tons	SN@5.50	Ea	15,800.00	213.00	104.00	16,117.00
20 tons	SN@6.00	Ea	20,800.00	232.00	113.00	21,145.00
25 tons	SN@7.00	Ea	26,200.00	271.00	132.00	26,603.00
30 tons	SN@8.00	Ea	31,400.00	309.00	151.00	31,860.00
40 tons	SN@9.00	Ea	41,000.00	348.00	170.00	41,518.00
50 tons	SN@10.0	Ea	51,500.00	387.00	189.00	52,076.00

Galvanized steel cooling tower. Assemble and set in place only. Add for pipe connections.

Description	Craft@Hrs	Unit	Material $	Labor $	Equipment $	Total $
10 tons	SN@16.0	Ea	4,000.00	618.00	302.00	4,920.00
15 tons	SN@16.0	Ea	4,540.00	618.00	302.00	5,460.00
20 tons	SN@18.0	Ea	7,120.00	696.00	340.00	8,156.00
25 tons	SN@20.0	Ea	8,790.00	773.00	378.00	9,941.00
30 tons	SN@25.0	Ea	10,000.00	966.00	472.00	11,438.00
40 tons	SN@30.0	Ea	12,800.00	1,160.00	566.00	14,526.00
50 tons	SN@35.0	Ea	14,000.00	1,350.00	661.00	16,011.00
60 tons	SN@40.0	Ea	14,900.00	1,550.00	755.00	17,205.00
80 tons	SN@45.0	Ea	16,100.00	1,740.00	849.00	18,689.00
100 tons	SN@50.0	Ea	19,100.00	1,930.00	944.00	21,974.00
125 tons	SN@60.0	Ea	23,300.00	2,320.00	1,130.00	26,750.00
150 tons	SN@65.0	Ea	26,800.00	2,510.00	1,230.00	30,540.00
175 tons	SN@70.0	Ea	30,200.00	2,710.00	1,320.00	34,230.00
200 tons	SN@70.0	Ea	33,400.00	2,710.00	1,320.00	37,430.00
300 tons	SN@75.0	Ea	48,000.00	2,900.00	1,420.00	52,320.00
400 tons	SN@80.0	Ea	60,900.00	3,090.00	1,510.00	65,500.00
500 tons	SN@80.0	Ea	72,900.00	3,090.00	1,510.00	77,500.00

Redwood or fir induced-draft cooling tower. Assemble and set in place only. Add for pipe connections.

Description	Craft@Hrs	Unit	Material $	Labor $	Equipment $	Total $
100 tons	SN@75.0	Ea	12,200.00	2,900.00	1,420.00	16,520.00
200 tons	SN@80.0	Ea	20,700.00	3,090.00	1,510.00	25,300.00
300 tons	SN@85.0	Ea	29,900.00	3,290.00	1,600.00	34,790.00
400 tons	SN@90.0	Ea	34,000.00	3,480.00	1,700.00	39,180.00
500 tons	SN@90.0	Ea	39,700.00	3,480.00	1,700.00	44,880.00

Description	Craft@Hrs	Unit	Material $	Labor $	Equipment $	Total $

Redwood or fir forced-draft cooling tower. Assemble and set in place only. Add for pipe connections.

Description	Craft@Hrs	Unit	Material $	Labor $	Equipment $	Total $
100 tons	SN@75.0	Ea	19,800.00	2,900.00	1,420.00	24,120.00
200 tons	SN@80.0	Ea	38,000.00	3,090.00	1,510.00	42,600.00
300 tons	SN@85.0	Ea	44,300.00	3,290.00	1,600.00	49,190.00
400 tons	SN@90.0	Ea	58,600.00	3,480.00	1,700.00	63,780.00
500 tons	SN@90.0	Ea	71,200.00	3,480.00	1,700.00	76,380.00

Cooling tower connection assembly. Includes condenser water pipe, fittings, valves, thermometers and gauges.

Description	Craft@Hrs	Unit	Material $	Labor $	Equipment $	Total $
2½" connection	SN@42.0	Ea	5,520.00	1,620.00	—	7,140.00
3" connection	SN@47.0	Ea	6,260.00	1,820.00	—	8,080.00
4" connection	SN@62.0	Ea	9,010.00	2,400.00	—	11,410.00
6" connection	SN@96.0	Ea	14,200.00	3,710.00	59.40	17,969.40
8" connection	SN@118.	Ea	20,300.00	4,560.00	73.00	24,933.00
10" connection	SN@150.	Ea	28,100.00	5,800.00	92.80	33,992.80
12" connection	SN@215.	Ea	34,300.00	8,310.00	133.00	42,743.00
14" connection	SN@260.	Ea	39,300.00	10,000.00	161.00	49,461.00

Carbon Steel, Schedule 40 with 150# Fittings & Butt-Welded Joints

Schedule 40 carbon steel (ASTM A-53 and A-120) pipe with 150 pound carbon steel welding fittings is commonly used for heating hot water, chilled water, glycol and low pressure steam systems where operating pressures don't exceed 250 PSIG.

Schedule 40 (ASTM A-53B SML) steel pipe with carbon steel welding fittings can also be used for refrigerant piping systems.

The cost estimates in this section are based on the conditions, limitations and wage rates described in the section "How to Use This Book" beginning on page 5.

Equipment cost, where shown, is $4.33 per hour for a 2-ton chain hoist and $6.47 per hour for 200 amp gas welder, trailer mounted.

Description	Craft@Hrs	Unit	Material $	Labor $	Equipment $	Total $

Schedule 40 carbon steel butt-welded horizontal pipe assembly.

Horizontally hung in a building. Assembly includes fittings and hanger assemblies. Based on a reducing tee and a 45-degree elbow every 16 feet for ½" pipe. A reducing tee and a 45-degree elbow every 50 feet for 12" pipe. Hangers spaced to meet plumbing code. *(Distance between fittings in the pipe assembly increases as pipe diameter increases.)* Use these figures for preliminary estimates.

Description	Craft@Hrs	Unit	Material $	Labor $	Equipment $	Total $
½"	P1@.144	LF	1.75	5.28	.47	7.50
¾"	P1@.170	LF	1.90	6.24	.55	8.69
1"	P1@.206	LF	2.33	7.56	.67	10.56
1¼"	P1@.208	LF	3.10	7.63	.67	11.40
1½"	P1@.238	LF	3.33	8.73	.77	12.83
2"	P1@.286	LF	4.40	10.50	.93	15.83
2½"	P1@.344	LF	7.73	12.60	1.11	21.44
3"	P1@.390	LF	9.66	14.30	1.26	25.22
4"	P1@.507	LF	13.40	18.60	1.64	33.64
6"	ER@.712	LF	21.60	28.10	1.10	50.80
8"	ER@.847	LF	37.30	33.40	1.31	72.01
10"	ER@1.02	LF	54.30	40.20	1.57	96.07
12"	ER@1.25	LF	69.60	49.30	1.93	120.83

Schedule 40 carbon steel butt-welded vertical pipe assembly.

Riser assembly, including fittings and riser clamps. Based on a reducing tee, and a riser clamp on every floor. Use these figures for preliminary estimates.

Description	Craft@Hrs	Unit	Material $	Labor $	Equipment $	Total $
½"	P1@.104	LF	1.33	3.82	.34	5.49
¾"	P1@.128	LF	1.49	4.70	.41	6.60
1"	P1@.162	LF	1.92	5.95	.52	8.39
1¼"	P1@.184	LF	3.33	6.75	.60	10.68
1½"	P1@.204	LF	3.72	7.49	.66	11.87
2"	P1@.247	LF	4.94	9.06	.80	14.80
2½"	P1@.304	LF	8.06	11.20	.98	20.24
3"	P1@.351	LF	9.73	12.90	1.14	23.77
4"	P1@.505	LF	12.90	18.50	1.63	33.03
6"	ER@.737	LF	23.10	29.00	1.14	53.24
8"	ER@.878	LF	41.80	34.60	1.35	77.75
10"	ER@1.07	LF	63.60	42.20	1.65	107.45
12"	ER@1.32	LF	83.80	52.00	2.04	137.84

Carbon Steel, Schedule 40 with 150# Fittings & Butt-Welded Joints

Description	Craft@Hrs	Unit	Material $	Labor $	Equipment $	Total $

Schedule 40 carbon steel pipe, plain end, butt-welded joints, pipe only

Description	Craft@Hrs	Unit	Material $	Labor $	Equipment $	Total $
½"	P1@.070	LF	.79	2.57	.23	3.59
¾"	P1@.080	LF	.90	2.94	.26	4.10
1"	P1@.100	LF	1.33	3.67	.32	5.32
1¼"	P1@.110	LF	1.72	4.04	.36	6.12
1½"	P1@.120	LF	2.07	4.40	.39	6.86
2"	P1@.140	LF	2.80	5.14	.45	8.39
2½"	P1@.170	LF	5.22	6.24	.55	12.01
3"	P1@.190	LF	6.86	6.97	.61	14.44
4"	P1@.300	LF	9.76	11.00	.97	21.73
6"	ER@.420	LF	16.90	16.50	.65	34.05
8"	ER@.510	LF	22.00	20.10	.79	42.89
10"	ER@.600	LF	31.30	23.60	.93	55.83
12"	ER@.760	LF	37.70	29.90	1.17	68.77

Schedule 40 carbon steel 45-degree ell, butt-welded joints

Description	Craft@Hrs	Unit	Material $	Labor $	Equipment $	Total $
½"	P1@.220	Ea	2.79	8.07	.71	11.57
¾"	P1@.330	Ea	2.79	12.10	1.07	15.96
1"	P1@.440	Ea	2.79	16.10	1.42	20.31
1¼"	P1@.560	Ea	2.94	20.60	1.81	25.35
1½"	P1@.710	Ea	2.94	26.10	2.30	31.34
2"	P1@.890	Ea	3.05	32.70	2.88	38.63
2½"	P1@1.11	Ea	4.40	40.70	3.59	48.69
3"	P1@1.33	Ea	5.31	48.80	4.30	58.41
4"	P1@1.78	Ea	7.83	65.30	5.76	78.89
6"	ER@2.67	Ea	20.30	105.00	4.12	129.42
8"	ER@3.20	Ea	47.80	126.00	4.94	178.74
10"	ER@4.00	Ea	73.20	158.00	6.17	237.37
12"	ER@4.80	Ea	107.00	189.00	7.41	303.41

Schedule 40 carbon steel 90-degree ell, butt-welded joints

Description	Craft@Hrs	Unit	Material $	Labor $	Equipment $	Total $
½"	P1@.220	Ea	2.72	8.07	.71	11.50
¾"	P1@.330	Ea	2.72	12.10	1.07	15.89
1"	P1@.440	Ea	2.72	16.10	1.42	20.24
1¼"	P1@.560	Ea	2.90	20.60	1.81	25.31
1½"	P1@.670	Ea	2.90	24.60	2.30	29.80
2"	P1@.890	Ea	3.18	32.70	2.88	38.76
2½"	P1@1.11	Ea	5.04	40.70	3.59	49.33
3"	P1@1.33	Ea	6.75	48.80	4.30	59.85
4"	P1@1.78	Ea	10.90	65.30	5.76	81.96
6"	ER@2.67	Ea	30.80	105.00	4.12	139.92
8"	ER@3.20	Ea	70.30	126.00	4.94	201.24
10"	ER@4.00	Ea	126.00	158.00	6.17	290.17
12"	ER@4.80	Ea	186.00	189.00	7.41	382.41

Carbon Steel, Schedule 40 with 150# Fittings & Butt-Welded Joints

Description	Craft@Hrs	Unit	Material $	Labor $	Equipment $	Total $

Schedule 40 carbon steel tee, butt-welded joints

Description	Craft@Hrs	Unit	Material $	Labor $	Equipment $	Total $
½"	P1@.330	Ea	8.63	12.10	1.07	21.80
¾"	P1@.500	Ea	8.63	18.40	1.62	28.65
1"	P1@.670	Ea	8.63	24.60	2.17	35.40
1¼"	P1@.830	Ea	9.60	30.50	2.68	42.78
1½"	P1@1.00	Ea	10.10	36.70	3.23	50.03
2"	P1@1.33	Ea	11.10	48.80	4.30	64.20
2½"	P1@1.67	Ea	15.40	61.30	5.40	82.10
3"	P1@2.00	Ea	17.10	73.40	6.47	96.97
4"	P1@2.67	Ea	23.50	98.00	8.64	130.14
6"	ER@4.00	Ea	42.90	158.00	6.17	207.07
8"	ER@4.80	Ea	99.90	189.00	7.41	296.31
10"	ER@6.00	Ea	171.00	236.00	9.26	416.26
12"	ER@7.20	Ea	257.00	284.00	11.10	552.10

Schedule 40 carbon steel reducing tee, butt-welded joints

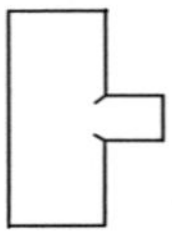

Description	Craft@Hrs	Unit	Material $	Labor $	Equipment $	Total $
1½" x ½"	P1@.940	Ea	13.10	34.50	3.04	50.64
1½" x ¾"	P1@.940	Ea	13.10	34.50	3.04	50.64
1½" x 1"	P1@.940	Ea	13.10	34.50	3.04	50.64
1½" x 1¼"	P1@.940	Ea	13.10	34.50	3.04	50.64
2" x ¾"	P1@1.22	Ea	17.60	44.80	3.04	65.44
2" x 1"	P1@1.22	Ea	17.60	44.80	3.95	66.35
2" x 1¼"	P1@1.22	Ea	17.60	44.80	3.95	66.35
2" x 1½"	P1@1.22	Ea	17.60	44.80	3.95	66.35
2½" x 1"	P1@1.56	Ea	21.80	57.30	5.05	84.15
2½" x 1¼"	P1@1.56	Ea	21.80	57.30	5.05	84.15
2½" x 1½"	P1@1.56	Ea	21.80	57.30	5.05	84.15
2½" x 2"	P1@1.56	Ea	21.80	57.30	5.05	84.15
3" x 1¼"	P1@1.89	Ea	22.40	69.40	6.11	97.91
3" x 2"	P1@1.89	Ea	22.40	69.40	6.11	97.91
3" x 2½"	P1@1.89	Ea	22.40	69.40	6.11	97.91
4" x 1½"	P1@2.44	Ea	23.90	89.50	7.89	121.29
4" x 2"	P1@2.44	Ea	23.90	89.50	7.89	121.29
4" x 2½"	P1@2.44	Ea	23.90	89.50	7.89	121.29
4" x 3"	P1@2.44	Ea	23.90	89.50	7.89	121.29
6" x 2½"	ER@3.78	Ea	54.00	149.00	5.83	208.83
6" x 3"	ER@3.78	Ea	54.00	149.00	5.83	208.83
6" x 4"	ER@3.78	Ea	54.00	149.00	5.83	208.83
8" x 3"	ER@4.40	Ea	133.00	173.00	6.79	312.79
8" x 4"	ER@4.40	Ea	133.00	173.00	6.79	312.79
8" x 6"	ER@4.40	Ea	133.00	173.00	6.79	312.79
10" x 4"	ER@5.60	Ea	226.00	221.00	8.64	455.64
10" x 6"	ER@5.60	Ea	226.00	221.00	8.64	455.64
10" x 8"	ER@5.60	Ea	226.00	221.00	8.64	455.64
12" x 6"	ER@6.80	Ea	335.00	268.00	10.50	613.50
12" x 8"	ER@6.80	Ea	335.00	268.00	10.50	613.50
12" x 10"	ER@6.80	Ea	335.00	268.00	10.50	613.50

Carbon Steel, Schedule 40 with 150# Fittings & Butt-Welded Joints

Description	Craft@Hrs	Unit	Material $	Labor $	Equipment $	Total $

Schedule 40 carbon steel concentric reducer, butt-welded joints

Description	Craft@Hrs	Unit	Material $	Labor $	Equipment $	Total $
¾"	P1@.295	Ea	5.83	10.80	.95	17.58
1"	P1@.395	Ea	6.97	14.50	1.28	22.75
1¼"	P1@.500	Ea	7.37	18.40	1.62	27.39
1½"	P1@.600	Ea	8.21	22.00	1.94	32.15
2"	P1@.800	Ea	7.65	29.40	2.59	39.64
2½"	P1@1.00	Ea	12.10	36.70	3.23	52.03
3"	P1@1.20	Ea	9.66	44.00	3.88	57.54
4"	P1@1.60	Ea	13.80	58.70	5.18	77.68
6"	ER@2.40	Ea	25.40	94.60	3.70	123.70
8"	ER@2.88	Ea	48.40	113.00	4.44	165.84
10"	ER@3.60	Ea	74.10	142.00	5.55	221.65
12"	ER@4.30	Ea	114.00	169.00	6.63	289.63

Schedule 40 carbon steel eccentric reducer, butt-welded joints

Description	Craft@Hrs	Unit	Material $	Labor $	Equipment $	Total $
¾"	P1@.295	Ea	7.16	10.80	.95	18.91
1"	P1@.395	Ea	9.67	14.50	1.28	25.45
1¼"	P1@.500	Ea	11.20	18.40	1.62	31.22
1½"	P1@.600	Ea	11.80	22.00	1.94	35.74
2"	P1@.800	Ea	9.15	29.40	2.59	41.14
2½"	P1@1.00	Ea	14.70	36.70	3.23	54.63
3"	P1@1.20	Ea	13.00	44.00	3.88	60.88
4"	P1@1.60	Ea	18.40	58.70	5.18	82.28
6"	ER@2.40	Ea	36.70	94.60	3.70	135.00
8"	ER@2.88	Ea	67.50	113.00	4.44	184.94
10"	ER@3.60	Ea	116.00	142.00	5.55	263.55
12"	ER@4.30	Ea	177.00	169.00	6.63	352.63

Schedule 40 carbon steel cap, butt-welded joints

Description	Craft@Hrs	Unit	Material $	Labor $	Equipment $	Total $
½"	P1@.150	Ea	6.61	5.51	.49	12.61
¾"	P1@.230	Ea	7.08	8.44	.74	16.26
1"	P1@.300	Ea	7.56	11.00	.96	19.52
1¼"	P1@.390	Ea	9.44	14.30	1.26	25.00
1½"	P1@.470	Ea	9.44	17.20	1.52	28.16
2"	P1@.620	Ea	9.44	22.80	2.01	34.25
2½"	P1@.770	Ea	11.70	28.30	2.49	42.49
3"	P1@.930	Ea	11.20	34.10	3.01	48.31
4"	P1@1.25	Ea	14.30	45.90	4.04	64.24
6"	ER@1.87	Ea	16.60	73.70	2.88	93.18
8"	ER@2.24	Ea	30.00	88.30	3.46	121.76
10"	ER@2.80	Ea	50.70	110.00	4.32	165.02
12"	ER@3.36	Ea	63.60	132.00	5.18	200.78

Carbon Steel, Schedule 40 with 150# Fittings & Butt-Welded Joints

Description	Craft@Hrs	Unit	Material $	Labor $	Equipment $	Total $

Schedule 40 carbon steel union, butt-welded joints

Description	Craft@Hrs	Unit	Material $	Labor $	Equipment $	Total $
½"	P1@.220	Ea	4.77	8.07	.71	13.55
¾"	P1@.330	Ea	6.15	12.10	1.07	19.32
1"	P1@.440	Ea	8.50	16.10	1.42	26.02
1¼"	P1@.560	Ea	17.90	20.60	1.81	40.31
1½"	P1@.670	Ea	19.60	24.60	2.17	46.37
2"	P1@.890	Ea	26.30	32.70	2.88	61.88
2½"	P1@1.11	Ea	70.30	40.70	3.59	114.59
3"	P1@1.33	Ea	83.30	48.80	4.30	136.40

Carbon steel weldolet, Schedule 40

Description	Craft@Hrs	Unit	Material $	Labor $	Equipment $	Total $
½"	P1@.330	Ea	6.63	12.10	1.07	19.80
¾"	P1@.500	Ea	7.32	18.40	1.62	27.34
1"	P1@.670	Ea	8.17	24.60	2.17	34.94
1¼"	P1@.830	Ea	12.40	30.50	2.68	45.58
1½"	P1@1.00	Ea	12.40	36.70	3.23	52.33
2"	P1@1.33	Ea	13.50	48.80	4.30	66.60
2½"	P1@1.67	Ea	36.90	61.30	5.40	103.60
3"	P1@2.00	Ea	39.80	73.40	6.47	119.67
4"	P1@2.67	Ea	51.80	98.00	8.64	158.44
6"	P1@4.00	Ea	211.00	147.00	12.90	370.90
8"	P1@4.80	Ea	402.00	176.00	15.50	593.50
10"	P1@6.00	Ea	552.00	220.00	19.40	791.40
12"	P1@7.20	Ea	675.00	264.00	23.30	962.30

Carbon steel threadolet, Schedule 40

Description	Craft@Hrs	Unit	Material $	Labor $	Equipment $	Total $
¾"	P1@.330	Ea	4.42	12.10	1.07	17.59
1"	P1@.440	Ea	5.23	16.10	1.42	22.75
1¼"	P1@.560	Ea	9.04	20.60	1.81	31.45
1½"	P1@.670	Ea	10.20	24.60	2.17	36.97
2"	P1@.890	Ea	11.80	32.70	2.88	47.38
2½"	P1@1.11	Ea	45.50	40.70	3.59	89.79
3"	P1@1.33	Ea	49.80	48.80	4.30	102.90
4"	P1@1.78	Ea	97.80	65.30	5.76	168.86

150# forged steel companion flange, weld neck

Description	Craft@Hrs	Unit	Material $	Labor $	Equipment $	Total $
2½"	P1@.610	Ea	26.20	22.40	1.97	50.57
3"	P1@.730	Ea	24.40	26.80	2.36	53.56
4"	P1@.980	Ea	33.80	36.00	3.17	72.97
6"	ER@1.47	Ea	54.30	57.90	2.27	114.47
8"	ER@1.77	Ea	110.00	69.70	2.73	182.43
10"	ER@2.20	Ea	180.00	86.70	3.39	270.09
12"	ER@2.64	Ea	262.00	104.00	4.07	370.07

Carbon Steel, Schedule 40 with 150# Fittings & Butt-Welded Joints

Description	Craft@Hrs	Unit	Material $	Labor $	Equipment $	Total $

Class 125 bronze body gate valve, threaded ends

Description	Craft@Hrs	Unit	Material $	Labor $	Equipment $	Total $
½"	P1@.210	Ea	19.20	7.71	—	26.91
¾"	P1@.250	Ea	24.00	9.18	—	33.18
1"	P1@.300	Ea	33.70	11.00	—	44.70
1¼"	P1@.400	Ea	42.50	14.70	—	57.20
1½"	P1@.450	Ea	57.30	16.50	—	73.80
2"	P1@.500	Ea	95.90	18.40	—	114.30
2½"	P1@.750	Ea	158.00	27.50	—	185.50
3"	P1@.950	Ea	227.00	34.90	—	261.90

Class 125 iron body gate valve, flanged ends

Description	Craft@Hrs	Unit	Material $	Labor $	Equipment $	Total $
2"	P1@.500	Ea	367.00	18.40	—	385.40
2½"	P1@.600	Ea	495.00	22.00	—	517.00
3"	P1@.750	Ea	538.00	27.50	—	565.50
4"	P1@1.35	Ea	788.00	49.50	—	837.50
6"	ER@2.50	Ea	1,500.00	98.50	1.55	1,600.05
8"	ER@3.00	Ea	2,470.00	118.00	1.86	2,589.86
10"	ER@4.00	Ea	4,040.00	158.00	2.47	4,200.47
12"	ER@4.50	Ea	5,590.00	177.00	2.78	5,769.78

Class 125 bronze body globe valve, threaded ends

Description	Craft@Hrs	Unit	Material $	Labor $	Equipment $	Total $
½"	P1@.210	Ea	35.70	7.71	—	43.41
¾"	P1@.250	Ea	47.40	9.18	—	56.58
1"	P1@.300	Ea	67.90	11.00	—	78.90
1¼"	P1@.400	Ea	95.60	14.70	—	110.30
1½"	P1@.450	Ea	128.00	16.50	—	144.50
2"	P1@.500	Ea	209.00	18.40	—	227.40

Class 125 iron body globe valve, flanged ends

Description	Craft@Hrs	Unit	Material $	Labor $	Equipment $	Total $
2½"	P1@.600	Ea	416.00	22.00	—	438.00
3"	P1@.750	Ea	498.00	27.50	—	525.50
4"	P1@1.35	Ea	669.00	49.50	—	718.50
6"	ER@2.50	Ea	1,180.00	98.50	1.55	1,280.05
8"	ER@3.00	Ea	1,800.00	118.00	1.86	1,919.86
10"	ER@4.00	Ea	2,750.00	158.00	2.47	2,910.47

200 PSIG iron body butterfly valve, lug-type, lever operated

Description	Craft@Hrs	Unit	Material $	Labor $	Equipment $	Total $
2"	P1@.450	Ea	163.00	16.50	—	179.50
2½"	P1@.450	Ea	169.00	16.50	—	185.50
3"	P1@.550	Ea	177.00	20.20	—	197.20
4"	P1@.550	Ea	221.00	20.20	—	241.20
6"	ER@.800	Ea	356.00	31.50	.49	387.99
8"	ER@.800	Ea	487.00	31.50	.49	518.99
10"	ER@.900	Ea	677.00	35.50	.56	713.06
12"	ER@1.00	Ea	884.00	39.40	.62	924.02

Carbon Steel, Schedule 40 with 150# Fittings & Butt-Welded Joints

Description	Craft@Hrs	Unit	Material $	Labor $	Equipment $	Total $

200 PSIG iron body butterfly valve, wafer-type, lever operated

Description	Craft@Hrs	Unit	Material $	Labor $	Equipment $	Total $
2"	P1@.450	Ea	148.00	16.50	—	164.50
2½"	P1@.450	Ea	151.00	16.50	—	167.50
3"	P1@.550	Ea	163.00	20.20	—	183.20
4"	P1@.550	Ea	197.00	20.20	—	217.20
6"	ER@.800	Ea	330.00	31.50	.49	361.99
8"	ER@.800	Ea	457.00	31.50	.49	488.99
10"	ER@.900	Ea	639.00	35.50	.56	675.06
12"	ER@1.00	Ea	839.00	39.40	.62	879.02

Class 125 bronze body swing check valve, threaded

Description	Craft@Hrs	Unit	Material $	Labor $	Equipment $	Total $
½"	P1@.210	Ea	20.40	7.71	—	28.11
¾"	P1@.250	Ea	29.20	9.18	—	38.38
1"	P1@.300	Ea	38.10	11.00	—	49.10
1¼"	P1@.400	Ea	54.40	14.70	—	69.10
1½"	P1@.450	Ea	76.70	16.50	—	93.20
2"	P1@.500	Ea	127.00	18.40	—	145.40

Class 125 iron body swing check valve, flanged joints

Description	Craft@Hrs	Unit	Material $	Labor $	Equipment $	Total $
2"	P1@.500	Ea	179.00	18.40	—	197.40
2½"	P1@.600	Ea	226.00	22.00	—	248.00
3"	P1@.750	Ea	282.00	27.50	—	309.50
4"	P1@1.35	Ea	414.00	49.50	—	463.50
6"	ER@2.50	Ea	803.00	98.50	1.55	903.05
8"	ER@3.00	Ea	1,440.00	118.00	1.86	1,559.86
10"	ER@4.00	Ea	2,350.00	158.00	2.47	2,510.47
12"	ER@4.50	Ea	3,220.00	177.00	2.78	3,399.78

Class 125 iron body silent check valve, flanged joints

Description	Craft@Hrs	Unit	Material $	Labor $	Equipment $	Total $
2"	P1@.500	Ea	131.00	18.40	—	149.40
2½"	P1@.600	Ea	150.00	22.00	—	172.00
3"	P1@.750	Ea	170.00	27.50	—	197.50
4"	P1@1.35	Ea	222.00	49.50	—	271.50
6"	ER@2.50	Ea	399.00	98.50	1.55	499.05
8"	ER@3.00	Ea	719.00	118.00	1.86	838.86
10"	ER@4.00	Ea	1,110.00	158.00	2.47	1,270.47

Class 125 bronze body strainer, threaded ends

Description	Craft@Hrs	Unit	Material $	Labor $	Equipment $	Total $
½"	P1@.210	Ea	30.10	7.71	—	37.81
¾"	P1@.250	Ea	39.60	9.18	—	48.78
1"	P1@.300	Ea	48.50	11.00	—	59.50
1¼"	P1@.400	Ea	67.80	14.70	—	82.50
1½"	P1@.450	Ea	102.00	16.50	—	118.50
2"	P1@.500	Ea	177.00	18.40	—	195.40

Carbon Steel, Schedule 40 with 150# Fittings & Butt-Welded Joints

Description	Craft@Hrs	Unit	Material $	Labor $	Equipment $	Total $
Class 125 iron body strainer, flanged						
2"	P1@.500	Ea	135.00	18.40	—	153.40
2½"	P1@.600	Ea	151.00	22.00	—	173.00
3"	P1@.750	Ea	175.00	27.50	—	202.50
4"	P1@1.35	Ea	297.00	49.50	—	346.50
6"	ER@2.50	Ea	604.00	98.50	1.55	704.05
8"	ER@3.00	Ea	1,020.00	118.00	1.86	1,139.86
Installation of 2-way control valve, threaded joints						
½"	P1@.210	Ea	—	7.71	—	7.71
¾"	P1@.250	Ea	—	9.18	—	9.18
1"	P1@.300	Ea	—	11.00	—	11.00
1¼"	P1@.400	Ea	—	14.70	—	14.70
1½"	P1@.450	Ea	—	16.50	—	16.50
2"	P1@.500	Ea	—	18.40	—	18.40
2½"	P1@.830	Ea	—	30.50	—	30.50
3"	P1@.990	Ea	—	36.30	—	36.30
Installation of 2-way control valve, flanged joints						
2"	P1@.500	Ea	—	18.40	—	18.40
2½"	P1@.600	Ea	—	22.00	—	22.00
3"	P1@.750	Ea	—	27.50	—	27.50
4"	P1@1.35	Ea	—	49.50	—	49.50
6"	P1@2.50	Ea	—	91.80	—	91.80
8"	P1@3.00	Ea	—	110.00	—	110.00
10"	P1@4.00	Ea	—	147.00	—	147.00
12"	P1@4.50	Ea	—	165.00	—	165.00
Installation of 3-way control valve, threaded joints						
½"	P1@.260	Ea	—	9.54	—	9.54
¾"	P1@.365	Ea	—	13.40	—	13.40
1"	P1@.475	Ea	—	17.40	—	17.40
1¼"	P1@.575	Ea	—	21.10	—	21.10
1½"	P1@.680	Ea	—	25.00	—	25.00
2"	P1@.910	Ea	—	33.40	—	33.40
2½"	P1@1.12	Ea	—	41.10	—	41.10
3"	P1@1.33	Ea	—	48.80	—	48.80
Installation of 3-way control valve, flanged joints						
2"	P1@.910	Ea	—	33.40	—	33.40
2½"	P1@1.12	Ea	—	41.10	—	41.10
3"	P1@1.33	Ea	—	48.80	—	48.80
4"	P1@2.00	Ea	—	73.40	—	73.40
6"	P1@3.70	Ea	—	136.00	—	136.00
8"	P1@4.40	Ea	—	161.00	—	161.00
10"	P1@5.90	Ea	—	217.00	—	217.00
12"	P1@6.50	Ea	—	239.00	—	239.00

Carbon Steel, Schedule 40 with 150# Fittings & Butt-Welded Joints

Description	Craft@Hrs	Unit	Material $	Labor $	Equipment $	Total $
Bolt and gasket sets						
2"	P1@.500	Ea	4.13	18.40	—	22.53
2½"	P1@.650	Ea	4.82	23.90	—	28.72
3"	P1@.750	Ea	8.15	27.50	—	35.65
4"	P1@1.00	Ea	13.60	36.70	—	50.30
6"	P1@1.20	Ea	22.70	44.00	—	66.70
8"	P1@1.25	Ea	25.40	45.90	—	71.30
10"	P1@1.70	Ea	43.70	62.40	—	106.10
12"	P1@2.20	Ea	49.30	80.70	—	130.00
Thermometer with well						
7"	P1@.250	Ea	185.00	9.18	—	194.18
9"	P1@.250	Ea	247.00	9.18	—	256.18
Dial-type pressure gauge						
2½"	P1@.200	Ea	37.70	7.34	—	45.04
3½"	P1@.200	Ea	49.70	7.34	—	57.04
Pressure/temperature tap						
Tap	P1@.150	Ea	17.30	5.51	—	22.81
Hanger with swivel assembly						
2"	P1@.300	Ea	6.59	11.00	—	17.59
2½"	P1@.350	Ea	8.53	12.80	—	21.33
3"	P1@.350	Ea	10.50	12.80	—	23.30
4"	P1@.350	Ea	16.30	12.80	—	29.10
5"	P1@.450	Ea	18.30	16.50	—	34.80
6"	P1@.450	Ea	23.80	16.50	—	40.30
8"	P1@.450	Ea	31.60	16.50	—	48.10
10"	P1@.550	Ea	43.40	20.20	—	63.60
12"	P1@.600	Ea	57.60	22.00	—	79.60
Riser clamp						
2"	P1@.115	Ea	5.35	4.22	—	9.57
2½"	P1@.120	Ea	5.62	4.40	—	10.02
3"	P1@.120	Ea	6.08	4.40	—	10.48
4"	P1@.125	Ea	7.75	4.59	—	12.34
5"	P1@.180	Ea	11.10	6.61	—	17.71
6"	P1@.200	Ea	13.40	7.34	—	20.74
8"	P1@.200	Ea	21.80	7.34	—	29.14
10"	P1@.250	Ea	32.40	9.18	—	41.58
12"	P1@.250	Ea	38.50	9.18	—	47.68

Carbon Steel, Schedule 40 with 150# M.I. Fittings & Threaded Joints

Schedule 40 carbon steel (ASTM A-53 and A-120) pipe with 150 pound malleable iron threaded fittings is commonly used for heating hot water, chilled water, glycol and low pressure steam systems where operating pressures do not exceed 125 PSIG and temperatures are below 250 degrees F., for pipe sizes 2" and smaller. For pipe sizes 2½" and larger, wrought steel fittings with either welded or grooved joints should be used.

Schedule 40 (ASTM A-53B SML) steel pipe with carbon steel welding fittings can also be used for refrigerant piping systems.

This section has been arranged to save the estimator's time by including all normally-used system components such as pipe, fittings, valves, hanger assemblies, riser clamps and miscellaneous items under one heading. Additional items can be found under "Plumbing and Piping Specialties." The cost estimates in this section are based on the conditions, limitations and wage rates described in the section "How to Use This Book" beginning on page 5.

Description	Craft@Hrs	Unit	Material $	Labor $	Equipment $	Total $

Schedule 40 carbon steel threaded horizontal pipe assembly.
Horizontally hung in a building. Assembly includes fittings and hanger assemblies. Based on a reducing tee and a 45-degree elbow every 16 feet for ½" pipe. A reducing tee and a 45-degree elbow every 30 feet for 4" pipe. Hangers spaced to meet plumbing code. *(Distance between fittings in the pipe assembly increases as pipe diameter increases.)* Use these figures for preliminary estimates.

½"	P1@.119	LF	1.72	4.37	—	6.09
¾"	P1@.132	LF	2.02	4.84	—	6.86
1"	P1@.144	LF	2.78	5.28	—	8.06
1¼"	P1@.160	LF	3.63	5.87	—	9.50
1½"	P1@.184	LF	4.03	6.75	—	10.78
2"	P1@.200	LF	6.11	7.34	—	13.45
2½"	P1@.236	LF	12.00	8.66	—	20.66
3"	P1@.271	LF	16.50	9.95	—	26.45
4"	P1@.398	LF	24.60	14.60	—	39.20

Schedule 40 carbon steel threaded vertical pipe assembly.
Riser assembly, including fittings and riser clamps. Based on a reducing tee, and a riser clamp on every floor. Use these figures for preliminary estimates.

½"	P1@.082	LF	1.32	3.01	—	4.33
¾"	P1@.096	LF	1.67	3.52	—	5.19
1"	P1@.121	LF	2.37	4.44	—	6.81
1¼"	P1@.138	LF	3.39	5.06	—	8.45
1½"	P1@.156	LF	3.85	5.73	—	9.58
2"	P1@.173	LF	5.74	6.35	—	12.09
2½"	P1@.200	LF	11.80	7.34	—	19.14
3"	P1@.236	LF	17.20	8.66	—	25.86
4"	P1@.354	LF	22.70	13.00	—	35.70

Carbon Steel, Schedule 40 with 150# M.I. Fittings & Threaded Joints

Description	Craft@Hrs	Unit	Material $	Labor $	Equipment $	Total $

Schedule 40 carbon steel pipe, threaded and coupled, pipe only

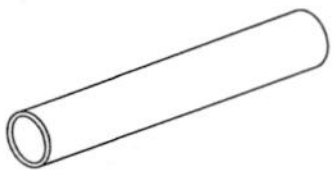

Description	Craft@Hrs	Unit	Material $	Labor $	Equipment $	Total $
½"	P1@.060	LF	1.19	2.20	—	3.39
¾"	P1@.070	LF	1.32	2.57	—	3.89
1"	P1@.090	LF	2.01	3.30	—	5.31
1¼"	P1@.100	LF	2.61	3.67	—	6.28
1½"	P1@.110	LF	3.10	4.04	—	7.14
2"	P1@.120	LF	4.60	4.40	—	9.00
2½"	P1@.130	LF	7.07	4.77	—	11.84
3"	P1@.150	LF	9.38	5.51	—	14.89
4"	P1@.260	LF	13.40	9.54	—	22.94

150# malleable iron 45-degree ell, threaded joints

Description	Craft@Hrs	Unit	Material $	Labor $	Equipment $	Total $
½"	P1@.120	Ea	1.87	4.40	—	6.27
¾"	P1@.130	Ea	2.28	4.77	—	7.05
1"	P1@.180	Ea	2.90	6.61	—	9.51
1¼"	P1@.240	Ea	5.14	8.81	—	13.95
1½"	P1@.300	Ea	6.35	11.00	—	17.35
2"	P1@.380	Ea	9.49	13.90	—	23.39
2½"	P1@.510	Ea	27.40	18.70	—	46.10
3"	P1@.640	Ea	35.60	23.50	—	59.10
4"	P1@.850	Ea	69.40	31.20	—	100.60

150# malleable iron 90-degree ell, threaded joints

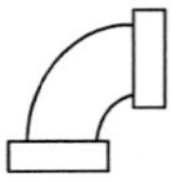

Description	Craft@Hrs	Unit	Material $	Labor $	Equipment $	Total $
½"	P1@.120	Ea	1.16	4.40	—	5.56
¾"	P1@.130	Ea	1.42	4.77	—	6.19
1"	P1@.180	Ea	2.43	6.61	—	9.04
1¼"	P1@.240	Ea	3.96	8.81	—	12.77
1½"	P1@.300	Ea	5.25	11.00	—	16.25
2"	P1@.380	Ea	8.94	13.90	—	22.84
2½"	P1@.510	Ea	19.60	18.70	—	38.30
3"	P1@.640	Ea	29.00	23.50	—	52.50
4"	P1@.850	Ea	67.80	31.20	—	99.00

150# malleable iron tee, threaded joints

Description	Craft@Hrs	Unit	Material $	Labor $	Equipment $	Total $
½"	P1@.180	Ea	1.54	6.61	—	8.15
¾"	P1@.190	Ea	2.23	6.97	—	9.20
1"	P1@.230	Ea	3.78	8.44	—	12.22
1¼"	P1@.310	Ea	6.19	11.40	—	17.59
1½"	P1@.390	Ea	7.60	14.30	—	21.90
2"	P1@.490	Ea	12.90	18.00	—	30.90
2½"	P1@.660	Ea	27.30	24.20	—	51.50
3"	P1@.830	Ea	39.90	30.50	—	70.40
4"	P1@1.10	Ea	95.70	40.40	—	136.10

Carbon Steel, Schedule 40 with 150# M.I. Fittings & Threaded Joints

Description	Craft@Hrs	Unit	Material $	Labor $	Equipment $	Total $

150# malleable iron reducing tee, threaded joints

Description	Craft@Hrs	Unit	Material $	Labor $	Equipment $	Total $
¾" x ½"	P1@.180	Ea	3.30	6.61	—	9.91
1" x ½"	P1@.220	Ea	4.19	8.07	—	12.26
1" x ¾"	P1@.220	Ea	4.19	8.07	—	12.26
1¼" x ½"	P1@.290	Ea	7.60	10.60	—	18.20
1¼" x ¾"	P1@.290	Ea	6.69	10.60	—	17.29
1¼" x 1"	P1@.290	Ea	7.25	10.60	—	17.85
1½" x ½"	P1@.370	Ea	8.92	13.60	—	22.52
1½" x ¾"	P1@.370	Ea	8.42	13.60	—	22.02
1½" x 1"	P1@.370	Ea	8.42	13.60	—	22.02
1½" x 1¼"	P1@.370	Ea	10.40	13.60	—	24.00
2" x 1"	P1@.460	Ea	13.20	16.90	—	30.10
2" x 1¼"	P1@.460	Ea	14.30	16.90	—	31.20
2" x 1½"	P1@.460	Ea	14.30	16.90	—	31.20
4" x 1½"	P1@1.05	Ea	99.10	38.50	—	137.60
4" x 2"	P1@1.05	Ea	99.10	38.50	—	137.60
4" x 3"	P1@1.05	Ea	99.10	38.50	—	137.60

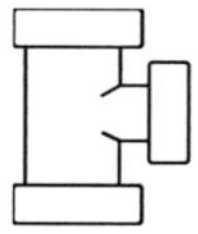

150# malleable iron reducer, threaded joints

Description	Craft@Hrs	Unit	Material $	Labor $	Equipment $	Total $
¾" x ½"	P1@.130	Ea	2.09	4.77	—	6.86
1" x ½"	P1@.160	Ea	3.27	5.87	—	9.14
1" x ¾"	P1@.160	Ea	3.27	5.87	—	9.14
1¼" x ¾"	P1@.210	Ea	3.27	7.71	—	10.98
1¼" x 1"	P1@.210	Ea	4.05	7.71	—	11.76
1½" x 1"	P1@.270	Ea	5.94	9.91	—	15.85
1½" x 1¼"	P1@.270	Ea	5.14	9.91	—	15.05
2" x 1¼"	P1@.340	Ea	8.50	12.50	—	21.00
2" x 1½"	P1@.340	Ea	7.46	12.50	—	19.96
3" x 2"	P1@.530	Ea	22.20	19.50	—	41.70
3" x 2½"	P1@.530	Ea	25.90	19.50	—	45.40
4" x 3"	P1@.750	Ea	25.90	27.50	—	53.40

150# malleable iron cross, threaded joints

Description	Craft@Hrs	Unit	Material $	Labor $	Equipment $	Total $
½"	P1@.280	Ea	5.62	10.30	—	15.92
¾"	P1@.320	Ea	6.86	11.70	—	18.56
1"	P1@.360	Ea	8.42	13.20	—	21.62
1¼"	P1@.480	Ea	13.50	17.60	—	31.10
1½"	P1@.600	Ea	16.70	22.00	—	38.70
2"	P1@.700	Ea	27.80	25.70	—	53.50
2½"	P1@1.02	Ea	51.30	37.40	—	88.70
3"	P1@1.28	Ea	71.10	47.00	—	118.10

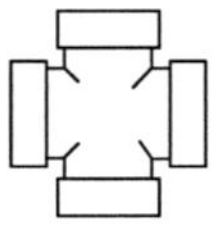

Carbon Steel, Schedule 40 with 150# M.I. Fittings & Threaded Joints

Description	Craft@Hrs	Unit	Material $	Labor $	Equipment $	Total $

150# malleable iron cap, threaded joint

Description	Craft@Hrs	Unit	Material $	Labor $	Equipment $	Total $
½"	P1@.090	Ea	1.17	3.30	—	4.47
¾"	P1@.100	Ea	1.56	3.67	—	5.23
1"	P1@.140	Ea	1.93	5.14	—	7.07
1¼"	P1@.180	Ea	2.49	6.61	—	9.10
1½"	P1@.230	Ea	3.39	8.44	—	11.83
2"	P1@.290	Ea	4.97	10.60	—	15.57
2½"	P1@.380	Ea	11.10	13.90	—	25.00
3"	P1@.480	Ea	16.40	17.60	—	34.00
4"	P1@.640	Ea	27.90	23.50	—	51.40

150# malleable iron plug, threaded joint

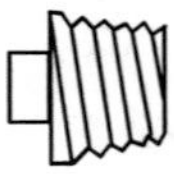

Description	Craft@Hrs	Unit	Material $	Labor $	Equipment $	Total $
½"	P1@.090	Ea	.92	3.30	—	4.22
¾"	P1@.100	Ea	.99	3.67	—	4.66
1"	P1@.140	Ea	.99	5.14	—	6.13
1¼"	P1@.180	Ea	1.57	6.61	—	8.18
1½"	P1@.230	Ea	2.26	8.44	—	10.70
2"	P1@.290	Ea	2.63	10.60	—	13.23
2½"	P1@.380	Ea	6.37	13.90	—	20.27
3"	P1@.480	Ea	7.41	17.60	—	25.01
4"	P1@.640	Ea	13.50	23.50	—	37.00

150# malleable iron union, threaded joints

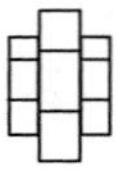

Description	Craft@Hrs	Unit	Material $	Labor $	Equipment $	Total $
½"	P1@.140	Ea	5.10	5.14	—	10.24
¾"	P1@.150	Ea	5.85	5.51	—	11.36
1"	P1@.210	Ea	7.61	7.71	—	15.32
1¼"	P1@.280	Ea	10.90	10.30	—	21.20
1½"	P1@.360	Ea	13.50	13.20	—	26.70
2"	P1@.450	Ea	15.70	16.50	—	32.20
2½"	P1@.610	Ea	46.90	22.40	—	69.30
3"	P1@.760	Ea	56.40	27.90	—	84.30

150# malleable iron coupling, threaded joints

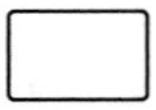

Description	Craft@Hrs	Unit	Material $	Labor $	Equipment $	Total $
½"	P1@.120	Ea	1.56	4.40	—	5.96
¾"	P1@.130	Ea	1.85	4.77	—	6.62
1"	P1@.180	Ea	2.78	6.61	—	9.39
1¼"	P1@.240	Ea	3.59	8.81	—	12.40
1½"	P1@.300	Ea	4.79	11.00	—	15.79
2"	P1@.380	Ea	6.96	13.90	—	20.86

Carbon Steel, Schedule 40 with 150# M.I. Fittings & Threaded Joints

Description	Craft@Hrs	Unit	Material $	Labor $	Equipment $	Total $

Steel pipe nipples — Close nipple, standard right-hand thread

Description	Craft@Hrs	Unit	Material $	Labor $	Equipment $	Total $
½"	P1@.025	Ea	.59	.92	—	1.51
¾"	P1@.025	Ea	.70	.92	—	1.62
1"	P1@.025	Ea	1.09	.92	—	2.01
1¼"	P1@.025	Ea	1.33	.92	—	2.25
1½"	P1@.025	Ea	1.59	.92	—	2.51
2"	P1@.025	Ea	2.05	.92	—	2.97
2½"	P1@.030	Ea	7.99	1.10	—	9.09
3"	P1@.030	Ea	8.65	1.10	—	9.75
4"	P1@.040	Ea	13.80	1.47	—	15.27
6"	P1@.050	Ea	69.70	1.84	—	71.54
8"	P1@.065	Ea	122.00	2.39	—	124.39

Steel pipe nipples — 1½" long, standard right-hand thread

Description	Craft@Hrs	Unit	Material $	Labor $	Equipment $	Total $
½"	P1@.025	Ea	.62	.92	—	1.54
¾"	P1@.025	Ea	.71	.92	—	1.63

Steel pipe nipples — 2" long, standard right-hand thread

Description	Craft@Hrs	Unit	Material $	Labor $	Equipment $	Total $
½"	P1@.025	Ea	.62	.92	—	1.54
¾"	P1@.025	Ea	.71	.92	—	1.63
1"	P1@.025	Ea	1.09	.92	—	2.01
1¼"	P1@.025	Ea	1.33	.92	—	2.25
1½"	P1@.025	Ea	1.59	.92	—	2.51

Steel pipe nipples — 2½" long, standard right-hand thread

Description	Craft@Hrs	Unit	Material $	Labor $	Equipment $	Total $
½"	P1@.025	Ea	.64	.92	—	1.56
¾"	P1@.025	Ea	.75	.92	—	1.67
1"	P1@.025	Ea	1.23	.92	—	2.15
1¼"	P1@.025	Ea	1.57	.92	—	2.49
1½"	P1@.025	Ea	1.96	.92	—	2.88
2"	P1@.025	Ea	2.31	.92	—	3.23

Steel pipe nipples — 3" long, standard right-hand thread

Description	Craft@Hrs	Unit	Material $	Labor $	Equipment $	Total $
½"	P1@.025	Ea	.74	.92	—	1.66
¾"	P1@.025	Ea	.88	.92	—	1.80
1"	P1@.025	Ea	1.28	.92	—	2.20
1¼"	P1@.025	Ea	1.66	.92	—	2.58
1½"	P1@.025	Ea	1.99	.92	—	2.91
2"	P1@.030	Ea	2.47	1.10	—	3.57
2½"	P1@.030	Ea	8.30	1.10	—	9.40
3"	P1@.035	Ea	9.35	1.28	—	10.63

Carbon Steel, Schedule 40 with 150# M.I. Fittings & Threaded Joints

Description	Craft@Hrs	Unit	Material $	Labor $	Equipment $	Total $

Steel pipe nipples — 3½" long, standard right-hand thread

Description	Craft@Hrs	Unit	Material $	Labor $	Equipment $	Total $
½"	P1@.025	Ea	.85	.92	—	1.77
¾"	P1@.025	Ea	1.02	.92	—	1.94
1"	P1@.025	Ea	1.47	.92	—	2.39
1¼"	P1@.025	Ea	1.88	.92	—	2.80
1½"	P1@.030	Ea	2.37	1.10	—	3.47
2"	P1@.030	Ea	2.91	1.10	—	4.01
2½"	P1@.035	Ea	11.00	1.28	—	12.28
3"	P1@.035	Ea	12.70	1.28	—	13.98
4"	P1@.040	Ea	17.40	1.47	—	18.87

Steel pipe nipples — 4" long, standard right-hand thread

Description	Craft@Hrs	Unit	Material $	Labor $	Equipment $	Total $
½"	P1@.025	Ea	.91	.92	—	1.83
¾"	P1@.025	Ea	1.09	.92	—	2.01
1"	P1@.025	Ea	1.55	.92	—	2.47
1¼"	P1@.030	Ea	1.93	1.10	—	3.03
1½"	P1@.030	Ea	2.43	1.10	—	3.53
2"	P1@.030	Ea	3.07	1.10	—	4.17
2½"	P1@.035	Ea	9.55	1.28	—	10.83
3"	P1@.040	Ea	10.70	1.47	—	12.17
4"	P1@.040	Ea	15.50	1.47	—	16.97

Steel pipe nipples — 4½" long, standard right-hand thread

Description	Craft@Hrs	Unit	Material $	Labor $	Equipment $	Total $
½"	P1@.025	Ea	1.09	.92	—	2.01
¾"	P1@.025	Ea	1.22	.92	—	2.14
1"	P1@.025	Ea	1.66	.92	—	2.58
1¼"	P1@.030	Ea	2.62	1.10	—	3.72
1½"	P1@.030	Ea	3.17	1.10	—	4.27
2"	P1@.030	Ea	3.40	1.10	—	4.50
2½"	P1@.035	Ea	12.20	1.28	—	13.48
3"	P1@.040	Ea	15.30	1.47	—	16.77
4"	P1@.050	Ea	19.30	1.84	—	21.14
6"	P1@.070	Ea	83.80	2.57	—	86.37

Steel pipe nipples — 5" long, standard right-hand thread

Description	Craft@Hrs	Unit	Material $	Labor $	Equipment $	Total $
½"	P1@.025	Ea	1.10	.92	—	2.02
¾"	P1@.025	Ea	1.22	.92	—	2.14
1"	P1@.030	Ea	1.81	1.10	—	2.91
1¼"	P1@.030	Ea	2.35	1.10	—	3.45
1½"	P1@.030	Ea	2.81	1.10	—	3.91
2"	P1@.035	Ea	3.53	1.28	—	4.81
2½"	P1@.040	Ea	12.90	1.47	—	14.37
3"	P1@.045	Ea	13.40	1.65	—	15.05
4"	P1@.050	Ea	21.00	1.84	—	22.84
6"	P1@.075	Ea	85.00	2.75	—	87.75
8"	P1@.100	Ea	142.00	3.67	—	145.67

Carbon Steel, Schedule 40 with 150# M.I. Fittings & Threaded Joints

Description	Craft@Hrs	Unit	Material $	Labor $	Equipment $	Total $

Steel pipe nipples — 6" long, standard right-hand thread

Description	Craft@Hrs	Unit	Material $	Labor $	Equipment $	Total $
½"	P1@.025	Ea	1.22	.92	—	2.14
¾"	P1@.025	Ea	1.49	.92	—	2.41
1"	P1@.030	Ea	1.99	1.10	—	3.09
1¼"	P1@.030	Ea	2.48	1.10	—	3.58
1½"	P1@.030	Ea	2.99	1.10	—	4.09
2"	P1@.035	Ea	4.03	1.28	—	5.31
2½"	P1@.040	Ea	11.60	1.47	—	13.07
3"	P1@.045	Ea	12.90	1.65	—	14.55
4"	P1@.060	Ea	19.30	2.20	—	21.50
6"	P1@.085	Ea	95.10	3.12	—	98.22
8"	P1@.120	Ea	152.00	4.40	—	156.40

Steel pipe nipples — 8" long, standard right-hand thread

Description	Craft@Hrs	Unit	Material $	Labor $	Equipment $	Total $
2½"	P1@.050	Ea	15.20	1.84	—	17.04
3"	P1@.055	Ea	18.60	2.02	—	20.62
4"	P1@.070	Ea	29.80	2.57	—	32.37
6"	P1@.100	Ea	99.80	3.67	—	103.47
8"	P1@.150	Ea	170.00	5.51	—	175.51

Steel pipe nipples — 10" long, standard right-hand thread

Description	Craft@Hrs	Unit	Material $	Labor $	Equipment $	Total $
2½"	P1@.055	Ea	18.30	2.02	—	20.32
3"	P1@.060	Ea	22.20	2.20	—	24.40
4"	P1@.080	Ea	36.20	2.94	—	39.14
6"	P1@.115	Ea	111.00	4.22	—	115.22
8"	P1@.165	Ea	192.00	6.06	—	198.06

Steel pipe nipples — 12" long, standard right-hand thread

Description	Craft@Hrs	Unit	Material $	Labor $	Equipment $	Total $
2½"	P1@.060	Ea	17.70	2.20	—	19.90
3"	P1@.065	Ea	21.80	2.39	—	24.19
4"	P1@.080	Ea	35.40	2.94	—	38.34
6"	P1@.130	Ea	127.00	4.77	—	131.77
8"	P1@.190	Ea	210.00	6.97	—	216.97

Class 125 bronze body gate valve, threaded ends

Description	Craft@Hrs	Unit	Material $	Labor $	Equipment $	Total $
½"	P1@.210	Ea	19.20	7.71	—	26.91
¾"	P1@.250	Ea	24.00	9.18	—	33.18
1"	P1@.300	Ea	33.70	11.00	—	44.70
1¼"	P1@.400	Ea	42.50	14.70	—	57.20
1½"	P1@.450	Ea	57.30	16.50	—	73.80
2"	P1@.500	Ea	95.90	18.40	—	114.30
2½"	P1@.750	Ea	158.00	27.50	—	185.50
3"	P1@.950	Ea	226.00	34.90	—	260.90

Carbon Steel, Schedule 40 with 150# M.I. Fittings & Threaded Joints

Description	Craft@Hrs	Unit	Material $	Labor $	Equipment $	Total $

Class 125 iron body gate valve, flanged ends

2"	P1@.500	Ea	328.00	18.40	—	346.40
2½"	P1@.600	Ea	495.00	22.00	—	517.00
3"	P1@.750	Ea	538.00	27.50	—	565.50
4"	P1@1.35	Ea	788.00	49.50	—	837.50

Class 125 bronze body globe valve, threaded ends

½"	P1@.210	Ea	35.70	7.71	—	43.41
¾"	P1@.250	Ea	47.40	9.18	—	56.58
1"	P1@.300	Ea	67.90	11.00	—	78.90
1¼"	P1@.400	Ea	95.60	14.70	—	110.30
1½"	P1@.450	Ea	128.00	16.50	—	144.50
2"	P1@.500	Ea	209.00	18.40	—	227.40

Class 125 iron body globe valve, flanged ends

2"	P1@.500	Ea	267.00	18.40	—	285.40
2½"	P1@.600	Ea	416.00	22.00	—	438.00
3"	P1@.750	Ea	498.00	27.50	—	525.50
4"	P1@1.35	Ea	669.00	49.50	—	718.50

200 PSIG iron body butterfly valve, lug-type, lever operated

2"	P1@.450	Ea	163.00	16.50	—	179.50
2½"	P1@.450	Ea	169.00	16.50	—	185.50
3"	P1@.550	Ea	177.00	20.20	—	197.20
4"	P1@.550	Ea	221.00	20.20	—	241.20

200 PSIG iron body butterfly valve, wafer-type, lever operated

2"	P1@.450	Ea	148.00	16.50	—	164.50
2½"	P1@.450	Ea	151.00	16.50	—	167.50
3"	P1@.550	Ea	163.00	20.20	—	183.20
4"	P1@.550	Ea	197.00	20.20	—	217.20

Class 125 bronze body 2-piece ball valve, threaded ends

½"	P1@.210	Ea	10.10	7.71	—	17.81
¾"	P1@.250	Ea	13.40	9.18	—	22.58
1"	P1@.300	Ea	23.30	11.00	—	34.30
1¼"	P1@.400	Ea	38.80	14.70	—	53.50
1½"	P1@.450	Ea	52.40	16.50	—	68.90
2"	P1@.500	Ea	66.50	18.40	—	84.90
3"	P1@.625	Ea	455.00	22.90	—	477.90
4"	P1@.690	Ea	595.00	25.30	—	620.30

Carbon Steel, Schedule 40 with 150# M.I. Fittings & Threaded Joints

Description	Craft@Hrs	Unit	Material $	Labor $	Equipment $	Total $

Class 125 bronze body swing check valve, threaded

½"	P1@.210	Ea	20.40	7.71	—	28.11
¾"	P1@.250	Ea	29.20	9.18	—	38.38
1"	P1@.300	Ea	38.10	11.00	—	49.10
1¼"	P1@.400	Ea	54.40	14.70	—	69.10
1½"	P1@.450	Ea	76.70	16.50	—	93.20
2"	P1@.500	Ea	127.00	18.40	—	145.40

Class 125 iron body swing check valve, flanged ends

2"	P1@.500	Ea	179.00	18.40	—	197.40
2½"	P1@.600	Ea	226.00	22.00	—	248.00
3"	P1@.750	Ea	282.00	27.50	—	309.50
4"	P1@1.35	Ea	414.00	49.50	—	463.50

Class 125 iron body silent check valve, wafer-type

2"	P1@.500	Ea	131.00	18.40	—	149.40
2½"	P1@.600	Ea	145.00	22.00	—	167.00
3"	P1@.750	Ea	167.00	27.50	—	194.50
4"	P1@1.35	Ea	223.00	49.50	—	272.50

Class 125 bronze body strainer, threaded ends

½"	P1@.210	Ea	29.70	7.71	—	37.41
¾"	P1@.250	Ea	39.00	9.18	—	48.18
1"	P1@.300	Ea	47.80	11.00	—	58.80
1¼"	P1@.400	Ea	66.80	14.70	—	81.50
1½"	P1@.450	Ea	100.00	16.50	—	116.50
2"	P1@.500	Ea	174.00	18.40	—	192.40

Class 125 iron body strainer, flanged ends

2"	P1@.500	Ea	133.00	18.40	—	151.40
2½"	P1@.600	Ea	149.00	22.00	—	171.00
3"	P1@.750	Ea	172.00	27.50	—	199.50
4"	P1@1.35	Ea	516.00	49.50	—	565.50

Installation of 2-way control valve, threaded joints

½"	P1@.210	Ea	—	7.71	—	7.71
¾"	P1@.275	Ea	—	10.10	—	10.10
1"	P1@.350	Ea	—	12.80	—	12.80
1¼"	P1@.430	Ea	—	15.80	—	15.80
1½"	P1@.505	Ea	—	18.50	—	18.50
2"	P1@.675	Ea	—	24.80	—	24.80
2½"	P1@.830	Ea	—	30.50	—	30.50
3"	P1@.990	Ea	—	36.30	—	36.30

Carbon Steel, Schedule 40 with 150# M.I. Fittings & Threaded Joints

Description	Craft@Hrs	Unit	Material $	Labor $	Equipment $	Total $
Installation of 3-way control valve, threaded joints						
½"	P1@.260	Ea	—	9.54	—	9.54
¾"	P1@.365	Ea	—	13.40	—	13.40
1"	P1@.475	Ea	—	17.40	—	17.40
1¼"	P1@.575	Ea	—	21.10	—	21.10
1½"	P1@.680	Ea	—	25.00	—	25.00
2"	P1@.910	Ea	—	33.40	—	33.40
2½"	P1@1.12	Ea	—	41.10	—	41.10
3"	P1@1.33	Ea	—	48.80	—	48.80
Companion flange, threaded						
2"	P1@.290	Ea	42.60	10.60	—	53.20
2½"	P1@.380	Ea	65.60	13.90	—	79.50
3"	P1@.460	Ea	71.00	16.90	—	87.90
4"	P1@.600	Ea	97.60	22.00	—	119.60
Bolt and gasket sets						
2"	P1@.500	Ea	4.17	18.40	—	22.57
2½"	P1@.650	Ea	4.87	23.90	—	28.77
3"	P1@.750	Ea	8.24	27.50	—	35.74
4"	P1@1.00	Ea	13.70	36.70	—	50.40
Thermometer with well						
7"	P1@.250	Ea	184.00	9.18	—	193.18
9"	P1@.250	Ea	190.00	9.18	—	199.18
Dial-type pressure gauge						
2½"	P1@.200	Ea	38.50	7.34	—	45.84
3½"	P1@.200	Ea	50.70	7.34	—	58.04
Pressure/temperature tap						
Tap	P1@.150	Ea	17.70	5.51	—	23.21
Hanger with swivel assembly						
½"	P1@.250	Ea	4.74	9.18	—	13.92
¾"	P1@.250	Ea	4.95	9.18	—	14.13
1"	P1@.250	Ea	5.21	9.18	—	14.39
1¼"	P1@.300	Ea	5.45	11.00	—	16.45
1½"	P1@.300	Ea	6.29	11.00	—	17.29
2"	P1@.300	Ea	6.90	11.00	—	17.90
2½"	P1@.350	Ea	8.94	12.80	—	21.74
3"	P1@.350	Ea	11.00	12.80	—	23.80
4"	P1@.350	Ea	17.40	12.80	—	30.20

Carbon Steel, Schedule 40 with 150# M.I. Fittings & Threaded Joints

Description	Craft@Hrs	Unit	Material $	Labor $	Equipment $	Total $
Riser clamp						
½"	P1@.100	Ea	2.76	3.67	—	6.43
¾"	P1@.100	Ea	4.06	3.67	—	7.73
1"	P1@.100	Ea	4.11	3.67	—	7.78
1¼"	P1@.105	Ea	4.95	3.85	—	8.80
1½"	P1@.110	Ea	5.22	4.04	—	9.26
2"	P1@.115	Ea	5.53	4.22	—	9.75
2½"	P1@.120	Ea	5.83	4.40	—	10.23
3"	P1@.120	Ea	6.30	4.40	—	10.70
4"	P1@.125	Ea	8.03	4.59	—	12.62

Carbon Steel, Schedule 5 with Pressfit Fittings

Schedule 5 carbon steel (ASTM A-53, A135 and A-795) pipe with pressfit fittings. The pressfit jointing method offers economy, speed and reliability for joining ¾" to 2" diameter pipe for fire protection, heating and cooling water systems. *(Use roll-grooved pipe and fittings for 2½" and larger pipe diameters.)*

The pressfit system for carbon steel is BOCA approved. Listed by SBCCI PST and ESI and is UL/ULC listed and FM approved for 175 PSI (1,200 kPa) on fire protection services and is rated to 300 PSI (2,065 kPa) for heating water systems.

The pressfit system requires no special preparation of the pipe ends before assembly. Pipe should be square cut (+/−0.030") and de-burred. Pressfit system products are designed only for use on approved Schedule 5 carbon steel pipe.

O-ring selection as specified by the mechanical design engineer. Type O O-rings are rated for application temperatures of +20 degrees F to +300 degrees F. Type O O-rings are recommended for many oxidizing acids, petroleum oils, halogenated hydrocarbons, lubricants, hydraulic fluids, organic liquids and air with hydrocarbons within the specified temperature range. Not recommended for steam services.

Standard type E O-rings are rated for application temperatures of −30 degrees F to +230 degrees F. Type E O-rings are recommended for hot water service within the specified temperature range plus a variety of dilute acids, oil free air and many chemical services.

Type T O-rings are rated for application temperatures of −20 degrees F to +180 degrees F. Type T O-rings are recommended for petroleum products, air with oil vapors, vegetable or mineral oils within the specified temperature range. Not recommended for hot water services over 150 degrees F., or steam, or hot dry air over 140 degrees F.

A specialized hand-held electric pressfit tool is required to make the pipe/fitting connections. Equipment cost, where shown, is $2.09 per hour for this electric pressfit tool.

Description	Craft@Hrs	Unit	Material $	Labor $	Equipment $	Total $

Schedule 5 carbon steel pipe, plain end, pressfit joints, pipe only. No hangers or fittings.

Description	Craft@Hrs	Unit	Material $	Labor $	Equipment $	Total $
¾"	P1@.030	LF	1.98	1.10	.03	3.11
1"	P1@.040	LF	2.69	1.47	.04	4.20
1¼"	P1@.040	LF	3.44	1.47	.04	4.95
1½"	P1@.050	LF	4.10	1.84	.05	5.99
2"	P1@.050	LF	4.92	1.84	.05	6.81

45-degree pressfit ell, with Type O O-ring (P x P)

Description	Craft@Hrs	Unit	Material $	Labor $	Equipment $	Total $
¾"	P1@.170	Ea	30.20	6.24	.18	36.62
1"	P1@.180	Ea	36.40	6.61	.19	43.20
1¼"	P1@.190	Ea	50.00	6.97	.20	57.17
1½"	P1@.200	Ea	63.70	7.34	.21	71.25
2"	P1@.200	Ea	86.30	7.34	.21	93.85

Description	Craft@Hrs	Unit	Material $	Labor $	Equipment $	Total $

90-degree standard radius pressfit ell, with Type O O-ring (P x P)

Description	Craft@Hrs	Unit	Material $	Labor $	Equipment $	Total $
¾"	P1@.170	Ea	31.80	6.24	.18	38.22
1"	P1@.180	Ea	35.80	6.61	.19	42.60
1¼"	P1@.190	Ea	50.00	6.97	.20	57.17
1½"	P1@.200	Ea	61.20	7.34	.21	68.75
2"	P1@.200	Ea	84.70	7.34	.21	92.25

90-degree short radius pressfit ell, with Type O O-ring (P x P)

Description	Craft@Hrs	Unit	Material $	Labor $	Equipment $	Total $
¾"	P1@.170	Ea	36.40	6.24	.18	42.82
1"	P1@.180	Ea	44.00	6.61	.19	50.80
1¼"	P1@.190	Ea	66.80	6.97	.20	73.97
1½"	P1@.200	Ea	82.70	7.34	.21	90.25
2"	P1@.200	Ea	108.00	7.34	.21	115.55

90-degree standard radius reducing pressfit x FIP ell, with Type O O-ring (P x F)

Description	Craft@Hrs	Unit	Material $	Labor $	Equipment $	Total $
1" x ½"	P1@.185	Ea	41.00	6.79	.19	47.98
1" x ¾"	P1@.185	Ea	41.40	6.79	.19	48.38
1¼" x ½"	P1@.195	Ea	54.60	7.16	.20	61.96
1¼" x ½"	P1@.195	Ea	56.20	7.16	.20	63.56
1½" x ½"	P1@.215	Ea	68.60	7.89	.22	76.71
1½" x ¾"	P1@.215	Ea	69.30	7.89	.22	77.41

90-degree short radius reducing pressfit x FIP ell, with Type O O-ring (P x F)

Description	Craft@Hrs	Unit	Material $	Labor $	Equipment $	Total $
1" x ¾"	P1@.185	Ea	37.30	6.79	.19	44.28
¾" x 1"	P1@.185	Ea	37.80	6.79	.19	44.78
1¼" x ¾"	P1@.195	Ea	52.30	7.16	.20	59.66
1½" x ¾"	P1@.215	Ea	66.40	7.89	.22	74.51

Standard coupling, pressfit, with Type O O-ring (P x P)

Description	Craft@Hrs	Unit	Material $	Labor $	Equipment $	Total $
¾"	P1@.170	Ea	23.70	6.24	.18	30.12
1"	P1@.180	Ea	30.80	6.61	.19	37.60
1¼"	P1@.190	Ea	42.60	6.97	.20	49.77
1½"	P1@.200	Ea	52.40	7.34	.21	59.95
2"	P1@.200	Ea	70.40	7.34	.21	77.95

Carbon Steel, Schedule 5 with Pressfit Fittings

Description	Craft@Hrs	Unit	Material $	Labor $	Equipment $	Total $

Slip coupling, pressfit, with Type O O-ring (P x P)

Description	Craft@Hrs	Unit	Material $	Labor $	Equipment $	Total $
¾"	P1@.170	Ea	32.00	6.24	.18	38.42
1"	P1@.180	Ea	41.00	6.61	.19	47.80
1¼"	P1@.190	Ea	54.60	6.97	.20	61.77
1½"	P1@.200	Ea	67.60	7.34	.21	75.15
2"	P1@.200	Ea	89.90	7.34	.21	97.45

Tee, pressfit, with Type O O-ring (P x P)

Description	Craft@Hrs	Unit	Material $	Labor $	Equipment $	Total $
¾"	P1@.190	Ea	36.30	6.97	.20	43.47
1"	P1@.220	Ea	42.80	8.07	.23	51.10
1¼"	P1@.240	Ea	56.60	8.81	.25	65.66
1½"	P1@.255	Ea	69.10	9.36	.26	78.72
2"	P1@.270	Ea	94.50	9.91	.28	104.69

Reducing tee, pressfit, with Type O O-ring (P x P x P)

Description	Craft@Hrs	Unit	Material $	Labor $	Equipment $	Total $
1" x ¾"	P1@.220	Ea	28.70	8.07	.23	37.00
1¼" x ¾"	P1@.240	Ea	58.20	8.81	.25	67.26
1¼" x 1"	P1@.240	Ea	69.30	8.81	.25	78.36
1½" x ¾"	P1@.255	Ea	67.60	9.36	.26	77.22
1½" x 1"	P1@.255	Ea	78.30	9.36	.26	87.92
2" x ¾"	P1@.270	Ea	89.30	9.91	.28	99.49
2" x 1"	P1@.270	Ea	103.00	9.91	.28	113.19
2" x 1½"	P1@.270	Ea	108.00	9.91	.28	118.19

Male threaded pressfit adapter with Type O O-ring (P x M)

Description	Craft@Hrs	Unit	Material $	Labor $	Equipment $	Total $
¾" x ½"	P1@.180	Ea	22.10	6.61	.19	28.90
¾" x ¾"	P1@.190	Ea	23.70	6.97	.20	30.87
¾" x 1"	P1@.190	Ea	26.20	6.97	.20	33.37
1" x ¾"	P1@.190	Ea	30.70	6.97	.20	37.87
1" x 1"	P1@.190	Ea	32.90	6.97	.20	40.07
1¼" x 1¼"	P1@.200	Ea	45.40	7.34	.21	52.95
1½" x 1½"	P1@.215	Ea	56.80	7.89	.23	64.92
2" x 2"	P1@.225	Ea	73.60	8.26	.24	82.10

Female threaded pressfit adapter with Type O O-ring (P x F)

Description	Craft@Hrs	Unit	Material $	Labor $	Equipment $	Total $
¾" x ½"	P1@.180	Ea	22.10	6.61	.19	28.90
¾" x ¾"	P1@.190	Ea	23.70	6.97	.20	30.87
1" x ½"	P1@.190	Ea	24.60	6.97	.20	31.77
1" x ¾"	P1@.190	Ea	32.90	6.97	.20	40.07
1" x 1"	P1@.200	Ea	33.70	7.34	.21	41.25

Carbon Steel, Schedule 80 with 300# Fittings & Butt-Welded Joints

Schedule 80 carbon steel (ASTM A-53 and A-120) pipe with 300 pound carbon steel welding fittings is commonly used for steam condensate and re-circulating water systems where operating pressures do not exceed 700 PSIG.

The cost estimates in this section are based on the conditions, limitations and wage rates described in the section "How to Use This Book" beginning on page 5.

Equipment cost, where shown, is $4.33 per hour for a 2-ton chain hoist and $6.47 per hour for 200 amp gas welder, trailer mounted.

Description	Craft@Hrs	Unit	Material $	Labor $	Equipment $	Total $

Schedule 80 carbon steel butt-welded horizontal pipe assembly.
Horizontally hung in a building. Assembly includes fittings and hanger assemblies. Based on a reducing tee and a 45-degree elbow every 16 feet for ½" pipe. A reducing tee and a 45-degree elbow every 50 feet for 12" pipe. Hangers spaced to meet plumbing code. *(Distance between fittings in the pipe assembly increases as pipe diameter increases.)* Use these figures for preliminary estimates.

Description	Craft@Hrs	Unit	Material $	Labor $	Equipment $	Total $
½"	P1@.183	LF	4.52	6.72	.59	11.83
¾"	P1@.203	LF	5.05	7.45	.66	13.16
1"	P1@.241	LF	5.77	8.84	.78	15.39
1¼"	P1@.246	LF	6.47	9.03	.80	16.30
1½"	P1@.271	LF	7.02	9.95	.88	17.85
2"	P1@.350	LF	9.14	12.80	1.13	23.07
2½"	P1@.437	LF	13.60	16.00	1.41	31.01
3"	P1@.532	LF	17.70	19.50	1.72	38.92
4"	P1@.601	LF	25.60	22.10	1.94	49.64
6"	ER@.833	LF	35.80	32.80	1.29	69.89
8"	ER@.994	LF	55.30	39.20	1.53	96.03
10"	ER@1.20	LF	93.70	47.30	1.85	142.85
12"	ER@1.43	LF	110.00	56.30	2.21	168.51

Schedule 80 carbon steel butt-welded vertical pipe assembly.
Riser assembly, including fittings and riser clamps. Based on a reducing tee, and a riser clamp on every floor. Use these figures for preliminary estimates.

Description	Craft@Hrs	Unit	Material $	Labor $	Equipment $	Total $
½"	P1@.153	LF	4.45	5.62	.49	10.56
¾"	P1@.163	LF	4.99	5.98	.53	11.50
1"	P1@.196	LF	5.72	7.19	.63	13.54
1¼"	P1@.229	LF	7.13	8.40	.74	16.27
1½"	P1@.240	LF	7.61	8.81	.78	17.20
2"	P1@.310	LF	9.51	11.40	1.00	21.91
2½"	P1@.395	LF	13.80	14.50	1.28	29.58
3"	P1@.516	LF	17.80	18.90	1.67	38.37
4"	P1@.610	LF	25.60	22.40	1.97	49.97
6"	ER@.877	LF	37.80	34.60	1.35	73.75
8"	ER@1.04	LF	60.40	41.00	1.60	103.00
10"	ER@1.28	LF	101.00	50.40	1.97	153.37
12"	ER@1.58	LF	125.00	62.30	2.44	189.74

Carbon Steel, Schedule 80 with 300# Fittings & Butt-Welded Joints

Description	Craft@Hrs	Unit	Material $	Labor $	Equipment $	Total $

Schedule 80 carbon steel pipe, plain end, butt-welded

Description	Craft@Hrs	Unit	Material $	Labor $	Equipment $	Total $
½"	P1@.080	LF	1.80	2.94	.26	5.00
¾"	P1@.090	LF	2.24	3.30	.29	5.83
1"	P1@.110	LF	3.07	4.04	.36	7.47
1¼"	P1@.120	LF	4.09	4.40	.39	8.88
1½"	P1@.130	LF	4.92	4.77	.42	10.11
2"	P1@.170	LF	6.83	6.24	.55	13.62
2½"	P1@.220	LF	10.50	8.07	.71	19.28
3"	P1@.280	LF	13.80	10.30	.91	25.01
4"	P1@.340	LF	20.40	12.50	1.10	34.00
6"	ER@.460	LF	30.80	18.10	.71	49.61
8"	ER@.560	LF	46.40	22.10	.86	69.36
10"	ER@.660	LF	80.40	26.00	1.02	107.42
12"	ER@.840	LF	92.50	33.10	1.30	126.90

Schedule 80 carbon steel 45-degree ell, butt-welded

Description	Craft@Hrs	Unit	Material $	Labor $	Equipment $	Total $
½"	P1@.300	Ea	6.53	11.00	.97	18.50
¾"	P1@.440	Ea	6.53	16.10	1.42	24.05
1"	P1@.590	Ea	6.53	21.70	1.91	30.14
1¼"	P1@.740	Ea	7.20	27.20	2.39	36.79
1½"	P1@.890	Ea	7.50	32.70	2.88	43.08
2"	P1@1.18	Ea	7.04	43.30	3.82	54.16
2½"	P1@1.49	Ea	9.61	54.70	4.82	69.13
3"	P1@1.77	Ea	12.10	65.00	5.73	82.83
4"	P1@2.37	Ea	18.30	87.00	7.67	112.97
6"	ER@3.55	Ea	51.40	140.00	5.48	196.88
8"	ER@4.26	Ea	83.30	168.00	6.57	257.87
10"	ER@5.32	Ea	133.00	210.00	8.21	351.21
12"	ER@6.38	Ea	201.00	251.00	9.84	461.84

Schedule 80 carbon steel 90-degree ell, butt-welded

Description	Craft@Hrs	Unit	Material $	Labor $	Equipment $	Total $
½"	P1@.300	Ea	6.60	11.00	.97	18.57
¾"	P1@.440	Ea	6.60	16.10	1.42	24.12
1"	P1@.590	Ea	6.60	21.70	1.91	30.21
1¼"	P1@.740	Ea	7.59	27.20	2.39	37.18
1½"	P1@.890	Ea	8.14	32.70	2.88	43.72
2"	P1@1.18	Ea	12.20	43.30	3.82	59.32
2½"	P1@1.49	Ea	12.20	54.70	4.82	71.72
3"	P1@1.77	Ea	15.90	65.00	5.73	86.63
4"	P1@2.37	Ea	37.90	87.00	7.67	132.57
6"	ER@3.55	Ea	72.20	140.00	5.48	217.68
8"	ER@4.26	Ea	132.00	168.00	6.57	306.57
10"	ER@5.32	Ea	241.00	210.00	8.21	459.21
12"	ER@6.38	Ea	339.00	251.00	9.84	599.84

Carbon Steel, Schedule 80 with 300# Fittings & Butt-Welded Joints

Description	Craft@Hrs	Unit	Material $	Labor $	Equipment $	Total $

Schedule 80 carbon steel tee, butt-welded

Description	Craft@Hrs	Unit	Material $	Labor $	Equipment $	Total $
½"	P1@.440	Ea	22.20	16.10	1.42	39.72
¾"	P1@.670	Ea	22.20	24.60	2.17	48.97
1"	P1@.890	Ea	22.20	32.70	2.88	57.78
1¼"	P1@1.11	Ea	22.20	40.70	3.59	66.49
1½"	P1@1.33	Ea	25.00	48.80	4.30	78.10
2"	P1@1.77	Ea	21.40	65.00	5.73	92.13
2½"	P1@2.22	Ea	51.60	81.50	7.18	140.28
3"	P1@2.66	Ea	56.70	97.60	8.60	162.90
4"	P1@3.55	Ea	59.80	130.00	11.50	201.30
6"	ER@5.32	Ea	96.10	210.00	8.21	314.31
8"	ER@6.38	Ea	193.00	251.00	9.84	453.84
10"	ER@7.98	Ea	311.00	314.00	12.30	637.30
12"	ER@9.58	Ea	450.00	377.00	14.80	841.80

Carbon Steel, Schedule 80 with 300# Fittings & Butt-Welded Joints

Description	Craft@Hrs	Unit	Material $	Labor $	Equipment $	Total $

Schedule 80 carbon steel reducing tee, butt-welded

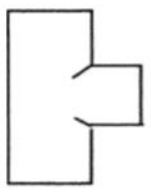

Description	Craft@Hrs	Unit	Material $	Labor $	Equipment $	Total $
1" x ¾"	P1@.810	Ea	27.00	29.70	2.62	59.32
1¼" x ½"	P1@1.04	Ea	27.00	38.20	3.36	68.56
1¼" x ¾"	P1@1.04	Ea	27.00	38.20	3.36	68.56
1¼" x 1"	P1@1.04	Ea	27.00	38.20	3.36	68.56
1½" x ½"	P1@1.26	Ea	27.00	46.20	4.08	77.28
1½" x ¾"	P1@1.26	Ea	27.00	46.20	4.08	77.28
1½" x 1"	P1@1.26	Ea	27.00	46.20	4.08	77.28
1½" x 1¼"	P1@1.26	Ea	27.00	46.20	4.08	77.28
2" x ¾"	P1@1.63	Ea	29.30	59.80	5.27	94.37
2" x 1"	P1@1.63	Ea	29.30	59.80	5.27	94.37
2" x 1¼"	P1@1.63	Ea	29.30	59.80	5.27	94.37
2" x 1½"	P1@1.63	Ea	29.30	59.80	5.27	94.37
2½" x 1"	P1@2.07	Ea	40.70	76.00	6.70	123.40
2½" x 1¼"	P1@2.07	Ea	40.70	76.00	6.70	123.40
2½" x 1½"	P1@2.07	Ea	40.70	76.00	6.70	123.40
2½" x 2"	P1@2.07	Ea	40.70	76.00	6.70	123.40
3" x 1¼"	P1@2.51	Ea	40.70	92.10	8.12	140.92
3" x 2"	P1@2.51	Ea	40.70	92.10	8.12	140.92
3" x 2½"	P1@2.51	Ea	40.70	92.10	8.12	140.92
4" x 1½"	P1@3.25	Ea	63.90	119.00	10.50	193.40
4" x 2"	P1@3.25	Ea	63.90	119.00	10.50	193.40
4" x 2½"	P1@3.25	Ea	63.90	119.00	10.50	193.40
4" x 3"	P1@3.25	Ea	63.90	119.00	10.50	193.40
6" x 2½"	ER@5.03	Ea	118.00	198.00	7.76	323.76
6" x 3"	ER@5.03	Ea	118.00	198.00	7.76	323.76
6" x 4"	ER@5.03	Ea	118.00	198.00	7.76	323.76
8" x 3"	ER@5.85	Ea	210.00	230.00	9.02	449.02
8" x 4"	ER@5.85	Ea	210.00	230.00	9.02	449.02
8" x 6"	ER@5.85	Ea	210.00	230.00	9.02	449.02
10" x 4"	ER@7.45	Ea	334.00	294.00	11.50	639.50
10" x 6"	ER@7.45	Ea	334.00	294.00	11.50	639.50
10" x 8"	ER@7.45	Ea	334.00	294.00	11.50	639.50
12" x 6"	ER@9.04	Ea	612.00	356.00	14.00	982.00
12" x 8"	ER@9.04	Ea	612.00	356.00	14.00	982.00
12" x 10"	ER@9.04	Ea	612.00	356.00	14.00	982.00

Schedule 80 carbon steel concentric reducer, butt-welded

Description	Craft@Hrs	Unit	Material $	Labor $	Equipment $	Total $
¾"	P1@.360	Ea	13.10	13.20	1.16	27.46
1"	P1@.500	Ea	14.10	18.40	1.62	34.12
1¼"	P1@.665	Ea	12.20	24.40	2.15	38.75
1½"	P1@.800	Ea	13.00	29.40	2.59	44.99
2"	P1@1.00	Ea	11.20	36.70	3.23	51.13
2½"	P1@1.33	Ea	14.80	48.80	4.30	67.90
3"	P1@1.63	Ea	14.10	59.80	5.27	79.17
4"	P1@2.00	Ea	18.60	73.40	6.47	98.47
6"	ER@2.96	Ea	40.70	117.00	2.74	160.44
8"	ER@3.90	Ea	60.30	154.00	3.60	217.90
10"	ER@4.79	Ea	83.30	189.00	4.43	276.73
12"	ER@5.85	Ea	122.00	230.00	5.41	357.41

Carbon Steel, Schedule 80 with 300# Fittings & Butt-Welded Joints

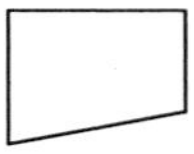

Description	Craft@Hrs	Unit	Material $	Labor $	Equipment $	Total $
Schedule 80 carbon steel eccentric reducer, butt-welded						
¾"	P1@.360	Ea	16.10	13.20	1.16	30.46
1"	P1@.500	Ea	21.00	18.40	1.62	41.02
1¼"	P1@.665	Ea	16.90	24.40	2.15	43.45
1½"	P1@.800	Ea	20.10	29.40	2.59	52.09
2"	P1@1.00	Ea	15.90	36.70	3.23	55.83
2½"	P1@1.33	Ea	19.30	48.80	4.30	72.40
3"	P1@1.63	Ea	18.60	59.80	5.27	83.67
4"	P1@2.00	Ea	26.20	73.40	6.47	106.07
6"	ER@2.96	Ea	59.90	117.00	2.74	179.64
8"	ER@3.90	Ea	89.30	154.00	3.60	246.90
10"	ER@4.79	Ea	156.00	189.00	4.43	349.43
12"	ER@5.85	Ea	243.00	230.00	5.41	478.41
Schedule 80 carbon steel cap, butt-welded						
½"	P1@.180	Ea	13.00	6.61	.58	20.19
¾"	P1@.270	Ea	18.10	9.91	.87	28.88
1"	P1@.360	Ea	8.42	13.20	1.16	22.78
1¼"	P1@.460	Ea	8.42	16.90	1.49	26.81
1½"	P1@.560	Ea	9.21	20.60	1.81	31.62
2"	P1@.740	Ea	7.98	27.20	2.39	37.57
2½"	P1@.920	Ea	8.93	33.80	2.98	45.71
3"	P1@1.11	Ea	9.51	40.70	3.59	53.80
4"	P1@1.50	Ea	12.70	55.10	4.85	72.65
6"	ER@2.24	Ea	23.70	88.30	2.07	114.07
8"	ER@2.68	Ea	36.70	106.00	2.48	145.18
10"	ER@3.36	Ea	54.50	132.00	3.11	189.61
12"	ER@4.00	Ea	72.00	158.00	3.70	233.70
Schedule 80 carbon steel union, butt-welded						
½"	P1@.300	Ea	11.00	11.00	.97	22.97
¾"	P1@.440	Ea	13.10	16.10	1.42	30.62
1"	P1@.590	Ea	17.40	21.70	1.91	41.01
1¼"	P1@.740	Ea	29.30	27.20	2.39	58.89
1½"	P1@.890	Ea	31.80	32.70	2.88	67.38
2"	P1@1.18	Ea	42.40	43.30	3.82	89.52
2½"	P1@1.49	Ea	92.30	54.70	4.82	151.82
3"	P1@1.77	Ea	115.00	65.00	5.73	185.73
Carbon steel weldolet						
½"	P1@.440	Ea	12.00	16.10	1.42	29.52
¾"	P1@.670	Ea	12.30	24.60	2.17	39.07
1"	P1@.890	Ea	13.30	32.70	2.88	48.88
1¼"	P1@1.11	Ea	17.40	40.70	3.59	61.69
1½"	P1@1.33	Ea	17.40	48.80	4.30	70.50
2"	P1@1.77	Ea	17.80	65.00	5.73	88.53
2½"	P1@2.22	Ea	40.60	81.50	7.18	129.28
3"	P1@2.66	Ea	42.10	97.60	8.60	148.30
4"	P1@3.55	Ea	93.40	130.00	11.50	234.90

Description	Craft@Hrs	Unit	Material $	Labor $	Equipment $	Total $

Carbon steel threadolet

½"	P1@.300	Ea	8.78	11.00	.97	20.75
¾"	P1@.440	Ea	9.70	16.10	1.42	27.22
1"	P1@.590	Ea	12.00	21.70	1.91	35.61
1¼"	P1@.740	Ea	53.80	27.20	2.39	83.39
1½"	P1@.890	Ea	53.80	32.70	2.88	89.38
2"	P1@1.18	Ea	70.90	43.30	3.82	118.02

300# forged steel slip-on companion flange, welding-type

2½"	P1@.810	Ea	29.90	29.70	2.62	62.22
3"	P1@.980	Ea	29.90	36.00	3.17	69.07
4"	P1@1.30	Ea	44.20	47.70	4.21	96.11
6"	ER@1.95	Ea	80.00	76.80	1.80	158.60
8"	ER@2.35	Ea	113.00	92.60	2.17	207.77
10"	ER@2.90	Ea	186.00	114.00	2.68	302.68
12"	ER@3.50	Ea	298.00	138.00	3.23	439.23

Class 300 bronze body gate valve, threaded joints

½"	P1@.210	Ea	44.90	7.71	—	52.61
¾"	P1@.250	Ea	58.60	9.18	—	67.78
1"	P1@.300	Ea	71.00	11.00	—	82.00
1¼"	P1@.400	Ea	96.90	14.70	—	111.60
1½"	P1@.450	Ea	122.00	16.50	—	138.50
2"	P1@.500	Ea	180.00	18.40	—	198.40
2½"	P1@.750	Ea	368.00	27.50	—	395.50
3"	P1@.950	Ea	522.00	34.90	—	556.90

Class 250 iron body gate valve, flanged

2½"	P1@.600	Ea	112.00	22.00	—	134.00
3"	P1@.750	Ea	807.00	27.50	—	834.50
4"	P1@1.35	Ea	1,180.00	49.50	—	1,229.50
6"	ER@2.50	Ea	1,590.00	98.50	1.55	1,690.05
8"	ER@3.00	Ea	5,440.00	118.00	1.86	5,559.86
10"	ER@4.00	Ea	8,870.00	158.00	2.47	9,030.47
12"	ER@4.50	Ea	14,500.00	177.00	2.78	14,679.78

Class 300 bronze body globe valve, threaded

½"	P1@.210	Ea	66.90	7.71	—	74.61
¾"	P1@.250	Ea	84.90	9.18	—	94.08
1"	P1@.300	Ea	121.00	11.00	—	132.00
1¼"	P1@.400	Ea	176.00	14.70	—	190.70
1½"	P1@.450	Ea	234.00	16.50	—	250.50
2"	P1@.500	Ea	324.00	18.40	—	342.40

Carbon Steel, Schedule 80 with 300# Fittings & Butt-Welded Joints

Description	Craft@Hrs	Unit	Material $	Labor $	Equipment $	Total $

Class 250 iron body globe valve, flanged

Description	Craft@Hrs	Unit	Material $	Labor $	Equipment $	Total $
2"	P1@.500	Ea	588.00	18.40	—	606.40
2½"	P1@.600	Ea	730.00	22.00	—	752.00
3"	P1@.750	Ea	870.00	27.50	—	897.50
4"	P1@1.35	Ea	1,200.00	49.50	—	1,249.50
6"	ER@2.50	Ea	2,370.00	98.50	1.55	2,470.05
8"	ER@3.00	Ea	5,230.00	118.00	1.86	5,349.86

200 PSIG iron body butterfly valve, lug-type, lever operated

Description	Craft@Hrs	Unit	Material $	Labor $	Equipment $	Total $
2"	P1@.450	Ea	163.00	16.50	—	179.50
2½"	P1@.450	Ea	169.00	16.50	—	185.50
3"	P1@.550	Ea	177.00	20.20	—	197.20
4"	P1@.550	Ea	221.00	20.20	—	241.20
6"	ER@.800	Ea	356.00	31.50	.49	387.99
8"	ER@.800	Ea	487.00	31.50	.49	518.99
10"	ER@.900	Ea	677.00	35.50	.56	713.06
12"	ER@1.00	Ea	884.00	39.40	.62	924.02

200 PSIG iron body butterfly valve, wafer-type, lever operated

Description	Craft@Hrs	Unit	Material $	Labor $	Equipment $	Total $
2"	P1@.450	Ea	148.00	16.50	—	164.50
2½"	P1@.450	Ea	151.00	16.50	—	167.50
3"	P1@.550	Ea	163.00	20.20	—	183.20
4"	P1@.550	Ea	197.00	20.20	—	217.20
6"	ER@.800	Ea	330.00	31.50	.49	361.99
8"	ER@.800	Ea	457.00	31.50	.49	488.99
10"	ER@.900	Ea	639.00	35.50	.56	675.06
12"	ER@1.00	Ea	839.00	39.40	.62	879.02

Class 300 bronze body swing check valve, threaded

Description	Craft@Hrs	Unit	Material $	Labor $	Equipment $	Total $
½"	P1@.210	Ea	139.00	7.71	—	146.71
¾"	P1@.250	Ea	162.00	9.18	—	171.18
1"	P1@.300	Ea	251.00	11.00	—	262.00
1¼"	P1@.400	Ea	288.00	14.70	—	302.70
1½"	P1@.450	Ea	411.00	16.50	—	427.50
2"	P1@.500	Ea	575.00	18.40	—	593.40

Class 250 iron body swing check valve, flanged

Description	Craft@Hrs	Unit	Material $	Labor $	Equipment $	Total $
2"	P1@.500	Ea	461.00	18.40	—	479.40
2½"	P1@.600	Ea	537.00	22.00	—	559.00
3"	P1@.750	Ea	667.00	27.50	—	694.50
4"	P1@1.35	Ea	818.00	49.50	—	867.50
6"	ER@2.50	Ea	1,580.00	98.50	1.55	1,680.05
8"	ER@3.00	Ea	2,970.00	118.00	1.86	3,089.86

Carbon Steel, Schedule 80 with 300# Fittings & Butt-Welded Joints

Description	Craft@Hrs	Unit	Material $	Labor $	Equipment $	Total $
Class 250 iron body silent check valve, flanged						
2"	P1@.500	Ea	193.00	18.40	—	211.40
2½"	P1@.600	Ea	222.00	22.00	—	244.00
3"	P1@.750	Ea	247.00	27.50	—	274.50
4"	P1@1.35	Ea	335.00	49.50	—	384.50
6"	ER@2.50	Ea	588.00	98.50	1.55	688.05
8"	ER@3.00	Ea	1,070.00	118.00	1.86	1,189.86
10"	ER@4.00	Ea	1,700.00	158.00	2.47	1,860.47
Class 250 bronze body strainer, threaded						
½"	P1@.210	Ea	40.00	7.71	—	47.71
¾"	P1@.250	Ea	54.10	9.18	—	63.28
1"	P1@.300	Ea	68.70	11.00	—	79.70
1¼"	P1@.400	Ea	97.30	14.70	—	112.00
1½"	P1@.450	Ea	127.00	16.50	—	143.50
2"	P1@.500	Ea	216.00	18.40	—	234.40
Class 250 iron body strainer, flanged						
2"	P1@.500	Ea	231.00	18.40	—	249.40
2½"	P1@.600	Ea	253.00	22.00	—	275.00
3"	P1@.750	Ea	295.00	27.50	—	322.50
4"	P1@1.35	Ea	524.00	49.50	—	573.50
6"	ER@2.50	Ea	839.00	98.50	1.55	939.05
8"	ER@3.00	Ea	1,530.00	118.00	1.86	1,649.86
Installation of 2-way control valve, threaded joints						
½"	P1@.210	Ea	—	7.71	—	7.71
¾"	P1@.250	Ea	—	9.18	—	9.18
1"	P1@.300	Ea	—	11.00	—	11.00
1¼"	P1@.400	Ea	—	14.70	—	14.70
1½"	P1@.450	Ea	—	16.50	—	16.50
2"	P1@.500	Ea	—	18.40	—	18.40
2½"	P1@.830	Ea	—	30.50	—	30.50
3"	P1@.990	Ea	—	36.30	—	36.30
Installation of 2-way control valve, flanged joints						
2"	P1@.500	Ea	—	18.40	—	18.40
2½"	P1@.600	Ea	—	22.00	—	22.00
3"	P1@.750	Ea	—	27.50	—	27.50
4"	P1@1.35	Ea	—	49.50	—	49.50
6"	P1@2.50	Ea	—	91.80	—	91.80
8"	P1@3.00	Ea	—	110.00	—	110.00
10"	P1@4.00	Ea	—	147.00	—	147.00
12"	P1@4.50	Ea	—	165.00	—	165.00

Carbon Steel, Schedule 80 with 300# Fittings & Butt-Welded Joints

Description	Craft@Hrs	Unit	Material $	Labor $	Equipment $	Total $

Installation of 3-way control valve, threaded joints

Description	Craft@Hrs	Unit	Material $	Labor $	Equipment $	Total $
½"	P1@.260	Ea	—	9.54	—	9.54
¾"	P1@.365	Ea	—	13.40	—	13.40
1"	P1@.475	Ea	—	17.40	—	17.40
1¼"	P1@.575	Ea	—	21.10	—	21.10
1½"	P1@.680	Ea	—	25.00	—	25.00
2"	P1@.910	Ea	—	33.40	—	33.40
2½"	P1@1.12	Ea	—	41.10	—	41.10
3"	P1@1.33	Ea	—	48.80	—	48.80

Installation of 3-way control valve, flanged joints

Description	Craft@Hrs	Unit	Material $	Labor $	Equipment $	Total $
2"	P1@.910	Ea	—	33.40	—	33.40
2½"	P1@1.12	Ea	—	41.10	—	41.10
3"	P1@1.33	Ea	—	48.80	—	48.80
4"	P1@2.00	Ea	—	73.40	—	73.40
6"	P1@3.70	Ea	—	136.00	—	136.00
8"	P1@4.40	Ea	—	161.00	—	161.00
10"	P1@5.90	Ea	—	217.00	—	217.00
12"	P1@6.50	Ea	—	239.00	—	239.00

Bolt and gasket set

Description	Craft@Hrs	Unit	Material $	Labor $	Equipment $	Total $
2"	P1@.500	Ea	4.17	18.40	—	22.57
2½"	P1@.650	Ea	4.87	23.90	—	28.77
3"	P1@.750	Ea	8.24	27.50	—	35.74
4"	P1@1.00	Ea	13.70	36.70	—	50.40
6"	P1@1.20	Ea	22.90	44.00	—	66.90
8"	P1@1.25	Ea	25.70	45.90	—	71.60
10"	P1@1.70	Ea	44.20	62.40	—	106.60
12"	P1@2.20	Ea	49.80	80.70	—	130.50

Thermometer with well

Description	Craft@Hrs	Unit	Material $	Labor $	Equipment $	Total $
7"	P1@.250	Ea	184.00	9.18	—	193.18
9"	P1@.250	Ea	190.00	9.18	—	199.18

Dial-type pressure gauge

Description	Craft@Hrs	Unit	Material $	Labor $	Equipment $	Total $
2½"	P1@.200	Ea	38.50	7.34	—	45.84
3½"	P1@.200	Ea	50.70	7.34	—	58.04

Pressure/temperature tap

Description	Craft@Hrs	Unit	Material $	Labor $	Equipment $	Total $
Tap	P1@.150	Ea	17.70	5.51	—	23.21

Carbon Steel, Schedule 80 with 300# Fittings & Butt-Welded Joints

Description	Craft@Hrs	Unit	Material $	Labor $	Equipment $	Total $
Hanger with swivel assembly						
½"	P1@.250	Ea	6.41	9.18	—	15.59
¾"	P1@.250	Ea	7.05	9.18	—	16.23
1"	P1@.250	Ea	7.62	9.18	—	16.80
1½"	P1@.300	Ea	7.98	11.00	—	18.98
2"	P1@.300	Ea	10.40	11.00	—	21.40
2½"	P1@.350	Ea	13.30	12.80	—	26.10
3"	P1@.350	Ea	16.90	12.80	—	29.70
4"	P1@.350	Ea	28.80	12.80	—	41.60
5"	P1@.450	Ea	33.70	16.50	—	50.20
6"	P1@.450	Ea	37.50	16.50	—	54.00
8"	P1@.450	Ea	49.50	16.50	—	66.00
10"	P1@.550	Ea	67.80	20.20	—	88.00
12"	P1@.600	Ea	89.70	22.00	—	111.70

Description	Craft@Hrs	Unit	Material $	Labor $	Equipment $	Total $
Riser clamp						
½"	P1@.100	Ea	2.76	3.67	—	6.43
¾"	P1@.100	Ea	4.06	3.67	—	7.73
1"	P1@.100	Ea	4.11	3.67	—	7.78
1¼"	P1@.105	Ea	4.95	3.85	—	8.80
1½"	P1@.110	Ea	5.22	4.04	—	9.26
2"	P1@.115	Ea	5.53	4.22	—	9.75
2½"	P1@.120	Ea	5.83	4.40	—	10.23
3"	P1@.120	Ea	6.30	4.40	—	10.70
4"	P1@.125	Ea	8.03	4.59	—	12.62
5"	P1@.180	Ea	11.50	6.61	—	18.11
6"	P1@.200	Ea	13.80	7.34	—	21.14
8"	P1@.200	Ea	22.60	7.34	—	29.94
10"	P1@.250	Ea	33.60	9.18	—	42.78
12"	P1@.250	Ea	39.90	9.18	—	49.08

Carbon Steel, Schedule 80 with 300# M.I. Fittings & Threaded Joints

Schedule 80 carbon steel (ASTM A-53 and A-120) pipe with 300 pound malleable iron threaded fittings is commonly used for heating hot water, chilled water, steam and steam condensate systems where operating pressures do not exceed 250 PSIG for pipe sized 2" and smaller. For pipe sizes 2½" and larger wrought steel fittings either welded or grooved should be used.

The cost estimates in this section are based on the conditions, limitations and wage rates described in the section "How to Use This Book" beginning on page 5.

Equipment cost, where shown, is $4.33 per hour for a 2-ton chain hoist.

Description	Craft@Hrs	Unit	Material $	Labor $	Equipment $	Total $

Schedule 80 carbon steel threaded horizontal pipe assembly.

Horizontally hung in a building. Assembly includes fittings and hanger assemblies. Based on a reducing tee and a 45-degree elbow every 16 feet for ½" pipe and a reducing tee and a 45-degree elbow every 30 feet for 4" pipe. Hangers spaced to meet plumbing code. *(Distance between fittings in the pipe assembly increases as pipe diameter increases.)* Use these figures for preliminary estimates.

Description	Craft@Hrs	Unit	Material $	Labor $	Equipment $	Total $
½"	P1@.121	LF	3.74	4.44	—	8.18
¾"	P1@.125	LF	4.27	4.59	—	8.86
1"	P1@.149	LF	5.35	5.47	—	10.82
1¼"	P1@.163	LF	6.16	5.98	—	12.14
1½"	P1@.186	LF	6.38	6.83	—	13.21
2"	P1@.210	LF	7.82	7.71	—	15.53
2½"	P1@.297	LF	10.40	10.90	—	21.30
3"	P1@.340	LF	22.20	12.50	—	34.70
4"	P1@.409	LF	33.50	15.00	—	48.50

Schedule 80 carbon steel threaded vertical pipe assembly.
Riser assembly, including fittings and riser clamps. Based on a reducing tee, and a riser clamp on every floor. Use these figures for preliminary estimates.

Description	Craft@Hrs	Unit	Material $	Labor $	Equipment $	Total $
½"	P1@.094	LF	3.07	3.45	—	6.52
¾"	P1@.104	LF	3.51	3.82	—	7.33
1"	P1@.124	LF	4.62	4.55	—	9.17
1¼"	P1@.143	LF	5.39	5.25	—	10.64
1½"	P1@.161	LF	5.95	5.91	—	11.86
2"	P1@.179	LF	7.09	6.57	—	13.66
2½"	P1@.257	LF	9.93	9.43	—	19.36
3"	P1@.296	LF	29.80	10.90	—	40.70
4"	P1@.373	LF	49.50	13.70	—	63.20

Carbon Steel, Schedule 80 with 300# M.I. Fittings & Threaded Joints

Description	Craft@Hrs	Unit	Material $	Labor $	Equipment $	Total $

Schedule 80 carbon steel pipe, threaded and coupled joints

Description	Craft@Hrs	Unit	Material $	Labor $	Equipment $	Total $
½"	P1@.070	LF	1.89	2.57	—	4.46
¾"	P1@.080	LF	2.37	2.94	—	5.31
1"	P1@.100	LF	3.23	3.67	—	6.90
1¼"	P1@.110	LF	4.24	4.04	—	8.28
1½"	P1@.120	LF	5.01	4.40	—	9.41
2"	P1@.130	LF	6.84	4.77	—	11.61
2½"	P1@.190	LF	10.70	6.97	—	17.67
3"	P1@.220	LF	14.20	8.07	—	22.27
4"	P1@.290	LF	20.40	10.60	—	31.00

300# malleable iron 45-degree ell, threaded joints

Description	Craft@Hrs	Unit	Material $	Labor $	Equipment $	Total $
½"	P1@.132	Ea	8.96	4.84	—	13.80
¾"	P1@.143	Ea	9.96	5.25	—	15.21
1"	P1@.198	Ea	11.10	7.27	—	18.37
1¼"	P1@.264	Ea	17.00	9.69	—	26.69
1½"	P1@.330	Ea	22.30	12.10	—	34.40
2"	P1@.418	Ea	33.20	15.30	—	48.50
2½"	P1@.561	Ea	52.30	20.60	—	72.90
3"	P1@.704	Ea	69.70	25.80	—	95.50
4"	P1@.935	Ea	134.00	34.30	—	168.30

300# malleable iron 90-degree ell, threaded joints

Description	Craft@Hrs	Unit	Material $	Labor $	Equipment $	Total $
½"	P1@.132	Ea	6.90	4.84	—	11.74
¾"	P1@.143	Ea	7.98	5.25	—	13.23
1"	P1@.198	Ea	10.20	7.27	—	17.47
1¼"	P1@.264	Ea	14.20	9.69	—	23.89
1½"	P1@.330	Ea	17.00	12.10	—	29.10
2"	P1@.418	Ea	23.80	15.30	—	39.10
2½"	P1@.561	Ea	45.30	20.60	—	65.90
3"	P1@.704	Ea	61.90	25.80	—	87.70
4"	P1@.935	Ea	124.00	34.30	—	158.30

Carbon Steel, Schedule 80 with 300# M.I. Fittings & Threaded Joints

Description	Craft@Hrs	Unit	Material $	Labor $	Equipment $	Total $

300# malleable iron tee, threaded joints

Description	Craft@Hrs	Unit	Material $	Labor $	Equipment $	Total $
½"	P1@.198	Ea	9.70	7.27	—	16.97
¾"	P1@.209	Ea	10.70	7.67	—	18.37
1"	P1@.253	Ea	12.80	9.29	—	22.09
1¼"	P1@.341	Ea	17.00	12.50	—	29.50
1½"	P1@.429	Ea	21.10	15.70	—	36.80
2"	P1@.539	Ea	30.90	19.80	—	50.70
2½"	P1@.726	Ea	58.90	26.60	—	85.50
3"	P1@.913	Ea	98.60	33.50	—	132.10
4"	P1@1.21	Ea	168.00	44.40	—	212.40

300# malleable iron reducing tee, threaded joints

Description	Craft@Hrs	Unit	Material $	Labor $	Equipment $	Total $
¾" x ½"	P1@.205	Ea	13.30	7.52	—	20.82
1" x ½"	P1@.205	Ea	16.80	7.52	—	24.32
1" x ¾"	P1@.205	Ea	16.80	7.52	—	24.32
1¼" x ½"	P1@.310	Ea	21.60	11.40	—	33.00
1¼" x ¾"	P1@.310	Ea	21.60	11.40	—	33.00
1¼" x 1"	P1@.310	Ea	21.60	11.40	—	33.00
1½" x ½"	P1@.400	Ea	26.70	14.70	—	41.40
1½" x ¾"	P1@.400	Ea	26.70	14.70	—	41.40
1½" x 1"	P1@.400	Ea	26.70	14.70	—	41.40
1½" x 1¼"	P1@.400	Ea	26.70	14.70	—	41.40
2" x 1"	P1@.500	Ea	39.60	18.40	—	58.00
2" x 1¼"	P1@.500	Ea	39.60	18.40	—	58.00
2" x 1½"	P1@.500	Ea	39.60	18.40	—	58.00
3" x 1¼"	P1@.830	Ea	206.00	30.50	—	236.50
3" x 1½"	P1@.830	Ea	206.00	30.50	—	236.50
3" x 2"	P1@.830	Ea	206.00	30.50	—	236.50

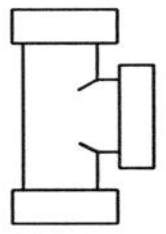

300# malleable iron reducer, threaded joints

Description	Craft@Hrs	Unit	Material $	Labor $	Equipment $	Total $
¾" x ½"	P1@.135	Ea	7.92	4.95	—	12.87
1" x ½"	P1@.170	Ea	9.47	6.24	—	15.71
1" x ¾"	P1@.170	Ea	9.47	6.24	—	15.71
1¼" x ¾"	P1@.220	Ea	12.00	8.07	—	20.07
1¼" x 1"	P1@.220	Ea	12.00	8.07	—	20.07
1½" x 1"	P1@.285	Ea	16.40	10.50	—	26.90
1½" x 1¼"	P1@.285	Ea	16.40	10.50	—	26.90
2" x 1¼"	P1@.360	Ea	23.70	13.20	—	36.90
2" x 1½"	P1@.360	Ea	23.70	13.20	—	36.90
3" x 2"	P1@.550	Ea	57.60	20.20	—	77.80
3" x 2½"	P1@.550	Ea	57.60	20.20	—	77.80

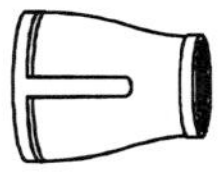

Carbon Steel, Schedule 80 with 300# M.I. Fittings & Threaded Joints

Description	Craft@Hrs	Unit	Material $	Labor $	Equipment $	Total $

300# malleable iron cross, threaded joints

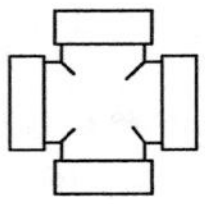

Description	Craft@Hrs	Unit	Material $	Labor $	Equipment $	Total $
½"	P1@.310	Ea	23.30	11.40	—	34.70
¾"	P1@.350	Ea	25.40	12.80	—	38.20
1"	P1@.400	Ea	28.80	14.70	—	43.50
1¼"	P1@.530	Ea	35.60	19.50	—	55.10
1½"	P1@.660	Ea	46.50	24.20	—	70.70
2"	P1@.770	Ea	66.70	28.30	—	95.00

300# malleable iron cap, threaded joint

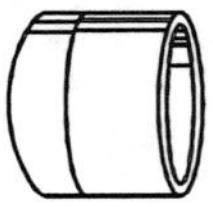

Description	Craft@Hrs	Unit	Material $	Labor $	Equipment $	Total $
½"	P1@.100	Ea	4.65	3.67	—	8.32
¾"	P1@.110	Ea	6.19	4.04	—	10.23
1"	P1@.155	Ea	8.00	5.69	—	13.69
1¼"	P1@.200	Ea	10.90	7.34	—	18.24
1½"	P1@.250	Ea	12.90	9.18	—	22.08
2"	P1@.320	Ea	17.80	11.70	—	29.50
3"	P1@.530	Ea	37.10	19.50	—	56.60
4"	P1@.705	Ea	68.20	25.90	—	94.10

300# malleable iron plug, threaded joint

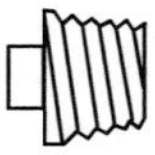

Description	Craft@Hrs	Unit	Material $	Labor $	Equipment $	Total $
½"	P1@.090	Ea	1.15	3.30	—	4.45
¾"	P1@.100	Ea	1.56	3.67	—	5.23
1"	P1@.140	Ea	1.86	5.14	—	7.00
1¼"	P1@.180	Ea	2.01	6.61	—	8.62
1½"	P1@.230	Ea	2.56	8.44	—	11.00
2"	P1@.290	Ea	3.96	10.60	—	14.56
2½"	P1@.380	Ea	5.59	13.90	—	19.49
3"	P1@.480	Ea	9.34	17.60	—	26.94
4"	P1@.640	Ea	26.70	23.50	—	50.20

300# malleable iron union, threaded joints

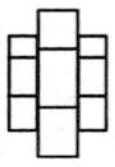

Description	Craft@Hrs	Unit	Material $	Labor $	Equipment $	Total $
½"	P1@.150	Ea	8.55	5.51	—	14.06
¾"	P1@.165	Ea	9.63	6.06	—	15.69
1"	P1@.230	Ea	12.60	8.44	—	21.04
1¼"	P1@.310	Ea	19.30	11.40	—	30.70
1½"	P1@.400	Ea	21.00	14.70	—	35.70
2"	P1@.500	Ea	26.70	18.40	—	45.10
2½"	P1@.670	Ea	58.90	24.60	—	83.50
3"	P1@.840	Ea	99.20	30.80	—	130.00
4"	P1@.960	Ea	277.00	35.20	—	312.20

Carbon Steel, Schedule 80 with 300# M.I. Fittings & Threaded Joints

Description	Craft@Hrs	Unit	Material $	Labor $	Equipment $	Total $

300# malleable iron coupling, threaded joints

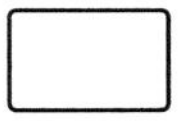

Description	Craft@Hrs	Unit	Material $	Labor $	Equipment $	Total $
½"	P1@.132	Ea	5.19	4.84	—	10.03
¾"	P1@.143	Ea	6.07	5.25	—	11.32
1"	P1@.198	Ea	7.14	7.27	—	14.41
1¼"	P1@.264	Ea	8.55	9.69	—	18.24
1½"	P1@.330	Ea	12.80	12.10	—	24.90
2"	P1@.418	Ea	17.80	15.30	—	33.10
2½"	P1@.561	Ea	22.90	20.60	—	43.50
3"	P1@.704	Ea	37.10	25.80	—	62.90
4"	P1@.935	Ea	88.70	34.30	—	123.00

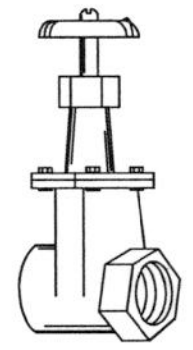

Class 300 bronze body gate valve, threaded joints

Description	Craft@Hrs	Unit	Material $	Labor $	Equipment $	Total $
½"	P1@.210	Ea	44.90	7.71	—	52.61
¾"	P1@.250	Ea	58.30	9.18	—	67.48
1"	P1@.300	Ea	70.90	11.00	—	81.90
1¼"	P1@.400	Ea	96.70	14.70	—	111.40
1½"	P1@.450	Ea	122.00	16.50	—	138.50
2"	P1@.500	Ea	179.00	18.40	—	197.40

Class 250 iron body gate valve, flanged joints

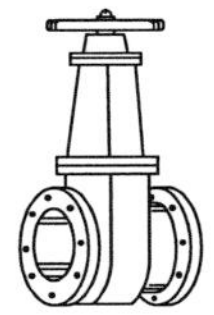

Description	Craft@Hrs	Unit	Material $	Labor $	Equipment $	Total $
2"	P1@.500	Ea	112.00	18.40	—	130.40
2½"	P1@.600	Ea	807.00	22.00	—	829.00
3"	P1@.750	Ea	1,020.00	27.50	—	1,047.50
4"	P1@1.35	Ea	1,740.00	49.50	—	1,789.50

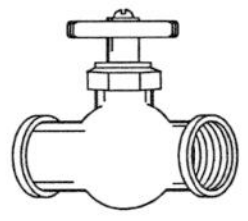

Class 300 bronze body globe valve, threaded joints

Description	Craft@Hrs	Unit	Material $	Labor $	Equipment $	Total $
½"	P1@.210	Ea	66.90	7.71	—	74.61
¾"	P1@.250	Ea	84.90	9.18	—	94.08
1"	P1@.300	Ea	121.00	11.00	—	132.00
1¼"	P1@.400	Ea	176.00	14.70	—	190.70
1½"	P1@.450	Ea	234.00	16.50	—	250.50
2"	P1@.500	Ea	324.00	18.40	—	342.40

Class 250 iron body globe valve, flanged joints

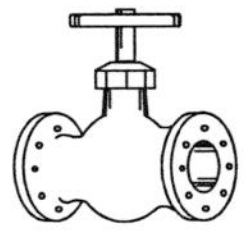

Description	Craft@Hrs	Unit	Material $	Labor $	Equipment $	Total $
2"	P1@.500	Ea	588.00	18.40	—	606.40
2½"	P1@.600	Ea	730.00	22.00	—	752.00
3"	P1@.750	Ea	870.00	27.50	—	897.50
4"	P1@1.35	Ea	1,200.00	49.50	—	1,249.50

200 PSIG iron body butterfly valve, lug-type, lever operated

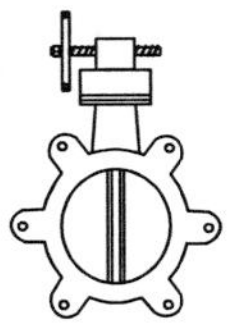

Description	Craft@Hrs	Unit	Material $	Labor $	Equipment $	Total $
2"	P1@.450	Ea	163.00	16.50	—	179.50
2½"	P1@.450	Ea	169.00	16.50	—	185.50
3"	P1@.550	Ea	177.00	20.20	—	197.20
4"	P1@.550	Ea	221.00	20.20	—	241.20

Carbon Steel, Schedule 80 with 300# M.I. Fittings & Threaded Joints

Description	Craft@Hrs	Unit	Material $	Labor $	Equipment $	Total $

200 PSIG iron body butterfly valve, wafer-type, lever operated

Description	Craft@Hrs	Unit	Material $	Labor $	Equipment $	Total $
2"	P1@.450	Ea	148.00	16.50	—	164.50
2½"	P1@.450	Ea	151.00	16.50	—	167.50
3"	P1@.550	Ea	163.00	20.20	—	183.20
4"	P1@.550	Ea	197.00	20.20	—	217.20

Class 150 bronze body 2-piece ball valve, threaded joints

Description	Craft@Hrs	Unit	Material $	Labor $	Equipment $	Total $
½"	P1@.210	Ea	15.20	7.71	—	22.91
¾"	P1@.250	Ea	24.50	9.18	—	33.68
1"	P1@.300	Ea	31.10	11.00	—	42.10
1¼"	P1@.400	Ea	38.80	14.70	—	53.50
1½"	P1@.450	Ea	52.40	16.50	—	68.90
2"	P1@.500	Ea	66.50	18.40	—	84.90
3"	P1@.625	Ea	455.00	22.90	—	477.90
4"	P1@.690	Ea	595.00	25.30	—	620.30

Class 300 bronze body swing check valve, threaded joints

Description	Craft@Hrs	Unit	Material $	Labor $	Equipment $	Total $
½"	P1@.210	Ea	139.00	7.71	—	146.71
¾"	P1@.250	Ea	162.00	9.18	—	171.18
1"	P1@.300	Ea	251.00	11.00	—	262.00
1¼"	P1@.400	Ea	288.00	14.70	—	302.70
1½"	P1@.450	Ea	411.00	16.50	—	427.50
2"	P1@.500	Ea	575.00	18.40	—	593.40

Class 250 iron body swing check valve, flanged joints

Description	Craft@Hrs	Unit	Material $	Labor $	Equipment $	Total $
2"	P1@.500	Ea	461.00	18.40	—	479.40
2½"	P1@.600	Ea	537.00	22.00	—	559.00
3"	P1@.750	Ea	667.00	27.50	—	694.50
4"	P1@1.35	Ea	818.00	49.50	—	867.50

Class 250 iron body silent check valve, wafer-type

Description	Craft@Hrs	Unit	Material $	Labor $	Equipment $	Total $
2"	P1@.500	Ea	193.00	18.40	—	211.40
2½"	P1@.600	Ea	222.00	22.00	—	244.00
3"	P1@.750	Ea	247.00	27.50	—	274.50
4"	P1@1.35	Ea	335.00	49.50	—	384.50

Class 250 bronze body wye strainer, threaded joints

Description	Craft@Hrs	Unit	Material $	Labor $	Equipment $	Total $
½"	P1@.210	Ea	40.00	7.71	—	47.71
¾"	P1@.250	Ea	54.10	9.18	—	63.28
1"	P1@.300	Ea	68.70	11.00	—	79.70
1¼"	P1@.400	Ea	97.30	14.70	—	112.00
1½"	P1@.450	Ea	127.00	16.50	—	143.50
2"	P1@.500	Ea	216.00	18.40	—	234.40

Carbon Steel, Schedule 80 with 300# M.I. Fittings & Threaded Joints

Description	Craft@Hrs	Unit	Material $	Labor $	Equipment $	Total $

Class 250 iron body strainer, flanged ends

Description	Craft@Hrs	Unit	Material $	Labor $	Equipment $	Total $
2"	P1@.500	Ea	231.00	18.40	—	249.40
2½"	P1@.600	Ea	253.00	22.00	—	275.00
3"	P1@.750	Ea	295.00	27.50	—	322.50
4"	P1@1.35	Ea	524.00	49.50	—	573.50

Installation of 2-way control valve, threaded joints

Description	Craft@Hrs	Unit	Material $	Labor $	Equipment $	Total $
½"	P1@.210	Ea	—	7.71	—	7.71
¾"	P1@.275	Ea	—	10.10	—	10.10
1"	P1@.350	Ea	—	12.80	—	12.80
1¼"	P1@.430	Ea	—	15.80	—	15.80
1½"	P1@.505	Ea	—	18.50	—	18.50
2"	P1@.675	Ea	—	24.80	—	24.80
2½"	P1@.830	Ea	—	30.50	—	30.50
3"	P1@.990	Ea	—	36.30	—	36.30

Installation of 3-way control valve, threaded joints

Description	Craft@Hrs	Unit	Material $	Labor $	Equipment $	Total $
½"	P1@.260	Ea	—	9.54	—	9.54
¾"	P1@.365	Ea	—	13.40	—	13.40
1"	P1@.475	Ea	—	17.40	—	17.40
1¼"	P1@.575	Ea	—	21.10	—	21.10
1½"	P1@.680	Ea	—	25.00	—	25.00
2"	P1@.910	Ea	—	33.40	—	33.40
2½"	P1@1.12	Ea	—	41.10	—	41.10
3"	P1@1.33	Ea	—	48.80	—	48.80

Threaded companion flange, 300# malleable iron

Description	Craft@Hrs	Unit	Material $	Labor $	Equipment $	Total $
2"	P1@.290	Ea	104.00	10.60	—	114.60
2½"	P1@.380	Ea	142.00	13.90	—	155.90
3"	P1@.460	Ea	135.00	16.90	—	151.90
4"	P1@.600	Ea	250.00	22.00	—	272.00

Bolt and gasket sets

Description	Craft@Hrs	Unit	Material $	Labor $	Equipment $	Total $
2"	P1@.500	Ea	4.17	18.40	—	22.57
2½"	P1@.650	Ea	4.87	23.90	—	28.77
3"	P1@.750	Ea	8.27	27.50	—	35.77
4"	P1@1.00	Ea	13.80	36.70	—	50.50

Carbon Steel, Schedule 80 with 300# M.I. Fittings & Threaded Joints

Description	Craft@Hrs	Unit	Material $	Labor $	Equipment $	Total $
Thermometer with well						
7"	P1@.250	Ea	184.00	9.18	—	193.18
9"	P1@.250	Ea	190.00	9.18	—	199.18
Dial-type pressure gauge						
2½"	P1@.200	Ea	38.50	7.34	—	45.84
3½"	P1@.200	Ea	50.70	7.34	—	58.04
Pressure/temperature tap						
Tap	P1@.150	Ea	17.70	5.51	—	23.21
Hanger with swivel assembly						
½"	P1@.250	Ea	4.74	9.18	—	13.92
¾"	P1@.250	Ea	4.95	9.18	—	14.13
1"	P1@.250	Ea	5.21	9.18	—	14.39
1¼"	P1@.300	Ea	5.45	11.00	—	16.45
1½"	P1@.300	Ea	6.29	11.00	—	17.29
2"	P1@.300	Ea	6.90	11.00	—	17.90
2½"	P1@.350	Ea	8.94	12.80	—	21.74
3"	P1@.350	Ea	11.00	12.80	—	23.80
4"	P1@.350	Ea	17.10	12.80	—	29.90
Riser clamp						
½"	P1@.100	Ea	2.76	3.67	—	6.43
¾"	P1@.100	Ea	4.06	3.67	—	7.73
1"	P1@.100	Ea	4.11	3.67	—	7.78
1¼"	P1@.105	Ea	4.95	3.85	—	8.80
1½"	P1@.110	Ea	5.22	4.04	—	9.26
2"	P1@.115	Ea	5.53	4.22	—	9.75
2½"	P1@.120	Ea	5.83	4.40	—	10.23
3"	P1@.120	Ea	6.30	4.40	—	10.70
4"	P1@.125	Ea	8.03	4.59	—	12.62

Carbon Steel, Schedule 160 with 3,000-6,000# Fittings

Schedule 160 carbon steel (ASTM A-106) pipe with steel fittings is commonly used for steam, steam condensate and recirculating systems.

The cost estimates in this section are based on the conditions, limitations and wage rates described in the section "How to Use This Book" beginning on page 5.

Description	Craft@Hrs	Unit	Material $	Labor $	Equipment $	Total $

Schedule 160 carbon steel threaded horizontal pipe assembly.

Horizontally hung in a building. Assembly includes fittings and hanger assemblies. Based on a reducing tee and a 45-degree elbow every 16 feet for ½" pipe. A reducing tee and a 45-degree elbow every 30 feet for 4" pipe. Hangers spaced to meet plumbing code. *(Distance between fittings in the pipe assembly increases as pipe diameter increases.)* Use these figures for preliminary estimates.

Description	Craft@Hrs	Unit	Material $	Labor $	Equipment $	Total $
½"	P1@.140	LF	5.45	5.14	—	10.59
¾"	P1@.145	LF	6.39	5.32	—	11.71
1"	P1@.172	LF	8.76	6.31	—	15.07
1¼"	P1@.184	LF	9.61	6.75	—	16.36
1½"	P1@.207	LF	11.10	7.60	—	18.70
2"	P1@.272	LF	16.80	9.98	—	26.78
2½"	P1@.388	LF	25.70	14.20	—	39.90
3"	P1@.434	LF	38.30	15.90	—	54.20
4"	P1@.530	LF	62.90	19.50	—	82.40

Schedule 160 carbon steel socket weld horizontal pipe assembly.

Horizontally hung in a building. Assembly includes fittings and hanger assemblies. Based on a reducing tee and a 45-degree elbow every 16 feet for ½" pipe. A reducing tee and a 45-degree elbow every 30 feet for 4" pipe Hangers spaced to meet plumbing code. *(Distance between fittings in the pipe assembly increases as pipe diameter increases.)* Use these figures for preliminary estimates.

Description	Craft@Hrs	Unit	Material $	Labor $	Equipment $	Total $
½"	P1@.175	LF	5.31	6.42	.57	12.30
¾"	P1@.206	LF	6.14	7.56	.67	14.37
1"	P1@.247	LF	8.45	9.06	.80	18.31
1¼"	P1@.250	LF	9.37	9.18	.81	19.36
1½"	P1@.284	LF	10.20	10.40	.92	21.52
2"	P1@.375	LF	15.90	13.80	1.21	30.91
2½"	P1@.503	LF	24.00	18.50	1.63	44.13
3"	P1@.568	LF	37.30	20.80	1.84	59.94
4"	P1@.709	LF	61.60	26.00	2.29	89.89

Carbon Steel, Schedule 160 with 3,000-6,000# Fittings

Description	Craft@Hrs	Unit	Material $	Labor $	Equipment $	Total $

Schedule 160 carbon steel threaded vertical pipe assembly. Riser assembly, including fittings and riser clamps. Based on a reducing tee, and a riser clamp on every floor. Use these figures for preliminary estimates.

Description	Craft@Hrs	Unit	Material $	Labor $	Equipment $	Total $
½"	P1@.114	LF	4.80	4.18	—	8.98
¾"	P1@.125	LF	5.74	4.59	—	10.33
1"	P1@.148	LF	7.94	5.43	—	13.37
1¼"	P1@.165	LF	9.72	6.06	—	15.78
1½"	P1@.183	LF	10.70	6.72	—	17.42
2"	P1@.242	LF	16.00	8.88	—	24.88
2½"	P1@.348	LF	24.00	12.80	—	36.80
3"	P1@.393	LF	36.90	14.40	—	51.30
4"	P1@.487	LF	59.10	17.90	—	77.00

Schedule 160 carbon steel socket weld vertical pipe assembly. Riser assembly, including fittings and riser clamps. Based on a reducing tee, and a riser clamp on every floor. Use these figures for preliminary estimates.

Description	Craft@Hrs	Unit	Material $	Labor $	Equipment $	Total $
½"	P1@.133	LF	4.67	4.88	.43	9.98
¾"	P1@.162	LF	5.52	5.95	.52	11.99
1"	P1@.199	LF	7.75	7.30	.64	15.69
1¼"	P1@.227	LF	9.03	8.33	.73	18.09
1½"	P1@.255	LF	10.00	9.36	.82	20.18
2"	P1@.341	LF	15.20	12.50	1.10	28.80
2½"	P1@.467	LF	22.30	17.10	1.51	40.91
3"	P1@.532	LF	37.30	19.50	1.72	58.52
4"	P1@.674	LF	58.30	24.70	2.18	85.18

Schedule 160 carbon steel plain end pipe

Description	Craft@Hrs	Unit	Material $	Labor $	Equipment $	Total $
½"	P1@.090	LF	4.43	3.30	—	7.73
¾"	P1@.100	LF	5.14	3.67	—	8.81
1"	P1@.120	LF	7.31	4.40	—	11.71
1¼"	P1@.130	LF	8.59	4.77	—	13.36
1½"	P1@.140	LF	13.40	5.14	—	18.54
2"	P1@.190	LF	11.00	6.97	—	17.97
3"	P1@.310	LF	21.50	11.40	—	32.90
4"	P1@.380	LF	33.40	13.90	—	47.30

Carbon Steel, Schedule 160 with 3,000-6,000# Fittings

Description	Craft@Hrs	Unit	Material $	Labor $	Equipment $	Total $

3,000# carbon steel 45-degree ell, threaded joints

Description	Craft@Hrs	Unit	Material $	Labor $	Equipment $	Total $
½"	P1@.132	Ea	8.61	4.84	—	13.45
¾"	P1@.143	Ea	9.83	5.25	—	15.08
1"	P1@.198	Ea	13.60	7.27	—	20.87
1¼"	P1@.264	Ea	18.70	9.69	—	28.39
1½"	P1@.330	Ea	28.30	12.10	—	40.40
2"	P1@.418	Ea	38.60	15.30	—	53.90
2½"	P1@.561	Ea	95.70	20.60	—	116.30
3"	P1@.704	Ea	155.00	25.80	—	180.80
4"	P1@.935	Ea	286.00	34.30	—	320.30

3,000# carbon steel 45-degree ell, socket-welded joints

Description	Craft@Hrs	Unit	Material $	Labor $	Equipment $	Total $
½"	P1@.300	Ea	6.98	11.00	.97	18.95
¾"	P1@.440	Ea	8.45	16.10	1.42	25.97
1"	P1@.590	Ea	11.00	21.70	1.91	34.61
1¼"	P1@.740	Ea	15.00	27.20	2.39	44.59
1½"	P1@.890	Ea	18.20	32.70	2.88	53.78
2"	P1@1.18	Ea	31.60	43.30	3.82	78.72
2½"	P1@1.49	Ea	81.60	54.70	4.82	141.12
3"	P1@1.77	Ea	137.00	65.00	5.73	207.73
4"	P1@2.37	Ea	271.00	87.00	7.67	365.67

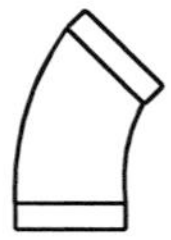

6,000# carbon steel 45-degree ell, threaded joints

Description	Craft@Hrs	Unit	Material $	Labor $	Equipment $	Total $
½"	P1@.132	Ea	14.60	4.84	—	19.44
¾"	P1@.143	Ea	19.90	5.25	—	25.15
1"	P1@.198	Ea	22.30	7.27	—	29.57
1¼"	P1@.264	Ea	59.20	9.69	—	68.89
1½"	P1@.330	Ea	59.20	12.10	—	71.30
2"	P1@.418	Ea	82.70	15.30	—	98.00
2½"	P1@.561	Ea	314.00	20.60	—	334.60
3"	P1@.704	Ea	323.00	25.80	—	348.80

6,000# carbon steel 45-degree ell, socket-welded joints

Description	Craft@Hrs	Unit	Material $	Labor $	Equipment $	Total $
½"	P1@.300	Ea	11.30	11.00	.97	23.27
¾"	P1@.440	Ea	13.00	16.10	1.42	30.52
1"	P1@.590	Ea	17.70	21.70	1.91	41.31
1¼"	P1@.740	Ea	63.50	27.20	2.39	93.09
1½"	P1@.890	Ea	42.90	32.70	2.88	78.48
2"	P1@1.18	Ea	53.30	43.30	3.82	100.42
2½"	P1@1.49	Ea	212.00	54.70	4.82	271.52
3"	P1@1.77	Ea	249.00	65.00	5.73	319.73
4"	P1@2.37	Ea	298.00	87.00	7.67	392.67

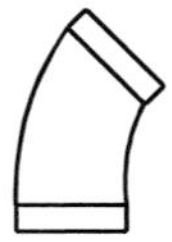

Carbon Steel, Schedule 160 with 3,000-6,000# Fittings

Description	Craft@Hrs	Unit	Material $	Labor $	Equipment $	Total $

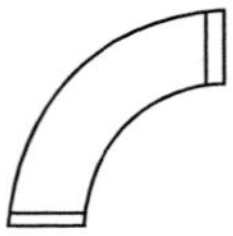

3,000# carbon steel 90-degree ell, threaded joints

Description	Craft@Hrs	Unit	Material $	Labor $	Equipment $	Total $
½"	P1@.132	Ea	6.73	4.84	—	11.57
¾"	P1@.143	Ea	8.10	5.25	—	13.35
1"	P1@.198	Ea	11.70	7.27	—	18.97
1¼"	P1@.264	Ea	19.90	9.69	—	29.59
1½"	P1@.330	Ea	29.80	12.10	—	41.90
2"	P1@.418	Ea	35.80	15.30	—	51.10
2½"	P1@.561	Ea	88.90	20.60	—	109.50
3"	P1@.704	Ea	136.00	25.80	—	161.80
4"	P1@.935	Ea	298.00	34.30	—	332.30

3,000# carbon steel 90-degree ell, socket-welded joints

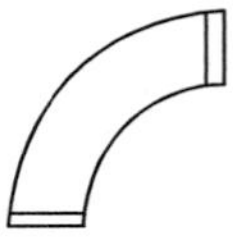

Description	Craft@Hrs	Unit	Material $	Labor $	Equipment $	Total $
½"	P1@.300	Ea	6.67	11.00	.97	18.64
¾"	P1@.440	Ea	6.89	16.10	1.42	24.41
1"	P1@.590	Ea	8.86	21.70	1.91	32.47
1¼"	P1@.740	Ea	14.60	27.20	2.39	44.19
1½"	P1@.890	Ea	18.60	32.70	2.88	54.18
2"	P1@1.18	Ea	28.30	43.30	3.82	75.42
2½"	P1@1.49	Ea	68.00	54.70	4.82	127.52
3"	P1@1.77	Ea	112.00	65.00	5.73	182.73
4"	P1@2.37	Ea	330.00	87.00	7.67	424.67

6,000# carbon steel 90-degree ell, threaded joints

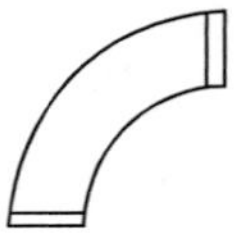

Description	Craft@Hrs	Unit	Material $	Labor $	Equipment $	Total $
½"	P1@.132	Ea	10.90	4.84	—	15.74
¾"	P1@.143	Ea	13.80	5.25	—	19.05
1"	P1@.198	Ea	17.90	7.27	—	25.17
1¼"	P1@.264	Ea	30.10	9.69	—	39.79
1½"	P1@.330	Ea	46.60	12.10	—	58.70
2"	P1@.418	Ea	82.00	15.30	—	97.30
2½"	P1@.561	Ea	122.00	20.60	—	142.60
3"	P1@.704	Ea	276.00	25.80	—	301.80

6,000# carbon steel 90-degree ell, socket-welded joints

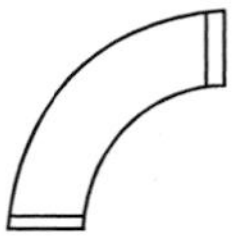

Description	Craft@Hrs	Unit	Material $	Labor $	Equipment $	Total $
½"	P1@.300	Ea	9.24	11.00	.97	21.21
¾"	P1@.440	Ea	10.80	16.10	1.42	28.32
1"	P1@.590	Ea	15.00	21.70	1.91	38.61
1¼"	P1@.740	Ea	22.00	27.20	2.39	51.59
1½"	P1@.890	Ea	33.10	32.70	2.88	68.68
2"	P1@1.18	Ea	48.50	43.30	3.82	95.62
2½"	P1@1.49	Ea	101.00	54.70	4.82	160.52
3"	P1@1.77	Ea	178.00	65.00	5.73	248.73
4"	P1@2.37	Ea	392.00	87.00	7.67	486.67

Description	Craft@Hrs	Unit	Material $	Labor $	Equipment $	Total $

3,000# carbon steel tee, threaded joints

Description	Craft@Hrs	Unit	Material $	Labor $	Equipment $	Total $
½"	P1@.198	Ea	8.89	7.27	—	16.16
¾"	P1@.209	Ea	12.70	7.67	—	20.37
1"	P1@.253	Ea	16.10	9.29	—	25.39
1¼"	P1@.341	Ea	26.80	12.50	—	39.30
1½"	P1@.429	Ea	36.50	15.70	—	52.20
2"	P1@.539	Ea	48.60	19.80	—	68.40
2½"	P1@.726	Ea	117.00	26.60	—	143.60
3"	P1@.913	Ea	220.00	33.50	—	253.50
4"	P1@1.21	Ea	364.00	44.40	—	408.40

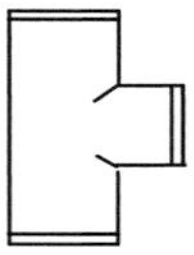

3,000# carbon steel tee, socket-welded joints

Description	Craft@Hrs	Unit	Material $	Labor $	Equipment $	Total $
½"	P1@.440	Ea	7.66	16.10	1.42	25.18
¾"	P1@.670	Ea	9.67	24.60	2.17	36.44
1"	P1@.890	Ea	13.40	32.70	2.88	48.98
1¼"	P1@1.11	Ea	17.90	40.70	3.59	62.19
1½"	P1@1.33	Ea	28.00	48.80	4.30	81.10
2"	P1@1.77	Ea	39.10	65.00	5.73	109.83
2½"	P1@2.22	Ea	94.60	81.50	7.18	183.28
3"	P1@2.66	Ea	223.00	97.60	8.60	329.20
4"	P1@3.55	Ea	357.00	130.00	11.50	498.50

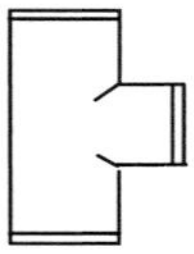

6,000# carbon steel tee, threaded joints

Description	Craft@Hrs	Unit	Material $	Labor $	Equipment $	Total $
½"	P1@.198	Ea	14.20	7.27	—	21.47
¾"	P1@.209	Ea	18.90	7.67	—	26.57
1"	P1@.253	Ea	31.60	9.29	—	40.89
1¼"	P1@.341	Ea	39.30	12.50	—	51.80
1½"	P1@.429	Ea	53.70	15.70	—	69.40
2"	P1@.539	Ea	105.00	19.80	—	124.80
2½"	P1@.726	Ea	377.00	26.60	—	403.60
3"	P1@.913	Ea	381.00	33.50	—	414.50

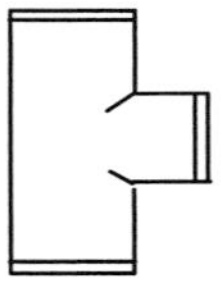

6,000# carbon steel tee, socket-welded joints

Description	Craft@Hrs	Unit	Material $	Labor $	Equipment $	Total $
½"	P1@.440	Ea	12.30	16.10	1.42	29.82
¾"	P1@.670	Ea	16.10	24.60	2.17	42.87
1"	P1@.890	Ea	20.30	32.70	2.88	55.88
1¼"	P1@1.11	Ea	57.70	40.70	3.59	101.99
1½"	P1@1.33	Ea	51.70	48.80	4.30	104.80
2"	P1@1.77	Ea	65.50	65.00	5.73	136.23
2½"	P1@2.22	Ea	138.00	81.50	7.18	226.68
3"	P1@2.66	Ea	233.00	97.60	8.60	339.20
4"	P1@3.55	Ea	442.00	130.00	11.50	583.50

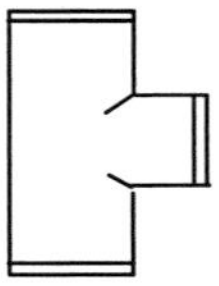

Carbon Steel, Schedule 160 with 3,000-6,000# Fittings

Description	Craft@Hrs	Unit	Material $	Labor $	Equipment $	Total $

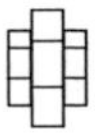

3,000# carbon steel union, threaded joints

Description	Craft@Hrs	Unit	Material $	Labor $	Equipment $	Total $
½"	P1@.150	Ea	9.75	5.51	—	15.26
¾"	P1@.165	Ea	11.70	6.06	—	17.76
1"	P1@.230	Ea	15.40	8.44	—	23.84
1¼"	P1@.310	Ea	25.90	11.40	—	37.30
1½"	P1@.400	Ea	28.30	14.70	—	43.00
2"	P1@.500	Ea	37.70	18.40	—	56.10
2½"	P1@.670	Ea	82.10	24.60	—	106.70
3"	P1@.840	Ea	102.00	30.80	—	132.80

3,000# carbon steel union, socket-welded joints

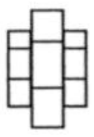

Description	Craft@Hrs	Unit	Material $	Labor $	Equipment $	Total $
½"	P1@.300	Ea	11.70	11.00	.97	23.67
¾"	P1@.440	Ea	13.40	16.10	1.42	30.92
1"	P1@.590	Ea	17.10	21.70	1.91	40.71
1¼"	P1@.740	Ea	27.60	27.20	2.39	57.19
1½"	P1@.890	Ea	28.60	32.70	2.88	64.18
2"	P1@1.18	Ea	40.50	43.30	3.82	87.62
2½"	P1@1.49	Ea	94.30	54.70	4.82	153.82
3"	P1@1.77	Ea	120.00	65.00	5.73	190.73

6,000# carbon steel union, threaded joints

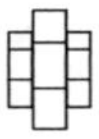

Description	Craft@Hrs	Unit	Material $	Labor $	Equipment $	Total $
½"	P1@.150	Ea	22.90	5.51	—	28.41
¾"	P1@.165	Ea	25.90	6.06	—	31.96
1"	P1@.230	Ea	34.10	8.44	—	42.54
1¼"	P1@.310	Ea	55.80	11.40	—	67.20
1½"	P1@.400	Ea	69.00	14.70	—	83.70
2"	P1@.500	Ea	98.30	18.40	—	116.70

6,000# carbon steel union, socket-welded joints

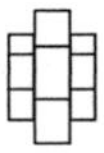

Description	Craft@Hrs	Unit	Material $	Labor $	Equipment $	Total $
½"	P1@.300	Ea	23.60	11.00	.97	35.57
¾"	P1@.440	Ea	27.00	16.10	1.42	44.52
1"	P1@.590	Ea	39.40	21.70	1.19	62.29
1¼"	P1@.740	Ea	60.60	27.20	2.39	90.19
1½"	P1@.890	Ea	70.80	32.70	2.88	106.38
2"	P1@1.18	Ea	107.00	43.30	3.82	154.12

3,000# carbon steel reducer, socket-welded joints

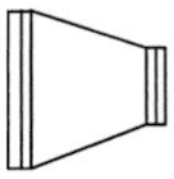

Description	Craft@Hrs	Unit	Material $	Labor $	Equipment $	Total $
¾"	P1@.360	Ea	6.37	13.20	1.16	20.73
1"	P1@.500	Ea	7.74	18.40	1.62	27.76
1¼"	P1@.665	Ea	8.96	24.40	2.15	35.51
1½"	P1@.800	Ea	11.20	29.40	2.59	43.19
2"	P1@1.00	Ea	15.40	36.70	3.23	55.33
2½"	P1@1.33	Ea	106.00	48.80	4.30	159.10
3"	P1@1.63	Ea	134.00	59.80	5.27	199.07

Description	Craft@Hrs	Unit	Material $	Labor $	Equipment $	Total $

6,000# carbon steel reducer, socket-welded joints

Description	Craft@Hrs	Unit	Material $	Labor $	Equipment $	Total $
¾"	P1@.360	Ea	24.50	13.20	1.15	38.85
1"	P1@.500	Ea	25.30	18.40	1.59	45.29
1¼"	P1@.665	Ea	54.20	24.40	2.12	80.72
1½"	P1@.800	Ea	54.20	29.40	2.55	86.15
2"	P1@1.00	Ea	72.10	36.70	3.18	111.98

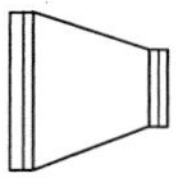

3,000# carbon steel cap, socket-welded joint

Description	Craft@Hrs	Unit	Material $	Labor $	Equipment $	Total $
½"	P1@.180	Ea	3.81	6.61	.58	11.00
¾"	P1@.270	Ea	4.43	9.91	.87	15.21
1"	P1@.360	Ea	6.75	13.20	1.16	21.11
1¼"	P1@.460	Ea	7.94	16.90	1.49	26.33
1½"	P1@.560	Ea	10.80	20.60	1.81	33.21
2"	P1@.740	Ea	16.90	27.20	2.39	46.49
2½"	P1@.920	Ea	90.80	33.80	2.98	127.58
3"	P1@1.11	Ea	93.70	40.70	3.59	137.99
4"	P1@1.50	Ea	142.00	55.10	4.85	201.95

6,000# carbon steel cap, socket-welded joint

Description	Craft@Hrs	Unit	Material $	Labor $	Equipment $	Total $
½"	P1@.180	Ea	7.31	6.61	.58	14.50
¾"	P1@.270	Ea	10.60	9.91	.87	21.38
1"	P1@.360	Ea	11.20	13.20	1.16	25.56
1¼"	P1@.460	Ea	26.30	16.90	1.49	44.69
1½"	P1@.560	Ea	34.10	20.60	1.81	56.51
2"	P1@.740	Ea	50.70	27.20	2.39	80.29

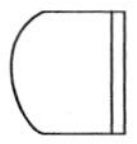

3,000# carbon steel weldolet

Description	Craft@Hrs	Unit	Material $	Labor $	Equipment $	Total $
½"	P1@.580	Ea	13.50	21.30	1.88	36.68
¾"	P1@.890	Ea	13.80	32.70	2.88	49.38
1"	P1@1.18	Ea	15.10	43.30	3.82	62.22
1¼"	P1@1.48	Ea	18.90	54.30	4.79	77.99
1½"	P1@1.75	Ea	18.90	64.20	5.66	88.76
2"	P1@2.35	Ea	20.40	86.20	7.60	114.20
2½"	P1@2.90	Ea	46.20	106.00	9.38	161.58
3"	P1@3.50	Ea	47.50	128.00	11.30	186.80
4"	P1@4.75	Ea	58.80	174.00	15.40	248.20

6,000# carbon steel weldolet

Description	Craft@Hrs	Unit	Material $	Labor $	Equipment $	Total $
½"	P1@.690	Ea	58.80	25.30	2.23	86.33
¾"	P1@1.05	Ea	61.30	38.50	3.40	103.20
1"	P1@1.40	Ea	64.30	51.40	4.53	120.23
1¼"	P1@1.75	Ea	77.40	64.20	5.66	147.26
1½"	P1@2.10	Ea	86.90	77.10	6.79	170.79
2"	P1@2.80	Ea	100.00	103.00	9.06	212.06
2½"	P1@3.50	Ea	147.00	128.00	11.30	286.30
3"	P1@4.20	Ea	163.00	154.00	13.60	330.60
4"	P1@5.70	Ea	279.00	209.00	18.40	506.40

Carbon Steel, Schedule 160 with 3,000-6,000# Fittings

Description	Craft@Hrs	Unit	Material $	Labor $	Equipment $	Total $

3,000# carbon steel threadolet

Description	Craft@Hrs	Unit	Material $	Labor $	Equipment $	Total $
½"	P1@.400	Ea	10.20	14.70	—	24.90
¾"	P1@.580	Ea	11.30	21.30	—	32.60
1"	P1@.790	Ea	13.80	29.00	—	42.80
1¼"	P1@.980	Ea	62.70	36.00	—	98.70
1½"	P1@1.20	Ea	62.70	44.00	—	106.70
2"	P1@1.60	Ea	82.80	58.70	—	141.50

6,000# carbon steel threadolet

Description	Craft@Hrs	Unit	Material $	Labor $	Equipment $	Total $
½"	P1@.480	Ea	15.60	17.60	—	33.20
¾"	P1@.690	Ea	17.00	25.30	—	42.30
1"	P1@.950	Ea	24.40	34.90	—	59.30
1¼"	P1@1.20	Ea	90.80	44.00	—	134.80
1½"	P1@1.45	Ea	90.80	53.20	—	144.00
2"	P1@1.90	Ea	148.00	69.70	—	217.70

Class 300 forged steel companion flange, weld neck

Description	Craft@Hrs	Unit	Material $	Labor $	Equipment $	Total $
½"	P1@.240	Ea	18.40	8.81	.78	27.99
¾"	P1@.320	Ea	18.40	11.70	1.04	31.14
1"	P1@.340	Ea	18.40	12.50	1.10	32.00
1¼"	P1@.425	Ea	23.90	15.60	1.37	40.87
1½"	P1@.560	Ea	23.90	20.60	1.81	46.31
2"	P1@.740	Ea	26.10	27.20	2.39	55.69
2½"	P1@.810	Ea	38.30	29.70	2.62	70.62
3"	P1@.980	Ea	39.20	36.00	3.17	78.37
4"	P1@1.30	Ea	57.80	47.70	4.21	109.71

Class 300 forged steel companion flange, slip on

Description	Craft@Hrs	Unit	Material $	Labor $	Equipment $	Total $
½"	P1@.240	Ea	14.70	8.81	—	23.51
¾"	P1@.320	Ea	14.70	11.70	—	26.40
1"	P1@.340	Ea	14.70	12.50	—	27.20
1½"	P1@.425	Ea	22.00	15.60	—	37.60
2"	P1@.560	Ea	20.70	20.60	—	41.30
2½"	P1@.740	Ea	35.30	27.20	—	62.50
3"	P1@.810	Ea	35.30	29.70	—	65.00
4"	P1@.980	Ea	52.10	36.00	—	88.10

Class 300 cast steel gate valve, flanged

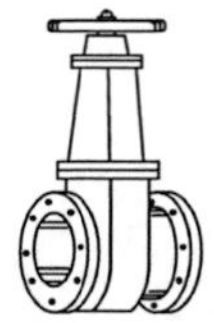

Description	Craft@Hrs	Unit	Material $	Labor $	Equipment $	Total $
2"	P1@.500	Ea	2,040.00	18.40	—	2,058.40
2½"	P1@.600	Ea	2,760.00	22.00	—	2,782.00
3"	P1@.750	Ea	2,760.00	27.50	—	2,787.50
4"	P1@1.35	Ea	3,850.00	49.50	—	3,899.50

Description	Craft@Hrs	Unit	Material $	Labor $	Equipment $	Total $

Class 600 cast steel gate valve, flanged

2"	P1@.500	Ea	3,390.00	18.40	—	3,408.40
2½"	P1@.600	Ea	4,880.00	22.00	—	4,902.00
3"	P1@.750	Ea	4,880.00	27.50	—	4,907.50
4"	P1@1.35	Ea	7,820.00	49.50	—	7,869.50

Class 300 cast steel body globe valve, flanged

2"	P1@.500	Ea	1,420.00	18.40	—	1,438.40
2½"	P1@.600	Ea	1,940.00	22.00	—	1,962.00
3"	P1@.750	Ea	1,940.00	27.50	—	1,967.50
4"	P1@1.35	Ea	2,660.00	49.50	—	2,709.50

Class 600 cast steel body globe valve, flanged

2"	P1@.500	Ea	1,990.00	18.40	—	2,008.40
2½"	P1@.600	Ea	3,010.00	22.00	—	3,032.00
3"	P1@.750	Ea	3,010.00	27.50	—	3,037.50
4"	P1@1.35	Ea	4,860.00	49.50	—	4,909.50

Class 300 cast steel body swing check valve, flanged

2"	P1@.500	Ea	1,350.00	18.40	—	1,368.40
2½"	P1@.600	Ea	2,030.00	22.00	—	2,052.00
3"	P1@.750	Ea	2,030.00	27.50	—	2,057.50
4"	P1@1.35	Ea	2,880.00	49.50	—	2,929.50

Class 600 cast steel body swing check valve, flanged

2"	P1@.500	Ea	2,070.00	18.40	—	2,088.40
2½"	P1@.600	Ea	3,270.00	22.00	—	3,292.00
3"	P1@.750	Ea	3,270.00	27.50	—	3,297.50
4"	P1@1.35	Ea	5,170.00	49.50	—	5,219.50

Installation of cast steel body 2-way control valve, threaded joints

½"	P1@.210	Ea	—	7.71	—	7.71
¾"	P1@.250	Ea	—	9.18	—	9.18
1"	P1@.300	Ea	—	11.00	—	11.00
1¼"	P1@.400	Ea	—	14.70	—	14.70
1½"	P1@.450	Ea	—	16.50	—	16.50
2"	P1@.500	Ea	—	18.40	—	18.40
2½"	P1@.830	Ea	—	30.50	—	30.50
3"	P1@.990	Ea	—	36.30	—	36.30

Installation of cast steel body 2-way control valve, flanged joints

2"	P1@.500	Ea	—	18.40	—	18.40
2½"	P1@.600	Ea	—	22.00	—	22.00
3"	P1@.750	Ea	—	27.50	—	27.50
4"	P1@1.35	Ea	—	49.50	—	49.50

Carbon Steel, Schedule 160 with 3,000-6,000# Fittings

Description	Craft@Hrs	Unit	Material $	Labor $	Equipment $	Total $

Installation of cast steel body 3-way control valve, threaded joints

Description	Craft@Hrs	Unit	Material $	Labor $	Equipment $	Total $
½"	P1@.260	Ea	—	9.54	—	9.54
¾"	P1@.365	Ea	—	13.40	—	13.40
1"	P1@.475	Ea	—	17.40	—	17.40
1¼"	P1@.575	Ea	—	21.10	—	21.10
1½"	P1@.680	Ea	—	25.00	—	25.00
2"	P1@.910	Ea	—	33.40	—	33.40
2½"	P1@1.12	Ea	—	41.10	—	41.10
3"	P1@1.33	Ea	—	48.80	—	48.80

Installation of cast steel body 3-way control valve, flanged joints

Description	Craft@Hrs	Unit	Material $	Labor $	Equipment $	Total $
2"	P1@.910	Ea	—	33.40	—	33.40
2½"	P1@1.12	Ea	—	41.10	—	41.10
3"	P1@1.33	Ea	—	48.80	—	48.80
4"	P1@2.00	Ea	—	73.40	—	73.40

300# bolt and gasket set, ring face

Description	Craft@Hrs	Unit	Material $	Labor $	Equipment $	Total $
2"	P1@.500	Ea	8.27	18.40	—	26.67
2½"	P1@.650	Ea	9.67	23.90	—	33.57
3"	P1@.750	Ea	16.60	27.50	—	44.10
4"	P1@1.00	Ea	27.00	36.70	—	63.70

300# bolt and gasket set, full face

Description	Craft@Hrs	Unit	Material $	Labor $	Equipment $	Total $
2"	P1@.500	Ea	16.60	18.40	—	35.00
2½"	P1@.650	Ea	19.40	23.90	—	43.30
3"	P1@.750	Ea	33.20	27.50	—	60.70
4"	P1@1.00	Ea	54.30	36.70	—	91.00

Thermometer with well

Description	Craft@Hrs	Unit	Material $	Labor $	Equipment $	Total $
7"	P1@.250	Ea	184.00	9.18	—	193.18
9"	P1@.250	Ea	190.00	9.18	—	199.18

Dial-type pressure gauge

Description	Craft@Hrs	Unit	Material $	Labor $	Equipment $	Total $
2½"	P1@.200	Ea	38.50	7.34	—	45.84
3½"	P1@.200	Ea	50.70	7.34	—	58.04

Pressure/temperature tap

Description	Craft@Hrs	Unit	Material $	Labor $	Equipment $	Total $
Tap	P1@.150	Ea	17.70	5.51	—	23.21

Carbon Steel, Schedule 160 with 3,000-6,000# Fittings

Description	Craft@Hrs	Unit	Material $	Labor $	Equipment $	Total $
Hanger with swivel assembly						
½"	P1@.250	Ea	6.19	9.18	—	15.37
¾"	P1@.250	Ea	6.80	9.18	—	15.98
1"	P1@.250	Ea	7.13	9.18	—	16.31
1¼"	P1@.300	Ea	7.69	11.00	—	18.69
1½"	P1@.300	Ea	10.00	11.00	—	21.00
2"	P1@.300	Ea	12.90	11.00	—	23.90
2½"	P1@.350	Ea	16.30	12.80	—	29.10
3"	P1@.350	Ea	27.80	12.80	—	40.60
4"	P1@.350	Ea	36.20	12.80	—	49.00
Riser clamp						
½"	P1@.100	Ea	2.66	3.67	—	6.33
¾"	P1@.100	Ea	3.91	3.67	—	7.58
1¼"	P1@.105	Ea	4.78	3.85	—	8.63
1½"	P1@.110	Ea	5.04	4.04	—	9.08
2"	P1@.115	Ea	5.35	4.22	—	9.57
2½"	P1@.120	Ea	5.62	4.40	—	10.02
3"	P1@.120	Ea	6.08	4.40	—	10.48
4"	P1@.125	Ea	7.75	4.59	—	12.34

Carbon Steel, Schedule 40 with Roll-Grooved Joints

Schedule 40 carbon steel (ASTM A-53 and A-120) roll-grooved pipe with factory grooved malleable iron, ductile iron, or steel fittings is commonly used for heating hot water, chilled water and condenser water systems.

Consult the manufacturers for maximum operating temperature/pressure ratings for the various combinations of pipe and fittings and system applications.

The cost estimates in this section are based on the conditions, limitations and wage rates described in the section "How to Use This Book" beginning on page 5.

Equipment cost, where shown, is $4.33 per hour for a 2-ton chain hoist.

Description	Craft@Hrs	Unit	Material $	Labor $	Equipment $	Total $

Schedule 40 carbon steel roll-grooved horizontal pipe assembly.
Horizontally hung in a building. Assembly includes fittings, couplings and hanger assemblies. Based on a reducing tee and a 45-degree elbow every 16 feet for ½" pipe. A reducing tee and a 45-degree elbow every 50 feet for 12" pipe. Hangers spaced to meet plumbing code. *(Distance between fittings in the pipe assembly increases as pipe diameter increases.)* Use these figures for preliminary estimates.

Description	Craft@Hrs	Unit	Material $	Labor $	Equipment $	Total $
2"	P1@.183	LF	9.28	6.72	—	16.00
2½"	P1@.192	LF	12.20	7.05	—	19.25
3"	P1@.220	LF	13.80	8.07	—	21.87
4"	P1@.262	LF	20.00	9.62	—	29.62
6"	ER@.347	LF	36.90	13.70	.21	50.81
8"	ER@.414	LF	53.80	16.30	.26	70.36
10"	ER@.530	LF	80.00	20.90	.33	101.23
12"	ER@.617	LF	143.00	24.30	.38	167.68

Schedule 40 carbon steel roll-grooved vertical pipe assembly.
Riser assembly, including fittings, couplings, and riser clamps. Based on a reducing tee, and a riser clamp on every floor. Use these figures for preliminary estimates.

Description	Craft@Hrs	Unit	Material $	Labor $	Equipment $	Total $
2"	P1@.120	LF	8.14	4.40	—	12.54
2½"	P1@.123	LF	11.60	4.51	—	16.11
3"	P1@.150	LF	12.00	5.51	—	17.51
4"	P1@.209	LF	18.10	7.67	—	25.77
6"	ER@.294	LF	35.20	11.60	.18	46.98
8"	ER@.395	LF	57.70	15.60	.24	73.54
10"	ER@.523	LF	80.00	20.60	.32	100.92
12"	ER@.608	LF	130.00	24.00	.38	154.38

Schedule 40 carbon steel pipe with roll-grooved joints

Description	Craft@Hrs	Unit	Material $	Labor $	Equipment $	Total $
2"	P1@.100	LF	3.68	3.67	—	7.35
2½"	P1@.115	LF	5.78	4.22	—	10.00
3"	P1@.135	LF	7.89	4.95	—	12.84
4"	P1@.190	LF	11.30	6.97	—	18.27
5"	ER@.240	LF	21.20	9.46	.15	30.81
6"	ER@.315	LF	32.10	12.40	.19	44.69
8"	ER@.370	LF	32.10	14.60	.23	46.93
10"	ER@.460	LF	45.00	18.10	.28	63.38
12"	ER@.550	LF	71.00	21.70	.34	93.04

Carbon Steel, Schedule 40 with Roll-Grooved Joints

Description	Craft@Hrs	Unit	Material $	Labor $	Equipment $	Total $

Schedule 40 carbon steel 45-degree ell with roll-grooved joints

Description	Craft@Hrs	Unit	Material $	Labor $	Equipment $	Total $
2"	P1@.170	Ea	27.00	6.24	—	33.24
2½"	P1@.185	Ea	36.70	6.79	—	43.49
3"	P1@.200	Ea	48.40	7.34	—	55.74
4"	P1@.230	Ea	71.40	8.44	—	79.84
5"	ER@.250	Ea	174.00	9.85	.15	184.00
6"	ER@.350	Ea	204.00	13.80	.22	218.02
8"	ER@.500	Ea	424.00	19.70	.31	444.01
10"	ER@.750	Ea	573.00	29.60	.46	603.06
12"	ER@1.00	Ea	1,000.00	39.40	.62	1,040.02

Schedule 40 carbon steel 90-degree ell with roll-grooved joints

Description	Craft@Hrs	Unit	Material $	Labor $	Equipment $	Total $
2"	P1@.170	Ea	27.00	6.24	—	33.24
2½"	P1@.185	Ea	36.70	6.79	—	43.49
3"	P1@.200	Ea	48.40	7.34	—	55.74
4"	P1@.230	Ea	71.40	8.44	—	79.84
5"	ER@.250	Ea	174.00	9.85	.15	184.00
6"	ER@.350	Ea	204.00	13.80	.22	218.02
8"	ER@.500	Ea	424.00	19.70	.31	444.01
10"	ER@.750	Ea	658.00	29.60	.46	688.06
12"	ER@1.00	Ea	1,050.00	39.40	.62	1,090.02

Schedule 40 carbon steel tee with roll-grooved joints

Description	Craft@Hrs	Unit	Material $	Labor $	Equipment $	Total $
2"	P1@.200	Ea	41.50	7.34	—	48.84
2½"	P1@.210	Ea	56.50	7.71	—	64.21
3"	P1@.230	Ea	78.80	8.44	—	87.24
4"	P1@.250	Ea	121.00	9.18	—	130.18
5"	ER@.300	Ea	285.00	11.80	.19	296.99
6"	ER@.500	Ea	331.00	19.70	.31	351.01
8"	ER@.750	Ea	725.00	29.60	.46	755.06
10"	ER@1.00	Ea	1,290.00	39.40	.62	1,330.02
12"	ER@1.50	Ea	1,780.00	59.10	.93	1,840.03

Carbon Steel, Schedule 40 with Roll-Grooved Joints

Description	Craft@Hrs	Unit	Material $	Labor $	Equipment $	Total $

Schedule 40 carbon steel reducing tee with roll-grooved joints

Description	Craft@Hrs	Unit	Material $	Labor $	Equipment $	Total $
3" x 2"	P1@.230	Ea	107.00	8.44	—	115.44
4" x 2"	P1@.240	Ea	145.00	8.81	—	153.81
4" x 2½"	P1@.245	Ea	145.00	8.99	—	153.99
4" x 3"	P1@.250	Ea	145.00	9.18	—	154.18
5" x 2"	P1@.260	Ea	265.00	9.54	—	274.54
5" x 3"	P1@.280	Ea	291.00	10.30	—	301.30
5" x 4"	P1@.300	Ea	314.00	11.00	—	325.00
6" x 3"	ER@.400	Ea	345.00	15.80	.25	361.05
6" x 4"	ER@.450	Ea	345.00	17.70	.28	362.98
6" x 5"	ER@.500	Ea	345.00	19.70	.31	365.01
8" x 3"	ER@.600	Ea	497.00	23.60	.37	520.97
8" x 4"	ER@.650	Ea	725.00	25.60	.40	751.00
8" x 5"	ER@.700	Ea	725.00	27.60	.43	753.03
8" x 6"	ER@.750	Ea	725.00	29.60	.46	755.06
10" x 4"	ER@.800	Ea	773.00	31.50	.49	804.99
10" x 5"	ER@.900	Ea	788.00	35.50	.56	824.06
10" x 6"	ER@1.00	Ea	788.00	39.40	.62	828.02
10" x 8"	ER@1.10	Ea	788.00	43.30	.68	831.98
12" x 6"	ER@1.20	Ea	115.00	47.30	.74	163.04
12" x 8"	ER@1.35	Ea	1,180.00	53.20	.84	1,234.04
12" x 10"	ER@1.50	Ea	1,240.00	59.10	.93	1,300.03

Schedule 40 carbon steel male adapter

Description	Craft@Hrs	Unit	Material $	Labor $	Equipment $	Total $
2"	P1@.170	Ea	17.60	6.24	—	23.84
2½"	P1@.185	Ea	20.90	6.79	—	27.69
3"	P1@.200	Ea	26.00	7.34	—	33.34
4"	P1@.230	Ea	43.50	8.44	—	51.94
5"	ER@.290	Ea	91.70	11.40	.18	103.28
6"	ER@.400	Ea	215.00	15.80	.25	231.05

Schedule 40 carbon steel female adapter

Description	Craft@Hrs	Unit	Material $	Labor $	Equipment $	Total $
2"	P1@.170	Ea	39.80	6.24	—	46.04
3"	P1@.200	Ea	61.60	7.34	—	68.94
4"	P1@.230	Ea	86.20	8.44	—	94.64

Schedule 40 carbon steel reducer, roll-grooved joints

Description	Craft@Hrs	Unit	Material $	Labor $	Equipment $	Total $
2"	P1@.160	Ea	37.20	5.87	—	43.07
3"	P1@.170	Ea	51.60	6.24	—	57.84
4"	P1@.210	Ea	62.80	7.71	—	70.51
5"	ER@.260	Ea	87.40	10.20	.16	97.76
6"	ER@.310	Ea	102.00	12.20	.19	114.39
8"	ER@.450	Ea	265.00	17.70	.28	282.98
10"	ER@.690	Ea	461.00	27.20	.43	488.63
12"	ER@.920	Ea	801.00	36.20	.57	837.77

Carbon Steel, Schedule 40 with Roll-Grooved Joints

Description	Craft@Hrs	Unit	Material $	Labor $	Equipment $	Total $

Schedule 40 carbon steel cap, roll-grooved joint

Description	Craft@Hrs	Unit	Material $	Labor $	Equipment $	Total $
2"	P1@.110	Ea	19.30	4.04	—	23.34
2½"	P1@.120	Ea	30.30	4.40	—	34.70
3"	P1@.130	Ea	30.30	4.77	—	35.07
4"	P1@.140	Ea	32.70	5.14	—	37.84
5"	ER@.150	Ea	75.90	5.91	.09	81.90
6"	ER@.170	Ea	78.80	6.70	.11	85.61
8"	ER@.200	Ea	152.00	7.88	.12	160.00
10"	ER@.225	Ea	274.00	8.87	.14	283.01
12"	ER@.250	Ea	444.00	9.85	.15	454.00

Flange with gasket, grooved joint

Description	Craft@Hrs	Unit	Material $	Labor $	Equipment $	Total $
2"	P1@.280	Ea	57.90	10.30	—	68.20
2½"	P1@.340	Ea	68.30	12.50	—	80.80
3"	P1@.390	Ea	77.30	14.30	—	91.60
4"	P1@.560	Ea	112.00	20.60	—	132.60
6"	ER@.780	Ea	140.00	30.70	—	170.70
8"	ER@1.10	Ea	164.00	43.30	—	207.30
10"	ER@1.40	Ea	209.00	55.20	—	264.20
12"	ER@1.70	Ea	288.00	67.00	—	355.00

Roll-grooved coupling with gasket

Description	Craft@Hrs	Unit	Material $	Labor $	Equipment $	Total $
2"	P1@.300	Ea	29.90	11.00	—	40.90
2½"	P1@.350	Ea	36.00	12.80	—	48.80
3"	P1@.400	Ea	39.80	14.70	—	54.50
4"	P1@.500	Ea	57.90	18.40	—	76.30
5"	ER@.600	Ea	88.70	23.60	—	112.30
6"	ER@.700	Ea	104.00	27.60	—	131.60
8"	ER@.900	Ea	174.00	35.50	—	209.50
10"	ER@1.10	Ea	241.00	43.30	—	284.30
12"	ER@1.35	Ea	274.00	53.20	—	327.20

Class 125 iron body gate valve, flanged ends

Description	Craft@Hrs	Unit	Material $	Labor $	Equipment $	Total $
2"	P1@.500	Ea	377.00	18.40	—	395.40
2½"	P1@.600	Ea	509.00	22.00	—	531.00
3"	P1@.750	Ea	552.00	27.50	—	579.50
4"	P1@1.35	Ea	811.00	49.50	—	860.50
6"	ER@2.50	Ea	1,540.00	98.50	1.55	1,640.05
8"	ER@3.00	Ea	2,550.00	118.00	1.86	2,669.86
10"	ER@4.00	Ea	4,160.00	158.00	2.47	4,320.47
12"	ER@4.50	Ea	5,770.00	177.00	2.78	5,949.78

Carbon Steel, Schedule 40 with Roll-Grooved Joints

Description	Craft@Hrs	Unit	Material $	Labor $	Equipment $	Total $

Class 125 iron body globe valve, flanged joints

Description	Craft@Hrs	Unit	Material $	Labor $	Equipment $	Total $
2"	P1@.500	Ea	275.00	18.40	—	293.40
2½"	P1@.600	Ea	427.00	22.00	—	449.00
3"	P1@.750	Ea	512.00	27.50	—	539.50
4"	P1@1.35	Ea	688.00	49.50	—	737.50
6"	ER@2.50	Ea	1,210.00	98.50	1.55	1,310.05
8"	ER@3.00	Ea	1,850.00	118.00	1.86	1,969.86
10"	ER@4.00	Ea	2,820.00	158.00	2.47	2,980.47

200 PSIG iron body butterfly valve, lug-type, lever operated

Description	Craft@Hrs	Unit	Material $	Labor $	Equipment $	Total $
2"	P1@.450	Ea	163.00	16.50	—	179.50
2½"	P1@.450	Ea	169.00	16.50	—	185.50
3"	P1@.550	Ea	177.00	20.20	—	197.20
4"	P1@.550	Ea	221.00	20.20	—	241.20
6"	ER@.800	Ea	356.00	31.50	.49	387.99
8"	ER@.800	Ea	487.00	31.50	.49	518.99
10"	ER@.900	Ea	677.00	35.50	.56	713.06
12"	ER@1.00	Ea	884.00	39.40	.62	924.02

200 PSIG iron body butterfly valve, wafer-type, lever operated

Description	Craft@Hrs	Unit	Material $	Labor $	Equipment $	Total $
2"	P1@.450	Ea	148.00	16.50	—	164.50
2½"	P1@.450	Ea	151.00	16.50	—	167.50
3"	P1@.550	Ea	163.00	20.20	—	183.20
4"	P1@.550	Ea	197.00	20.20	—	217.20
6"	ER@.800	Ea	330.00	31.50	.49	361.99
8"	ER@.800	Ea	457.00	31.50	.49	488.99
10"	ER@.900	Ea	639.00	35.50	.56	675.06
12"	ER@1.00	Ea	839.00	39.40	.62	879.02

Class 125 iron body swing check valve, flanged ends

Description	Craft@Hrs	Unit	Material $	Labor $	Equipment $	Total $
2"	P1@.500	Ea	179.00	18.40	—	197.40
2½"	P1@.600	Ea	226.00	22.00	—	248.00
3"	P1@.750	Ea	282.00	27.50	—	309.50
4"	P1@1.35	Ea	414.00	49.50	—	463.50
6"	ER@2.50	Ea	803.00	98.50	1.55	903.05
8"	ER@3.00	Ea	1,440.00	118.00	1.86	1,559.86
10"	ER@4.00	Ea	2,350.00	158.00	2.47	2,510.47
12"	ER@4.50	Ea	3,220.00	177.00	2.78	3,399.78

Class 125 iron body silent check valve, flanged joints

Description	Craft@Hrs	Unit	Material $	Labor $	Equipment $	Total $
2"	P1@.500	Ea	131.00	18.40	—	149.40
2½"	P1@.600	Ea	150.00	22.00	—	172.00
3"	P1@.750	Ea	170.00	27.50	—	197.50
4"	P1@1.35	Ea	222.00	49.50	—	271.50
5"	ER@2.00	Ea	322.00	78.80	1.24	402.04
6"	ER@2.50	Ea	399.00	98.50	1.55	499.05
8"	ER@3.00	Ea	719.00	118.00	1.86	838.86
10"	ER@4.00	Ea	1,110.00	158.00	2.47	1,270.47

Description	Craft@Hrs	Unit	Material $	Labor $	Equipment $	Total $

Class 125 iron body strainer, flanged

Description	Craft@Hrs	Unit	Material $	Labor $	Equipment $	Total $
2"	P1@.500	Ea	135.00	18.40	—	153.40
2½"	P1@.600	Ea	151.00	22.00	—	173.00
3"	P1@.750	Ea	175.00	27.50	—	202.50
4"	P1@1.35	Ea	297.00	49.50	—	346.50
6"	ER@2.50	Ea	604.00	98.50	1.55	704.05
8"	ER@3.00	Ea	996.00	118.00	1.86	1,115.86

Installation of 2-way control valve, flanged joints

Description	Craft@Hrs	Unit	Material $	Labor $	Equipment $	Total $
2"	P1@.500	Ea	—	18.40	—	18.40
2½"	P1@.600	Ea	—	22.00	—	22.00
3"	P1@.750	Ea	—	27.50	—	27.50
4"	P1@1.35	Ea	—	49.50	—	49.50
6"	P1@2.50	Ea	—	91.80	—	91.80
8"	P1@3.00	Ea	—	110.00	—	110.00
10"	P1@4.00	Ea	—	147.00	—	147.00
12"	P1@4.50	Ea	—	165.00	—	165.00

Installation of 3-way control valve, flanged joints

Description	Craft@Hrs	Unit	Material $	Labor $	Equipment $	Total $
2"	P1@.910	Ea	—	33.40	—	33.40
2½"	P1@1.12	Ea	—	41.10	—	41.10
3"	P1@1.33	Ea	—	48.80	—	48.80
4"	P1@2.00	Ea	—	73.40	—	73.40
6"	P1@3.70	Ea	—	136.00	—	136.00
8"	P1@4.40	Ea	—	161.00	—	161.00
10"	P1@5.90	Ea	—	217.00	—	217.00
12"	P1@6.50	Ea	—	239.00	—	239.00

Bolt and gasket sets

Description	Craft@Hrs	Unit	Material $	Labor $	Equipment $	Total $
2"	P1@.500	Ea	4.17	18.40	—	22.57
2½"	P1@.650	Ea	4.87	23.90	—	28.77
3"	P1@.750	Ea	8.27	27.50	—	35.77
4"	P1@1.00	Ea	13.80	36.70	—	50.50
6"	P1@1.20	Ea	22.90	44.00	—	66.90
8"	P1@1.25	Ea	25.70	45.90	—	71.60
10"	P1@1.70	Ea	44.50	62.40	—	106.90
12"	P1@2.20	Ea	49.80	80.70	—	130.50

Thermometer with well

Description	Craft@Hrs	Unit	Material $	Labor $	Equipment $	Total $
7"	P1@.250	Ea	182.00	9.18	—	191.18
9"	P1@.250	Ea	187.00	9.18	—	196.18

Dial-type pressure gauge

Description	Craft@Hrs	Unit	Material $	Labor $	Equipment $	Total $
2½"	P1@.200	Ea	38.10	7.34	—	45.44
3½"	P1@.200	Ea	50.20	7.34	—	57.54

Carbon Steel, Schedule 40 with Roll-Grooved Joints

Description	Craft@Hrs	Unit	Material $	Labor $	Equipment $	Total $

Pressure/temperature tap

Description	Craft@Hrs	Unit	Material $	Labor $	Equipment $	Total $
Tap	P1@.150	Ea	17.50	5.51	—	23.01

Hanger with swivel assembly

Description	Craft@Hrs	Unit	Material $	Labor $	Equipment $	Total $
2"	P1@.300	Ea	6.66	11.00	—	17.66
2½"	P1@.350	Ea	8.62	12.80	—	21.42
3"	P1@.350	Ea	10.60	12.80	—	23.40
4"	P1@.350	Ea	16.50	12.80	—	29.30
5"	P1@.450	Ea	18.50	16.50	—	35.00
6"	P1@.450	Ea	24.00	16.50	—	40.50
8"	P1@.450	Ea	31.90	16.50	—	48.40
10"	P1@.550	Ea	43.90	20.20	—	64.10
12"	P1@.600	Ea	58.20	22.00	—	80.20

Riser clamp

Description	Craft@Hrs	Unit	Material $	Labor $	Equipment $	Total $
2"	P1@.115	Ea	5.35	4.22	—	9.57
2½"	P1@.120	Ea	5.62	4.40	—	10.02
3"	P1@.120	Ea	6.08	4.40	—	10.48
4"	P1@.125	Ea	7.75	4.59	—	12.34
5"	P1@.180	Ea	11.10	6.61	—	17.71
6"	P1@.200	Ea	13.40	7.34	—	20.74
8"	P1@.200	Ea	21.80	7.34	—	29.14
10"	P1@.250	Ea	32.90	9.18	—	42.08
12"	P1@.250	Ea	38.50	9.18	—	47.68

Carbon Steel, Schedule 10 with Roll-Grooved Joints

Schedule 10 carbon steel (ASTM A-53) roll-grooved pipe with factory grooved malleable iron, ductile iron, or steel fittings is commonly used for sprinkler systems, heating hot water, chilled water and condenser water systems.

Consult the manufacturers for maximum operating temperature/pressure ratings for the various combinations of pipe and fittings and system applications.

The cost estimates in this section are based on the conditions, limitations and wage rates described in the section "How to Use This Book" beginning on page 5.

Equipment cost, where shown, is $4.33 per hour for a 2-ton chain hoist.

Description	Craft@Hrs	Unit	Material $	Labor $	Equipment $	Total $

Schedule 10 carbon steel roll-grooved horizontal pipe assembly.

Horizontally hung in a building. Assembly includes fittings, couplings and hanger assemblies. Based on a reducing tee and a 45-degree elbow every 16 feet for ½" pipe. A reducing tee and a 45-degree elbow every 50 feet for 12" pipe. Hangers spaced to meet plumbing code. *(Distance between fittings in the pipe assembly increases as pipe diameter increases.)* Use these figures for preliminary estimates.

Description	Craft@Hrs	Unit	Material $	Labor $	Equipment $	Total $
2"	P1@.178	LF	8.24	6.53	—	14.77
2½"	P1@.188	LF	10.30	6.90	—	17.20
3"	P1@.215	LF	11.90	7.89	—	19.79
4"	P1@.255	LF	17.00	9.36	—	26.36
6"	ER@.339	LF	32.10	13.40	.21	45.71
8"	ER@.404	LF	48.90	15.90	.25	65.05
10"	ER@.517	LF	73.30	20.40	.32	94.02
12"	ER@.602	LF	131.00	23.70	.37	155.07

Schedule 10 carbon steel roll-grooved vertical pipe assembly.
Riser assembly, including fittings, couplings, and riser clamps. Based on a reducing tee, and a riser clamp on every floor. Use these figures for preliminary estimates.

Description	Craft@Hrs	Unit	Material $	Labor $	Equipment $	Total $
2"	P1@.117	LF	7.90	4.29	—	12.19
2½"	P1@.120	LF	10.70	4.40	—	15.10
3"	P1@.146	LF	11.00	5.36	—	16.36
4"	P1@.204	LF	16.90	7.49	—	24.39
6"	ER@.286	LF	33.30	11.30	.18	44.78
8"	ER@.385	LF	58.60	15.20	.24	74.04
10"	ER@.510	LF	81.30	20.10	.32	101.72
12"	ER@.593	LF	131.00	23.40	.37	154.77

Carbon Steel, Schedule 10 with Roll-Grooved Joints

Description	Craft@Hrs	Unit	Material $	Labor $	Equipment $	Total $

Schedule 10 carbon steel pipe, roll-grooved joints

Description	Craft@Hrs	Unit	Material $	Labor $	Equipment $	Total $
2"	P1@.080	LF	2.41	2.94	—	5.35
2½"	P1@.100	LF	3.28	3.67	—	6.95
3"	P1@.115	LF	3.98	4.22	—	8.20
4"	P1@.160	LF	5.29	5.87	—	11.16
5"	ER@.200	LF	9.04	7.88	.12	17.04
6"	ER@.260	LF	9.54	10.20	.16	19.90
8"	ER@.300	LF	13.10	11.80	.19	25.09
10"	ER@.380	LF	17.90	15.00	.24	33.14
12"	ER@.450	LF	25.30	17.70	.28	43.28

Schedule 10 carbon steel 45-degree ell, roll-grooved joints

Description	Craft@Hrs	Unit	Material $	Labor $	Equipment $	Total $
2"	P1@.170	Ea	27.00	6.24	—	33.24
2½"	P1@.185	Ea	36.70	6.79	—	43.49
3"	P1@.200	Ea	48.40	7.34	—	55.74
4"	P1@.230	Ea	71.40	8.44	—	79.84
5"	ER@.250	Ea	174.00	9.85	.15	184.00
6"	ER@.350	Ea	204.00	13.80	.22	218.02
8"	ER@.500	Ea	424.00	19.70	.31	444.01
10"	ER@.750	Ea	573.00	29.60	.46	603.06
12"	ER@1.00	Ea	1,000.00	39.40	.62	1,040.02

Schedule 10 carbon steel 90-degree ell, roll-grooved joints

Description	Craft@Hrs	Unit	Material $	Labor $	Equipment $	Total $
2"	P1@.170	Ea	27.00	6.24	—	33.24
2½"	P1@.185	Ea	36.70	6.79	—	43.49
3"	P1@.200	Ea	48.40	7.34	—	55.74
4"	P1@.230	Ea	71.40	8.44	—	79.84
5"	ER@.250	Ea	174.00	9.85	.15	184.00
6"	ER@.350	Ea	204.00	13.80	.22	218.02
8"	ER@.500	Ea	424.00	19.70	.31	444.01
10"	ER@.750	Ea	658.00	29.60	.46	688.06
12"	ER@1.00	Ea	1,050.00	39.40	.62	1,090.02

Schedule 10 carbon steel tee, roll-grooved joints

Description	Craft@Hrs	Unit	Material $	Labor $	Equipment $	Total $
2"	P1@.200	Ea	41.50	7.34	—	48.84
2½"	P1@.210	Ea	56.50	7.71	—	64.21
3"	P1@.230	Ea	78.80	8.44	—	87.24
4"	P1@.250	Ea	121.00	9.18	—	130.18
5"	ER@.300	Ea	285.00	11.80	.19	296.99
6"	ER@.500	Ea	331.00	19.70	.31	351.01
8"	ER@.750	Ea	725.00	29.60	.46	755.06
10"	ER@1.00	Ea	1,290.00	39.40	.62	1,330.02
12"	ER@1.50	Ea	1,780.00	59.10	.93	1,840.03

Description	Craft@Hrs	Unit	Material $	Labor $	Equipment $	Total $

Schedule 10 carbon steel reducing tee, roll-grooved joints

Description	Craft@Hrs	Unit	Material $	Labor $	Equipment $	Total $
3" x 2"	P1@.230	Ea	107.00	8.44	—	115.44
4" x 2"	P1@.240	Ea	145.00	8.81	—	153.81
4" x 2½"	P1@.245	Ea	145.00	8.99	—	153.99
4" x 3"	P1@.250	Ea	145.00	9.18	—	154.18
5" x 2"	P1@.260	Ea	265.00	9.54	—	274.54
5" x 3"	P1@.280	Ea	291.00	10.30	—	301.30
5" x 4"	P1@.300	Ea	314.00	11.00	—	325.00
6" x 3"	ER@.400	Ea	345.00	15.80	.25	361.05
6" x 4"	ER@.450	Ea	345.00	17.70	.28	362.98
6" x 5"	ER@.500	Ea	345.00	19.70	.31	365.01
8" x 3"	ER@.600	Ea	497.00	23.60	.37	520.97
8" x 4"	ER@.650	Ea	725.00	25.60	.40	751.00
8" x 5"	ER@.700	Ea	725.00	27.60	.43	753.03
8" x 6"	ER@.750	Ea	725.00	29.60	.46	755.06
10" x 4"	ER@.800	Ea	773.00	31.50	.49	804.99
10" x 5"	ER@.900	Ea	788.00	35.50	.56	824.06
10" x 6"	ER@1.00	Ea	788.00	39.40	.62	828.02
10" x 8"	ER@1.10	Ea	788.00	43.30	.68	831.98
12" x 6"	ER@1.20	Ea	115.00	47.30	.74	163.04
12" x 8"	ER@1.35	Ea	1,180.00	53.20	.84	1,234.04
12" x 10"	ER@1.50	Ea	1,240.00	59.10	.93	1,300.03

Schedule 10 carbon steel male adapter

Description	Craft@Hrs	Unit	Material $	Labor $	Equipment $	Total $
2"	P1@.170	Ea	17.60	6.24	—	23.84
2½"	P1@.185	Ea	20.90	6.79	—	27.69
3"	P1@.200	Ea	26.00	7.34	—	33.34
4"	P1@.230	Ea	43.50	8.44	—	51.94
5"	ER@.290	Ea	91.70	11.40	—	103.10
6"	ER@.400	Ea	215.00	15.80	—	230.80

Schedule 10 carbon steel female adapter

Description	Craft@Hrs	Unit	Material $	Labor $	Equipment $	Total $
2"	P1@.170	Ea	39.80	6.24	—	46.04
3"	P1@.200	Ea	61.60	7.34	—	68.94
4"	P1@.230	Ea	86.20	8.44	—	94.64

Schedule 10 carbon steel reducer, roll-grooved joints

Description	Craft@Hrs	Unit	Material $	Labor $	Equipment $	Total $
2"	P1@.160	Ea	37.20	5.87	—	43.07
3"	P1@.170	Ea	51.60	6.24	—	57.84
4"	P1@.210	Ea	62.80	7.71	—	70.51
5"	ER@.260	Ea	87.40	10.20	—	97.60
6"	ER@.310	Ea	102.00	12.20	—	114.20
8"	ER@.450	Ea	265.00	17.70	—	282.70
10"	ER@.690	Ea	461.00	27.20	—	488.20
12"	ER@.920	Ea	801.00	36.20	—	837.20

Carbon Steel, Schedule 10 with Roll-Grooved Joints

Description	Craft@Hrs	Unit	Material $	Labor $	Equipment $	Total $
Schedule 10 carbon steel cap, roll-grooved joints						
2"	P1@.110	Ea	19.30	4.04	—	23.34
2½"	P1@.120	Ea	30.30	4.40	—	34.70
3"	P1@.130	Ea	30.30	4.77	—	35.07
4"	P1@.140	Ea	32.70	5.14	—	37.84
5"	ER@.150	Ea	75.90	5.91	—	81.81
6"	ER@.170	Ea	78.80	6.70	—	85.50
8"	ER@.200	Ea	152.00	7.88	—	159.88
10"	ER@.225	Ea	274.00	8.87	—	282.87
12"	ER@.250	Ea	444.00	9.85	—	453.85
Flange with gasket, grooved joint						
2"	P1@.280	Ea	57.90	10.30	—	68.20
2½"	P1@.340	Ea	68.30	12.50	—	80.80
3"	P1@.390	Ea	77.30	14.30	—	91.60
4"	P1@.560	Ea	112.00	20.60	—	132.60
6"	ER@.780	Ea	140.00	30.70	—	170.70
8"	ER@1.10	Ea	164.00	43.30	—	207.30
10"	ER@1.40	Ea	209.00	55.20	—	264.20
12"	ER@1.70	Ea	288.00	67.00	—	355.00
Roll-grooved coupling with gasket						
2"	P1@.300	Ea	29.90	11.00	—	40.90
2½"	P1@.350	Ea	36.00	12.80	—	48.80
3"	P1@.400	Ea	39.80	14.70	—	54.50
4"	P1@.500	Ea	57.90	18.40	—	76.30
5"	ER@.600	Ea	88.70	23.60	—	112.30
6"	ER@.700	Ea	104.00	27.60	—	131.60
8"	ER@.900	Ea	174.00	35.50	—	209.50
10"	ER@1.10	Ea	241.00	43.30	—	284.30
12"	ER@1.35	Ea	274.00	53.20	—	327.20
Class 125 iron body gate valve, flanged joints						
2"	P1@.500	Ea	488.00	18.40	—	506.40
2½"	P1@.600	Ea	662.00	22.00	—	684.00
3"	P1@.750	Ea	717.00	27.50	—	744.50
4"	P1@1.35	Ea	1,040.00	49.50	—	1,089.50
5"	ER@2.00	Ea	2,010.00	78.80	1.24	2,090.04
6"	ER@2.50	Ea	2,010.00	98.50	1.55	2,110.05
8"	ER@3.00	Ea	3,280.00	118.00	1.86	3,399.86
10"	ER@4.00	Ea	5,450.00	158.00	2.47	5,610.47
12"	ER@4.50	Ea	7,440.00	177.00	2.78	7,619.78

Carbon Steel, Schedule 10 with Roll-Grooved Joints

Description	Craft@Hrs	Unit	Material $	Labor $	Equipment $	Total $

Class 125 iron body globe valve, flanged joints

Description	Craft@Hrs	Unit	Material $	Labor $	Equipment $	Total $
2"	P1@.500	Ea	267.00	18.40	—	285.40
2½"	P1@.600	Ea	413.00	22.00	—	435.00
3"	P1@.750	Ea	498.00	27.50	—	525.50
4"	P1@1.35	Ea	665.00	49.50	—	714.50
5"	ER@2.00	Ea	1,180.00	78.80	1.24	1,260.04
6"	ER@2.50	Ea	1,180.00	98.50	1.55	1,280.05
8"	ER@3.00	Ea	1,790.00	118.00	1.86	1,909.86
10"	ER@4.00	Ea	2,750.00	158.00	2.47	2,910.47

200 PSIG iron body butterfly valve, lug-type, lever operated

Description	Craft@Hrs	Unit	Material $	Labor $	Equipment $	Total $
2"	P1@.450	Ea	163.00	16.50	—	179.50
2½"	P1@.450	Ea	169.00	16.50	—	185.50
3"	P1@.550	Ea	177.00	20.20	—	197.20
4"	P1@.550	Ea	221.00	20.20	—	241.20
5"	ER@.800	Ea	291.00	31.50	.49	322.99
6"	ER@.800	Ea	356.00	31.50	.49	387.99
8"	ER@.800	Ea	487.00	31.50	.49	518.99
10"	ER@.900	Ea	677.00	35.50	.56	713.06
12"	ER@1.00	Ea	884.00	39.40	.62	924.02

200 PSIG iron body butterfly valve, wafer-type, lever operated

Description	Craft@Hrs	Unit	Material $	Labor $	Equipment $	Total $
2"	P1@.450	Ea	148.00	16.50	—	164.50
2½"	P1@.450	Ea	151.00	16.50	—	167.50
3"	P1@.550	Ea	163.00	20.20	—	183.20
4"	P1@.550	Ea	197.00	20.20	—	217.20
5"	ER@.800	Ea	263.00	31.50	.49	294.99
6"	ER@.800	Ea	330.00	31.50	.49	361.99
8"	ER@.800	Ea	457.00	31.50	.49	488.99
10"	ER@.900	Ea	639.00	35.50	.56	675.06
12"	ER@1.00	Ea	839.00	39.40	.62	879.02

Class 125 iron body swing check valve, flanged joints

Description	Craft@Hrs	Unit	Material $	Labor $	Equipment $	Total $
2"	P1@.500	Ea	179.00	18.40	—	197.40
2½"	P1@.600	Ea	226.00	22.00	—	248.00
3"	P1@.750	Ea	282.00	27.50	—	309.50
4"	P1@1.35	Ea	414.00	49.50	—	463.50
5"	ER@2.00	Ea	797.00	78.80	1.24	877.04
6"	ER@2.50	Ea	803.00	98.50	1.55	903.05
8"	ER@3.00	Ea	1,440.00	118.00	1.86	1,559.86
10"	ER@4.00	Ea	2,350.00	158.00	2.47	2,510.47
12"	ER@4.50	Ea	3,220.00	177.00	2.78	3,399.78

Carbon Steel, Schedule 10 with Roll-Grooved Joints

Description	Craft@Hrs	Unit	Material $	Labor $	Equipment $	Total $

Class 125 iron body silent check valve, flanged joints

2"	P1@.500	Ea	131.00	18.40	—	149.40
2½"	P1@.600	Ea	150.00	22.00	—	172.00
3"	P1@.750	Ea	170.00	27.50	—	197.50
4"	P1@1.35	Ea	222.00	49.50	—	271.50
5"	ER@2.00	Ea	322.00	78.80	1.24	402.04
6"	ER@2.50	Ea	399.00	98.50	1.55	499.05
8"	ER@3.00	Ea	719.00	118.00	1.86	838.86
10"	ER@4.00	Ea	1,110.00	158.00	2.47	1,270.47

Class 125 iron body strainer, flanged

2"	P1@.500	Ea	135.00	18.40	—	153.40
2½"	P1@.600	Ea	151.00	22.00	—	173.00
3"	P1@.750	Ea	174.00	27.50	—	201.50
4"	P1@1.35	Ea	297.00	49.50	—	346.50
5"	ER@2.00	Ea	603.00	78.80	1.24	683.04
6"	ER@2.50	Ea	603.00	98.50	1.55	703.05
8"	ER@3.00	Ea	1,020.00	118.00	1.86	1,139.86

Installation of 2-way control valve, flanged joints

2"	P1@.500	Ea	—	18.40	—	18.40
2½"	P1@.600	Ea	—	22.00	—	22.00
3"	P1@.750	Ea	—	27.50	—	27.50
4"	P1@1.35	Ea	—	49.50	—	49.50
6"	P1@2.50	Ea	—	91.80	—	91.80
8"	P1@3.00	Ea	—	110.00	—	110.00
10"	P1@4.00	Ea	—	147.00	—	147.00
12"	P1@4.50	Ea	—	165.00	—	165.00

Installation of 3-way control valve, flanged joints

2"	P1@.910	Ea	—	33.40	—	33.40
2½"	P1@1.12	Ea	—	41.10	—	41.10
3"	P1@1.33	Ea	—	48.80	—	48.80
4"	P1@2.00	Ea	—	73.40	—	73.40
6"	P1@3.70	Ea	—	136.00	—	136.00
8"	P1@4.40	Ea	—	161.00	—	161.00
10"	P1@5.90	Ea	—	217.00	—	217.00
12"	P1@6.50	Ea	—	239.00	—	239.00

Description	Craft@Hrs	Unit	Material $	Labor $	Equipment $	Total $
Bolt and gasket sets						
2"	P1@.500	Ea	4.17	18.40	—	22.57
2½"	P1@.650	Ea	4.87	23.90	—	28.77
3"	P1@.750	Ea	8.24	27.50	—	35.74
4"	P1@1.00	Ea	13.70	36.70	—	50.40
5"	P1@1.10	Ea	22.90	40.40	—	63.30
6"	P1@1.20	Ea	22.90	44.00	—	66.90
8"	P1@1.25	Ea	25.70	45.90	—	71.60
10"	P1@1.70	Ea	44.20	62.40	—	106.60
12"	P1@2.20	Ea	49.80	80.70	—	130.50
Thermometer with well						
7"	P1@.250	Ea	184.00	9.18	—	193.18
9"	P1@.250	Ea	190.00	9.18	—	199.18
Dial-type pressure gauge						
2½"	P1@.200	Ea	38.50	7.34	—	45.84
3½"	P1@.200	Ea	50.70	7.34	—	58.04
Pressure/temperature tap						
Tap	P1@.150	Ea	17.70	5.51	—	23.21
Hanger with swivel assembly						
2"	P1@.300	Ea	6.90	11.00	—	17.90
2½"	P1@.350	Ea	8.94	12.80	—	21.74
3"	P1@.350	Ea	11.00	12.80	—	23.80
4"	P1@.350	Ea	17.10	12.80	—	29.90
5"	P1@.450	Ea	19.30	16.50	—	35.80
6"	P1@.450	Ea	24.80	16.50	—	41.30
8"	P1@.450	Ea	33.10	16.50	—	49.60
10"	P1@.550	Ea	45.50	20.20	—	65.70
12"	P1@.600	Ea	60.40	22.00	—	82.40
Riser clamp						
2"	P1@.115	Ea	5.53	4.22	—	9.75
2½"	P1@.120	Ea	5.83	4.40	—	10.23
3"	P1@.120	Ea	6.30	4.40	—	10.70
4"	P1@.125	Ea	8.03	4.59	—	12.62
5"	P1@.180	Ea	11.50	6.61	—	18.11
6"	P1@.200	Ea	13.80	7.34	—	21.14
8"	P1@.200	Ea	22.60	7.34	—	29.94
10"	P1@.250	Ea	33.60	9.18	—	42.78
12"	P1@.250	Ea	39.90	9.18	—	49.08

Carbon Steel, Schedule 40 with Cut-Grooved Joints

Schedule 40 carbon steel (ASTM A-53 and A-120) cut-grooved pipe with factory-grooved malleable iron, ductile iron or steel fittings is commonly used for heating hot water, chilled water and condenser water systems.

Consult the manufacturers for maximum operating temperature/pressure ratings for various combinations of pipe and fittings.

This section has been arranged to save the estimator's time by including all normally-used system components such as pipe, fittings, valves, hanger assemblies, riser clamps and miscellaneous items under one heading. Additional items can be found under "Plumbing and Piping Specialties." The cost estimates in this section are based on the conditions, limitations and wage rates described in the section "How to Use This Book" beginning on page 5.

Equipment cost, where shown, is $4.33 per hour for a 2-ton chain hoist.

Description	Craft@Hrs	Unit	Material $	Labor $	Equipment $	Total $

Schedule 40 carbon steel pipe with cut-grooved joints

Description	Craft@Hrs	Unit	Material $	Labor $	Equipment $	Total $
2"	P1@.110	LF	4.10	4.04	—	8.14
2½"	P1@.130	LF	6.44	4.77	—	11.21
3"	P1@.150	LF	8.74	5.51	—	14.25
4"	P1@.210	LF	12.60	7.71	—	20.31
5"	ER@.270	LF	23.70	10.60	.17	34.47
6"	ER@.350	LF	35.70	13.80	.22	49.72
8"	ER@.410	LF	35.70	16.20	.25	52.15
10"	ER@.510	LF	50.10	20.10	.32	70.52
12"	ER@.610	LF	79.00	24.00	.38	103.38

Schedule 40 carbon steel 45-degree ell with cut-grooved joints

Description	Craft@Hrs	Unit	Material $	Labor $	Equipment $	Total $
2"	P1@.170	Ea	27.00	6.24	—	33.24
2½"	P1@.185	Ea	36.90	6.79	—	43.69
3"	P1@.200	Ea	48.40	7.34	—	55.74
4"	P1@.230	Ea	71.40	8.44	—	79.84
5"	ER@.250	Ea	174.00	9.85	.15	184.00
6"	ER@.350	Ea	204.00	13.80	.22	218.02
8"	ER@.500	Ea	424.00	19.70	.31	444.01
10"	ER@.750	Ea	573.00	29.60	.46	603.06
12"	ER@1.00	Ea	1,000.00	39.40	.62	1,040.02

Schedule 40 carbon steel 90-degree ell with cut-grooved joints

Description	Craft@Hrs	Unit	Material $	Labor $	Equipment $	Total $
2"	P1@.170	Ea	27.00	6.24	—	33.24
2½"	P1@.185	Ea	36.90	6.79	—	43.69
3"	P1@.200	Ea	48.40	7.34	—	55.74
4"	P1@.230	Ea	71.40	8.44	—	79.84
5"	ER@.250	Ea	174.00	9.85	.15	184.00
6"	ER@.350	Ea	204.00	13.80	.22	218.02
8"	ER@.500	Ea	424.00	19.70	.31	444.01
10"	ER@.750	Ea	658.00	29.60	.46	688.06
12"	ER@1.00	Ea	1,780.00	39.40	.62	1,820.02

Carbon Steel, Schedule 40 with Cut-Grooved Joints

Description	Craft@Hrs	Unit	Material $	Labor $	Equipment $	Total $

Schedule 40 carbon steel tee with cut-grooved joints

Description	Craft@Hrs	Unit	Material $	Labor $	Equipment $	Total $
2"	P1@.200	Ea	41.50	7.34	—	48.84
2½"	P1@.210	Ea	56.50	7.71	—	64.21
3"	P1@.230	Ea	78.80	8.44	—	87.24
4"	P1@.250	Ea	121.00	9.18	—	130.18
5"	ER@.300	Ea	285.00	11.80	.19	296.99
6"	ER@.500	Ea	331.00	19.70	.31	351.01
8"	ER@.750	Ea	725.00	29.60	.46	755.06
10"	ER@1.00	Ea	1,290.00	39.40	.62	1,330.02
12"	ER@1.50	Ea	1,780.00	59.10	.93	1,840.03

Schedule 40 carbon steel reducing tee with cut-grooved joints

Description	Craft@Hrs	Unit	Material $	Labor $	Equipment $	Total $
3" x 2"	P1@.230	Ea	107.00	8.44	—	115.44
4" x 2"	P1@.240	Ea	145.00	8.81	—	153.81
4" x 2½"	P1@.245	Ea	145.00	8.99	—	153.99
4" x 3"	P1@.250	Ea	145.00	9.18	—	154.18
5" x 2"	P1@.260	Ea	265.00	9.54	—	274.54
5" x 3"	P1@.280	Ea	291.00	10.30	—	301.30
5" x 4"	P1@.300	Ea	314.00	11.00	—	325.00
6" x 3"	ER@.400	Ea	345.00	15.80	.25	361.05
6" x 4"	ER@.450	Ea	345.00	17.70	.28	362.98
6" x 5"	ER@.500	Ea	345.00	19.70	.31	365.01
8" x 3"	ER@.600	Ea	497.00	23.60	.37	520.97
8" x 4"	ER@.650	Ea	725.00	25.60	.40	751.00
8" x 5"	ER@.700	Ea	725.00	27.60	.43	753.03
8" x 6"	ER@.750	Ea	725.00	29.60	.46	755.06
10" x 4"	ER@.800	Ea	773.00	31.50	.49	804.99
10" x 5"	ER@.900	Ea	788.00	35.50	.56	824.06
10" x 6"	ER@1.00	Ea	788.00	39.40	.62	828.02
10" x 8"	ER@1.10	Ea	788.00	43.30	.68	831.98
12" x 6"	ER@1.20	Ea	115.00	47.30	.74	163.04
12" x 8"	ER@1.35	Ea	1,180.00	53.20	.84	1,234.04
12" x 10"	ER@1.50	Ea	1,240.00	59.10	.93	1,300.03

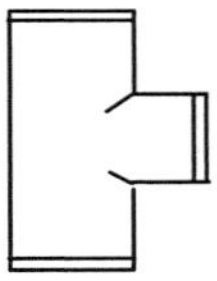

Schedule 40 carbon steel male adapter

Description	Craft@Hrs	Unit	Material $	Labor $	Equipment $	Total $
2"	P1@.170	Ea	17.60	6.24	—	23.84
2½"	P1@.185	Ea	20.90	6.79	—	27.69
3"	P1@.200	Ea	26.00	7.34	—	33.34
4"	P1@.230	Ea	43.20	8.44	—	51.64
5"	ER@.290	Ea	91.70	11.40	—	103.10
6"	ER@.400	Ea	215.00	15.80	—	230.80

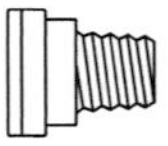

Schedule 40 carbon steel female adapter

Description	Craft@Hrs	Unit	Material $	Labor $	Equipment $	Total $
2"	P1@.170	Ea	39.80	6.24	—	46.04
3"	P1@.200	Ea	61.60	7.34	—	68.94
4"	P1@.230	Ea	86.20	8.44	—	94.64

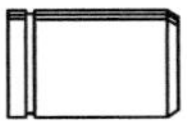

Carbon Steel, Schedule 40 with Cut-Grooved Joints

Description	Craft@Hrs	Unit	Material $	Labor $	Equipment $	Total $
Schedule 40 carbon steel reducer						
2"	P1@.160	Ea	37.20	5.87	—	43.07
3"	P1@.170	Ea	51.60	6.24	—	57.84
4"	P1@.210	Ea	62.80	7.71	—	70.51
5"	ER@.260	Ea	87.40	10.20	—	97.60
6"	ER@.310	Ea	102.00	12.20	—	114.20
8"	ER@.450	Ea	265.00	17.70	—	282.70
10"	ER@.690	Ea	461.00	27.20	—	488.20
12"	ER@.920	Ea	801.00	36.20	—	837.20
Schedule 40 carbon steel cap						
2"	P1@.110	Ea	19.30	4.04	—	23.34
2½"	P1@.120	Ea	30.30	4.40	—	34.70
3"	P1@.130	Ea	30.30	4.77	—	35.07
4"	P1@.140	Ea	32.70	5.14	—	37.84
5"	ER@.150	Ea	75.90	5.91	—	81.81
6"	ER@.170	Ea	78.80	6.70	—	85.50
8"	ER@.200	Ea	152.00	7.88	—	159.88
10"	ER@.225	Ea	274.00	8.87	—	282.87
12"	ER@.250	Ea	444.00	9.85	—	453.85
Flange with gasket, grooved joint						
2"	P1@.280	Ea	57.90	10.30	—	68.20
2½"	P1@.340	Ea	68.30	12.50	—	80.80
3"	P1@.390	Ea	77.30	14.30	—	91.60
4"	P1@.560	Ea	112.00	20.60	—	132.60
6"	ER@.780	Ea	140.00	30.70	—	170.70
8"	ER@1.10	Ea	164.00	43.30	—	207.30
10"	ER@1.40	Ea	209.00	55.20	—	264.20
12"	ER@1.70	Ea	288.00	67.00	—	355.00
Cut-grooved joint coupling with gasket						
2"	P1@.300	Ea	29.90	11.00	—	40.90
2½"	P1@.350	Ea	36.00	12.80	—	48.80
3"	P1@.400	Ea	39.80	14.70	—	54.50
4"	P1@.500	Ea	57.90	18.40	—	76.30
5"	ER@.600	Ea	88.70	23.60	—	112.30
6"	ER@.700	Ea	104.00	27.60	—	131.60
8"	ER@.900	Ea	174.00	35.50	—	209.50
10"	ER@1.10	Ea	241.00	43.30	—	284.30
12"	ER@1.35	Ea	274.00	53.20	—	327.20

Description	Craft@Hrs	Unit	Material $	Labor $	Equipment $	Total $

Class 125 iron body gate valve, flanged ends

2"	P1@.500	Ea	523.00	18.40	—	541.40
2½"	P1@.600	Ea	708.00	22.00	—	730.00
3"	P1@.750	Ea	768.00	27.50	—	795.50
4"	P1@1.35	Ea	1,130.00	49.50	—	1,179.50
6"	ER@2.50	Ea	2,170.00	98.50	1.55	2,270.05
8"	ER@3.00	Ea	3,520.00	118.00	1.86	3,639.86
10"	ER@4.00	Ea	5,810.00	158.00	2.47	5,970.47
12"	ER@4.50	Ea	7,960.00	177.00	2.78	8,139.78

Class 125 iron body globe valve, flanged joints

2"	P1@.500	Ea	267.00	18.40	—	285.40
2½"	P1@.600	Ea	416.00	22.00	—	438.00
3"	P1@.750	Ea	498.00	27.50	—	525.50
4"	ER@1.35	Ea	669.00	53.20	—	722.20
6"	ER@2.50	Ea	1,180.00	98.50	1.55	1,280.05
8"	ER@3.00	Ea	1,800.00	118.00	1.86	1,919.86
10"	ER@4.00	Ea	2,750.00	158.00	2.47	2,910.47

200 PSIG iron body butterfly valve, lug-type, lever operated

2"	P1@.450	Ea	163.00	16.50	—	179.50
2½"	P1@.450	Ea	169.00	16.50	—	185.50
3"	P1@.550	Ea	177.00	20.20	—	197.20
4"	P1@.550	Ea	221.00	20.20	—	241.20
6"	ER@.800	Ea	356.00	31.50	.49	387.99
8"	ER@.800	Ea	487.00	31.50	.49	518.99
10"	ER@.900	Ea	677.00	35.50	.56	713.06
12"	ER@1.00	Ea	884.00	39.40	.62	924.02

200 PSIG iron body butterfly valve, wafer-type, lever operated

2"	P1@.450	Ea	148.00	16.50	—	164.50
2½"	P1@.450	Ea	151.00	16.50	—	167.50
3"	P1@.550	Ea	163.00	20.20	—	183.20
4"	P1@.550	Ea	197.00	20.20	—	217.20
6"	ER@.800	Ea	330.00	31.50	.49	361.99
8"	ER@.800	Ea	457.00	31.50	.49	488.99
10"	ER@.900	Ea	639.00	35.50	.56	675.06
12"	ER@1.00	Ea	839.00	39.40	.62	879.02

Class 125 iron body swing check valve, flanged ends

2"	P1@.500	Ea	179.00	18.40	—	197.40
2½"	P1@.600	Ea	226.00	22.00	—	248.00
3"	P1@.750	Ea	282.00	27.50	—	309.50
4"	P1@1.35	Ea	414.00	49.50	—	463.50
6"	ER@2.50	Ea	803.00	98.50	1.55	903.05
8"	ER@3.00	Ea	1,440.00	118.00	1.86	1,559.86
10"	ER@4.00	Ea	2,350.00	158.00	2.47	2,510.47
12"	ER@4.50	Ea	3,220.00	177.00	2.78	3,399.78

Carbon Steel, Schedule 40 with Cut-Grooved Joints

Description	Craft@Hrs	Unit	Material $	Labor $	Equipment $	Total $

Class 125 iron body silent check valve, flanged joints

2"	P1@.500	Ea	131.00	18.40	—	149.40
2½"	P1@.600	Ea	150.00	22.00	—	172.00
3"	P1@.750	Ea	170.00	27.50	—	197.50
4"	P1@1.35	Ea	222.00	49.50	—	271.50
5"	ER@2.00	Ea	322.00	78.80	1.24	402.04
6"	ER@2.50	Ea	399.00	98.50	1.55	499.05
8"	ER@3.00	Ea	719.00	118.00	1.86	838.86
10"	ER@4.00	Ea	1,110.00	158.00	2.47	1,270.47

Class 125 iron body strainer, flanged

2"	P1@.500	Ea	135.00	18.40	—	153.40
2½"	P1@.600	Ea	151.00	22.00	—	173.00
3"	P1@.750	Ea	175.00	27.50	—	202.50
4"	P1@1.35	Ea	297.00	49.50	—	346.50
6"	ER@2.50	Ea	604.00	98.50	1.55	704.05
8"	ER@3.00	Ea	1,020.00	118.00	1.86	1,139.86

Installation of 2-way control valve, flanged joints

2"	P1@.500	Ea	—	18.40	—	18.40
2½"	P1@.600	Ea	—	22.00	—	22.00
3"	P1@.750	Ea	—	27.50	—	27.50
4"	P1@1.35	Ea	—	49.50	—	49.50
6"	P1@2.50	Ea	—	91.80	—	91.80
8"	P1@3.00	Ea	—	110.00	—	110.00
10"	P1@4.00	Ea	—	147.00	—	147.00
12"	P1@4.50	Ea	—	165.00	—	165.00

Installation of 3-way control valve, flanged joints

2"	P1@.910	Ea	—	33.40	—	33.40
2½"	P1@1.12	Ea	—	41.10	—	41.10
3"	P1@1.33	Ea	—	48.80	—	48.80
4"	P1@2.00	Ea	—	73.40	—	73.40
6"	P1@3.70	Ea	—	136.00	—	136.00
8"	P1@4.40	Ea	—	161.00	—	161.00
10"	P1@5.90	Ea	—	217.00	—	217.00
12"	P1@6.50	Ea	—	239.00	—	239.00

Bolt and gasket sets

2"	P1@.500	Ea	4.17	18.40	—	22.57
2½"	P1@.650	Ea	4.87	23.90	—	28.77
3"	P1@.750	Ea	8.24	27.50	—	35.74
4"	P1@1.00	Ea	13.70	36.70	—	50.40
6"	P1@1.20	Ea	22.90	44.00	—	66.90
8"	P1@1.25	Ea	25.70	45.90	—	71.60
10"	P1@1.70	Ea	44.20	62.40	—	106.60
12"	P1@2.20	Ea	49.80	80.70	—	130.50

Carbon Steel, Schedule 40 with Cut-Grooved Joints

Description	Craft@Hrs	Unit	Material $	Labor $	Equipment $	Total $
Thermometer with well						
7"	P1@.250	Ea	184.00	9.18	—	193.18
9"	P1@.250	Ea	190.00	9.18	—	199.18
Dial-type pressure gauge						
2½"	P1@.200	Ea	38.50	7.34	—	45.84
3½"	P1@.200	Ea	50.70	7.34	—	58.04
Pressure/temperature tap						
Tap	P1@.150	Ea	17.60	5.51	—	23.11
Hanger with swivel assembly						
2"	P1@.300	Ea	6.87	11.00	—	17.87
2½"	P1@.350	Ea	8.88	12.80	—	21.68
3"	P1@.350	Ea	11.00	12.80	—	23.80
4"	P1@.350	Ea	17.00	12.80	—	29.80
5"	P1@.450	Ea	19.20	16.50	—	35.70
6"	P1@.450	Ea	24.70	16.50	—	41.20
8"	P1@.450	Ea	32.90	16.50	—	49.40
10"	P1@.550	Ea	45.30	20.20	—	65.50
12"	P1@.600	Ea	60.20	22.00	—	82.20
Riser clamp						
2"	P1@.115	Ea	5.50	4.22	—	9.72
2½"	P1@.120	Ea	5.79	4.40	—	10.19
3"	P1@.120	Ea	6.27	4.40	—	10.67
4"	P1@.125	Ea	7.99	4.59	—	12.58
5"	P1@.180	Ea	11.40	6.61	—	18.01
6"	P1@.200	Ea	13.80	7.34	—	21.14
8"	P1@.200	Ea	22.50	7.34	—	29.84
10"	P1@.250	Ea	33.50	9.18	—	42.68
12"	P1@.250	Ea	39.70	9.18	—	48.88
Galvanized steel pipe sleeves						
2"	P1@.130	Ea	5.56	4.77	—	10.33
2½"	P1@.150	Ea	5.64	5.51	—	11.15
3"	P1@.180	Ea	5.78	6.61	—	12.39
4"	P1@.220	Ea	6.59	8.07	—	14.66
5"	P1@.250	Ea	8.15	9.18	—	17.33
6"	P1@.270	Ea	8.80	9.91	—	18.71
8"	P1@.270	Ea	10.10	9.91	—	20.01
10"	P1@.290	Ea	11.70	10.60	—	22.30
12"	P1@.310	Ea	16.10	11.40	—	27.50
14"	P1@.330	Ea	16.10	12.10	—	28.20

Description	Craft@Hrs	Unit	Material $	Labor $	Equipment $	Total $

Leadership in Energy & Environmental Design (LEED).

LEED offers an incentive to building owners to install energy-efficient and environmentally sensitive HVAC equipment. The LEED program awards points for application of the best available technology. The HVAC system can earn a maximum of 17 points toward LEED certification. A LEED-certified air conditioning system must comply with ASHRAE 90.1-2004 (efficiency) and ASHRAE 62.1-2004 (indoor air quality) standards. LEED certification also requires recording controls (Carrier ComfortView 3 or equal) to measure system performance. Zone sensor thermostats (Carrier Debonair or equal) are required to provide performance and efficiency data to the digital recorder. An indoor air quality CO_2 sensor, duct-mounted aspirator modules and a refrigerant such as Puron R-410A are also required.

Description	Craft@Hrs	Unit	Material $	Labor $	Equipment $	Total $
Add for a LEED-certified system with USGBC rating	—	%	20.0	—	—	—
Add for LEED registration and inspection	—	Ea	2,050.00	—	—	2,050.00
Add for LEED central digital performance monitor/recorder, ComfortView 3 or equal	—	Ea	2,410.00	—	—	2,410.00
Add for LEED-certified zone recording thermostats, Carrier DebonAir, USB remote-ready	—	Ea	358.00	—	—	358.00
Add for IAQ (Indoor Air Quality) CO_2 sensor and duct-mounted aspirator box	—	Ea	436.00	—	—	436.00

Residential A/C Cooling System.

With remote exterior condenser/compressor, field installed DX coil in existing supply air plenum. Includes electrical & control wiring, R-410A refrigerant piping & connections, programmable digital thermostat, start-up and testing. Costs based on a maximum distance between outdoor condenser and coil of 60' and include a contractor mark up of 25%. Use these figures for preliminary budget purposes. Costs are based on using 650 SF per ton as the cooling requirement with a thermal envelope of R20 or greater, i.e., a 1.5-ton A/C unit will serve a 975 SF home. Typical subcontract prices.

Description	Craft@Hrs	Unit	Material $	Labor $	Equipment $	Total $
1.5 tons	—	Ea	—	—	—	3,040.00
2 tons	—	Ea	—	—	—	3,270.00
2.5 tons	—	Ea	—	—	—	4,060.00
3 tons	—	Ea	—	—	—	5,520.00
5 tons	—	Ea	—	—	—	8,340.00
Add for high heat loads	—	%	—	—	—	12.0

Description	Craft@Hrs	Unit	Material $	Labor $	Equipment $	Total $

Residential forced-air heating, cooling & ventilation system.

Including a remote exterior condenser/compressor, field installed R-410A refrigerant DX "A" coil in supply air plenum, a gas-fired high efficiency furnace, programmable digital thermostat, ductwork, grilles & registers, HRV (heat recovery ventilator), kitchen and bathroom exhaust fans, ventilation and exhaust ductwork and wall hoods, electrical & control wiring, gas, refrigerant & flue piping & associated connections, start-up, testing, and a contractor markup of 25%. Use these figures for preliminary budget purposes. Heating & cooling load calculations based on a thermal envelope of R20 or greater. Typical subcontract prices.

Description	Craft@Hrs	Unit	Material $	Labor $	Equipment $	Total $
900 – 1,000 SF	—	Ea	—	—	—	15,800.00
1,100 – 1,400 SF	—	Ea	—	—	—	18,000.00
1,500 – 1,750 SF	—	Ea	—	—	—	22,200.00
1,800 – 2000 SF	—	Ea	—	—	—	23,800.00
2,200 – 2,500 SF	—	Ea	—	—	—	26,700.00
2,600 – 3,000 SF	—	Ea	—	—	—	32,200.00
Deduct for mid-efficient unit	—	Ea	—	—	—	-283.00
Deduct for electric furnace	—	Ea	—	—	—	-283.00

Description	Craft@Hrs	Unit	Material $	Labor $	Equipment $	Total $

Packaged self-contained rooftop DX air conditioning units.

EnergySmart certified, ASHRAE, U.S. Green Building Council and Underwriters Laboratories approved. Includes cooling coils, compressor, heat rejection coils, regulator valves, refrigerant tank and remote digital single-zone control package. Set in place with a 13,000 lb. truck crane. Add installation costs from the section that follows.

Description	Craft@Hrs	Unit	Material $	Labor $	Equipment $	Total $
2-T, 800 CFM	SN@5.00	Ea	3,890.00	193.00	50.50	4,133.50
2½-T, 1,000 CFM	SN@6.00	Ea	4,660.00	232.00	60.60	4,952.60
3-T, 1,200 CFM	SN@7.00	Ea	5,570.00	271.00	70.70	5,911.70
4-T, 1,600 CFM	SN@8.00	Ea	6,820.00	309.00	80.80	7,209.80
5-T, 2,000 CFM	SN@9.00	Ea	8,520.00	348.00	90.90	8,958.90
7½-T, 3,000 CFM	SN@10.0	Ea	11,800.00	387.00	101.00	12,288.00
10-T, 4,000 CFM	SN@12.0	Ea	14,700.00	464.00	121.00	15,285.00
12-T, 5,000 CFM	SN@14.0	Ea	17,600.00	541.00	141.00	18,282.00
15-T, 6,000 CFM	SN@15.0	Ea	21,500.00	580.00	151.00	22,231.00
20-T, 8,000 CFM	SN@16.0	Ea	28,500.00	618.00	162.00	29,280.00
25-T, 10,000 CFM	SN@17.0	Ea	34,700.00	657.00	172.00	35,529.00
30-T, 12,000 CFM	SN@19.0	Ea	39,700.00	734.00	192.00	40,626.00
40-T, 16,000 CFM	SN@22.0	Ea	49,600.00	850.00	222.00	50,672.00

Rooftop DX air conditioning unit, hot water coil.

EnergySmart certified, ASHRAE, U.S. Green Building Council, and Underwriters Laboratories approved. Add for the hot water coil. Single zone controls. Add installation costs from the section that follows.

Description	Craft@Hrs	Unit	Material $	Labor $	Equipment $	Total $
2-T, 800 CFM	SN@5.00	Ea	3,790.00	193.00	50.50	4,033.50
2½-T, 1,000 CFM	SN@6.00	Ea	4,560.00	232.00	60.60	4,852.60
3-T, 1,200 CFM	SN@7.00	Ea	5,450.00	271.00	70.70	5,791.70
4-T, 1,600 CFM	SN@8.00	Ea	6,650.00	309.00	80.80	7,039.80
5-T, 2,000 CFM	SN@9.00	Ea	8,330.00	348.00	90.90	8,768.90
7½-T, 3,000 CFM	SN@10.0	Ea	12,600.00	387.00	101.00	13,088.00
10-T, 4,000 CFM	SN@12.0	Ea	14,400.00	464.00	121.00	14,985.00
12-T, 5,000 CFM	SN@14.0	Ea	17,100.00	541.00	141.00	17,782.00
15-T, 6,000 CFM	SN@15.0	Ea	21,000.00	580.00	151.00	21,731.00
20-T, 8,000 CFM	SN@16.0	Ea	27,900.00	618.00	162.00	28,680.00
25-T, 10,000 CFM	SN@17.0	Ea	33,700.00	657.00	172.00	34,529.00
30-T, 12,000 CFM	SN@19.0	Ea	39,700.00	734.00	192.00	40,626.00
40-T, 16,000 CFM	SN@22.0	Ea	49,600.00	850.00	222.00	50,672.00

Description	Craft@Hrs	Unit	Material $	Labor $	Equipment $	Total $

Installation costs for packaged rooftop cooling
Typical installation costs for packaged rooftop cooling. Costs will be higher for larger capacity units.

Description	Craft@Hrs	Unit	Material $	Labor $	Equipment $	Total $
Cut, frame and gasket downcomer hole in roof	S2@1.00	Ea	32.20	36.00	—	68.20
Mount duct hangers	S2@.250	Ea	4.15	9.01	—	13.16
Cut and mount sheet metal duct	S2@.350	LF	6.38	12.60	—	18.98
Apply and coat duct insulation	S2@.150	LF	2.76	5.41	—	8.17
Install piping for gas line	P1@.100	LF	4.30	3.67	—	7.97
Install piping for chilled or hot water/steam line	P1@.100	LF	3.26	3.67	—	6.93
Install electrical wiring	BE@.150	LF	1.98	6.06	—	8.04
Install HVAC controls	BE@.500	Ea	—	20.20	—	20.20
Commission and test	P1@4.00	Ea	—	147.00	—	147.00
Air balance and fine-tune	P1@4.00	Ea	—	147.00	—	147.00

Packaged air handler with chilled water and hot water/steam coil.
Costs shown based on 400 CFM per ton cooling. Unit includes insulated single wall casing, fan section, cooling coil section, heating coil section, mixing plenum section, bag filter section, fan motor, variable pitch drive, vibration isolators and drain pan. Set in place only. Make additional allowances for coil connections, controls, motor starters and power wiring. (12,000 BTUs equals 1 ton cooling.) Use these costs for preliminary estimates.

Description	Craft@Hrs	Unit	Material $	Labor $	Equipment $	Total $
3-T, 1,200 CFM	SN@4.00	Ea	3,130.00	155.00	40.40	3,325.40
4-T, 1,600 CFM	SN@5.50	Ea	4,320.00	213.00	55.50	4,588.50
5-T, 2,000 CFM	SN@7.00	Ea	5,140.00	271.00	70.70	5,481.70
7½-T, 3,000 CFM	SN@9.00	Ea	7,250.00	348.00	90.90	7,688.90
10-T, 4,000 CFM	SN@11.0	Ea	9,190.00	425.00	111.00	9,726.00
12-T, 5,000 CFM	SN@13.0	Ea	10,900.00	502.00	131.00	11,533.00
15-T, 6,000 CFM	SN@14.0	Ea	12,300.00	541.00	141.00	12,982.00
20-T, 8,000 CFM	SN@15.0	Ea	14,600.00	580.00	151.00	15,331.00
25-T, 10,000 CFM	SN@16.0	Ea	14,500.00	618.00	162.00	15,280.00
30-T, 12,000 CFM	SN@18.0	Ea	18,200.00	696.00	182.00	19,078.00
40-T, 16,000 CFM	SN@21.0	Ea	22,400.00	812.00	212.00	23,424.00

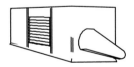

Air Handling Unit Accessories

Description	Craft@Hrs	Unit	Material $	Labor $	Equipment $	Total $

Air handling unit accessories and options. Use these costs for preliminary estimates.

Description	Craft@Hrs	Unit	Material $	Labor $	Equipment $	Total $
DDC controls per zone complete	SN@2.75	Ea	1,020.00	106.00	—	1,126.00
Electric controls per zone complete	SN@2.75	Ea	545.00	106.00	—	651.00
Pneumatic controls per zone complete	SN@2.75	Ea	513.00	106.00	—	619.00
Variable speed drive 5 HP	SN@4.00	Ea	4,110.00	155.00	—	4,265.00
Variable speed drive 7.5HP	SN@4.00	Ea	4,790.00	155.00	—	4,945.00
Variable speed drive 10 HP	SN@6.00	Ea	5,450.00	232.00	—	5,682.00
Variable speed drive 15 HP	SN@8.00	Ea	6,150.00	309.00	—	6,459.00
Variable speed drive 20 HP	SN@12.0	Ea	6,850.00	464.00	—	7,314.00
Variable speed drive 25 HP	SN@16.0	Ea	7,530.00	618.00	—	8,148.00
Variable speed drive 30 HP	SN@18.0	Ea	8,880.00	696.00	—	9,576.00
Variable speed drive 40 HP	SN@22.0	Ea	11,000.00	850.00	—	11,850.00
Variable speed drive 50 HP	SN@26.0	Ea	13,100.00	1,000.00	—	14,100.00
Variable inlet vanes 1,000-1,500 CFM	SN@4.00	Ea	205.00	155.00	2.47	362.47
Variable inlet vanes 1,600-2,500 CFM	SN@4.00	Ea	445.00	155.00	2.47	602.47
Variable inlet vanes 2,600-5,000 CFM	SN@6.00	Ea	1,250.00	232.00	3.71	1,485.71
Variable inlet vanes 6,000-10,000 CFM	SN@8.00	Ea	2,050.00	309.00	4.95	2,363.95
Variable inlet vanes 11,000-20,000 CFM	SN@10.0	Ea	3,410.00	387.00	6.19	3,803.19

Description	Craft@Hrs	Unit	Material $	Labor $	Equipment $	Total $

Heat Recovery Ventilators provide a fresh air supply to tightly sealed building envelopes. An HRV extracts heat from the stale indoor air being exhausted and transfers the heat to the fresh air being drawn into the building through the HRV. Heat recovery ventilators are also excellent dehumidifiers.

HRV unit features include energy-efficient defrost cycle, cross-flow polypropylene heat exchanger, acoustically lined cabinet, outdoor air filter.

Allow 15-30 cfm per person or .04 to.05 cfm per square foot when sizing HRV unit. (ASHRAE62-19890) (cfm = cubic feet per minute)

Description	Craft@Hrs	Unit	Material $	Labor $	Equipment $	Total $

Commercial heat recovery ventilator. Set/hang in place only. Make additional allowances for duct, diffusers, controls, air balancing, electrical connections and condensate drain.

HRV 700 cfm	SN@4.75	Ea	5,960.00	184.00	89.70	6,233.70
HRV 1,200 cfm	SN@5.25	Ea	6,420.00	203.00	99.10	6,722.10
HRV 2,500 cfm	SN@6.75	Ea	26,400.00	261.00	127.00	26,788.00

Swimming pool heat recovery ventilator. Set/hang in place only. Make additional allowances for duct, diffusers, controls, air balancing, electrical connections and condensate drain.

| HRV 700 cfm | SN@5.00 | Ea | 6,910.00 | 193.00 | 94.40 | 7,197.40 |
| HRV 1,200 cfm | SN@5.75 | Ea | 7,830.00 | 222.00 | 109.00 | 8,161.00 |

Conventional heat recovery ventilator. Hang in place only. Make additional allowances for duct, diffusers, controls, air balancing, electrical connections and condensate drain.

HRV 65–150 cfm	S2@2.45	Ea	1,100.00	88.30	—	1,188.30
HRV 115–200 cfm	S2@2.65	Ea	1,300.00	95.50	—	1,395.50

Compact heat recovery ventilator. Hang in place only. Make additional allowances for duct, diffusers, controls, air balancing, electrical connections and condensate drain.

HRV 65–127 cfm	S2@2.45	Ea	1,170.00	88.30	—	1,258.30
HRV 115–195 cfm	S2@2.65	Ea	1,280.00	95.50	—	1,375.50

High-efficiency heat recovery ventilator. Hang in place only. Make additional allowances for duct, diffusers, controls, air balancing, electrical connections and condensate drain.

HRV 65–127 cfm	S2@2.65	Ea	1,290.00	95.50	—	1,385.50
HRV 115–180cfm	S2@2.90	Ea	1,970.00	105.00	—	2,075.00
HRV 180–265cfm	S2@3.25	Ea	2,050.00	117.00	—	2,167.00

Heat recovery ventilator controls

HRV basic control	S2@1.00	Ea	99.50	36.00	—	135.50
HRV std. control	S2@1.00	Ea	159.00	36.00	—	195.00
HRV auto control	S2@1.00	Ea	215.00	36.00	—	251.00
60 minute timer	S2@.600	Ea	66.80	21.60	—	88.40
Interlock relay	S2@.600	Ea	79.40	21.60	—	101.00

Heat recovery ventilator accessories

6" diffusers	S2@.500	Ea	17.00	18.00	—	35.00
6" wall hoods	S2@1.50	Ea	29.30	54.10	—	83.40
6" tee fittings	S2@.350	Ea	3.92	12.60	—	16.52
6" flex duct	S2@.025	LF	.93	.90	—	1.83
6" flex insul duct	S2@.025	LF	1.79	.90	—	2.69
HRV filters	S2@.450	Ea	24.00	16.20	—	40.20

Installation costs for heat recovery ventilators

Description	Craft@Hrs	Unit	Material $	Labor $	Equipment $	Total $
Hang heat recovery ventilator, 150 CFM	S2@2.00	Ea	5.11	72.10	—	77.21
Hang heat recovery ventilator, 250 CMF	S2@2.25	LF	7.67	81.10	—	88.77
Heat recovery ventilator ducting, 150 CFM	S2@4.00	LF	342.00	144.00	—	486.00
Heat recovery ventilator ducting, 250 CFM	S2@4.75	LF	383.00	171.00	—	554.00
Heat recovery ventilator drain line, gravity	P1@1.50	LF	25.60	55.10	—	80.70
Heat recovery ventilator drain line, pumped	P1@2.00	LF	230.00	73.40	—	303.40
Heat recovery ventilator power wiring, 115 volt	BE@2.00	Ea	86.90	80.80	—	167.70
Heat recovery ventilator power wiring, 24 volt	BE@2.00	Ea	20.40	80.80	—	101.20
Commission and test	P1@1.50	Ea	—	55.10	—	55.10
Run air balance and fine tune system	P1@1.75	Ea	—	64.20	—	64.20

Water Coil Piping

Engineering drawings of coil piping details have a bad reputation among HVAC contractors and estimators. They're notorious for what they leave out. They'll rarely show more than one coil bank, no matter how big the system. Furthermore, the drawings hardly ever call out sizes for either the piping or the control valves. Don't be taken in by the apparent simplicity of the system as shown in these drawings. It's likely to be only the tip of the iceberg. For example, unless the air handling capacity of the system is less than 16,000 CFM, the single coil bank shown won't be adequate. Add one or two more coil banks and you're looking at a lot more piping — and a more complex system that takes longer to install.

You probably haven't even decided which equipment supplier to use. This is hardly the time for you to start researching heating and cooling coils. Nevertheless, you need better, more complete and realistic data to come up with a competitive estimate.

I'll tell you how and where to track down the hard data that you need, and also pass along a few tips on estimating water coil piping. They'll help you avoid leaving something out of your estimates — a real pitfall for any beginner. It's those little (but essential) items, so easily overlooked, that are so deadly to a profit margin. Finally, check the next two pages of diagrams with tables. The data given there, combined with the data you collected earlier, forms the basis for informed guesswork.

The Hard Data

There are two things you absolutely must know to estimate water coil piping. First, the size of the branch piping to the coils; second, the CFM rating of the air handling units for the system.

To find the pipe sizes, look at either the floor plans or the details for the equipment room. They'll list the sizes of the branch run-out pipes. Once you know them, you can make a good guess at the right size for the control valves. (See the diagrams and tables on pages 295 and 296.) The only information you need is the CFM ratings of the air handling units and the branch piping sizes to the coils. If the system's capacity is over 16,000 CFM, you need two or three coil banks.

A Few Tips on Estimating Water Coil Piping

1) Coil connection sizes seldom match branch pipe sizes. Be sure to include the reducing fittings you'll need in any estimate.

2) Two-way and three-way control valves are both usually one pipe size smaller than the pipe where they're installed. That means you'll need either two or three reducing fittings per control valve. Be sure you include their cost in your estimates.

Using the Diagrams

In the following diagrams, for clarity, some items aren't included. These items are: balance valves, shut-off valves, reducers, strainers, gauges and gauge taps. Any details you need about these items for your estimate are in the engineer's coil piping details.

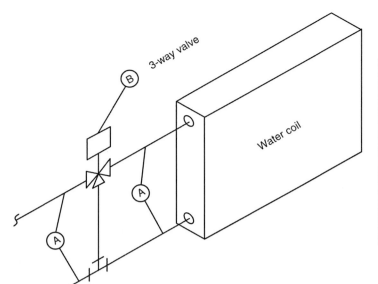

If Ⓐ is:	Then Ⓑ is:
1"	¾"
1¼"	1"
1½"	1¼"
2"	1½"
2½"	2"
3"	2½"

Typical water coil piping for A.H. units up to 16,000 CFM

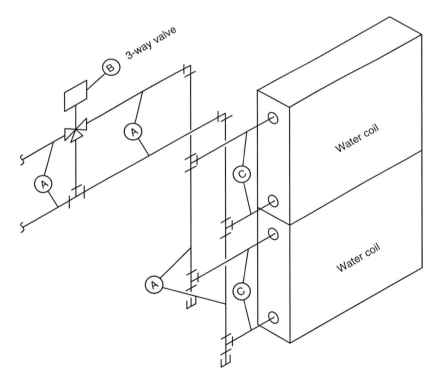

If Ⓐ is:	Then Ⓑ is:	And Ⓒ is:
2½"	2"	2"
3"	2½"	2½"
4"	3"	3"

Typical water coil piping for A.H. units from 16,000 to 26,000 CFM

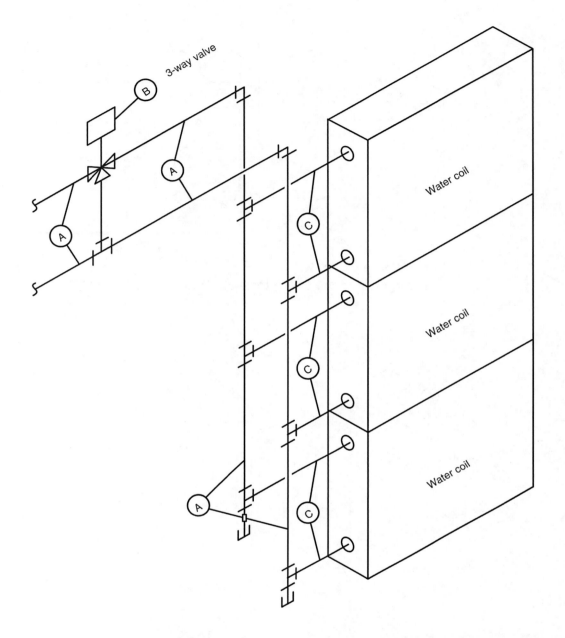

If Ⓐ is:	Then Ⓑ is:	And Ⓒ is:
4"	3"	2½"
6"	4"	3"
8"	6"	4"
10"	8"	6"

Typical water coil piping for A.H. units over 26,000 CFM

Description	Craft@Hrs	Unit	Material $	Labor $	Equipment $	Total $

Air handling unit coil connection, one row coil bank, non-regulated flow. Connection assembly includes pipe, fittings, pipe insulation, valves, gauges, thermometers and vents.

Description	Craft@Hrs	Unit	Material $	Labor $	Equipment $	Total $
1½" supply	SN@14.0	Ea	1,820.00	541.00	—	2,361.00
2" supply	SN@18.0	Ea	2,240.00	696.00	—	2,936.00
2½" supply	SN@28.0	Ea	2,990.00	1,080.00	—	4,070.00
3" supply	SN@31.5	Ea	3,280.00	1,220.00	—	4,500.00
4" supply	SN@34.0	Ea	3,520.00	1,310.00	—	4,830.00
6" supply	SN@41.0	Ea	5,090.00	1,580.00	25.40	6,695.40

Air handling unit coil connection, one row coil bank, 2-way control valve design. Connection assembly includes pipe and fittings, pipe insulation, valves, gauges, thermometers and vents.

Description	Craft@Hrs	Unit	Material $	Labor $	Equipment $	Total $
1½" supply	SN@16.5	Ea	2,050.00	638.00	—	2,688.00
2" supply	SN@21.0	Ea	2,650.00	812.00	—	3,462.00
2½" supply	SN@33.6	Ea	3,220.00	1,300.00	—	4,520.00
3" supply	SN@36.0	Ea	3,460.00	1,390.00	—	4,850.00
4" supply	SN@43.8	Ea	4,200.00	1,690.00	—	5,890.00
6" supply	SN@54.0	Ea	6,730.00	2,090.00	33.40	8,853.40

Air handling unit coil connection, one row coil bank, 3-way control valve design. Connection assembly includes pipe and fittings, pipe insulation, valves, gauges, thermometers and vents.

Description	Craft@Hrs	Unit	Material $	Labor $	Equipment $	Total $
1½" supply	SN@18.0	Ea	2,370.00	696.00	—	3,066.00
2" supply	SN@26.0	Ea	2,960.00	1,000.00	—	3,960.00
2½" supply	SN@41.5	Ea	3,710.00	1,600.00	—	5,310.00
3" supply	SN@43.8	Ea	4,090.00	1,690.00	—	5,780.00
4" supply	SN@54.6	Ea	5,270.00	2,110.00	—	7,380.00
6" supply	SN@61.4	Ea	7,500.00	2,370.00	38.00	9,908.00

Air handling unit coil connection, two row coil bank, non-regulated flow. Connection assembly includes pipe and fittings, pipe insulation, valves, gauges, thermometers and vents.

Description	Craft@Hrs	Unit	Material $	Labor $	Equipment $	Total $
2½" supply	SN@48.0	Ea	4,490.00	1,860.00	—	6,350.00
3" supply	SN@51.5	Ea	5,460.00	1,990.00	—	7,450.00
4" supply	SN@63.6	Ea	7,230.00	2,460.00	—	9,690.00
6" supply	SN@79.2	Ea	10,600.00	3,060.00	49.00	13,709.00

Air handling unit coil connection, two row coil bank, 2-way control valve design. Connection assembly includes pipe and fittings, pipe insulation, valves, gauges, thermometers and vents.

Description	Craft@Hrs	Unit	Material $	Labor $	Equipment $	Total $
2½" supply	SN@54.0	Ea	5,030.00	2,090.00	—	7,120.00
3" supply	SN@58.5	Ea	6,320.00	2,260.00	—	8,580.00
4" supply	SN@67.0	Ea	7,770.00	2,590.00	—	10,360.00
6" supply	SN@83.5	Ea	10,800.00	3,230.00	51.70	14,081.70

Air Handling Unit Coil Connections

Description	Craft@Hrs	Unit	Material $	Labor $	Equipment $	Total $

Air handling unit coil connection, two row coil bank, 3-way control valve design. Connection assembly includes pipe and fittings, pipe insulation, valves, gauges, thermometers and vents.

Description	Craft@Hrs	Unit	Material $	Labor $	Equipment $	Total $
2½" supply	SN@58.8	Ea	5,270.00	2,270.00	—	7,540.00
3" supply	SN@64.0	Ea	6,640.00	2,470.00	—	9,110.00
4" supply	SN@72.0	Ea	8,410.00	2,780.00	—	11,190.00
6" supply	SN@87.0	Ea	11,200.00	3,360.00	53.80	14,613.80

Air handling unit coil connection, three-row coil bank, non-regulated flow. Connection assembly includes pipe and fittings, pipe insulation, valves, gauges, thermometers and vents.

Description	Craft@Hrs	Unit	Material $	Labor $	Equipment $	Total $
4" supply	SN@72.0	Ea	7,610.00	2,780.00	—	10,390.00
6" supply	SN@94.0	Ea	13,300.00	3,630.00	—	16,930.00
8" supply	SN@120.	Ea	17,800.00	4,640.00	74.20	22,514.20

Air handling unit coil connection, three-row coil bank, 2-way control valve design. Connection assembly includes pipe and fittings, pipe insulation, valves, gauges, thermometers and vents.

Description	Craft@Hrs	Unit	Material $	Labor $	Equipment $	Total $
4" supply	SN@78.0	Ea	8,090.00	3,010.00	—	11,100.00
6" supply	SN@105.	Ea	14,600.00	4,060.00	—	18,660.00
8" supply	SN@132.	Ea	19,900.00	5,100.00	81.60	25,081.60

Air handling unit coil connection, three-row coil bank, 3-way control valve design. Connection assembly includes pipe and fittings, pipe insulation, valves, gauges, thermometers and vents.

Description	Craft@Hrs	Unit	Material $	Labor $	Equipment $	Total $
4" supply	SN@82.0	Ea	9,370.00	3,170.00	—	12,540.00
6" supply	SN@112.	Ea	15,000.00	4,330.00	—	19,330.00
8" supply	SN@140.	Ea	22,400.00	5,410.00	86.60	27,896.60

Description	Craft@Hrs	Unit	Material $	Labor $	Equipment $	Total $

Residential gas-fired upflow furnace. Set in place only. Make additional allowances for gas and electrical connections. Costs shown per thousand BTU per hour (MBH) Mid-efficiency, 80% AFUE.

Description	Craft@Hrs	Unit	Material $	Labor $	Equipment $	Total $
50 MBH input	SN@2.00	Ea	943.00	77.30	—	1,020.30
75 MBH input	SN@2.00	Ea	1,230.00	77.30	3.83	1,311.13
100 MBH input	SN@2.00	Ea	1,430.00	77.30	3.83	1,511.13
125 MBH input	SN@2.00	Ea	1,640.00	77.30	3.83	1,721.13

Residential combination gas-fired furnace and A/C unit. With remote exterior condenser/compressor. Set in place only. Make additional allowances for gas, electrical and DX connections. Costs shown per thousand BTU per hour (MBH) Mid-efficiency, 80% AFUE.

Description	Craft@Hrs	Unit	Material $	Labor $	Equipment $	Total $
30 MBH, 1.5 tons	SN@2.50	Ea	2,810.00	96.60	—	2,906.60
36 MBH, 2 tons	SN@2.75	Ea	3,330.00	106.00	—	3,436.00
42 MBH, 2 tons	SN@3.00	Ea	3,740.00	116.00	—	3,856.00
48 MBH, 2.5 tons	SN@3.50	Ea	4,360.00	135.00	—	4,495.00
60 MBH, 3 tons	SN@4.00	Ea	5,210.00	155.00	—	5,365.00

High-efficiency, condensing gas furnace. Direct drive, atmospheric draft, electronic ignition. Set in place only. Make additional allowances for flue, gas and electrical connections. Costs shown per thousand BTU per hour (MBH) High efficiency 94% AFUE.

Description	Craft@Hrs	Unit	Material $	Labor $	Equipment $	Total $
40 MBH input	S2@2.00	Ea	1,960.00	72.10	—	2,032.10
60 MBH input	S2@2.00	Ea	1,960.00	72.10	—	2,032.10
70 MBH input	S2@2.00	Ea	1,870.00	72.10	—	1,942.10
90 MBH input	S2@2.00	Ea	1,870.00	72.10	—	1,942.10
100 MBH input	S2@2.00	Ea	2,670.00	72.10	—	2,742.10
120 MBH input	S2@2.00	Ea	2,760.00	72.10	—	2,832.10
2 speed fan & limit control kit	S2@1.25	Ea	38.20	45.10	—	83.30
Propane conversion kit	S2@1.75	Ea	149.00	63.10	—	212.10

Gas-Fired Furnaces

Central dehumidification system. Upflow or counterflow, 400 CFM, 30 gallon per day baseload capacity at 0.5 water column pressure differential. With inlet filter, drain pan, drain line, heater plate, cooling draft exchanger, humidistat control sensor, control relay, actuator, cabinet, two 24" by 24" by 4' long duct transitions, gaskets, hangers and ground fault protection. Whirlpool UGD160UH or equal. Add the cost of control wiring and electrical wiring.

Central dehumidification system	SW@0.50	Ea	1,750.00	21.00	—	1,771.00
Cut, frame and gasket hole in wall, ceiling or roof	SW@1.00	Ea	32.70	41.90	—	74.60
Mount duct hangers	SW@0.25	Ea	4.13	10.50	—	14.63
Cut and mount sheet metal duct	SW@0.35	LF	6.36	14.70	—	21.06
Apply and coat duct insulation	SW@0.15	LF	2.75	6.29	—	9.04
Install electrical wiring	BE@0.15	LF	1.98	6.06	—	8.04
Install dehumidifier controls	BE@0.50	Ea	—	20.20	—	20.20
Test and balance	SW@1.50	Ea	—	62.90	—	62.90

Wall furnace, mid-efficiency, pilot light ignition. Gas-fired, atmospheric draft (radiant heat, no fan). Set in place only. Make additional allowances for gas and flue connections. Costs shown per thousand BTU per hour (MBH). Mid-efficiency 80% AFUE.

35 MBH input	S2@2.75	Ea	1,320.00	99.10	—	1,419.10
50 MBH input	S2@2.75	Ea	1,490.00	99.10	—	1,589.10
65 MBH input	S2@2.75	Ea	1,500.00	99.10	—	1,599.10
Fan blower kit	S2@1.75	Ea	213.00	63.10	—	276.10

Wall furnace, mid-efficiency, electronic ignition. Gas-fired, atmospheric draft (radiant heat, no fan). Set in place only. Make additional allowances for gas and flue connections. Costs shown per thousand BTU per hour (MBH). Mid-efficiency 80% AFUE.

35 MBH input	S2@2.75	Ea	1,650.00	99.10	—	1,749.10
65 MBH input	S2@2.75	Ea	1,850.00	99.10	—	1,949.10
Fan blower kit	S2@1.75	Ea	213.00	63.10	—	276.10

Wall furnace, direct vent (included). Gas-fired, pilot light ignition, mid-efficiency (radiant heat, no fan). Set in place only. Make additional allowances for vent, gas and electrical connections. Costs shown per thousand BTU per hour (MBH). Mid-efficiency 80% AFUE.

15 MBH input	S2@2.75	Ea	885.00	99.10	—	984.10
25 MBH input	S2@2.75	Ea	1,070.00	99.10	—	1,169.10
33 MBH input	S2@2.75	Ea	1,140.00	99.10	—	1,239.10
40 MBH input	S2@2.75	Ea	1,570.00	99.10	—	1,669.10
65 MBH input	S2@2.75	Ea	1,720.00	99.10	—	1,819.10
Direct vent kit	S2@2.50	Ea	150.00	90.10	—	240.10
Fan blower kit	S2@1.75	Ea	213.00	63.10	—	276.10

Description	Craft@Hrs	Unit	Material $	Labor $	Equipment $	Total $

Enthalpy energy recovery system. Steam-to-air or hot-water-to-air heat exchange. For boiler room downstream retrofits or new construction to qualify for EPA EnergyStar certification. Equipment cost is for a 40' rolling platform scissor lift. Add the cost of the energy recovery wheel ventilator.

Description	Craft@Hrs	Unit	Material $	Labor $	Equipment $	Total $
Reinforced attic joists	P1@3.00	Ea	69.00	110.00	—	179.00
Vibration isolators	S2@.750	LS	44.50	27.00	—	71.50
Drill and mount coil service drain	P1@.150	LF	3.32	5.51	—	8.83
Install electrical wiring	BE@.150	LF	1.98	6.06	—	8.04
Mount unit heater or enthalpy wheel	P1@.500	Ea	—	18.40	—	18.40
Connect ductwork mains to manifold	SW@4.50	Ea	169.00	189.00	—	358.00
Coil piping and condensate drain	P1@4.00	Ea	8.87	147.00	—	155.87
Zone control thermostat, manual	BE@.750	Ea	140.00	30.30	—	170.30
Master heat controller with auto change-over	BE@.750	Ea	328.00	30.30	—	358.30
Wiring of controls	BE@3.00	Ea	1.65	121.00	—	122.65
Calibration test	BE@2.00	Ea	—	80.80	—	80.80

Enthalpy energy recovery wheel. Modine Slimvent, Reznor XAWS or equal. For both heating and cooling applications in commercial and institutional buildings. Complies with the ARI Standard 1060-2000. Provides recovery of up to 75% of heat energy and 80% of cooling energy.

Description	Craft@Hrs	Unit	Material $	Labor $	Equipment $	Total $
800-1,000 CFM	SL@5.00	Ea	4,000.00	182.00	—	4,182.00
1,300-2,000 CFM	SL@5.00	Ea	5,020.00	182.00	—	5,202.00
2,400-3,000 CFM	SL@5.00	Ea	5,840.00	182.00	—	6,022.00
3,200-3,800 CFM	SL@5.00	Ea	7,670.00	182.00	—	7,852.00
4,000-4,800 CFM	SL@5.00	Ea	8,680.00	182.00	—	8,862.00
5,000-6,000 CFM	SL@5.00	Ea	9,720.00	182.00	—	9,902.00

Horizontal hot water unit heaters. Fan-driven, with integral check valves, bypass piping, drain line fitting and vacuum breaker, Modine HSB series or equal.

Description	Craft@Hrs	Unit	Material $	Labor $	Equipment $	Total $
12.5 MBtu, 200 CFM	SL@2.00	Ea	465.00	72.90	4.33	542.23
17 MBtu, 300 CFM	SL@2.00	Ea	513.00	72.90	4.33	590.23
25 MBtu, 500 CFM	SL@2.50	Ea	587.00	91.10	5.41	683.51
30 MBtu, 700 CFM	SL@3.00	Ea	619.00	109.00	6.49	734.49
50 MBtu, 1,000 CFM	SL@3.50	Ea	719.00	128.00	7.58	854.58
60 MBtu, 1,300 CFM	SL@4.00	Ea	853.00	146.00	8.66	1,007.66

Unit Heaters

Description	Craft@Hrs	Unit	Material $	Labor $	Equipment $	Total $

Vertical hot water unit heaters. Fan-driven, with integral check valves, bypass piping, drain line fitting and vacuum breaker, Modine V500 series or equal.

Description	Craft@Hrs	Unit	Material $	Labor $	Equipment $	Total $
12.5 MBtu, 200 CFM	SL@2.00	Ea	521.00	72.90	4.33	598.23
17 MBtu, 300 CFM	SL@2.00	Ea	533.00	72.90	4.33	610.23
25 MBtu, 500 CFM	SL@2.50	Ea	691.00	91.10	5.41	787.51
30 MBtu, 700 CFM	SL@3.00	Ea	721.00	109.00	6.49	836.49
50 MBtu, 1,000 CFM	SL@3.50	Ea	869.00	128.00	7.58	1,004.58
60 MBtu, 1,300 CFM	SL@4.00	Ea	954.00	146.00	8.66	1,108.66

Horizontal steam coil unit heaters. Fan-driven, with integral check valves, bypass piping, drain line fitting and vacuum breaker, Modine HSB series or equal.

Description	Craft@Hrs	Unit	Material $	Labor $	Equipment $	Total $
12.5 MBtu, 200 CFM	SL@2.00	Ea	758.00	72.90	4.33	835.23
17 MBtu, 300 CFM	SL@2.00	Ea	875.00	72.90	4.33	952.23
25 MBtu, 500 CFM	SL@2.50	Ea	960.00	91.10	5.41	1,056.51
30 MBtu, 700 CFM	SL@3.00	Ea	1,050.00	109.00	6.49	1,165.49
50 MBtu, 1,000 CFM	SL@3.50	Ea	1,270.00	128.00	7.58	1,405.58
60 MBtu, 1,300 CFM	SL@4.00	Ea	1,580.00	146.00	8.66	1,734.66

Vertical steam coil unit heaters. Fan-driven, with integral check valves, bypass piping, drain line fitting and vacuum breaker, Modine V500 series or equal.

Description	Craft@Hrs	Unit	Material $	Labor $	Equipment $	Total $
12.5 MBtu, 200 CFM	SL@2.00	Ea	817.00	72.90	4.33	894.23
17 MBtu, 300 CFM	SL@2.00	Ea	896.00	72.90	4.33	973.23
25 MBtu, 500 CFM	SL@2.50	Ea	1,050.00	91.10	5.41	1,146.51
30 MBtu, 700 CFM	SL@3.00	Ea	1,400.00	109.00	6.49	1,515.49
50 MBtu, 1,000 CFM	SL@3.50	Ea	1,640.00	128.00	7.58	1,775.58
60 MBtu, 1,300 CFM	SL@4.00	Ea	1,740.00	146.00	8.66	1,894.66

Description	Craft@Hrs	Unit	Material $	Labor $	Equipment $	Total $

Unit heater, gas, direct-fired, mid-efficiency, propeller fan, atmospheric draft, single-stage, standing pilot, aluminized burner and heat exchanger. Hang in place only. Make additional allowances for flue, gas and electrical connections. Costs shown per thousand BTU per hour (MBH). Mid-efficiency 78% AFUE.

Description	Craft@Hrs	Unit	Material $	Labor $	Equipment $	Total $
30 MBH input	S2@4.00	Ea	1,120.00	144.00	—	1,264.00
50 MBH input	S2@4.00	Ea	1,200.00	144.00	—	1,344.00
75 MBH input	S2@4.00	Ea	1,340.00	144.00	—	1,484.00
100 MBH input	S2@4.00	Ea	1,490.00	144.00	—	1,634.00
125 MBH input	S2@4.00	Ea	1,770.00	144.00	—	1,914.00
145 MBH input	S2@4.00	Ea	1,840.00	144.00	—	1,984.00
175 MBH input	S2@4.00	Ea	2,020.00	144.00	—	2,164.00
200 MBH input	S2@4.00	Ea	2,230.00	144.00	—	2,374.00
250 MBH input	S2@4.00	Ea	2,350.00	144.00	—	2,494.00
300 MBH input	S2@4.00	Ea	3,040.00	144.00	—	3,184.00
350 MBH input	S2@4.00	Ea	3,590.00	144.00	—	3,734.00
400 MBH input	S2@4.00	Ea	4,080.00	144.00	—	4,224.00

Unit heater, gas-fired, high-efficiency, propeller fan, power vented (included), direct-fired, single-stage, intermittent pilot ignition, aluminized burner and heat exchanger. Hang in place and install power vent. Make additional allowances for gas and electrical connections. Costs shown per thousand BTU per hour (MBH). High-efficiency 85% AFUE.

Description	Craft@Hrs	Unit	Material $	Labor $	Equipment $	Total $
30 MBH input	S2@5.25	Ea	1,610.00	189.00	—	1,799.00
50 MBH input	S2@5.25	Ea	1,710.00	189.00	—	1,899.00
75 MBH input	S2@5.25	Ea	2,740.00	189.00	—	2,929.00
100 MBH input	S2@5.25	Ea	2,070.00	189.00	—	2,259.00
125 MBH input	S2@5.25	Ea	2,310.00	189.00	—	2,499.00
145 MBH input	S2@5.25	Ea	2,440.00	189.00	—	2,629.00
175 MBH input	S2@5.25	Ea	2,590.00	189.00	—	2,779.00
200 MBH input	S2@5.25	Ea	2,760.00	189.00	—	2,949.00
250 MBH input	S2@5.25	Ea	3,200.00	189.00	—	3,389.00
300 MBH input	S2@5.25	Ea	3,760.00	189.00	—	3,949.00
350 MBH input	S2@5.25	Ea	4,210.00	189.00	—	4,399.00
400 MBH input	S2@5.25	Ea	4,610.00	189.00	—	4,799.00

Unit heater, gas, indirect-fired, mid-efficiency, propeller fan, power vented (included), single-stage, intermittent pilot ignition, aluminized burner and heat exchanger, Hang in place and install power vent. Make additional allowances for gas and electrical connections. Costs shown per thousand BTU per hour (MBH). Mid-efficiency 78% AFUE.

Description	Craft@Hrs	Unit	Material $	Labor $	Equipment $	Total $
130 MBH input	S2@5.25	Ea	2,940.00	189.00	—	3,129.00
150 MBH input	S2@5.25	Ea	3,200.00	189.00	—	3,389.00
170 MBH input	S2@5.25	Ea	3,300.00	189.00	—	3,489.00
225 MBH input	S2@5.25	Ea	3,700.00	189.00	—	3,889.00
280 MBH input	S2@5.25	Ea	4,400.00	189.00	—	4,589.00
340 MBH input	S2@5.25	Ea	4,980.00	189.00	—	5,169.00

Unit Heaters

Description	Craft@Hrs	Unit	Material $	Labor $	Equipment $	Total $

High-intensity infrared heater, gas-fired. Ceramic grid. Automatic spark ignition. Including reflector. Approved for unvented commercial indoor installations. Not for use in residential applications. Labor includes hang in place only. Make additional allowances for gas and 110 volt electrical connections. Refer to local gas code for proper installation requirements.

Description	Craft@Hrs	Unit	Material $	Labor $	Equipment $	Total $
30,000 BTU	P1@1.20	Ea	563.00	44.00	2.60	609.60
60,000 BTU	P1@1.80	Ea	719.00	66.10	3.90	789.00
100,000 BTU	P1@2.50	Ea	1,700.00	91.80	5.41	1,797.21
130,000 BTU	P1@2.75	Ea	2,110.00	101.00	5.95	2,216.95
160,000 BTU	P1@2.75	Ea	2,400.00	101.00	5.95	2,506.95
Add for propane	—	Ea	135.00	—	—	135.00

High-intensity infrared heater, gas-fired. Ceramic grid. Millivolt standing pilot ignition. Approved for unvented commercial indoor installations. Not for use in residential applications. Labor includes hang in place only. Make additional allowances for gas connections. Millivolt standing pilot does not require electrical source.

Description	Craft@Hrs	Unit	Material $	Labor $	Equipment $	Total $
30,000 BTU	P1@1.20	Ea	1,310.00	44.00	2.60	1,356.60
60,000 BTU	P1@1.80	Ea	1,700.00	66.10	3.90	1,770.00
100,000 BTU	P1@2.50	Ea	2,130.00	91.80	5.41	2,227.21
130,000 BTU	P1@2.75	Ea	2,540.00	101.00	5.95	2,646.95
160,000 BTU	P1@2.75	Ea	2,950.00	101.00	5.95	3,056.95
Add for propane	—	Ea	135.00	—	—	135.00

Infrared tube heater, burner unit only, gas-fired, negative pressure type. Labor includes hang in place only. Not for use in residential applications. Make additional allowances for burner tube, burner tube reflector, venting, gas piping and electrical requirements.

Description	Craft@Hrs	Unit	Material $	Labor $	Equipment $	Total $
30,000 BTU	P1@1.20	Ea	1,620.00	44.00	2.60	1,666.60
60,000 BTU	P1@1.80	Ea	1,530.00	66.10	3.90	1,600.00
100,000 BTU	P1@2.50	Ea	1,760.00	91.80	5.41	1,857.21
150,000 BTU	P1@2.75	Ea	1,870.00	101.00	5.95	1,976.95
175,000 BTU	P1@2.75	Ea	2,170.00	101.00	5.95	2,276.95
200,000 BTU	P1@2.75	Ea	2,510.00	101.00	5.95	2,616.95
Add for propane	—	Ea	135.00	—	—	135.00

Infrared tube heater, 4" diameter, 10' lengths. 16 gauge burner tube and .024" mill finish aluminum reflector. Labor includes hang in place only. Not for use in residential applications. Make additional allowances for burner unit, venting, gas piping and electrical requirements.

Description	Craft@Hrs	Unit	Material $	Labor $	Equipment $	Total $
Burner tubing	P1@.200	LF	6.67	7.34	—	14.01
Tubing U-bend	P1@.850	Ea	238.00	31.20	—	269.20
Reflector shield	P1@.200	LF	6.67	7.34	—	14.01
Vent terminal	P1@.600	Ea	105.00	22.00	—	127.00

Description	Craft@Hrs	Unit	Material $	Labor $	Equipment $	Total $

Heat Pumps

In general, heat pumps are more effective in milder winter climates. In colder winter climates, secondary/supplemental electric heating elements (or equivalent heating source) are recommended.

A geothermal heat pump system's efficient operation is largely dependent upon proper design, layout, sizing and installation of the ground loop piping system for closed loop systems. In the case of well-to-well, the efficient operation of open loop systems is largely dependent upon the ability of the supply and return, or dump wells to meet the water volume demand of the heat pumps.

Rules of thumb when sizing loop & well systems:

Heating capacity, when considered or measured in tons cooling or refrigeration, is generally 3 to 4 times that of required cooling capacity for climates within 100 miles of the 49th parallel, i.e., if 2 tons of cooling is required, allow 6-8 tons of heating capacity (1 ton = 12,000 Btu).

Use 250 square feet per ton, e.g., a 2,000-square-foot home requires a 6-ton heat pump (heating capacity).

(Calculation based on R-20 wall and R-40 attic insulation, minimum dual pane argon-charged windows.)

Closed Vertical Loop:

Vertical Bore Hole Loop — ¾" or 1" diameter, socket fusioned loop = 200'-250' total bore hole depth per ton. For example, a 6-ton unit requires between 1,200'-1,500' (twelve to fifteen 5" bore holes, each 100' deep) with the associated 12-15 loop drops, spaced a minimum of 20' apart, connected to an adequately sized header, immersed in a bentonite slurry and applied with a tremie tube.

Closed Horizontal Loop:

Straight run — ¾" or 1" diameter manifolded loop = 400 horizontal feet per ton. For example, a 6-ton unit requires 2,400 horizontal feet (eight 300' runs, 150' out, 150' back, minimum 2' radius return bend, placed a minimum 10' apart, minimum 4' buried depth, or buried below potential winter frost line).

Slinky — ¾" or 1" diameter manifolded loop = 400 horizontal feet per ton. For example, a 6-ton unit requires 2,400 horizontal feet (eight 300' runs, 50' out, 50' back, coiled, placed a minimum 20' apart, minimum 4' buried depth, or buried below potential winter frost line. (Below 7' at 100 miles, or farther, north of the 49th parallel.)

Open Loop:

Well-to-Well — 2.5 gallons per minute per ton, each well (two required: supply & return). For example, a 6-ton unit requires 12-15 gallons per minute capacity per well.

Leadership in Energy & Environmental Design (LEED)

LEED offers an incentive to building owners to install energy-efficient and environmentally sensitive HVAC equipment. The LEED program awards points for application of the best available technology. The HVAC system can earn a maximum of 17 points toward LEED certification. A LEED-certified air conditioning system must comply with ASHRAE 90.1-2004 (efficiency) and ASHRAE 62.1-2004 (indoor air quality) standards. LEED certification also requires recording controls (Carrier ComfortView 3 or equal) to measure system performance. Zone sensor thermostats (Carrier Debonair or equal) are required to provide performance and efficiency data to the digital recorder. An indoor air quality CO_2 sensor, duct-mounted aspirator modules and a refrigerant such as Puron R-410A are also required.

Heat Pump Systems

Description	Craft@Hrs	Unit	Material $	Labor $	Equipment $	Total $

Geothermal ground or water source heat pump (open or closed loop), w/active cooling. Glycol/water mixture or water-to-air thermal transfer. The refrigeration circuit operates in the heating and cooling mode as required (sometimes referred to as active heating & cooling or mechanical heating & cooling). This unit is suited for either open or closed loop systems. These units provide a higher degree of cooling capacity than the open loop, passive "free" cooling units. See table below. Optional de-superheater system provides domestic hot water temperature preheat or boost. Make additional allowances for the thermal source, i.e., horizontal ground loop, vertical bore hole ground loop, supply and return water wells, pond or lake source loop, etc. Make additional allowances for connections, ancillary components, start up and testing. Based on R-410A refrigerant.

Description	Craft@Hrs	Unit	Material $	Labor $	Equipment $	Total $
Heat pump, open or closed loop, ½ ton	P1@2.25	Ea	1,860.00	82.60	—	1,942.60
Heat pump, open or closed loop, ¾ ton	P1@2.25	Ea	1,910.00	82.60	—	1,992.60
Heat pump, open or closed loop, 1 ton	P1@2.25	Ea	2,000.00	82.60	—	2,082.60
Heat pump, open or closed loop, 1½ ton	P1@2.25	Ea	2,330.00	82.60	—	2,412.60
Heat pump, open or closed loop, 2 ton	P1@2.50	Ea	2,410.00	91.80	—	2,501.80
Heat pump, open or closed loop, 2½ ton	P1@2.50	Ea	2,980.00	91.80	—	3,071.80
Heat pump, open or closed loop, 3 ton	P1@3.00	Ea	3,100.00	110.00	—	3,210.00
Heat pump, open or closed loop, 4 ton	P1@3.50	Ea	3,540.00	128.00	—	3,668.00
Heat pump, open or closed loop, 5 ton	P1@4.00	Ea	6,980.00	147.00	—	7,127.00
Heat pump, open or closed loop, 7½ ton	P1@4.75	Ea	7,480.00	174.00	—	7,654.00
Optional de-superheater coil	—	Ea	841.00	—	—	841.00
Optional supplementary electric coil 5 KW	—	Ea	182.00	—	—	182.00
Optional supplementary electric coil 10 KW	—	Ea	215.00	—	—	215.00
Optional supplementary electric coil 15 KW	—	Ea	466.00	—	—	466.00
Optional supplementary electric coil 20 KW	—	Ea	517.00	—	—	517.00
Optional supplementary electric coil 25 KW	—	Ea	634.00	—	—	634.00
Water-to-water thermal transfer, additional	—	Ea	414.00	—	—	414.00

Description	Craft@Hrs	Unit	Material $	Labor $	Equipment $	Total $

Geothermal water source heat pump (open loop), w/free cooling.

Water-to-air thermal transfer. The refrigeration circuit is only operated in the heating mode while passive "free" cooling is provided by pumping cold well water directly to a unit-mounted coil that bypasses the refrigeration (compressor) unit. This unit is suited to areas where the temperature of ground water supply is below 50°F. Optional de-superheater system provides domestic hot water temperature preheat or boost. Make additional allowances for the water source, i.e., supply and return water wells, pond or lake source, etc. Make additional allowances for connections, ancillary components, start up and testing.

Description	Craft@Hrs	Unit	Material $	Labor $	Equipment $	Total $
Heat pump, open loop, free cooling, 2 ton	P1@2.50	Ea	3,280.00	91.80	—	3,371.80
Heat pump, open loop, free cooling, 3 ton	P1@3.00	Ea	4,030.00	110.00	—	4,140.00
Heat pump, open loop, free cooling, 4 ton	P1@3.50	Ea	4,520.00	128.00	—	4,648.00
Heat pump, open loop, free cooling, 5 ton	P1@4.00	Ea	7,950.00	147.00	—	8,097.00
Heat pump, open loop, free cooling, 7½ ton	P1@4.75	Ea	8,630.00	174.00	—	8,804.00
Optional de-superheater coil	—	Ea	841.00	—	—	841.00
Optional supplementary electric coil 5 KW	—	Ea	182.00	—	—	182.00
Optional supplementary electric coil 10 KW	—	Ea	215.00	—	—	215.00
Optional supplementary electric coil 15 KW	—	Ea	466.00	—	—	466.00
Optional supplementary electric coil 20 KW	—	Ea	518.00	—	—	518.00
Optional supplementary electric coil 25 KW	—	Ea	634.00	—	—	634.00
Water-to-water thermal transfer additional	—	Ea	414.00	—	—	414.00

Description	Craft@Hrs	Unit	Material $	Labor $	Equipment $	Total $

Geothermal water source heat pump (closed refrigerant loop). Water-to-air thermal transfer. Optional de-superheater system provides domestic hot water temperature preheat or boost. Make additional allowances for the heat transfer loop, i.e., refrigerant loop. Make additional allowances for connections, ancillary components, start up and testing.

Heat pump, 2 ton	P1@2.50	Ea	1,970.00	91.80	—	2,061.80
Heat pump, 3 ton	P1@3.00	Ea	2,560.00	110.00	—	2,670.00
Heat pump, 4 ton	P1@3.50	Ea	2,960.00	128.00	—	3,088.00
Heat pump, 5 ton	P1@4.00	Ea	6,240.00	147.00	—	6,387.00
Heat pump, 7½ ton	P1@4.75	Ea	5,920.00	173.00	—	5,803.00
Optional de-superheater coil	—	Ea	807.00	—	—	807.00
Optional supplementary electric coil 5 KW	—	Ea	188.00	—	—	188.00
Optional supplementary electric coil 10 KW	—	Ea	222.00	—	—	222.00
Optional supplementary electric coil 15 KW	—	Ea	484.00	—	—	484.00
Optional supplementary electric coil 20 KW	—	Ea	537.00	—	—	537.00
Optional supplementary electric coil 25 KW	—	Ea	658.00	—	—	658.00

LEED certification for geothermal heat pumps

Add for a LEED-certified heat pump with USGBC rating	—	%	20.0	—	—	—
Add for LEED registration and inspection	—	Ea	2,160.00	—	—	2,160.00
Add for LEED central digital performance monitor/recorder, ComfortView 3 or equal	—	Ea	2,530.00	—	—	2,530.00
Add for LEED-certified zone recording thermostats, Carrier DebonAir, USB remote-ready	—	Ea	377.00	—	—	377.00
Add for IAQ (Indoor Air Quality) CO_2 sensor and duct-mounted aspirator box	—	Ea	456.00	—	—	456.00

Description	Craft@Hrs	Unit	Material $	Labor $	Equipment $	Total $

Heat pump, residential air-to-air split system. Outdoor unit installed on pad or wall bracket. Indoor coil installed in supply air plenum upstream of supply air fan (air handler). Units rated at 13SEER (Seasonal Energy Efficiency Rating). System includes outdoor condenser/compressor unit, indoor evaporative coil, programmable thermostat, indoor/outdoor sensor. Condenser/compressor unit comes pre-charged from the factory with R410A refrigerant. 208/230 Volt, single phase. Rated in nominal tonnage. (12,000 BTU's equals 1 ton cooling).

Description	Craft@Hrs	Unit	Material $	Labor $	Equipment $	Total $
Setting outdoor unit on pad or wall bracket						
2.0-ton outdoor unit	P1@1.50	Ea	2,420.00	55.10	—	2,475.10
3.0-ton outdoor unit	P1@1.75	Ea	2,660.00	64.20	—	2,724.20
3.5-ton outdoor unit	P1@2.00	Ea	3,220.00	73.40	—	3,293.40
4.0-ton outdoor unit	P1@2.25	Ea	3,450.00	82.60	—	3,532.60
5.0-ton outdoor unit	P1@2.50	Ea	4,050.00	91.80	—	4,141.80
Mounting evaporative coil in air stream						
2.0-ton evap. coil	P1@1.15	Ea	26.30	42.20	—	68.50
3.0-ton evap. coil	P1@1.35	Ea	31.60	49.50	—	81.10
3.5-ton evap. coil	P1@1.50	Ea	36.80	55.10	—	91.90
4.0-ton evap. coil	P1@2.00	Ea	42.00	73.40	—	115.40
5.0-ton evap. coil	P1@2.25	Ea	47.30	82.60	—	129.90
Install and connect (braze) refrigerant line set – 50' (Liquid x Suction). Connect refrigerant line set. Condensor/compressor unit to evaporative coil.						
2.0-ton 3/8" - 5/8"	P1@2.00	Ea	231.00	73.40	—	304.40
3.0-ton 1/2" - 7/8"	P1@2.25	Ea	268.00	82.60	—	350.60
3.5-ton 5/8" - 1"	P1@3.75	Ea	405.00	138.00	—	543.00
4.0-ton 7/8" -1-1/8"	P1@4.50	Ea	484.00	165.00	—	649.00
5.0-ton 1" - 1-3/8"	P1@6.00	Ea	652.00	220.00	—	872.00
Control wiring						
24 volt	BE@3.00	Ea	30.00	121.00	—	151.00
Power wiring compressor						
208/230 Volt	BE@2.00	Ea	180.00	80.80	—	260.80
Evaporative coil drain	P1@1.00	Ea	20.30	36.70	—	57.00
Equipment pad, concrete slab,						
4' x 4' x 2"	CF@4.00	Ea	31.00	141.00	—	172.00
Equipment pad, patio stones,						
2' x 2' x 1–1/2" (4 ea)	CF@2.00	Ea	41.30	70.70	—	112.00

Heat Pump Systems

Description	Craft@Hrs	Unit	Material $	Labor $	Equipment $	Total $

Heat pump, split system, single-speed blower, single-phase scroll compressor – 208 volt, 13 SEER (Seasonal Energy Efficiency Rating)

Air-to-air thermal transfer. Material costs include outdoor stat / low ambient, short cycle timer, high-pressure cut out and digital heat/cool thermostat with manual change-over. Labor includes setting outdoor condenser, indoor blower and coil sections in place only. Make additional allowances for refrigerant piping / connections, electrical / temperature control wiring / connections and evaporator coil condensate drain.

Description	Craft@Hrs	Unit	Material $	Labor $	Equipment $	Total $
1.5 tons cooling	S2@4.00	Ea	3,150.00	144.00	—	3,294.00
2 tons cooling	S2@4.25	Ea	3,240.00	153.00	—	3,393.00
2.5 tons cooling	S2@4.50	Ea	3,760.00	162.00	—	3,922.00
3 tons cooling	S2@5.00	Ea	4,210.00	180.00	—	4,390.00
3.5 tons cooling	S2@5.75	Ea	4,620.00	207.00	—	4,827.00
4 tons cooling	S2@6.25	Ea	5,110.00	225.00	—	5,335.00
5 tons cooling	S2@7.50	Ea	5,920.00	270.00	—	6,190.00

Heat pump, split system, single-speed blower, three-phase scroll compressor – 230 volt, 13 SEER (Seasonal Energy Efficiency Rating)

Air-to-air thermal transfer. Material costs include outdoor stat / low ambient, short cycle timer, high-pressure cut out and digital heat/cool thermostat with manual change over. Labor includes setting outdoor condenser, indoor blower and coil sections in place only. Make additional allowances for refrigerant piping / connections, electrical / temperature control wiring / connections and evaporator coil condensate drain.

Description	Craft@Hrs	Unit	Material $	Labor $	Equipment $	Total $
3.5 tons cooling	S2@5.75	Ea	5,150.00	207.00	—	5,357.00
4 tons cooling	S2@6.25	Ea	5,610.00	225.00	—	5,835.00
5 tons cooling	S2@7.50	Ea	6,430.00	270.00	—	6,700.00

Heat pump, split system, variable-speed blower, single-phase scroll compressor – 208 volt, 13 SEER (Seasonal Energy Efficiency Rating)

Air-to-air thermal transfer. Material costs include outdoor stat / low ambient, short cycle timer, high-pressure cut out and digital heat/cool thermostat with manual change over. Labor includes setting outdoor condenser, indoor blower and coil sections in place only. Make additional allowances for refrigerant piping / connections, electrical / temperature control wiring / connections and evaporator coil condensate drain.

Description	Craft@Hrs	Unit	Material $	Labor $	Equipment $	Total $
1.5 tons cooling	S2@4.00	Ea	3,770.00	144.00	—	3,914.00
2 tons cooling	S2@4.25	Ea	3,830.00	153.00	—	3,983.00
2.5 tons cooling	S2@4.50	Ea	4,410.00	162.00	—	4,572.00
3 tons cooling	S2@5.00	Ea	4,870.00	180.00	—	5,050.00
3.5 tons cooling	S2@5.75	Ea	5,240.00	207.00	—	5,447.00
4 tons cooling	S2@6.25	Ea	5,920.00	225.00	—	6,145.00
5 tons cooling	S2@7.50	Ea	6,710.00	270.00	—	6,980.00

Description	Craft@Hrs	Unit	Material $	Labor $	Equipment $	Total $

Heat pump, split system, variable-speed blower, three-phase scroll compressor – 230 volt, 13 SEER (Seasonal Energy Efficiency Rating)

Air-to-air thermal transfer. Material costs include outdoor stat / low ambient, short cycle timer, high-pressure cut out and digital heat/cool thermostat with manual change over. Labor includes setting outdoor condenser, indoor blower and coil sections in place only. Make additional allowances for refrigerant piping / connections, electrical / temperature control wiring / connections and evaporator coil condensate drain.

Description	Craft@Hrs	Unit	Material $	Labor $	Equipment $	Total $
3.5 tons cooling	S2@5.75	Ea	5,810.00	207.00	—	6,017.00
4 tons cooling	S2@6.25	Ea	6,410.00	225.00	—	6,635.00
5 tons cooling	S2@7.50	Ea	7,280.00	270.00	—	7,550.00

Heat pump, split system, single-speed blower, single-phase scroll compressor – 208 volt, 14 SEER (Seasonal Energy Efficiency Rating)

Air-to-air thermal transfer. Material costs include outdoor stat / low ambient, short cycle timer, high-pressure cut out and digital heat/cool thermostat with manual change over. Labor includes setting outdoor condenser, indoor blower and coil sections in place only. Make additional allowances for refrigerant piping / connections, electrical / temperature control wiring / connections and evaporator coil condensate drain.

Description	Craft@Hrs	Unit	Material $	Labor $	Equipment $	Total $
2 tons cooling	S2@4.25	Ea	4,360.00	153.00	—	4,513.00
2.5 tons cooling	S2@4.50	Ea	4,930.00	162.00	—	5,092.00
3 tons cooling	S2@5.00	Ea	5,240.00	180.00	—	5,420.00
3.5 tons cooling	S2@5.75	Ea	5,990.00	207.00	—	6,197.00
4 tons cooling	S2@6.25	Ea	6,670.00	225.00	—	6,895.00
5 tons cooling	S2@7.50	Ea	7,300.00	270.00	—	7,570.00

Heat pump, split system, variable-speed blower, single-phase scroll compressor – 208 volt, 14 SEER (Seasonal Energy Efficiency Rating)

Air-to-air thermal transfer. Material costs include outdoor stat / low ambient, short cycle timer, high-pressure cut out and digital heat/cool thermostat with manual change over. Labor includes setting outdoor condenser, indoor blower and coil sections in place only. Make additional allowances for refrigerant piping / connections, electrical / temperature control wiring / connections and evaporator coil condensate drain.

Description	Craft@Hrs	Unit	Material $	Labor $	Equipment $	Total $
2 tons cooling	S2@4.25	Ea	4,960.00	153.00	—	5,113.00
2.5 tons cooling	S2@4.50	Ea	5,420.00	162.00	—	5,582.00
3 tons cooling	S2@5.00	Ea	5,820.00	180.00	—	6,000.00
3.5 tons cooling	S2@5.75	Ea	6,550.00	207.00	—	6,757.00
4 tons cooling	S2@6.25	Ea	7,100.00	225.00	—	7,325.00
5 tons cooling	S2@7.50	Ea	10,500.00	270.00	—	10,770.00

Heat Pump System Accessories

Description	Craft@Hrs	Unit	Material $	Labor $	Equipment $	Total $

Heat pump system accessories

Description	Craft@Hrs	Unit	Material $	Labor $	Equipment $	Total $
Fossil fuel kit	S2@2.50	Ea	214.00	90.10	—	304.10
High pressure cut out kit	S2@.500	Ea	74.90	18.00	—	92.90
Outdoor stat/low ambient sensors	S2@.500	Ea	70.40	18.00	—	88.40
Short cycle protection	S2@.500	Ea	55.60	18.00	—	73.60
Digital heat/cool thermostat w/manual change-over	S2@.750	Ea	158.00	27.00	—	185.00
Digital, programmable heat/cool thermostat w/auto change-over	S2@.750	Ea	496.00	27.00	—	523.00

Heat pump, supplemental electric heating coil

Set in place only. Make additional allowances for electrical connections.

Description	Craft@Hrs	Unit	Material $	Labor $	Equipment $	Total $
5 KW, 240 Volt	S2@1.50	Ea	179.00	54.10	—	233.10
7.5 KW, 240 Volt	S2@1.75	Ea	214.00	63.10	—	277.10
10 KW, 240 Volt	S2@2.00	Ea	250.00	72.10	—	322.10
12.5 KW, 240 V	S2@2.25	Ea	416.00	81.10	—	497.10
15 KW, 240 Volt	S2@2.75	Ea	526.00	99.10	—	625.10
20 KW, 240 Volt	S2@3.50	Ea	589.00	126.00	—	715.00
25 KW, 240 Volt	S2@4.25	Ea	708.00	153.00	—	861.00

LEED certification for geothermal heat pumps

Description	Craft@Hrs	Unit	Material $	Labor $	Equipment $	Total $
Add for a LEED-certified heat pump with USGBC rating	—	%	20.0	—	—	—
Add for LEED registration and inspection	—	Ea	2,160.00	—	—	2,160.00
Add for LEED central digital performance monitor/recorder, ComfortView 3 or equal	—	Ea	2,530.00	—	—	2,530.00
Add for LEED-certified zone recording thermostats, Carrier DebonAir, USB remote-ready	—	Ea	377.00	—	—	377.00
Add for IAQ (Indoor Air Quality) CO_2 sensor and duct-mounted aspirator box	—	Ea	456.00	—	—	456.00

Description	Craft@Hrs	Unit	Material $	Labor $	Equipment $	Total $

Water pump, submersible. Domestic water, geothermal open loop, well, lake or pond applications. Stainless steel jacket with multiple-stage impellers and integral foot valve.

Description	Craft@Hrs	Unit	Material $	Labor $	Equipment $	Total $
Submersible pump, 120/240 volt, ½ HP	P1@2.00	Ea	572.00	73.40	—	645.40
Submersible pump, 240 volt, ¾ HP	P1@2.00	Ea	712.00	73.40	—	785.40
Submersible pump, 240 volt, 1 HP	P1@2.00	Ea	956.00	73.40	—	1,029.40
Submersible pump, 240 volt, 1½ HP	P1@2.00	Ea	1,250.00	73.40	—	1,323.40
Submersible pump, 240 volt, 2 HP	P1@2.00	Ea	1,460.00	73.40	—	1,533.40
Submersible pump, 240 volt, 3 HP	P1@2.00	Ea	1,640.00	73.40	—	1,713.40
Submersible pump, 240 volt, 5 HP	P1@2.00	Ea	1,910.00	73.40	—	1,983.40
Pump controller – constant pressure, VSD	P1@1.50	Ea	967.00	55.10	—	1,022.10
Pump control box, 120 volt	P1@.600	Ea	57.20	22.00	—	79.20
Pump control box, 240 volt	P1@.600	Ea	70.10	22.00	—	92.10
Pump stand, PVC	P1@1.00	Ea	154.00	36.70	—	190.70
Pump stand, galvanized steel	P1@1.35	Ea	208.00	49.50	—	257.50

Water pump, jet. Domestic water well, lake and pond applications. 110 volt, cast iron body, single-stage impeller, including pressure switch & pressure gauge. Use for shallow depths to 25' maximum lift. Deep well applications to 120' maximum lift.

Description	Craft@Hrs	Unit	Material $	Labor $	Equipment $	Total $
Jet pump, shallow depth ½ HP	P1@2.00	Ea	242.00	73.40	—	315.40
Jet pump, shallow depth ¾ HP	P1@2.00	Ea	281.00	73.40	—	354.40
Jet pump, deep well ½ HP	P1@2.00	Ea	329.00	73.40	—	402.40
Jet pump, deep well ¾ HP	P1@2.00	Ea	382.00	73.40	—	455.40
Optional pump-mounted pressure tank 5 gallon	P1@.500	Ea	102.00	18.40	—	120.40

Water Pump System Accessories

Description	Craft@Hrs	Unit	Material $	Labor $	Equipment $	Total $

Water pump system accessories. Pressure tanks are factory pre-charged. VSD = variable speed drive.

Description	Craft@Hrs	Unit	Material $	Labor $	Equipment $	Total $
Pressure tank, bladder type, 5 gallon	P1@1.00	Ea	89.20	36.70	—	125.90
Pressure tank, bladder type, 10 gallon	P1@2.00	Ea	102.00	73.40	—	175.40
Pressure tank, bladder type, 20 gallon	P1@2.00	Ea	121.00	73.40	—	194.40
Pressure tank, bladder type, 30 gallon	P1@2.00	Ea	173.00	73.40	—	246.40
Pressure tank, bladder type, 40 gallon	P1@2.00	Ea	229.00	73.40	—	302.40
Brass tank tee, ¾" or 1" w/ union	P1@.300	Ea	53.60	11.00	—	64.60
Brass tank tee, ¾" or 1" w/o union	P1@.300	Ea	43.20	11.00	—	54.20
Constant pressure VSD pump controller	P1@1.50	Ea	967.00	55.10	—	1,022.10
Pressure switch, 110/220 volt	P1@.350	Ea	40.80	12.80	—	53.60
Pressure gauge, 0-100 psi, ¼"	P1@.200	Ea	10.20	7.34	—	17.54
Sediment faucet (hose bibb), ½"	P1@1.50	Ea	8.29	55.10	—	63.39
Double injector deep well venturi, 1", brass	P1@.600	Ea	217.00	22.00	—	239.00
Double injector deep well venturi, 1¼", brass	P1@.650	Ea	249.00	23.90	—	272.90
Double injector deep well venturi, 1½", brass	P1@.700	Ea	281.00	25.70	—	306.70
Double injector deep well venturi, 1", plastic	P1@.550	Ea	70.10	20.20	—	90.30
Double injector deep well venturi, 1¼", plastic	P1@.600	Ea	86.50	22.00	—	108.50
Double injector deep well venturi, 1½", plastic	P1@.650	Ea	94.20	23.90	—	118.10
Foot valve, 1", brass	P1@.400	Ea	73.80	14.70	—	88.50
Foot valve, 1¼", brass	P1@.400	Ea	76.30	14.70	—	91.00
Foot valve, 1½", brass	P1@.400	Ea	81.40	14.70	—	96.10
Foot valve, 1", plastic	P1@.400	Ea	22.90	14.70	—	37.60
Foot valve, 1¼", plastic	P1@.400	Ea	28.10	14.70	—	42.80
Foot valve, 1½", plastic	P1@.400	Ea	31.20	14.70	—	45.90
Barbed adapter, FIP/MIP, 1" brass	P1@.230	Ea	12.10	8.44	—	20.54
Barbed adapter, FIP/MIP, 1¼" brass	P1@.250	Ea	14.00	9.18	—	23.18

Description	Craft@Hrs	Unit	Material $	Labor $	Equipment $	Total $
Barbed adapter, FIP/MIP, 1½" brass	P1@.265	Ea	17.00	9.73	—	26.73
Barbed adapter, FIP/MIP, 1" plastic	P1@.215	Ea	11.10	7.89	—	18.99
Barbed adapter, FIP/MIP, 1¼" plastic	P1@.220	Ea	12.20	8.07	—	20.27
Barbed adapter, FIP/MIP, 1½" plastic	P1@.235	Ea	16.40	8.62	—	25.02
Pitless adapter, 1" x 5" brass	P1@1.25	Ea	179.00	45.90	—	224.90
Pitless adapter, 1¼" x 5" brass	P1@1.25	Ea	193.00	45.90	—	238.90
Pitless adapter, 1½" x 5" brass	P1@1.25	Ea	217.00	45.90	—	262.90
Pitless adapter, 2" x 6" brass	P1@1.80	Ea	329.00	66.10	—	395.10
Poly pipe, 75 psi, 1"	P1@.020	LF	.28	.73	—	1.01
Poly pipe, 75 psi, 1¼"	P1@.021	LF	.32	.77	—	1.09
Poly pipe, 75 psi, 1½"	P1@.022	LF	.34	.81	—	1.15
Poly pipe, 100 psi, 1"	P1@.021	LF	.28	.77	—	1.05
Poly pipe, 100 psi, 1¼"	P1@.022	LF	.32	.81	—	1.13
Poly pipe, 100 psi, 1½"	P1@.023	LF	.34	.84	—	1.18
Electrical wire (110 volt), #12/4	P1@.015	LF	.34	.55	—	.89
Electrical wire (220 volt), #10/4	P1@.015	LF	.36	.55	—	.91
Heat shrink kit	—	Ea	4.51	—	—	4.51

Geothermal/Domestic Water Wells

Geothermal water wells

A geothermal water well is constructed and operates exactly like a water well used for drinking water. In many residential applications, the supply well of the geothermal system also functions as the domestic water supply well.

The cost of a drilled well depends on many variables, such as costs to get the drilling rig and crew to the site, the type of geologic formation the well drilling rig must drill through, the depth the driller must achieve to find water in that particular location and the distance from an existing available water source to provide the water required to drill (to refill the water truck if necessary).

Well drillers charge by the foot and usually have a 100' minimum charge-out fee. They can only speculate to what depth they must drill to find water. Water being found at a particular depth nearby doesn't always translate to finding water at another given location at the same depth. It's important to remember that drilling a relatively small hole in the ground and hoping to hit what is described as a vein of water as part of an underground water table is always a gamble. As such, it is, without proper research, difficult to determine a realistic budget for well drilling. It's always prudent to call the water authority or a well driller in the area for the most accurate information on the water and geological characteristics of a given location.

Use the following to begin to formulate a budget exclusive of costs associated with factors listed above. Because of the inherent uncertainties of the process, these prices are given only as typical subcontract costs.

Description	Craft@Hrs	Unit	Material $	Labor $	Equipment $	Total $

Drilled well, 5" diameter, PVC casing, set into bedrock. Based on a 2-man crew, 1 rotary drill rig and a 1,500 gallon water truck. Minimum 100' charge in most cases. Typical subcontract costs.

Description	Craft@Hrs	Unit	Material $	Labor $	Equipment $	Total $
Drilled through till, clay, sand	—	LF	—	—	—	35.70
Drilled through till and boulder	—	LF	—	—	—	46.10
Drilled through limestone soft bedrock	—	LF	—	—	—	51.90
Drilled through granite hard bedrock	—	LF	—	—	—	104.00
Add for galvanized steel casing	—	LF	—	—	—	13.60

Drilled well, 6" diameter, PVC casing, set into bedrock. Based on a 2-man crew, 1 rotary drill rig and a 1,500 gallon water truck. Minimum 100' charge in most cases. Typical subcontract costs.

Description	Craft@Hrs	Unit	Material $	Labor $	Equipment $	Total $
Drilled through till, clay, sand	—	LF	—	—	—	39.70
Drilled through till and boulder	—	LF	—	—	—	50.50
Drilled through limestone soft bedrock	—	LF	—	—	—	54.70
Drilled through granite hard bedrock	—	LF	—	—	—	111.00
Add for galvanized steel casing	—	LF	—	—	—	16.60

Geothermal Water Wells/Geothermal Closed Loops

Description	Craft@Hrs	Unit	Material $	Labor $	Equipment $	Total $

Drilled well, 8" diameter, PVC casing, set into bedrock. Based on a 2-man crew, 1-rotary drill rig and a 1,500 gallon water truck. Minimum 100' charge in most cases. Typical subcontract costs.

Description	Craft@Hrs	Unit	Material $	Labor $	Equipment $	Total $
Drilled through till, clay, sand	—	LF	—	—	—	50.50
Drilled through till and boulder	—	LF	—	—	—	57.60
Drilled through limestone soft bedrock	—	LF	—	—	—	64.90
Drilled through granite hard bedrock	—	LF	—	—	—	131.00
Add for galvanized steel casing	—	LF	—	—	—	33.10

Drilled well, 10" diameter, PVC casing, set into bedrock. Based on a 2-man crew, 1 rotary drill rig and a 1,500 gallon water truck. Minimum 100' charge in most cases. Typical subcontract costs.

Description	Craft@Hrs	Unit	Material $	Labor $	Equipment $	Total $
Drilled through till, clay, sand	—	LF	—	—	—	60.50
Drilled through till and boulder	—	LF	—	—	—	67.70
Drilled through limestone soft bedrock	—	LF	—	—	—	74.80
Drilled through granite hard bedrock	—	LF	—	—	—	174.00
Add for galvanized steel casing	—	LF	—	—	—	42.20

Geothermal bore hole (vertical loop), 3" diameter, drilled to a maximum of 30'. Includes installation of a polypropylene fluid loop using a 180-degree return bend with socket fusion joints. Bore hole filled with bentonite slurry through a tremie tube and mud pump. Includes socket fusioned header and supply & return mains to heat pump. Based on a minimum of 1,000 vertical feet (2,000' of pipe) Based on a 2-man crew, 1 skid steer-mounted rotary drill rig and a 50 gallon water tank. Typical subcontract costs.

Description	Craft@Hrs	Unit	Material $	Labor $	Equipment $	Total $
¾" vertical loop	—	LF	—	—	—	12.20
1" vertical loop	—	LF	—	—	—	13.40
Add for thermally-enhanced Bentonite	—	LF	—	—	—	3.54

Geothermal horizontal loop, directionally bored. Includes installation of a polypropylene fluid loop. Based on a minimum of 2,000 horizontal feet (2,000' of pipe). Based on a 3-man crew, 1 directional boring/horizontal drill machine. Typical subcontract costs.

Description	Craft@Hrs	Unit	Material $	Labor $	Equipment $	Total $
¾" straight horizontal loop	—	LF	—	—	—	7.41
1" straight horizontal loop	—	LF	—	—	—	8.08

Description	Craft@Hrs	Unit	Material $	Labor $	Equipment $	Total $

Geothermal horizontal loop, slinky type. Includes installation of a polypropylene fluid loop. Based on a minimum of 2,000 horizontal feet (2,000' of pipe). Based on a 3-man crew, 1 excavator backhoe. Typical subcontract costs.

Description	Craft@Hrs	Unit	Material $	Labor $	Equipment $	Total $
¾" slinky loop	—	LF	—	—	—	5.61
1" slinky loop	—	LF	—	—	—	6.34
¾" slinky pond loop	—	LF	—	—	—	4.39
1" slinky pond loop	—	LF	—	—	—	5.11

Description	Craft@Hrs	Unit	Material $	Labor $	Equipment $	Total $

Biomass-fired hot water boilers for central heating systems. For residential, commercial and light industrial applications. 15 PSI, EPA, ASME and UL rated, with wood-pellet and/or corn fuel gravity-feed hopper, ash hood cleanout, secondary natural gas or #2 oil burner, platinum stack transition emissions catalyst, forced draft fan, factory-installed boiler trim, feedwater pump and valves, and tempering valve. Add the cost of the following if required: a concrete pad, optional stainless steel 2 GPM potable hot water coil insert, expansion tank and circulating pump, stack and stack liner, CO_2 fire extinguisher system and electrical connection. Add the cost of post-installation inspection by both the fire marshal and the mechanical inspector to validate federal, state and local energy tax credits or energy tax credit offsets. See pollution control stack costs below. Equipment cost is for an appliance dolly, 3,000 lb. come-along, and ½ ton chain hoist. Listed by boiler horsepower (BHP) rating and input capacity in thousands of BTU per hour (MBH).

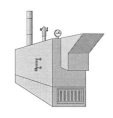

Description	Craft@Hrs	Unit	Material $	Labor $	Equipment $	Total $
3 BHP, 102 MBH	SN@24.0	Ea	11,400.00	928.00	82.40	12,410.40
3 BHP, 104 MBH	SN@24.0	Ea	10,400.00	928.00	82.40	11,410.40
6 BHP, 200 MBH	SN@28.0	Ea	13,400.00	1,080.00	96.10	14,576.10
14 BHP, 500 MBH	SN@30.0	Ea	15,900.00	1,160.00	103.00	17,163.00
27 BHP, 900 MBH	SN@44.0	Ea	23,100.00	1,700.00	151.00	24,951.00

LEED certification requires 88% boiler efficiency, recording controls, compliance with SCAQMD 1146.2 emission standards and zone thermostats.

Description	Craft@Hrs	Unit	Material $	Labor $	Equipment $	Total $
Add for LEED-certified boiler with USGBC rating	—	%	15.0	—	—	—
Add for LEED registration and inspection	—	Ea	—	—	—	2,130.00
Add for central performance monitor/ recorder	—	Ea	2,640.00	—	—	2,640.00
Add for zone recording thermostat	—	Ea	392.00	—	—	392.00

Retrofit pollution control stack for biomass hot water boilers. Stack-mounted platinum catalytic screen for biomass-fired hot water boilers for central radiant heating systems. E.P.A. and California Code standards compliant. UL rated. Includes stack mounting transition and extensions, baghouse structural steel frame, stack transition cleanout and inspection stack access bolted ports and an integral CO_2 fire extinguisher system. Add the cost of a concrete pad, electrical connection, post-installation calibration, inspection by both the fire marshal and the mechanical inspector to validate federal, state and local emission reduction tax credits and associated energy tax credit offsets. Equipment cost is for a 2-ton capacity hydraulic truck crane. Listed by boiler horsepower (BHP) rating and input capacity in thousands of BTU per hour (MBH). Use these figures to estimate the cost of adding a pollution control stack to an existing biomass boiler to meet environmental regulations.

Description	Craft@Hrs	Unit	Material $	Labor $	Equipment $	Total $
3 BHP, 102 MBH	SN@20.0	Ea	1,410.00	773.00	68.60	2,251.60
3 BHP, 104 MBH	SN@20.0	Ea	1,410.00	773.00	68.60	2,251.60
6 BHP, 200 MBH	SN@20.0	Ea	2,100.00	773.00	68.60	2,941.60
14 BHP, 500 MBH	SN@40.0	Ea	2,580.00	1,550.00	137.00	4,267.00
27 BHP, 900 MBH	SN@40.0	Ea	3,500.00	1,550.00	137.00	5,187.00

Biomass-Fired Boilers

Description	Craft@Hrs	Unit	Material $	Labor $	Equipment $	Total $

Biomass-fired water or steam boilers for multi-dwelling, process and industrial applications. 15 PSI, EPA, ASME and UL rated. Includes wood pellet and/or corn fuel auger feed hopper, ash auger cleanout, secondary natural gas or #2 oil burner, platinum stack transition emissions catalyst, forced draft fan, factory-installed boiler trim, feedwater pump and valves, and tempering valve. Add the following costs if required: a concrete pad, optional stainless steel 10 GPM potable hot water coil insert, circulating pump, expansion tank, shell and tube heat exchanger, stack and stack liner, CO_2 fire extinguisher system, pollution control baghouse and filter, electrical connection. Add the cost of post-installation inspection by both the fire marshal and the mechanical inspector to validate federal, state and local energy tax credits or energy tax credit offsets. See pollution control stack costs below. Equipment cost is for a 5-ton capacity forklift. Listed by boiler horsepower (BHP) rating and input capacity in thousands of BTU per hour (MBH).

Description	Craft@Hrs	Unit	Material $	Labor $	Equipment $	Total $
30 BHP, 1,000 MBH	SN@168	Ea	82,800.00	6,490.00	1,220.00	90,510.00
50 BHP, 1,700 MBH	SN@200	Ea	152,000.00	7,730.00	1,450.00	161,180.00
100 BHP, 3,400 MBH	SN@400	Ea	236,000.00	15,500.00	2,900.00	254,400.00
125 BHP, 4,100 MBH	SN@400	Ea	298,000.00	15,500.00	2,900.00	316,400.00
150 BHP, 5,000 MBH	SN@500	Ea	290,000.00	19,300.00	3,620.00	312,920.00
200 BHP, 6,700 MBH	SN@600	Ea	398,000.00	23,200.00	4,350.00	425,550.00

LEED certification requires 88% boiler efficiency, recording controls, compliance with SCAQMD 1146.2 emission standards and zone thermostats.

Description	Craft@Hrs	Unit	Material $	Labor $	Equipment $	Total $
Add for LEED-certified boiler with USGBC rating	—	%	6.0	—	—	—
Add for LEED registration and inspection	—	Ea	—	—	—	2,130.00
Add for central performance monitor/ recorder	—	Ea	2,640.00	—	—	2,640.00
Add for zone recording thermostat	—	Ea	392.00	—	—	392.00

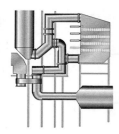

Biomass boiler stack tie-in to baghouse filter. Ionically bonded silicon-lined carbon steel. Costs assume a 20-foot vertical stack height, connection via T-shaped connector with cleanout to the baghouse inlet plenum. Includes rigging, breeching, connecting stack sections from back "turnaround" plenum of boiler to baghouse inlet plenum, vendor-supplied structural steel stack support brackets and aluminized external coating. Equipment cost assumes rental of a 2-ton hydraulic truck crane. Biomass boilers, stoves and heaters smaller than 30 BHP can use a conventional boiler stack (without a baghouse) and require only a platinum catalytic screen. Costs shown assume a 20-foot stack height and 10' horizontal run from T-connector. Cost per 10' length, including a bolt and gasket set on each end. Add 15% for each 10' section (or portion thereof) beyond 20' vertically or horizontally.

Description	Craft@Hrs	Unit	Material $	Labor $	Equipment $	Total $
10" diameter	SN@12.0	Ea	2,930.00	464.00	227.00	3,621.00
12" diameter	SN@12.0	Ea	3,270.00	464.00	227.00	3,961.00
16" diameter	SN@16.0	Ea	3,500.00	618.00	302.00	4,420.00
20" diameter	SN@20.0	Ea	4,430.00	773.00	378.00	5,581.00
24" diameter	SN@24.0	Ea	4,900.00	928.00	453.00	6,281.00
30" diameter	SN@32.0	Ea	7,250.00	1,240.00	604.00	9,094.00
56" diameter	SN@40.0	Ea	12,300.00	1,550.00	755.00	14,605.00

Biomass-Fired Boilers

Description	Craft@Hrs	Unit	Material $	Labor $	Equipment $	Total $

Retrofit pollution control stack for biomass water or steam boiler.

Stack-mounted platinum screen catalytic reactor with air filtration fabric baghouse. For use with biomass-fired hot water or low-pressure steam boilers for central radiant heating or process systems. Meets EPA, NFPA, California Code and UL standards for multi-dwelling, process and industrial applications. Cost includes equipment stack transitions and fittings, transition access bolted inspection and cleanout ports, structural steel frame equipment support and ladders, bottom ash cleanout for baghouse, alarmed stack gas composition monitor and recorder, urea tank and urea pump & regulating valve and an integral CO_2 fire extinguisher system. Costs do not include the following: concrete pad, electrical tie-in, calibration and commissioning, post-installation inspection by the fire marshal and the mechanical inspector to validate federal, state and local energy tax credits or energy tax credit offsets. Equipment cost is for a 2-ton capacity hydraulic truck crane. Listed by boiler horsepower (BHP) rating and input capacity in thousands of BTU per hour (MBH). Use these figures to estimate the cost of adding a pollution control stack to an existing biomass boiler to meet environmental regulations.

Description	Craft@Hrs	Unit	Material $	Labor $	Equipment $	Total $
30 BHP, 1,000 MBH	SN@40.0	Ea	40,700.00	1,550.00	755.00	43,005.00
50 BHP, 1,700 MBH	SN@80.0	Ea	72,200.00	3,090.00	1,510.00	76,800.00
100 BHP, 3,400 MBH	SN@80.0	Ea	127,000.00	3,090.00	1,510.00	131,600.00
125 BHP, 4,100 MBH	SN@120	Ea	135,000.00	4,640.00	2,270.00	141,910.00
150 BHP, 5,000 MBH	SN@120	Ea	144,000.00	4,640.00	2,270.00	150,910.00
200 BHP, 6,700 MBH	SN@160	Ea	153,000.00	6,180.00	3,020.00	162,200.00

Biomass-fired central air space heaters for residential, multi-dwelling, dwelling, and light commercial applications.

15 PSI, EPA, ASHRAE and UL rated. Includes a wood pellet and/or corn fuel gravity-feed hopper, ash hood cleanout, secondary natural gas or #2 oil burner, platinum stack transition emissions catalyst, forced draft fan, ducted hot-air circulation plenum and integral three- speed circulation duct blower. Add the cost of a concrete pad, optional stainless steel 2 GPM potable hot water coil insert, expansion tank and circulating pump, stack and stack liner, CO_2 fire extinguisher system and electrical connection. These costs do not include post-installation inspection by both the fire marshal and the mechanical inspector to validate federal, state and local energy tax credits or energy tax credit offsets. Equipment cost is for a 2-ton capacity forklift. Listed by input capacity in thousands of BTU per hour (MBH).

Description	Craft@Hrs	Unit	Material $	Labor $	Equipment $	Total $
90 MBH	SN@16.0	Ea	3,370.00	618.00	96.10	4,084.10
120 MBH	SN@16.0	Ea	4,360.00	618.00	96.10	5,074.10
170 MBH	SN@20.0	Ea	7,650.00	773.00	120.00	8,543.00
500 MBH	SN@30.0	Ea	11,900.00	1,160.00	180.00	13,240.00
950 MBH	SN@44.0	Ea	14,200.00	1,700.00	264.00	16,164.00

Description	Craft@Hrs	Unit	Material $	Labor $	Equipment $	Total $

Water mist fire extinguishing systems for biomass boilers. 20 minute full saturation cycle, with automatic reset to enable instantaneous reactivation for re-ignition. NFPA rated. Includes high pressure spray heads, piping, pumps, valves, alarm horns, emergency halogen lighting, calibration, training, insurance inspection and safety procedure documentation. Based on remote sensor relays for the fuel hopper, fuel conveyor, combustion chamber and baghouse filter.

Description	Craft@Hrs	Unit	Material $	Labor $	Equipment $	Total $
30 BHP boiler	ST@60.0	Ea	14,600.00	2,570.00	—	17,170.00
50 BHP boiler	ST@60.0	Ea	18,400.00	2,570.00	—	20,970.00
100 BHP boiler	ST@80.0	Ea	26,800.00	3,420.00	—	30,220.00
125 BHP boiler	ST@80.0	Ea	29,100.00	3,420.00	—	32,520.00
150 BHP boiler	ST@120	Ea	33,400.00	5,130.00	—	38,530.00
200 BHP boiler	ST@120	Ea	38,900.00	5,130.00	—	44,030.00
Add for CO_2 based extinguishing system	—	%	—	—	—	20.0
Add for Halon-based extinguishing system	—	%	—	—	—	30.0

Pulse-type boilers for central radiant heating. 15 PSI, EPA, ASME and UL rated. Modular design for multiple installations. Natural gas only. With forced draft fan, factory-installed boiler trim, feedwater pump and valves, and tempering valve. For residential, commercial and light industrial radiant heating and potable water applications. Add the cost of a concrete pad, circulating pump, stack and stack liner, CO_2 fire extinguisher system, electrical connection, and post-installation inspection by both the fire marshal and the mechanical inspector to validate federal, state and local energy tax credits or energy tax credit offsets. Equipment cost is for an appliance dolly, come-along and chain hoist. Listed by boiler horsepower (BHP) rating and input capacity in thousands of BTU per hour (MBH).

Description	Craft@Hrs	Unit	Material $	Labor $	Equipment $	Total $
3 BHP, 90 MBH	P1@24.0	Ea	4,680.00	881.00	288.00	5,849.00
3 BHP, 150 MBH	P1@24.0	Ea	6,770.00	881.00	288.00	7,939.00
6 BHP, 299 MBH	P1@28.0	Ea	9,310.00	1,030.00	336.00	10,676.00

LEED certification requires 88% boiler efficiency, recording controls, compliance with SCAQMD 1146.2 emission standards and zone thermostats.

Description	Craft@Hrs	Unit	Material $	Labor $	Equipment $	Total $
Add for LEED-certified boiler with USGBC rating	—	%	15.0	—	—	—
Add for LEED registration and inspection	—	Ea	—	—	—	2,130.00
Add for central performance monitor/ recorder	—	Ea	1,230.00	—	—	1,230.00
Add for zone recording thermostat	—	Ea	169.00	—	—	169.00

Description	Craft@Hrs	Unit	Material $	Labor $	Equipment $	Total $

Passive flat-panel solar water heater. Ambient pressure, modular design for for multiple installation, EPA, DOE and UL rated, with piping, tempering valve, circulating pump, controls, storage tank and heating coil. Cost does not include electrical connection, roofing modification (bracing) to accommodate the additional weight of water-filled panels, insulation or padding between roof and panels. Add the cost of post-installation inspection by the mechanical inspector to validate federal, state and local energy tax credits or energy tax credit offsets. For larger arrays (laundries, institutional facilities, food processing plants), develop an estimate based on required daily capacity and multiply these costs by the number of panels required. Equipment cost is for a 1-ton capacity hydraulic crane truck.

Description	Craft@Hrs	Unit	Material $	Labor $	Equipment $	Total $
One 32.3 SF solar panel, 80 gallons per day	P1@16.0	Ea	3,650.00	587.00	1,060.00	5,297.00
Two 32.3 SF solar panels, 120 gallons per day	P1@20.0	Ea	5,100.00	734.00	1,320.00	7,154.00

Fans and Blowers

Description	Craft@Hrs	Unit	Material $	Labor $	Equipment $	Total $

Centrifugal air-foil wheel blower, with motor, drive and isolation, set in place only.

Description	Craft@Hrs	Unit	Material $	Labor $	Equipment $	Total $
1/3 HP, 12" dia.	SN@2.50	Ea	1,650.00	96.60	—	1,746.60
1/2 HP, 15" dia.	SN@4.50	Ea	2,870.00	174.00	—	3,044.00
1 HP, 18" dia.	SN@7.00	Ea	3,320.00	271.00	—	3,591.00
2 HP, 18" dia.	SN@7.50	Ea	3,930.00	290.00	—	4,220.00
3 HP, 27" dia.	SN@9.00	Ea	6,210.00	348.00	5.57	6,563.57
5 HP, 30" dia.	SN@11.0	Ea	7,330.00	425.00	6.80	7,761.80
7½ HP, 36" dia.	SN@12.0	Ea	8,880.00	464.00	7.42	9,351.42
10 HP, 40" dia.	SN@14.0	Ea	12,300.00	541.00	8.66	12,849.66
15 HP, 44" dia.	SN@17.0	Ea	16,500.00	657.00	10.50	17,167.50
20 HP, 60" dia.	SN@20.0	Ea	18,700.00	773.00	12.40	19,485.40

Utility centrifugal blower, with motor, drive and isolation, set in place only.

Description	Craft@Hrs	Unit	Material $	Labor $	Equipment $	Total $
1/3 HP, 10" dia.	SN@2.50	Ea	1,070.00	96.60	—	1,166.60
1/2 HP, 12" dia.	SN@4.50	Ea	1,400.00	174.00	—	1,574.00
1 HP, 15" dia.	SN@6.50	Ea	2,280.00	251.00	—	2,531.00
2 HP, 18" dia.	SN@7.50	Ea	2,590.00	290.00	—	2,880.00
3 HP, 24" dia.	SN@9.00	Ea	3,740.00	348.00	5.57	4,093.57
5 HP, 27" dia.	SN@11.0	Ea	4,300.00	425.00	6.80	4,731.80
7½ HP, 36" dia.	SN@12.0	Ea	6,900.00	464.00	7.42	7,371.42
10 HP, 40" dia.	SN@14.0	Ea	8,330.00	541.00	8.66	8,879.66
15 HP, 44" dia.	SN@17.0	Ea	9,020.00	657.00	10.50	9,687.50

Vane-axial fan, set in place only.

Description	Craft@Hrs	Unit	Material $	Labor $	Equipment $	Total $
30 HP, 36" dia.	SN@16.0	Ea	21,900.00	618.00	302.00	22,820.00
50 HP, 42" dia.	SN@24.0	Ea	24,400.00	928.00	453.00	25,781.00
100 HP, 48" dia.	SN@32.0	Ea	30,400.00	1,240.00	604.00	32,244.00
200 HP, 54" dia.	SN@40.0	Ea	43,000.00	1,550.00	755.00	45,305.00

Description	Craft@Hrs	Unit	Material $	Labor $	Equipment $	Total $

Tube-axial fan, set in place only.

Description	Craft@Hrs	Unit	Material $	Labor $	Equipment $	Total $
15 HP, 24" dia.	SN@10.0	Ea	6,470.00	387.00	189.00	7,046.00
25 HP, 30" dia.	SN@13.0	Ea	10,000.00	502.00	245.00	10,747.00
30 HP, 36" dia.	SN@16.0	Ea	13,400.00	618.00	302.00	14,320.00
75 HP, 40" dia.	SN@22.0	Ea	20,000.00	850.00	415.00	21,265.00
100 HP, 48" dia.	SN@32.0	Ea	30,500.00	1,240.00	604.00	32,344.00

Roof exhaust fan, ASHRAE/American National Standards Institute specification 62.2-2007 and Air Conditioning Contractors of America Standard ACCA 5QI-2007, "HVAC Indoor Air Quality Installation Specification 2007." U.S. Green Building Council, Underwriters laboratories and CSA approved. Cost includes fan and bearing, shaft mount and power transmission assembly or direct drive, enclosure, mount plates, electric motor, 110 or 408, 3-phase, 60 Hz, switch or relay assembly, motor fusing and ground fault protection, fan blade guard. Set in place only. Add installation costs from the figures below.

Description	Craft@Hrs	Unit	Material $	Labor $	Equipment $	Total $
1/4 HP, 14" dia.	SN@2.50	Ea	1,700.00	96.60	—	1,796.60
1/3 HP, 22" dia.	SN@2.90	Ea	2,100.00	112.00	—	2,212.00
1/2 HP, 24" dia.	SN@3.50	Ea	2,400.00	135.00	—	2,535.00
1 HP, 30" dia.	SN@4.50	Ea	3,320.00	174.00	—	3,494.00
2 HP, 36" dia.	SN@5.00	Ea	4,750.00	193.00	94.40	5,037.40
3 HP, 40" dia.	SN@6.00	Ea	7,110.00	232.00	113.00	7,455.00
5 HP, 48" dia.	SN@7.00	Ea	9,870.00	271.00	132.00	10,273.00
Cut, frame and gasket downcomer hole in roof	S2@1.00	Ea	33.80	36.00	—	69.80
Mount duct hangers	S2@.250	Ea	4.35	9.01	—	13.36
Cut and mount sheet metal duct	S2@.350	LF	6.69	12.60	—	19.29
Apply and coat duct insulation	S2@.150	LF	2.90	5.41	—	8.31
Install electrical wiring	BE@.150	LF	2.09	6.06	—	8.15
Install HVAC controls	BE@.500	Ea	—	20.20	—	20.20
Commission and test	P1@2.00	Ea	—	73.40	—	73.40
Run air balance and fine-tune system	P1@1.00	Ea	—	36.70	—	36.70

Ventilators and Residential Exhaust Fans

Description	Craft@Hrs	Unit	Material $	Labor $	Equipment $	Total $

Ceiling-mounted commercial ventilator. Low-profile metal grille. Galvanized steel housing. Designed for continuous operation. Acoustically insulated. Can be installed horizontally, vertically or in-line. Labor includes layout and fasten in place only. Make additional allowances for vent duct and electrical requirements.

Description	Craft@Hrs	Unit	Material $	Labor $	Equipment $	Total $
100 CFM	P1@1.05	Ea	169.00	38.50	—	207.50
150 CFM	P1@1.15	Ea	213.00	42.20	—	255.20
200 CFM	P1@1.35	Ea	269.00	49.50	—	318.50
250 CFM	P1@1.45	Ea	339.00	53.20	—	392.20
300 CFM	P1@1.65	Ea	404.00	60.60	—	464.60
400 CFM	P1@2.15	Ea	520.00	78.90	—	598.90
500 CFM	P1@2.50	Ea	632.00	91.80	—	723.80
700 CFM	P1@2.75	Ea	880.00	101.00	—	981.00
900 CFM	P1@2.85	Ea	973.00	105.00	188.00	1,266.00
1,500 CFM	P1@3.50	Ea	1,350.00	128.00	231.00	1,709.00
2,000 CFM	P1@3.85	Ea	1,480.00	141.00	254.00	1,875.00
3,500 CFM	P1@4.25	Ea	1,690.00	156.00	281.00	2,127.00

Powered attic ventilator. Thermostatically controlled. 1/10 Hp, 120V/6-Hz motor. 8" x 24" metal housing. Labor includes layout and fasten in place only. Make additional allowances for electrical and roofing requirements.

Description	Craft@Hrs	Unit	Material $	Labor $	Equipment $	Total $
1,200 CFM	P1@2.00	Ea	213.00	73.40	—	286.40

Exterior wall/roof exhaust fan. Fully enclosed. Thermally protected. Lifetime lubricated 120V/6-Hz motor. Weather-resistant aluminum housing. Balance blower wheel. Neoprene vibration isolation. Includes backdraft damper, flashing plate and critter screen. 10" discharge outlet. Labor includes layout and fasten in place only. Make additional allowances for vent duct, electrical and roofing requirements.

Description	Craft@Hrs	Unit	Material $	Labor $	Equipment $	Total $
600 CFM	P1@2.50	Ea	632.00	91.80	—	723.80
900 CFM	P1@2.85	Ea	980.00	105.00	—	1,085.00

Room ventilator fan. Interior thru-wall room-to-room air transfer application. High efficiency. Manually switched operation with built-in variable speed control. 8" and 10" discharge outlets. Built in damper. Includes intake and discharge grilles. Labor includes layout and fasten in place only. Make additional allowances for electrical requirements.

Description	Craft@Hrs	Unit	Material $	Labor $	Equipment $	Total $
180 CFM	P1@1.25	Ea	131.00	45.90	—	176.90
270 CFM	P1@1.50	Ea	198.00	55.10	—	253.10
380 CFM	P1@1.75	Ea	269.00	64.20	—	333.20

Ceiling exhaust fan, with automatic motion response control. Adjustable (5-60 minutes) time interval operation. Manual override feature. 4" discharge outlet. Pre-wired outlet box. Balanced blower wheel. Permanently lubricated motor. Neoprene vibration isolators. Quiet operation. Suitable for washroom exhaust applications. Labor includes layout and fasten in place only. Make additional allowances for vent duct and electrical requirements.

Description	Craft@Hrs	Unit	Material $	Labor $	Equipment $	Total $
90 CFM	P1@1.00	Ea	203.00	36.70	—	239.70
130 CFM	P1@1.15	Ea	314.00	42.20	—	356.20

Description	Craft@Hrs	Unit	Material $	Labor $	Equipment $	Total $

Ceiling exhaust fan, with automatic humidity response control.

Adjustable time interval operation (5-60 minutes). Manual override feature. 4" discharge outlet. Pre-wired outlet box. Balanced blower wheel. Permanently lubricated motor. Neoprene vibration isolators. Quiet operation. Suitable for washroom exhaust applications. Labor includes layout and fasten in place only. Make additional allowances for vent duct and electrical requirements.

Description	Craft@Hrs	Unit	Material $	Labor $	Equipment $	Total $
90 CFM	P1@1.00	Ea	239.00	36.70	—	275.70
130 CFM	P1@1.15	Ea	350.00	42.20	—	392.20

Ceiling exhaust fan, manually-switched operation. 4" discharge outlet.
Pre-wired outlet box. 4 pole motor. Suitable for washroom exhaust applications. Labor includes layout and fasten in place only. Make additional allowances for vent duct and electrical requirements.

Description	Craft@Hrs	Unit	Material $	Labor $	Equipment $	Total $
50 CFM	P1@1.00	Ea	45.30	36.70	—	82.00
70 CFM	P1@1.00	Ea	67.70	36.70	—	104.40
90 CFM	P1@1.00	Ea	79.00	36.70	—	115.70
110 CFM	P1@1.05	Ea	94.60	38.50	—	133.10

Ceiling exhaust fan, high efficiency. Manually switched operation. 8" and
10" discharge outlets. Labor includes layout and fasten in place only. Suitable for kitchens, laundry rooms, rec rooms and workshops. Make additional allowances for vent duct and electrical requirements.

Description	Craft@Hrs	Unit	Material $	Labor $	Equipment $	Total $
160 CFM	P1@1.25	Ea	85.60	45.90	—	131.50
180 CFM	P1@1.25	Ea	123.00	45.90	—	168.90
270 CFM	P1@1.50	Ea	181.00	55.10	—	236.10
350 CFM	P1@1.75	Ea	227.00	64.20	—	291.20

Discharge hood ducting kit. Including discharge hood, 4" x 10' of flexible
duct, duct clamps, 3" x 4" transition coupling.

Description	Craft@Hrs	Unit	Material $	Labor $	Equipment $	Total $
Wall hood, 4"	P1@2.00	Ea	22.00	73.40	—	95.40
Roof hood, 4"	P1@2.25	Ea	43.90	82.60	—	126.50
Wall hood, 6"	P1@2.55	Ea	98.50	93.60	—	192.10

Fan controls. Switch only.

Description	Craft@Hrs	Unit	Material $	Labor $	Equipment $	Total $
1 hour timer	BE@.155	Ea	72.00	6.26	—	78.26
Humidistat	BE@.155	Ea	49.60	6.26	—	55.86
Speed controller	BE@.155	Ea	67.70	6.26	—	73.96

Ventilators and Residential Exhaust Fans

Description	Craft@Hrs	Unit	Material $	Labor $	Equipment $	Total $

Air admittance valve. For ventless sanitary plumbing. Provides a water seal in traps without roof penetration and vent piping. Complies with ASHRAE/American National Standards Institute spec. 62.2-2007 and American Society for Sanitary Engineering specs 1050 and 1051.

Description	Craft@Hrs	Unit	Material $	Labor $	Equipment $	Total $
Cut, frame and gasket for wall inset and vent valve access						
door	SW@1.00	Ea	34.70	41.90	—	76.60
Mount valve	P1@.500	Ea	—	18.40	—	18.40
Connect valve to sanitary S- or						
P-trap	P1@1.00	Ea	—	36.70	—	36.70

Bathroom fans and heaters. Comply with ASHRAE/American National Standards Institute spec 62.2-2007 and meet new indoor air quality code requirements for dehumidification, mold and mildew prevention and radon mitigation. Continental Fan, Tamarack, Broan or equal. Includes enclosure, mount plates, 110 or 408 volt electric motor, switch or relay assembly, motor fusing, ground fault protection. Add the following installation costs, if required.

Description	Craft@Hrs	Unit	Material $	Labor $	Equipment $	Total $
Cut, frame and gasket exhaust hole in						
sidewall	SW@1.00	Ea	34.70	41.90	—	76.60
Cut and mount sheet metal						
duct	SW@.150	LF	6.88	6.29	—	13.17
Apply and coat duct						
insulation	S2@.050	LF	3.00	1.80	—	4.80
Install electrical wiring						
(typical)	BE@.150	LF	2.15	6.06	—	8.21
Install controls	BE@.500	Ea	—	20.20	—	20.20
Test and adjust	P1@.250	Ea	—	9.18	—	9.18

Bathroom exhaust fan with heater, commercial grade, Continental Fan, Broan or equal.

Description	Craft@Hrs	Unit	Material $	Labor $	Equipment $	Total $
106 CFM fan,						
4" duct	BE@.500	Ea	243.00	20.20	—	263.20
152 CFM fan,						
4" duct	BE@.500	Ea	275.00	20.20	—	295.20
247 CFM fan,						
6" duct	BE@.500	Ea	459.00	20.20	—	479.20
418 CFM fan,						
8" duct	BE@.500	Ea	591.00	20.20	—	611.20

Infrared bulb heater and fan. 4-point adjustable mounting brackets span up to 24 inches. Uses 250 watt, 120 volt R-40 infrared bulbs (not included). Heater and fan units include 70 CFM, 3.5 sones ventilator fan. Damper and duct connector included. Plastic matte-white molded grille and compact housings. CFM (cubic feet of air per minute).

Description	Craft@Hrs	Unit	Material $	Labor $	Equipment $	Total $
1-bulb, 70 CFM,						
3.5 sones	BE@1.00	Ea	68.70	40.40	—	109.10
2-bulb heater,						
7" duct	BE@1.10	Ea	86.50	44.40	—	130.90

Description	Craft@Hrs	Unit	Material $	Labor $	Equipment $	Total $

Heat-A-Lamp® bulb heater and fan. NuTone. Swivel-mount. Torsion spring holds plates firmly to ceiling. 250 watt radiant heat from one R-40 infrared heat lamp. Uses 4-inch duct. Adjustable socket for ceilings up to 1-inch thick. Adjustable hanger bars. Automatic reset for thermal protection. White polymeric finish. UL listed.

2.6 amp, 1 lamp, 12-1/2" x 10"	BE@1.50	Ea	71.10	60.60	—	131.70
5.0 amp, 2 lamp, 15-3/8" x 11"	BE@1.50	Ea	91.50	60.60	—	152.10

Surface-mount ceiling bath resistance heater and fan. 1,250 watt. 4,266 BTU. 120 volt. Chrome alloy wire element for instant heat. Built-in fan. Automatic overheat protection. Low-profile housing. Permanently lubricated motor. Mounts to standard 3¼-inch round or 4-inch octagonal ceiling electrical box. Satin-finish aluminum grille extends 2¾ inch from ceiling. 11" diameter x 2¾" deep.

10.7 amp	BE@.750	Ea	72.50	30.30	—	102.80

Designer Series bath heater and exhaust fan. 1,500 watt fan-forced heater. 120 watt light capacity, 7 watt night-light (bulbs not included). Permanently lubricated motor. Polymeric damper. Torsion-spring grille mount. Non-glare light-diffusing glass lens. Fits single gang opening. Suitable for use with insulation. 100 CFM, 1.2 sones.

4" duct, 100 watt lamp	BE@1.00	Ea	294.00	40.40	—	334.40

Ventilation exhausters

Includes all-weather enclosure, air balance control, software module and NFPA-certified fusible plug-type vertical duct riser gravity dampers, and efficiency optimization arrays. Add the cost of roof curb, control wiring, electrical wiring, permits and final inspection. For wall- or roof-mount in single- and multi-unit residences, commercial (i.e. hotels), institutional (i.e. hospitals and schools) buildings. Set in place only. Add installation costs from the section below. Costs assume rental of an appliance dolly, a 3,000 lb. come-a-long and a ½-ton chain hoist.

Downblast ventilation exhausters. Rooftop mounted, direct drive, centrifugal type. Suitable for either ducted or nonducted applications.

1/4 HP, 14" dia.	SN@2.50	Ea	1,700.00	96.60	—	1,796.60
1/3 HP, 22" dia.	SN@2.90	Ea	2,100.00	112.00	—	2,212.00
1/2 HP, 24" dia.	SN@3.50	Ea	2,400.00	135.00	—	2,535.00
1 HP, 30" dia.	SN@4.50	Ea	3,320.00	174.00	—	3,494.00
2 HP, 36" dia.	SN@5.00	Ea	4,750.00	193.00	17.20	4,960.20
3 HP, 40" dia.	SN@6.00	Ea	7,110.00	232.00	20.60	7,362.60
5 HP, 48" dia.	SN@7.00	Ea	9,870.00	271.00	24.00	10,165.00

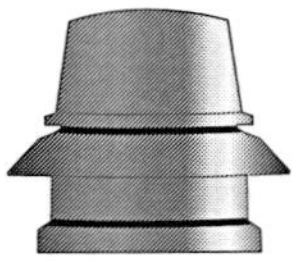

Ventilators and Residential Exhaust Fans

Description	Craft@Hrs	Unit	Material $	Labor $	Equipment $	Total $

Upblast ventilation exhausters. Direct drive, centrifugal, for roof or wall mount.

Description	Craft@Hrs	Unit	Material $	Labor $	Equipment $	Total $
1/4 HP, 14" dia.	SN@2.50	Ea	1,840.00	96.60	—	1,936.60
1/3 HP, 22" dia.	SN@2.90	Ea	2,220.00	112.00	—	2,332.00
1/2 HP, 24" dia.	SN@3.50	Ea	2,560.00	135.00	—	2,695.00
1 HP, 30" dia.	SN@4.50	Ea	3,550.00	174.00	—	3,724.00
2 HP, 36" dia.	SN@5.00	Ea	4,990.00	193.00	17.20	5,200.20
3 HP, 40" dia.	SN@6.00	Ea	7,340.00	232.00	20.60	7,592.60
5 HP, 48" dia.	SN@7.00	Ea	10,100.00	271.00	24.00	10,395.00

Energy-efficient exhauster arrays. Upblast belt-drive centrifugal UL762 roof exhausters. Low profile metal grille. Galvanized steel housing. Designed for continuous operation. Acoustically insulated. Can be installed horizontally, vertically or in-line. Labor includes layout and fasten in place only. Make additional allowances for vent duct and electrical requirements.

Description	Craft@Hrs	Unit	Material $	Labor $	Equipment $	Total $
100 CFM	P1@1.05	Ea	262.00	38.50	—	300.50
150 CFM	P1@1.15	Ea	289.00	42.20	—	331.20
200 CFM	P1@1.35	Ea	314.00	49.50	—	363.50
250 CFM	P1@1.45	Ea	406.00	53.20	—	459.20
300 CFM	P1@1.65	Ea	459.00	60.60	—	519.60
400 CFM	P1@2.15	Ea	576.00	78.90	—	654.90
500 CFM	P1@2.50	Ea	669.00	91.80	—	760.80
700 CFM	P1@2.75	Ea	957.00	101.00	—	1,058.00
900 CFM	P1@2.85	Ea	1,040.00	105.00	9.78	1,154.78
1,500 CFM	P1@3.50	Ea	1,590.00	128.00	12.00	1,730.00
2,000 CFM	P1@3.85	Ea	1,760.00	141.00	13.20	1,914.20
3,500 CFM	P1@4.25	Ea	1,970.00	156.00	14.60	2,140.60

Installation costs for ventilation exhausters.

Description	Craft@Hrs	Unit	Material $	Labor $	Equipment $	Total $
Cut, frame and gasket downcomer hole in sidewall or roof	S2@1.00	Ea	34.20	36.00	—	70.20
Mount duct hangers	S2@.250	Ea	4.35	9.01	—	13.36
Cut & mount sheet metal duct	S2@.350	LF	6.69	12.60	—	19.29
Apply and coat duct insulation	S2@.150	LF	2.90	5.41	—	8.31
Install electrical wiring	BE@.150	LF	2.09	6.06	—	8.15
Install exhauster controls	BE@.500	Ea	—	20.20	—	20.20
Commission and test	P1@4.00	Ea	—	147.00	—	147.00
Run air balance and fine-tune system	P1@4.00	Ea	—	147.00	—	147.00

Unitary or factory-packaged air conditioning units are often too small to meet the heating and cooling needs of a large single-zone space. In such cases, engineers design what's called a "built-up" system. They're custom fitted to the client's needs and the space. A typical "built-up" system includes the following equipment:

- centrifugal blowers or axial-flow fans
- damper assemblies
- coil banks
- filter banks

Once the system is assembled, it's enclosed in a shop-fabricated housing or casing. The illustration on page 332 shows a standard apparatus housing. The walls and roof are shop-built panels. These panels consist of a layer of thermal insulation covered on one or both sides with sheets of 18 gauge galvanized steel. An interior framework of steel angles or channels supports the panels. Seal all mating surfaces to minimize air leakage, using gaskets, caulking or sealant.

Apparatus Housing

There are two types of panels used in building apparatus housings: "double-skin" panels and "single-skin" panels. Double-skin panels have a layer of thermal insulation material sandwiched between two sheets of 18 gauge galvanized steel. Single-skin panels also have a layer of insulation. However, only the outside surface is clad in 18 gauge steel sheeting. On the inside the panel insulation is exposed.

The costs of constructing and installing apparatus housings vary greatly. This is due to the wide range of possible conditions and needs that you face on each job. Some of the many factors that influence costs include the following:

- internal operating pressures
- external wind loads
- seismic requirements
- physical locations

That's why I recommend that you always prepare a detailed cost estimate. Then you can take all the particulars into account. However, for budget estimating purposes only, use the following cost data:

Single-skin panels	$11.40 per SF*
Double-skin panels	$19.30 per SF*
Access doors (20" x 60")	$394.00 each

*Total area of walls and roof, less doors.

When you're estimating material costs for apparatus housings, remember to add in an extra 25 percent to cover the following costs: waste, seams, gaskets, sealants and miscellaneous assembly hardware.

Labor for Shop-Fabrication and Field Assembly of Apparatus Housings. The costs listed are manhours per square foot (SF) of panel. To find the total SF of panel, add up areas for all the panels.

Description	Shop Labor (MH per SF)	Field Labor (MH per SF)
Single-skin panels	.040	.250
Double-skin panels	.085	.300
Add for access doors	.400	.240

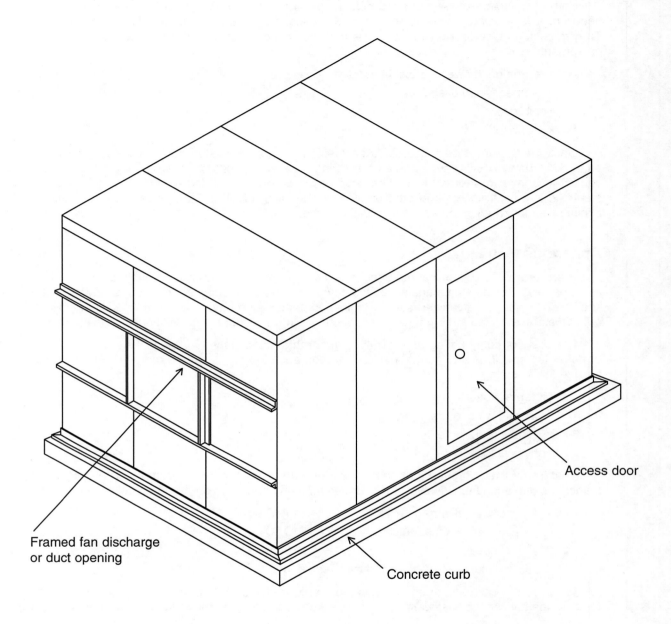

Framed fan discharge
or duct opening

Access door

Concrete curb

Figure 1 - Typical apparatus housing

Description	Craft@Hrs	Unit	Material $	Labor $	Equipment $	Total $

Supply air side-wall register, fixed pattern, with damper. Steel construction.

8" x 4"	S2@.250	Ea	10.60	9.01	—	19.61
8" x 6"	S2@.250	Ea	11.00	9.01	—	20.01
10" x 4"	S2@.250	Ea	12.00	9.01	—	21.01
10" x 6"	S2@.250	Ea	20.40	9.01	—	29.41
10" x 8"	S2@.250	Ea	19.50	9.01	—	28.51
12" x 6"	S2@.250	Ea	20.70	9.01	—	29.71
12" x 8"	S2@.250	Ea	16.80	9.01	—	25.81
14" x 6"	S2@.250	Ea	16.80	9.01	—	25.81
14" x 8"	S2@.250	Ea	19.30	9.01	—	28.31

Return air side-wall grille, fixed pattern, without damper. Steel construction.

14" x 6"	S2@.250	Ea	9.20	9.01	—	18.21
14" x 8"	S2@.250	Ea	8.90	9.01	—	17.91
16" x 16"	S2@.250	Ea	26.90	9.01	—	35.91
20" x 10"	S2@.300	Ea	40.00	10.80	—	50.80
20" x 16"	S2@.300	Ea	43.80	10.80	—	54.60
20" x 20"	S2@.300	Ea	58.00	10.80	—	68.80
24" x 24"	S2@.350	Ea	75.40	12.60	—	88.00

Supply air double-deflection side-wall register, adjustable pattern. Aluminum construction.

8" x 4"	S2@.250	Ea	34.50	9.01	—	43.51
8" x 6"	S2@.250	Ea	38.40	9.01	—	47.41
10" x 4"	S2@.250	Ea	44.10	9.01	—	53.11
10" x 6"	S2@.250	Ea	45.30	9.01	—	54.31
10" x 8"	S2@.250	Ea	48.70	9.01	—	57.71
12" x 6"	S2@.250	Ea	50.40	9.01	—	59.41
12" x 8"	S2@.250	Ea	56.40	9.01	—	65.41
12" x 10"	S2@.250	Ea	63.90	9.01	—	72.91
14" x 6"	S2@.250	Ea	63.90	9.01	—	72.91
14" x 8"	S2@.250	Ea	64.70	9.01	—	73.71
14" x 10"	S2@.250	Ea	68.90	9.01	—	77.91
16" x 8"	S2@.250	Ea	68.90	9.01	—	77.91
16" x 10"	S2@.250	Ea	75.40	9.01	—	84.41
18" x 8"	S2@.300	Ea	78.60	10.80	—	89.40
18" x 10"	S2@.300	Ea	84.30	10.80	—	95.10
18" x 12"	S2@.300	Ea	88.90	10.80	—	99.70

Air Devices, Diffusers and Grilles

Description	Craft@Hrs	Unit	Material $	Labor $	Equipment $	Total $

Return air single-deflection side-wall grille, adjustable pattern.
Aluminum construction.

Description	Craft@Hrs	Unit	Material $	Labor $	Equipment $	Total $
8" x 4"	S2@.250	Ea	31.20	9.01	—	40.21
8" x 6"	S2@.250	Ea	34.50	9.01	—	43.51
10" x 4"	S2@.250	Ea	37.80	9.01	—	46.81
10" x 6"	S2@.250	Ea	40.90	9.01	—	49.91
10" x 8"	S2@.250	Ea	43.60	9.01	—	52.61
12" x 6"	S2@.250	Ea	45.10	9.01	—	54.11
12" x 8"	S2@.250	Ea	50.70	9.01	—	59.71
12" x 10"	S2@.250	Ea	57.20	9.01	—	66.21
14" x 6"	S2@.250	Ea	57.20	9.01	—	66.21
14" x 8"	S2@.250	Ea	58.60	9.01	—	67.61
14" x 10"	S2@.250	Ea	61.50	9.01	—	70.51
16" x 8"	S2@.250	Ea	61.50	9.01	—	70.51
16" x 10"	S2@.300	Ea	68.00	10.80	—	78.80
18" x 8"	S2@.300	Ea	70.80	10.80	—	81.60
18" x 10"	S2@.300	Ea	76.00	10.80	—	86.80
18" x 12"	S2@.300	Ea	80.20	10.80	—	91.00

4-way adjustable pattern ceiling diffuser, 24" x 24" with collar and volume damper. Steel construction.

Description	Craft@Hrs	Unit	Material $	Labor $	Equipment $	Total $
6" neck	S2@.350	Ea	88.40	12.60	—	101.00
8" neck	S2@.350	Ea	92.90	12.60	—	105.50
10" neck	S2@.350	Ea	99.90	12.60	—	112.50
12" neck	S2@.350	Ea	104.00	12.60	—	116.60
14" neck	S2@.400	Ea	125.00	14.40	—	139.40

Fixed pattern ceiling diffuser, 24" x 24" with collar, without volume damper. Steel construction.

Description	Craft@Hrs	Unit	Material $	Labor $	Equipment $	Total $
6" neck	S2@.350	Ea	53.30	12.60	—	65.90
8" neck	S2@.350	Ea	57.50	12.60	—	70.10
10" neck	S2@.350	Ea	64.50	12.60	—	77.10
12" neck	S2@.350	Ea	68.40	12.60	—	81.00
14" neck	S2@.350	Ea	88.40	12.60	—	101.00

Perforated ceiling diffuser, 24" x 24" fixed pattern with collar, without volume damper. Steel construction.

Description	Craft@Hrs	Unit	Material $	Labor $	Equipment $	Total $
6" neck	S2@.350	Ea	70.90	12.60	—	83.50
8" neck	S2@.350	Ea	73.50	12.60	—	86.10
10" neck	S2@.350	Ea	78.80	12.60	—	91.40
12" neck	S2@.350	Ea	84.30	12.60	—	96.90
14" neck	S2@.400	Ea	94.20	14.40	—	108.60
16" neck	S2@.720	Ea	159.00	25.90	—	184.90

Description	Craft@Hrs	Unit	Material $	Labor $	Equipment $	Total $

Round 3-cone, 2-position ceiling diffuser with collar, without volume damper. Steel construction.

Description	Craft@Hrs	Unit	Material $	Labor $	Equipment $	Total $
6"	S2@.350	Ea	73.30	12.60	—	85.90
8"	S2@.350	Ea	84.30	12.60	—	96.90
10"	S2@.350	Ea	102.00	12.60	—	114.60
12"	S2@.350	Ea	120.00	12.60	—	132.60
14"	S2@.350	Ea	160.00	12.60	—	172.60

Round adjustable ceiling diffuser with collar and volume damper. Steel construction.

Description	Craft@Hrs	Unit	Material $	Labor $	Equipment $	Total $
6"	S2@.350	Ea	110.00	12.60	—	122.60
8"	S2@.350	Ea	131.00	12.60	—	143.60
10"	S2@.350	Ea	102.00	12.60	—	114.60
12"	S2@.350	Ea	160.00	12.60	—	172.60
14"	S2@.350	Ea	298.00	12.60	—	310.60
16"	S2@.450	Ea	349.00	16.20	—	365.20
18"	S2@.450	Ea	402.00	16.20	—	418.20
20"	S2@.550	Ea	475.00	19.80	—	494.80
24"	S2@.650	Ea	653.00	23.40	—	676.40

Makeup Air Unit

Description	Craft@Hrs	Unit	Material $	Labor $	Equipment $	Total $

Makeup air unit. Makeup air unit uses 100% direct-heated outdoor air for warehouse, factory and institutional applications to provide higher thermal efficiency and balanced distribution of system air while pre-heating air for use in boiler or other HVAC applications. ANSI standard Z83.4/CSA 3.7 certified. With weatherproof metal enclosure, burner, analog electric controls for burner and fan, thermostat, actuator, natural gas burner array, and either roof-mount supports or sidewall external mounts and transition and gaskets. Set in place only. Add installation costs.

Description	Craft@Hrs	Unit	Material $	Labor $	Equipment $	Total $
200 to 399 CFM	SN@2.25	Ea	650.00	87.00	—	737.00
400 to 599 CFM	SN@2.25	Ea	718.00	87.00	—	805.00
600 to 799 CFM	SN@2.75	Ea	826.00	106.00	51.90	983.90
800 to 999 CFM	SN@3.00	Ea	994.00	116.00	56.60	1,166.60
1,000 to 1,499 CFM	SN@3.50	Ea	1,320.00	135.00	66.10	1,521.10
1,500 to 1,999 CFM	SN@4.00	Ea	1,660.00	155.00	75.50	1,890.50
4,000 to 5,999 CFM	SN@6.00	Ea	4,210.00	232.00	113.00	4,555.00
7,500 to 10,000 CFM	SN@8.00	Ea	7,200.00	309.00	151.00	7,660.00
15,000 to 25,000 CFM	SN@10.0	Ea	10,200.00	387.00	189.00	10,776.00
30,000 to 40,000 CFM	SN@16.0	Ea	15,100.00	618.00	302.00	16,020.00
60,000 to 80,000 CFM	SN@24.0	Ea	26,400.00	928.00	453.00	27,781.00

Installation costs for makeup air units.

Description	Craft@Hrs	Unit	Material $	Labor $	Equipment $	Total $
Cut, frame and gasket downcomer/ sidewall hole	S2@1.00	Ea	31.40	36.00	—	67.40
Mount duct hangers	S2@.250	Ea	4.04	9.01	—	13.05
Cut and mount sheet metal duct	S2@.350	LF	6.24	12.60	—	18.84
Apply and coat duct insulation	S2@.150	LF	2.72	5.41	—	8.13
Install piping for gas line	P1@.100	LF	4.21	3.67	—	7.88
Install electrical wiring	BE@.150	LF	1.95	6.06	—	8.01
Install makeup air unit controls	BE@.500	Ea	—	20.20	—	20.20
Commission and test	P1@4.00	Ea	—	147.00	—	147.00
Run air balance and fine-tune system	P1@4.00	Ea	—	147.00	—	147.00

Variable-air volume units regulate supply air to ducted air conditioning systems. These systems are certified to ANSI standard Z83.4/CSA 3.7 and can be either roof-mounted or upper sidewall externally-mounted. Costs include a weatherproof metal enclosure, heater coil or electrical heater fixture, fan and fan controls, downcomer or sidewall entry transition and gaskets, and either roof-mount supports or sidewall external mounts. Includes set in place only. Add installation costs from the table that follows.

Description	Craft@Hrs	Unit	Material $	Labor $	Equipment $	Total $

Variable-air volume terminal unit with analog electric controls, thermostat, actuator and controller.

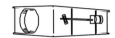

Description	Craft@Hrs	Unit	Material $	Labor $	Equipment $	Total $
200- 400 CFM	SN@2.00	Ea	581.00	77.30	37.80	696.10
400- 600 CFM	SN@2.00	Ea	647.00	77.30	37.80	762.10
600- 800 CFM	SN@2.50	Ea	681.00	96.60	47.20	824.80
800-1,000 CFM	SN@2.75	Ea	739.00	106.00	51.90	896.90
1,000-1,500 CFM	SN@3.00	Ea	1,070.00	116.00	56.60	1,242.60
1,500-2,000 CFM	SN@3.50	Ea	1,370.00	135.00	66.10	1,571.10

Variable-air volume terminal reheat unit with analog electric controls, thermostat, actuator, controller and a two-row hot water coil.

Description	Craft@Hrs	Unit	Material $	Labor $	Equipment $	Total $
200- 400 CFM	SN@2.25	Ea	670.00	87.00	42.50	799.50
400- 600 CFM	SN@2.25	Ea	739.00	87.00	42.50	868.50
600- 800 CFM	SN@2.75	Ea	851.00	106.00	51.90	1,008.90
800-1,000 CFM	SN@3.00	Ea	1,020.00	116.00	52.30	1,188.30
1,000-1,500 CFM	SN@3.50	Ea	1,370.00	135.00	56.60	1,561.60
1,500-2,000 CFM	SN@4.00	Ea	1,720.00	155.00	75.50	1,950.50

Installation costs for variable-air volume units.

Description	Craft@Hrs	Unit	Material $	Labor $	Equipment $	Total $
Cut, frame and gasket downcomer hole in roof	S2@1.00	Ea	30.70	36.00	—	66.70
Mount duct hangers	S2@.250	Ea	3.96	9.01	—	12.97
Cut and mount sheet metal duct	S2@.350	LF	6.11	12.60	—	18.71
Apply and coat duct insulation	S2@.150	LF	2.67	5.41	—	8.08
Install piping for chilled or hot water/steam	P1@.100	LF	3.14	3.67	—	6.81
Install electrical wiring	BE@.150	LF	1.91	6.06	—	7.97
Install VAV controls	BE@.500	Ea	—	20.20	—	20.20
Commission and test	P1@4.00	Ea	—	147.00	—	147.00
Run air balance and fine-tune system	P1@4.00	Ea	—	147.00	—	147.00

Terminal Units (VAV)

Air balance software and control modules for variable-air volume ducted zone control. For commercial, institutional and industrial applications. Johnson Controls or equal. The zone and duct-mounted sensor and response unit comprises a digital controller, actuator and differential pressure transducer. These components are wired together and securely housed inside a plenum-rated enclosure which mounts directly to the VAV "serviced zone" box. Complies with Underwriters Laboratories file E107041, CCN PAZX, UL916 Energy Management Equipment. Interfaces with Windows-based commercial desktop computers per the digital control manufacturer's specifications.

Description	Craft@Hrs	Unit	Material $	Labor $	Equipment $	Total $

Fully programmable 28 point controller, 16 universal inputs, 12 universal outputs, 12 HOA switches, a battery-backed real-time clock, plus preconfigured function blocks for a real-time clock. Stores schedules and trending information.

Description	Craft@Hrs	Unit	Material $	Labor $	Equipment $	Total $
Programmable controller	BE@.500	Ea	1,010.00	20.20	—	1,030.20

Configurable variable air volume pressure independent terminal box controller with wireless transceiver (868.3 MHz)

Description	Craft@Hrs	Unit	Material $	Labor $	Equipment $	Total $
Terminal box controller	BE@.500	Ea	3,040.00	20.20	—	3,060.20

Configurable controller, preconfigured LNS plug-in, and all input devices required for typical rooftop unit application

Description	Craft@Hrs	Unit	Material $	Labor $	Equipment $	Total $
Preconfigured controller	BE@.500	Ea	2,010.00	20.20	—	2,030.20

Variable-air volume controller system software for a Windows computer. Includes a software license and support for two concurrent client browser connections with no limit on the number of concurrent users and no data tag limit.

Description	Craft@Hrs	Unit	Material $	Labor $	Equipment $	Total $
System software	BE@18.0	Ea	7,650.00	727.00	—	8,377.00

Description	Craft@Hrs	Unit	Material $	Labor $	Equipment $	Total $

Opposed-blade damper for rectangular ductwork

Description	Craft@Hrs	Unit	Material $	Labor $	Equipment $	Total $
12" x 6"	S2@.600	Ea	65.50	21.60	—	87.10
12" x 10"	S2@.650	Ea	77.30	23.40	—	100.70
12" x 12"	S2@.700	Ea	94.80	25.20	—	120.00
18" x 12"	S2@.800	Ea	109.00	28.80	—	137.80
18" x 18"	S2@.850	Ea	121.00	30.60	—	151.60
24" x 12"	S2@.850	Ea	128.00	30.60	—	158.60
24" x 18"	S2@.900	Ea	140.00	32.40	—	172.40
24" x 24"	S2@1.00	Ea	156.00	36.00	—	192.00
30" x 12"	S2@.900	Ea	146.00	32.40	—	178.40
30" x 18"	S2@1.00	Ea	163.00	36.00	—	199.00
30" x 24"	S2@1.10	Ea	185.00	39.60	—	224.60
30" x 30"	S2@1.20	Ea	198.00	43.20	—	241.20
36" x 18"	S2@1.10	Ea	191.00	39.60	—	230.60
36" x 36"	S2@1.40	Ea	261.00	50.50	—	311.50
42" x 24"	S2@1.30	Ea	266.00	46.90	—	312.90
42" x 42"	S2@1.50	Ea	364.00	54.10	—	418.10
48" x 24"	S2@1.40	Ea	276.00	50.50	—	326.50
48" x 48"	S2@1.60	Ea	383.00	57.70	—	440.70
54" x 24"	S2@1.45	Ea	350.00	52.30	—	402.30
54" x 54"	S2@1.70	Ea	437.00	61.30	—	498.30

Single-blade damper for rectangular ductwork

Description	Craft@Hrs	Unit	Material $	Labor $	Equipment $	Total $
12" x 6"	S2@.600	Ea	41.00	21.60	—	62.60
12" x 8"	S2@.625	Ea	44.90	22.50	—	67.40
12" x 10"	S2@.650	Ea	50.50	23.40	—	73.90
12" x 12"	S2@.700	Ea	59.20	25.20	—	84.40
18" x 12"	S2@.800	Ea	65.90	28.80	—	94.70
18" x 18"	S2@.850	Ea	74.80	30.60	—	105.40
24" x 12"	S2@.850	Ea	79.20	30.60	—	109.80
24" x 18"	S2@.900	Ea	82.40	32.40	—	114.80
24" x 24"	S2@1.00	Ea	90.50	36.00	—	126.50

Single-blade butterfly damper for round ductwork

Description	Craft@Hrs	Unit	Material $	Labor $	Equipment $	Total $
6"	S2@.600	Ea	41.00	21.60	—	62.60
8"	S2@.550	Ea	38.60	19.80	—	58.40
10"	S2@.600	Ea	44.80	21.60	—	66.40
12"	S2@.650	Ea	50.70	23.40	—	74.10
14"	S2@.700	Ea	57.70	25.20	—	82.90
16"	S2@.765	Ea	64.00	27.60	—	91.60
18"	S2@.835	Ea	70.90	30.10	—	101.00
20"	S2@.900	Ea	79.70	32.40	—	112.10

Ductwork Specialties

Description	Craft@Hrs	Unit	Material $	Labor $	Equipment $	Total $

Curtain-type duct fire damper, 1½-hour rating, U.L. label

Description	Craft@Hrs	Unit	Material $	Labor $	Equipment $	Total $
12" x 6"	S2@1.00	Ea	74.50	36.00	—	110.50
12" x 8"	S2@1.00	Ea	80.60	36.00	—	116.60
12" x 10"	S2@1.05	Ea	88.50	37.80	—	126.30
12" x 12"	S2@1.10	Ea	95.30	39.60	—	134.90
18" x 12"	S2@1.20	Ea	133.00	43.20	—	176.20
18" x 18"	S2@1.30	Ea	145.00	46.90	—	191.90
24" x 12"	S2@1.40	Ea	160.00	50.50	—	210.50
24" x 18"	S2@1.50	Ea	183.00	54.10	—	237.10
24" x 24"	S2@1.60	Ea	217.00	57.70	—	274.70
30" x 12"	S2@1.50	Ea	186.00	54.10	—	240.10
30" x 18"	S2@1.80	Ea	206.00	64.90	—	270.90
30" x 24"	S2@2.00	Ea	274.00	72.10	—	346.10
30" x 30"	S2@2.30	Ea	295.00	82.90	—	377.90
36" x 18"	S2@2.00	Ea	236.00	72.10	—	308.10
36" x 36"	S2@2.40	Ea	351.00	86.50	—	437.50
42" x 24"	S2@2.40	Ea	305.00	86.50	—	391.50
42" x 42"	S2@2.90	Ea	441.00	105.00	—	546.00
48" x 24"	S2@2.90	Ea	331.00	105.00	—	436.00
48" x 48"	S2@3.50	Ea	548.00	126.00	—	674.00
54" x 24"	S2@3.00	Ea	358.00	108.00	—	466.00
54" x 54"	S2@3.80	Ea	638.00	137.00	—	775.00

Correction factors:

(1) For 22 gauge break-away connection, multiply material cost by 1.30

(2) For blades out of air stream cap, multiply material cost by 1.25

Description	Craft@Hrs	Unit	Material $	Labor $	Equipment $	Total $

Fusible plug dampers for air duct risers. Vertical riser and header air duct fusible (melting) and heat-sensing automatic counterweight closure actuators and closure dampers. For large A/C systems in multi-story institutions such as schools, hospitals and office buildings. 1½-hour rating. The liner and insulation must meet the requirements of U.L. 181 and NFPA No. 90 A-96, as modified, "Standard for the Installation of Air Conditioning and Ventilation Systems."

Description	Craft@Hrs	Unit	Material $	Labor $	Equipment $	Total $
12" x 6"	S2@1.00	Ea	98.50	36.00	—	134.50
12" x 8"	S2@1.00	Ea	101.00	36.00	—	137.00
12" x 10"	S2@1.05	Ea	104.00	37.80	—	141.80
12" x 12"	S2@1.10	Ea	106.00	39.60	—	145.60
18" x 12"	S2@1.20	Ea	134.00	43.20	—	177.20
18" x 18"	S2@1.30	Ea	173.00	46.90	—	219.90
24" x 12"	S2@1.40	Ea	185.00	50.50	—	235.50
24" x 18"	S2@1.50	Ea	206.00	54.10	—	260.10
24" x 24"	S2@1.60	Ea	236.00	57.70	—	293.70
30" x 12"	S2@1.50	Ea	223.00	54.10	—	277.10
30" x 18"	S2@1.80	Ea	280.00	64.90	—	344.90
30" x 24"	S2@2.00	Ea	325.00	72.10	—	397.10
30" x 30"	S2@2.30	Ea	391.00	82.90	—	473.90
36" x 18"	S2@2.00	Ea	314.00	72.10	—	386.10
36" x 36"	S2@2.40	Ea	419.00	86.50	—	505.50
42" x 24"	S2@2.40	Ea	391.00	86.50	—	477.50
42" x 42"	S2@2.90	Ea	475.00	105.00	—	580.00
48" x 24"	S2@2.90	Ea	425.00	105.00	—	530.00
48" x 48"	S2@3.50	Ea	950.00	126.00	—	1,076.00
54" x 24"	S2@3.00	Ea	1,220.00	108.00	—	1,328.00
54" x 54"	S2@3.80	Ea	1,990.00	137.00	—	2,127.00
Cut, frame and gasket access hole in sidewall	S2@1.00	Ea	57.50	36.00	—	93.50
Mount duct hangers	S2@.250	Ea	3.99	9.01	—	13.00
Cut and mount sheet metal duct	S2@.350	LF	6.15	12.60	—	18.75
Apply and coat duct insulation	S2@.150	LF	2.67	5.41	—	8.08

Ductwork Specialties

Description	Craft@Hrs	Unit	Material $	Labor $	Equipment $	Total $

Single-skin 2" fixed-position turning vanes for duct

Description	Craft@Hrs	Unit	Material $	Labor $	Equipment $	Total $
12" x 6"	S2@.280	Ea	8.69	10.10	—	18.79
12" x 8"	S2@.285	Ea	10.80	10.30	—	21.10
12" x 10"	S2@.290	Ea	13.90	10.50	—	24.40
12" x 12"	S2@.300	Ea	17.20	10.80	—	28.00
18" x 12"	S2@.375	Ea	26.00	13.50	—	39.50
18" x 18"	S2@.450	Ea	31.60	16.20	—	47.80
24" x 12"	S2@.450	Ea	34.60	16.20	—	50.80
24" x 18"	S2@.500	Ea	37.60	18.00	—	55.60
24" x 24"	S2@.525	Ea	42.90	18.90	—	61.80
30" x 12"	S2@.550	Ea	43.30	19.80	—	63.10
30" x 18"	S2@.750	Ea	62.50	27.00	—	89.50
30" x 24"	S2@.800	Ea	83.00	28.80	—	111.80
30" x 30"	S2@.900	Ea	105.00	32.40	—	137.40
36" x 18"	S2@.750	Ea	74.10	27.00	—	101.10
36" x 36"	S2@.800	Ea	151.00	28.80	—	179.80
42" x 24"	S2@.900	Ea	117.00	32.40	—	149.40
42" x 42"	S2@1.00	Ea	206.00	36.00	—	242.00
48" x 24"	S2@1.10	Ea	133.00	39.60	—	172.60
48" x 48"	S2@1.50	Ea	266.00	54.10	—	320.10
54" x 24"	S2@1.15	Ea	149.00	41.40	—	190.40
54" x 54"	S2@1.25	Ea	339.00	45.10	—	384.10
60" x 24"	S2@1.30	Ea	168.00	46.90	—	214.90
60" x 36"	S2@1.40	Ea	252.00	50.50	—	302.50
60" x 60"	S2@1.50	Ea	417.00	54.10	—	471.10
72" x 36"	S2@1.65	Ea	305.00	59.50	—	364.50
72" x 54"	S2@2.40	Ea	449.00	86.50	—	535.50

Note: Labor includes cost to cut, assemble and install vanes and rails. For 4-inch air-foil vanes, multiply costs by 1.80 and labor costs by .90

Description	Craft@Hrs	Unit	Material $	Labor $	Equipment $	Total $

Spin-in galvanized steel collars without volume damper

Description	Craft@Hrs	Unit	Material $	Labor $	Equipment $	Total $
4"	S2@.200	Ea	6.48	7.21	—	13.69
5"	S2@.200	Ea	7.27	7.21	—	14.48
6"	S2@.200	Ea	8.01	7.21	—	15.22
7"	S2@.200	Ea	8.82	7.21	—	16.03
8"	S2@.250	Ea	10.20	9.01	—	19.21
9"	S2@.250	Ea	11.30	9.01	—	20.31
10"	S2@.300	Ea	12.10	10.80	—	22.90
12"	S2@.350	Ea	14.50	12.60	—	27.10
14"	S2@.400	Ea	17.30	14.40	—	31.70
16"	S2@.450	Ea	20.60	16.20	—	36.80
18"	S2@.500	Ea	25.70	18.00	—	43.70

Spin-in galvanized steel collar with volume damper

Description	Craft@Hrs	Unit	Material $	Labor $	Equipment $	Total $
4"	S2@.250	Ea	21.70	9.01	—	30.71
5"	S2@.250	Ea	22.10	9.01	—	31.11
6"	S2@.250	Ea	22.30	9.01	—	31.31
7"	S2@.250	Ea	22.80	9.01	—	31.81
8"	S2@.300	Ea	23.30	10.80	—	34.10
9"	S2@.300	Ea	23.60	10.80	—	34.40
10"	S2@.350	Ea	24.20	12.60	—	36.80
12"	S2@.400	Ea	25.90	14.40	—	40.30
14"	S2@.450	Ea	27.70	16.20	—	43.90
16"	S2@.500	Ea	29.40	18.00	—	47.40
18"	S2@.550	Ea	31.60	19.80	—	51.40

Duct-to-equipment flexible connection

Description	Craft@Hrs	Unit	Material $	Labor $	Equipment $	Total $
18" x 12"	S2@.650	Ea	22.00	23.40	—	45.40
18" x 18"	S2@.850	Ea	26.50	30.60	—	57.10
24" x 12"	S2@.850	Ea	26.50	30.60	—	57.10
24" x 18"	S2@1.00	Ea	30.90	36.00	—	66.90
24" x 24"	S2@1.30	Ea	35.40	46.90	—	82.30
30" x 12"	S2@1.00	Ea	30.90	36.00	—	66.90
30" x 18"	S2@1.30	Ea	35.40	46.90	—	82.30
30" x 30"	S2@1.60	Ea	44.30	57.70	—	102.00
36" x 12"	S2@1.30	Ea	35.40	46.90	—	82.30
36" x 24"	S2@1.60	Ea	44.30	57.70	—	102.00
36" x 36"	S2@1.80	Ea	52.50	64.90	—	117.40
48" x 24"	S2@1.80	Ea	52.50	64.90	—	117.40
48" x 48"	S2@2.00	Ea	69.60	72.10	—	141.70

Galvanized Steel Ductwork

The costs for fabricating and installing galvanized steel ductwork are usually based on the total weights of the duct and fittings. Because of cost differences, fitting weights should be kept separate from straight duct weights.

Round spiral duct: Normally-used duct gauge/size relationships for low-pressure systems, up to 2 inches static pressure, are as follows:

Duct Diameter (inches)	U.S. Standard Gauge
Up to 12	26
13 through 24	24
26 through 36	22

Rectangular duct: Normally-used duct gauge/size relationships for low-pressure systems, up to 2 inches static pressure, are as follows:

Duct Size (inches)	U.S. Standard Gauge
Up to 12	26
13 through 30	24
31 through 54	22
55 through 72	20

The fitting weights in the following tables are in accordance with the above gauges.

Other material costs: All weights in the tables are net weights; they do not include bracing, cleats, scrap, hangers, end closures, sealants and miscellaneous hardware. Add 15% to calculated duct/fitting weights to cover these items if ductwork is purchased. Add 20% if ducting is manufactured in contractor's shop. The additional 5% covers the costs for scrap which is already included in an outside vendor's price.

Ductwork costs: Typical costs for purchased ductwork are as follows:

Straight duct, less than 1,000 pounds per order	$3.13/lb.
Straight duct, over 1,000 pounds per order	$2.81/lb.
Fittings, less than 1,000 pounds per order	$6.31/lb.
Fittings, over 1,000 pounds per order	$5.66/lb.
Lined ducts and fittings, add	$1.68/lb.*
Delivery costs	$0.15/lb.

Duct sizes shown on drawings are always inside, or net, dimensions and must be increased in size when estimating lined ductwork.

If ductwork is manufactured in the contractor's fabrication shop, the above costs can be reduced by about 20%, depending on the shop's rate of productivity.

Installation costs: An average crew can install approximately 25 pounds of duct and fittings per manhour under normal conditions. See "Applying Correction Factors" on page 6 for situations that do not conform to the definition of a standard labor unit.

EXAMPLE:

What is the cost to furnish and install 2,585 pounds of unlined duct and 560 pounds of unlined fittings? The ductwork will be purchased from an outside vendor. A 15% allowance for miscellaneous material is included in the weights.

Material:	
2,585 pounds of straight duct x \$2.81/lb.	= \$7,263.85
560 pounds of fittings x \$6.31/lb.	= 3,533.60
Delivery cost: (2,585 + 560) x \$.15/lb.	= 471.75
Total material cost	= \$11,269.20*
Labor:	
$\frac{2{,}585 + 560}{25}$ =125.8 *MH*; 125.8 *MH* × \$36.04	= \$4,533.83
Total installed cost: \$11,152.45 + \$4,511.19	= \$15,803.03*

**Sales tax not included.*

Installed Ductwork Per Pound

Description	Craft@Hrs	Unit	Material $	Labor $	Equipment $	Total $

Installed ductwork per lb. (under 1,000 lbs.). Duct purchased from an independent fabrication shop, installed by this contractor's shop. Material price includes scrap, cleats, hangers, sealant, miscellaneous hardware, delivery and fabrication shop markups.

Description	Craft@Hrs	Unit	Material $	Labor $	Equipment $	Total $
Straight duct	S2@.042	Lb	3.10	1.51	—	4.61
Duct fittings	S2@.042	Lb	6.31	1.51	—	7.82
Duct & fittings	S2@.042	Lb	4.18	1.51	—	5.69

Installed ductwork per lb. (over 1,000 lbs.). Duct purchased from an independent fabrication shop, installed by this contractor's shop. Material price includes scrap, cleats, hangers, sealant, miscellaneous hardware, delivery and subcontractor markups.

Description	Craft@Hrs	Unit	Material $	Labor $	Equipment $	Total $
Straight duct	S2@.039	Lb	2.81	1.41	—	4.22
Duct fittings	S2@.039	Lb	5.66	1.41	—	7.07
Duct & fittings	S2@.039	Lb	3.77	1.41	—	5.18

Installed lined ductwork per lb. (under 1,000 lbs.). Duct purchased from an independent fabrication shop, installed by this contractor's shop. Material price includes scrap, cleats, hangers, sealant, miscellaneous hardware, delivery and subcontractor markups.

Description	Craft@Hrs	Unit	Material $	Labor $	Equipment $	Total $
Straight duct	S2@.042	Lb	4.80	1.51	—	6.31
Duct fittings	S2@.042	Lb	7.95	1.51	—	9.46
Duct & fittings	S2@.042	Lb	5.85	1.51	—	7.36

Installed lined ductwork per lb. (over 1,000 lbs.). Duct purchased from an independent fabrication shop, installed by this contractor's shop. Material price includes scrap, cleats, hangers, sealant, miscellaneous hardware, delivery and subcontractor markups.

Description	Craft@Hrs	Unit	Material $	Labor $	Equipment $	Total $
Straight duct	S2@.039	Lb	4.51	1.41	—	5.92
Duct fittings	S2@.039	Lb	7.37	1.41	—	8.78
Duct & fittings	S2@.039	Lb	5.43	1.41	—	6.84

Installed ductwork per lb. (under 1,000 lbs.). Duct fabricated and installed by this contractor's shop. Material price includes scrap, cleats, hangers, sealant, miscellaneous hardware and delivery.

Description	Craft@Hrs	Unit	Material $	Labor $	Equipment $	Total $
Straight duct	S2@.051	Lb	1.92	1.84	—	3.76
Duct fittings	S2@.074	Lb	2.20	2.67	—	4.87
Duct & fittings	S2@.059	Lb	2.02	2.13	—	4.15

Installed ductwork per lb. (over 1,000 lbs.). Duct fabricated and installed by this contractor's shop. Material price includes scrap, cleats, hangers, sealant, miscellaneous hardware and delivery.

Description	Craft@Hrs	Unit	Material $	Labor $	Equipment $	Total $
Straight duct	S2@.046	Lb	1.77	1.66	—	3.43
Duct fittings	S2@.067	Lb	2.03	2.41	—	4.44
Duct & fittings	S2@.053	Lb	1.84	1.91	—	3.75

Description	Craft@Hrs	Unit	Material $	Labor $	Equipment $	Total $

Installed lined ductwork per lb. (under 1,000 lbs.). Duct fabricated and installed by this contractor's shop. Material price includes scrap, cleats, hangers, sealant, miscellaneous hardware and delivery.

Description	Craft@Hrs	Unit	Material $	Labor $	Equipment $	Total $
Straight duct	S2@.062	Lb	2.83	2.23	—	5.06
Duct fittings	S2@.085	Lb	3.13	3.06	—	6.19
Duct & fittings	S2@.070	Lb	2.94	2.52	—	5.46

Installed lined ductwork per lb. (over 1,000 lbs.). Duct fabricated and installed by this contractor's shop. Material price includes scrap, cleats, hangers, sealant, miscellaneous hardware and delivery. Use for preliminary estimates.

Description	Craft@Hrs	Unit	Material $	Labor $	Equipment $	Total $
Straight duct	S2@.057	Lb	2.69	2.05	—	4.74
Duct fittings	S2@.078	Lb	3.00	2.81	—	5.81
Duct & fittings	S2@.064	Lb	5.21	2.31	—	7.52

Galvanized Steel Spiral Ductwork

Weights of Galvanized Steel Spiral Duct (pounds per LF)					
Diameter (inches)	U.S. Standard Gauge				
	26	24	22	20	18
3	.76	1.01	1.23	1.46	—
4	1.02	1.35	1.64	1.94	—
5	1.28	1.69	2.06	2.42	—
6	1.54	2.03	2.47	2.91	3.88
7	1.79	2.37	2.88	3.40	4.52
8	2.05	2.71	3.29	3.86	5.17
9	2.31	3.05	3.71	4.37	5.82
10	2.57	3.39	4.12	4.86	6.47
12	3.08	4.07	4.95	5.83	7.75
14	3.59	4.74	5.77	6.81	9.05
16	4.11	5.42	6.60	7.78	10.34
18	4.63	6.10	7.43	8.76	11.64
20	5.15	6.78	8.25	9.73	12.93
22	5.65	7.46	9.08	10.71	14.93
24	6.16	8.14	9.91	11.68	15.52
26	6.67	8.82	10.73	12.66	16.82
28	7.18	9.50	11.56	13.63	18.11
30	7.71	10.18	12.38	14.60	19.41
32	8.22	10.84	13.21	15.58	20.71
34	—	11.52	14.05	16.55	22.00
36	—	12.20	14.90	17.53	23.29

Weights of Galvanized Steel Round Spiral Fittings (pounds per piece)				
Diameter (inches)	90° elbow	45° elbow	Coupling	Reducer
3	1.3	1.0	0.5	1.0
4	2.2	1.3	0.6	1.2
5	3.3	1.9	0.7	1.4
6	4.3	2.5	0.9	1.8
7	5.8	3.3	1.0	2.0
8	7.3	4.3	1.2	2.4
9	8.8	5.3	2.6	3.3
10	11.8	7.5	2.9	4.4
12	16.3	10.0	3.5	5.3
14	22.0	13.0	4.1	6.2
16	28.3	15.8	4.6	6.9
18	34.5	19.0	5.2	7.8
20	41.5	23.5	5.8	8.7
22	48.3	27.5	6.4	9.6
24	57.5	32.0	6.9	10.3
26	68.8	37.5	7.5	11.2
28	76.8	42.8	8.1	12.1
30	87.0	48.0	8.7	13.0
32	99.5	55.0	9.2	13.8
34	112.0	61.0	9.8	14.7
36	161.0	89.5	10.4	15.6

Largest run diam. (inches)	Weights of Galvanized Steel Round Spiral Fittings (pounds per piece) Tee with reducing run and branch Branch diameter (inches)													
	3	4	5	6	7	8	9	10	12	14	16	18	20	22
4	2.3													
5	2.8	3.0												
6	3.2	3.5	3.7											
7	3.7	3.9	4.1	4.5										
8	4.1	4.4	4.7	5.0	5.3									
9	4.5	4.8	5.1	5.5	5.8	6.2								
10	6.4	6.7	7.1	7.5	7.8	8.2	8.6							
12	7.7	8.1	8.5	8.9	9.4	9.9	10.3	10.7						
14	8.9	9.4	10.0	10.5	11.0	11.5	12.1	12.6	13.6					
16	10.2	10.8	11.4	12.0	12.5	13.1	13.7	14.3	15.5	16.7				
18	11.5	12.1	12.7	13.5	14.1	14.8	15.5	16.1	17.5	18.8	19.2			
20	12.7	13.4	14.2	14.9	17.7	16.4	17.2	17.9	19.4	20.9	22.4	23.9		
22	14.0	14.8	15.6	16.4	17.2	18.1	18.9	19.7	21.3	23.0	24.6	26.2	27.9	
24	15.3	16.1	17.0	17.8	18.7	19.6	20.5	21.4	23.2	25.0	26.8	28.5	30.3	32.1
26	16.5	17.5	18.5	19.5	20.4	21.3	22.3	23.3	25.2	26.2	29.1	31.1	33.0	34.9
28	17.7	18.7	19.8	20.8	21.9	22.9	23.9	25.0	27.0	29.2	31.2	33.2	35.4	37.4
30	18.9	20.0	21.1	22.2	23.3	24.5	25.5	26.7	28.9	31.1	33.3	35.5	37.7	40.0
32	20.4	21.6	22.6	24.0	25.2	26.4	27.6	28.8	31.2	33.6	36.0	38.4	40.8	42.2
34	21.6	22.9	24.2	25.4	26.7	28.0	29.3	30.5	33.1	35.6	38.1	40.7	42.2	45.6
36	29.6	31.4	33.1	34.8	36.6	38.6	40.1	41.8	45.3	48.8	52.2	55.7	59.2	62.7

Galvanized Steel Round Spiral Fittings

Run diam. (inches)	Weights of Galvanized Steel Round Spiral Fittings (pounds per piece) Cross with reducing run and branch Branch diameter (inches)													
	3	4	5	6	7	8	9	10	12	14	16	18	20	22
4	2.6													
5	3.2	3.6												
6	3.8	4.2	4.6											
7	4.4	4.9	5.4	5.8										
8	4.9	5.4	6.0	6.5	7.0									
9	5.4	6.0	6.6	7.2	7.7	8.3								
10	7.6	8.2	8.9	9.5	10.2	10.8	11.4							
12	8.8	9.6	10.3	11.0	11.7	12.5	13.2	13.8						
14	10.1	10.9	11.7	12.6	13.4	14.2	15.1	15.9	17.5					
16	11.3	12.2	13.1	14.0	15.0	15.9	16.8	17.7	19.5	21.4				
18	12.5	13.5	14.5	15.5	16.5	17.5	18.5	19.5	21.5	23.5	25.5			
20	13.8	15.0	16.1	17.2	18.2	19.4	20.6	21.7	23.9	26.2	28.4	30.6		
22	15.2	16.4	17.7	18.9	20.1	21.4	22.6	23.8	26.2	28.7	31.2	33.6	36.1	
24	16.6	17.9	19.3	20.6	22.0	23.3	24.6	26.0	28.7	31.5	34.1	36.7	39.4	42.1
26	18.0	19.4	20.9	22.3	23.8	25.2	26.7	28.2	31.1	33.9	36.9	39.8	42.7	45.6
28	19.3	20.8	22.4	24.0	25.5	27.1	28.6	30.2	32.3	36.4	39.6	42.7	45.8	48.9
30	20.6	22.3	23.9	25.6	27.3	28.9	30.6	32.3	35.6	38.9	42.3	45.6	49.9	52.2
32	22.2	24.0	25.8	27.6	29.4	31.2	33.0	34.8	38.4	42.0	45.6	49.1	52.8	56.5
34	23.5	25.4	27.3	29.2	31.1	33.1	34.9	36.9	40.7	44.5	48.3	52.1	55.9	59.8
36	32.2	34.8	37.4	40.0	42.8	45.3	48.0	50.5	55.7	61.0	66.2	71.5	76.7	82.0

Size (inches)	Weights of 26 Gauge Galvanized Steel Rectangular Duct (pounds per LF)								
	4	5	6	7	8	9	10	11	12
4	1.21								
5	1.36	1.51							
6	1.51	1.66	1.81						
7	1.66	1.81	1.96	2.11					
8	1.81	1.96	2.11	2.27	2.42				
9	1.96	2.11	2.27	2.42	2.57	2.72			
10	2.11	2.27	2.42	2.57	2.72	2.87	3.02		
11	2.27	2.42	2.57	2.72	2.87	3.02	3.17	3.32	
12	2.42	2.57	2.72	2.87	3.02	3.17	3.32	3.47	3.62

Galvanized Steel Rectangular Ductwork

Size (inches)	14	16	18	20	22	24	26	28	30
	Weights of 24 Gauge Galvanized Steel Rectangular Duct (pounds per LF)								
6	3.85	4.24	4.62	5.01	5.40	5.78	6.17	6.55	6.94
8	4.24	4.62	5.01	5.40	5.78	6.17	6.55	6.94	7.32
10	4.62	5.01	5.40	5.78	6.17	6.55	6.94	7.32	7.71
12	5.01	5.40	5.78	6.17	6.55	6.94	7.32	7.71	8.09
14	5.40	5.78	6.17	6.55	6.94	7.32	7.71	8.09	8.48
16	5.78	6.17	6.55	6.94	7.32	7.71	8.09	8.48	8.86
18	6.17	6.55	6.94	7.32	7.71	8.09	8.48	8.86	9.25
20	6.55	6.94	7.32	7.71	8.09	8.48	8.86	9.25	9.64
22	6.94	7.32	7.71	8.09	8.48	8.86	9.25	9.64	10.00
24	7.32	7.71	8.09	8.48	8.86	9.25	9.64	10.00	10.40
26	7.71	8.09	8.48	8.86	9.25	9.64	10.00	10.40	10.80
28	8.09	8.48	8.86	9.25	9.64	10.00	10.40	10.80	11.20
30	8.48	8.86	9.25	9.64	10.00	10.40	10.80	11.20	11.60

Size (inches)	32	34	36	38	40	42	44	46	48	50	52	54
	Weights of 22 Gauge Galvanized Steel Rectangular Duct (pounds per LF)											
8	9.37	9.84	10.3	10.8	11.3	11.7	12.2	12.7	13.1	13.6	14.1	14.5
10	9.84	10.3	10.8	11.3	11.7	12.2	12.7	13.1	13.6	14.1	14.5	15.0
12	10.3	10.8	11.3	11.7	12.2	12.7	13.1	13.6	14.1	14.5	15.0	15.5
14	10.8	11.3	11.7	12.2	12.7	13.1	13.6	14.1	14.5	15.0	15.5	15.9
16	11.3	11.7	12.2	12.7	13.1	13.6	14.1	14.5	15.0	15.5	15.9	16.4
18	11.7	12.2	12.7	13.1	13.6	14.1	14.5	15.0	15.5	15.9	16.4	16.9
20	12.2	12.7	13.1	13.6	14.1	14.5	15.0	15.5	15.9	16.4	16.9	17.3
22	12.7	13.1	13.6	14.1	14.5	15.0	15.5	15.9	16.4	16.9	17.3	17.8
24	13.1	13.6	14.1	14.5	15.0	15.5	15.9	16.4	16.9	17.3	17.8	18.3
26	13.6	14.1	14.5	15.0	15.5	15.9	16.4	16.9	17.3	17.8	18.3	18.7
28	14.1	14.5	15.0	15.5	15.9	16.4	16.9	17.3	17.8	18.3	18.7	19.2
30	14.5	15.0	15.5	15.9	16.4	16.9	17.3	17.8	18.3	18.7	19.2	19.7
32	15.0	15.5	15.9	16.4	16.9	17.3	17.8	18.3	18.7	19.2	19.7	20.2
34	15.5	15.9	16.4	16.9	17.3	17.8	18.3	18.7	19.2	19.7	20.2	20.6
36	15.9	16.4	16.9	17.3	17.8	18.3	18.7	19.2	19.7	20.2	20.6	21.1
38	16.4	16.9	17.3	17.8	18.3	18.7	19.2	19.7	20.2	20.6	21.1	21.6
40	16.9	17.3	17.8	18.3	18.7	19.2	19.7	20.2	20.6	21.1	21.6	22.0
42	17.3	17.8	18.3	18.7	19.2	19.7	20.2	20.6	21.1	21.6	22.0	22.5
44	17.8	18.3	18.7	19.2	19.7	20.2	20.6	21.1	21.6	22.0	22.5	23.0
46	18.3	18.7	19.2	19.7	20.2	20.6	21.1	21.6	22.0	22.5	23.0	23.4
48	18.7	19.2	19.7	20.2	20.6	21.1	21.6	22.0	22.5	23.0	23.4	23.9
50	19.2	19.7	20.2	20.6	21.1	21.6	22.0	22.5	23.0	23.4	23.9	24.4
52	19.7	20.2	20.6	21.1	21.6	22.0	22.5	23.0	23.4	23.9	24.4	24.8
54	20.2	20.6	21.1	21.6	22.0	22.5	23.0	23.4	23.9	24.4	24.8	25.3

Galvanized Steel Rectangular Ductwork

Weights of 20 Gauge Galvanized Steel Rectangular Duct (pounds per LF)								
Size (inches)	**56**	**58**	**60**	**62**	**64**	**66**	**68**	**70**	**72**
18	20.4	21.0	21.5	22.1	22.6	23.2	23.7	24.3	24.8
20	21.0	21.5	22.1	22.6	23.2	23.7	24.3	24.8	25.4
22	21.5	22.1	22.6	23.2	23.7	24.3	24.8	25.4	25.9
24	22.1	22.6	23.2	23.7	24.3	24.8	25.4	25.9	26.5
26	22.6	23.2	23.7	24.3	24.8	25.4	25.9	26.5	27.0
28	23.2	23.7	24.3	24.8	25.4	25.9	26.5	27.0	27.6
30	23.7	24.3	24.8	25.4	25.9	26.5	27.0	27.6	28.2
32	24.3	24.8	25.4	25.9	26.5	27.0	27.6	28.2	28.7
34	24.8	25.4	25.9	26.5	27.0	27.6	28.2	28.7	29.3
36	25.4	25.9	26.5	27.0	27.6	28.2	28.7	29.3	29.8
38	25.9	26.5	27.0	27.6	28.2	28.7	29.3	29.8	30.4
40	26.5	27.0	27.6	28.2	28.7	29.3	29.8	30.4	30.9
42	27.0	27.6	28.2	28.7	29.3	29.8	30.4	30.9	31.5
44	27.6	28.2	28.7	29.3	29.8	30.4	30.9	31.5	32.0
46	28.2	28.7	29.3	29.8	30.4	30.9	31.5	32.0	32.6
48	28.7	29.3	29.8	30.4	30.9	31.5	32.0	32.6	33.1
50	29.3	29.8	30.4	30.9	31.5	32.0	32.6	33.1	33.7
52	29.8	30.4	30.9	31.5	32.0	32.6	33.1	33.7	34.2
54	30.4	30.9	31.5	32.0	32.6	33.1	33.7	34.2	34.8
56	30.9	31.5	32.0	32.6	33.1	33.7	34.2	34.8	35.3
58	31.5	32.0	32.6	33.1	33.7	34.2	34.8	35.3	35.9
60	32.0	32.6	33.1	33.7	34.2	34.8	35.3	35.9	36.4
62	32.6	33.1	33.7	34.2	34.8	35.3	35.9	36.4	37.0
64	33.1	33.7	34.2	34.8	35.3	35.9	36.4	37.0	37.5
66	33.7	34.2	34.8	35.3	35.9	36.4	37.0	37.5	38.1
68	34.2	34.8	35.3	35.9	36.4	37.0	37.5	38.1	38.6
70	34.8	35.3	35.9	36.4	37.0	37.5	38.1	38.6	39.2
72	35.3	35.9	36.4	37.0	37.5	38.1	38.6	39.2	39.7

Weights of Galvanized Steel Rectangular 90 Degree Elbow (pounds per piece)									
Size (inches)	4	6	8	10	12	14	16	18	20
4	1.05	1.45	1.95	2.60	3.30	5.30	6.50	7.80	9.25
6	1.45	1.95	2.60	3.30	3.70	5.80	7.75	8.75	9.75
8	1.95	2.60	3.30	3.70	4.25	6.50	8.50	9.75	10.4
10	2.60	3.30	3.70	4.25	4.85	7.30	9.50	10.7	11.2
12	3.30	3.70	4.25	4.85	5.60	8.20	10.5	11.7	12.2
14	5.30	5.80	6.50	7.30	8.20	9.25	11.5	12.6	13.2
16	6.50	7.50	8.50	9.50	10.5	11.5	12.5	13.4	14.4
18	7.80	8.75	9.75	10.7	11.7	12.6	13.4	14.5	15.7
20	9.25	9.75	10.4	11.2	12.2	13.8	14.4	15.7	17.2
22	10.8	11.8	12.8	13.8	14.8	25.8	16.8	17.8	18.8
24	12.6	13.6	14.6	15.6	16.6	17.6	18.6	19.6	20.5
26	14.4	15.4	16.4	17.4	18.4	19.4	20.4	21.4	22.3
28	16.4	17.4	18.4	19.4	20.4	21.4	22.4	23.4	24.3
30	18.5	19.5	20.5	21.5	22.5	23.5	24.5	25.5	26.4
32	24.3	25.7	27.0	28.4	29.7	31.1	32.4	33.7	35.0
34	28.2	29.4	30.6	31.8	33.0	34.3	35.5	36.7	37.9
36	31.3	32.5	33.7	34.9	36.1	37.3	38.5	39.7	40.9
38	34.5	35.7	36.9	38.1	39.3	40.5	41.7	42.9	44.1
40	37.9	39.1	40.3	41.5	42.7	43.9	45.1	46.3	47.5
42	41.4	42.6	43.8	45.0	46.2	47.4	48.6	49.8	51.0
44	45.1	46.4	47.6	48.8	50.0	51.2	52.4	53.6	54.8
46	49.0	50.2	51.4	52.6	53.8	55.0	56.2	57.4	58.6
48	53.0	54.2	55.4	56.6	57.8	59.0	60.2	61.4	62.6
50	57.2	58.5	59.7	60.9	62.1	63.3	64.5	65.7	66.9

Note: Elbow weights do not include turning vanes.

Galvanized Steel Rectangular 90-Degree Elbows

Weights of Galvanized Steel Rectangular 90 Degree Elbows (pounds per piece)									
Size (inches)	22	24	26	28	30	32	34	36	38
22	19.6	21.9	24.1	26.3	28.5	37.0	40.8	44.5	48.3
24	21.9	24.2	26.4	28.6	30.9	39.5	43.6	47.2	50.8
26	24.1	26.4	28.7	30.9	33.2	42.0	46.4	49.8	53.3
28	26.3	28.6	30.9	33.3	35.6	44.5	49.2	52.5	55.8
30	28.5	30.9	33.2	35.6	37.9	47.0	52.0	55.1	58.3
32	37.0	39.5	42.0	44.5	47.0	49.5	52.0	57.8	60.8
34	40.8	43.6	46.4	49.2	52.0	54.8	57.6	60.5	63.3
36	44.5	47.2	49.8	52.5	55.1	57.8	60.5	63.1	65.8
38	48.3	50.8	53.3	55.8	58.3	60.8	63.3	65.8	68.3
40	52.1	54.6	57.1	59.6	62.1	64.6	67.1	69.6	72.1
42	55.9	58.4	60.9	63.4	65.9	68.4	70.9	73.4	75.9
44	59.6	62.1	64.6	67.1	69.6	72.1	74.6	77.1	79.6
46	63.4	65.9	68.4	70.9	73.4	75.9	78.4	80.9	83.4
48	67.2	69.7	72.2	74.7	77.2	79.7	82.2	84.7	87.2
50	70.9	73.4	75.9	78.4	80.9	83.4	85.9	88.4	90.9
52	74.8	77.3	79.8	82.3	84.8	87.3	89.8	92.3	94.7
54	78.5	81.0	83.5	86.0	88.5	91.0	93.5	96.0	98.5
56	97.2	100	103	106	109	112	115	118	121
58	103	106	109	112	115	118	121	124	127
60	110	113	116	119	122	125	128	131	134
62	116	119	122	125	128	131	134	137	140
64	122	125	129	132	135	138	141	144	147
66	128	131	135	138	141	144	148	151	154
68	135	138	142	145	148	151	154	157	160
70	141	144	148	151	154	157	160	163	166
72	147	150	154	157	160	163	166	169	172

Note: Elbow weights do not include turning vanes.

Weights of Galvanized Steel Rectangular 90 Degree Elbows (pounds per piece)									
Size (inches)	40	42	44	46	48	50	52	54	56
40	74.6	78.4	82.1	85.9	89.7	93.4	97.2	102	125
42	78.4	82.3	86.2	90.1	94.0	97.9	102	106	130
44	82.1	86.2	90.3	94.4	98.5	102	106	110	135
46	85.9	90.1	94.4	98.7	102	106	110	113	140
48	89.7	94.0	98.5	102	106	110	113	117	144
50	93.4	97.9	102	106	110	113	117	122	149
52	97.2	102	106	110	113	117	122	126	154
54	102	106	110	113	117	122	126	131	159
56	126	130	134	139	145	150	155	160	165
58	132	137	142	147	151	156	161	166	170
60	137	142	147	152	157	162	167	172	176
62	143	148	153	158	163	168	173	178	182
64	150	155	160	165	169	174	179	184	188
66	156	161	166	171	176	181	186	191	195
68	163	168	173	178	182	187	192	197	201
70	170	175	180	185	189	194	199	204	208
72	176	181	185	190	194	199	204	210	215

Note: Elbow weights do not include turning vanes.

Galvanized Steel Rectangular 90-Degree Elbows

Weights of Galvanized Steel Rectangular 90 Degree Elbows (pounds per piece)								
Size (inches)	58	60	62	64	66	68	70	72
58	176	182	187	194	200	207	214	221
60	182	187	194	200	207	214	221	227
62	187	194	200	207	214	221	227	233
64	194	200	207	214	221	227	233	239
66	200	207	214	221	227	233	239	246
68	207	214	221	227	233	239	246	252
70	214	221	227	233	239	246	252	259
72	221	227	233	239	246	252	259	266

Note: Elbow weights do not include turning vanes.

Weights of Galvanized Steel Rectangular Drops and Tap-In Tees (pounds per piece)								
Size (inches)	6	8	10	12	16	20	24	28	32
6	1.81	2.11	2.42	2.72	4.24	5.01	5.78	6.55	8.90
8	2.11	2.42	2.72	3.02	4.62	5.40	6.17	6.94	9.37
10	2.42	2.72	3.02	3.32	5.01	5.78	6.55	7.32	9.84
12	2.72	3.02	3.32	3.62	5.40	6.17	6.94	7.71	10.3
16	4.24	4.62	5.01	5.40	6.17	6.94	7.71	8.48	11.3
20	5.01	5.40	5.78	6.17	6.94	7.71	8.48	9.25	12.2
24	5.78	6.17	6.55	6.94	7.71	8.48	9.25	10.0	13.1
28	6.55	6.94	7.32	7.71	8.48	9.25	10.0	10.8	14.1
32	8.90	9.37	9.84	10.3	11.3	12.2	13.1	14.1	15.0
36	9.84	10.3	10.8	11.3	12.2	13.1	14.1	15.0	15.9
40	10.8	11.3	11.7	12.2	13.1	14.1	15.0	15.9	16.9
44	11.7	12.2	12.7	13.1	14.1	15.0	15.9	16.9	17.8
48	12.7	13.1	13.6	14.1	15.0	15.9	16.9	17.8	18.7
52	13.6	14.1	14.5	15.0	15.9	16.9	17.8	18.7	19.7
56	16.0	17.6	18.2	18.8	19.6	21.0	22.1	23.2	24.3
60	18.7	19.0	19.3	19.9	21.0	22.1	23.2	24.3	25.4

Note: Weights of drops and tap-in tees do not include splitter dampers.

Galvanized Steel Spiral Duct

Costs for Steel Spiral Duct. Costs per linear foot of duct, based on quantities less than 1,000 pounds. Add 50% to the material cost for lined duct. These costs include delivery, typical bracing, cleats, scrap, hangers, end closures, sealants and miscellaneous hardware.

Description	Craft@Hrs	Unit	Material $	Labor $	Equipment $	Total $

Galvanized steel spiral duct 26 gauge

Description	Craft@Hrs	Unit	Material $	Labor $	Equipment $	Total $
3"	S2@.030	LF	2.61	1.08	—	3.69
4"	S2@.041	LF	3.53	1.48	—	5.01
5"	S2@.051	LF	4.44	1.84	—	6.28
6"	S2@.061	LF	5.29	2.20	—	7.49
7"	S2@.072	LF	6.15	2.59	—	8.74
8"	S2@.082	LF	7.05	2.96	—	10.01
9"	S2@.092	LF	7.95	3.32	—	11.27
10"	S2@.103	LF	8.81	3.71	—	12.52
12"	S2@.123	LF	10.50	4.43	—	14.93
14" - 16"	S2@.174	LF	13.20	6.27	—	19.47
18" - 20"	S2@.195	LF	16.80	7.03	—	23.83
22" - 24"	S2@.236	LF	20.40	8.51	—	28.91
26" - 28"	S2@.277	LF	24.00	9.98	—	33.98
30" - 32"	S2@.318	LF	27.60	11.50	—	39.10

Galvanized steel spiral duct 24 gauge

Description	Craft@Hrs	Unit	Material $	Labor $	Equipment $	Total $
3"	S2@.041	LF	3.52	1.48	—	5.00
4"	S2@.054	LF	4.61	1.95	—	6.56
5"	S2@.067	LF	5.79	2.41	—	8.20
6"	S2@.081	LF	6.97	2.92	—	9.89
7"	S2@.095	LF	8.10	3.42	—	11.52
8"	S2@.108	LF	9.32	3.89	—	13.21
9"	S2@.122	LF	10.50	4.40	—	14.90
10"	S2@.135	LF	11.60	4.87	—	16.47
12"	S2@.163	LF	13.90	5.87	—	19.77
14" - 16"	S2@.203	LF	17.40	7.32	—	24.72
18" - 20"	S2@.257	LF	22.10	9.26	—	31.36
22" - 24"	S2@.312	LF	26.90	11.20	—	38.10
26" - 28"	S2@.366	LF	31.60	13.20	—	44.80
30" - 32"	S2@.419	LF	36.10	15.10	—	51.20
34" - 36"	S2@.474	LF	41.10	17.10	—	58.20

Description	Craft@Hrs	Unit	Material $	Labor $	Equipment $	Total $

Galvanized steel spiral duct 22 gauge

Description	Craft@Hrs	Unit	Material $	Labor $	Equipment $	Total $
3"	S2@.049	LF	4.32	1.77	—	6.09
4"	S2@.065	LF	5.71	2.34	—	8.05
5"	S2@.082	LF	7.19	2.96	—	10.15
6"	S2@.099	LF	8.54	3.57	—	12.11
7"	S2@.115	LF	10.00	4.14	—	14.14
8"	S2@.132	LF	11.30	4.76	—	16.06
9"	S2@.148	LF	12.80	5.33	—	18.13
10"	S2@.165	LF	14.50	5.95	—	20.45
12"	S2@.198	LF	17.30	7.14	—	24.44
14" - 16"	S2@.247	LF	21.40	8.90	—	30.30
18" - 20"	S2@.313	LF	27.40	11.30	—	38.70
22" - 24"	S2@.380	LF	32.90	13.70	—	46.60
26" - 28"	S2@.444	LF	38.90	16.00	—	54.90
30" - 32"	S2@.491	LF	44.80	17.70	—	62.50
34" - 36"	S2@.576	LF	50.70	20.80	—	71.50

Galvanized steel spiral duct 20 gauge

Description	Craft@Hrs	Unit	Material $	Labor $	Equipment $	Total $
3"	S2@.058	LF	5.10	2.09	—	7.19
4"	S2@.078	LF	6.78	2.81	—	9.59
5"	S2@.097	LF	8.42	3.50	—	11.92
6"	S2@.116	LF	10.10	4.18	—	14.28
7"	S2@.136	LF	11.70	4.90	—	16.60
8"	S2@.154	LF	13.40	5.55	—	18.95
9"	S2@.175	LF	15.10	6.31	—	21.41
10"	S2@.194	LF	16.90	6.99	—	23.89
12"	S2@.233	LF	20.50	8.40	—	28.90
14" - 16"	S2@.291	LF	25.00	10.50	—	35.50
18" - 20"	S2@.369	LF	32.10	13.30	—	45.40
22" - 24"	S2@.446	LF	38.90	16.10	—	55.00
26" - 28"	S2@.525	LF	45.80	18.90	—	64.70
30" - 32"	S2@.604	LF	52.30	21.80	—	74.10
34" - 36"	S2@.681	LF	59.10	24.50	—	83.60

Galvanized steel spiral duct 18 gauge

Description	Craft@Hrs	Unit	Material $	Labor $	Equipment $	Total $
6"	S2@.155	LF	13.40	5.59	—	18.99
7"	S2@.181	LF	15.70	6.52	—	22.22
8"	S2@.207	LF	17.80	7.46	—	25.26
9"	S2@.233	LF	20.20	8.40	—	28.60
10"	S2@.259	LF	22.30	9.33	—	31.63
12"	S2@.310	LF	27.20	11.20	—	38.40
14" - 16"	S2@.387	LF	33.80	13.90	—	47.70
18" - 20"	S2@.491	LF	43.10	17.70	—	60.80
22" - 24"	S2@.608	LF	52.70	21.90	—	74.60
26" - 28"	S2@.696	LF	60.60	25.10	—	85.70
30" - 32"	S2@.803	LF	69.60	28.90	—	98.50
34" - 36"	S2@.905	LF	78.70	32.60	—	111.30

Galvanized Steel Spiral Duct Fittings

Costs for Steel Spiral Duct Fittings. Costs per fitting, based on quantities less than 1,000 pounds. For quantities over 1,000 pounds, deduct 15% from material costs. For lined duct add 50%. These costs include delivery, typical bracing, cleats, scrap, hangers, end closures, sealants and miscellaneous hardware.

Description	Craft@Hrs	Unit	Material $	Labor $	Equipment $	Total $
Galvanized steel spiral duct 90-degree elbows						
3"	S2@.052	Ea	3.26	1.87	—	5.13
4"	S2@.088	Ea	5.51	3.17	—	8.68
5"	S2@.132	Ea	8.25	4.76	—	13.01
6"	S2@.172	Ea	10.80	6.20	—	17.00
7"	S2@.232	Ea	14.70	8.36	—	23.06
8"	S2@.292	Ea	18.50	10.50	—	29.00
9"	S2@.352	Ea	22.20	12.70	—	34.90
10"	S2@.470	Ea	29.50	16.90	—	46.40
12"	S2@.653	Ea	40.90	23.50	—	64.40
14" - 16"	S2@1.01	Ea	62.90	36.40	—	99.30
18" - 20"	S2@1.52	Ea	95.10	54.80	—	149.90
22" - 24"	S2@2.11	Ea	132.00	76.00	—	208.00
26" - 28"	S2@2.91	Ea	185.00	105.00	—	290.00
30" - 32"	S2@3.73	Ea	235.00	134.00	—	369.00
34" - 36"	S2@5.47	Ea	343.00	197.00	—	540.00
Galvanized steel spiral duct 45-degree elbows						
3"	S2@.040	Ea	2.50	1.44	—	3.94
4"	S2@.052	Ea	3.26	1.87	—	5.13
5"	S2@.076	Ea	4.75	2.74	—	7.49
6"	S2@.100	Ea	6.29	3.60	—	9.89
7"	S2@.132	Ea	8.25	4.76	—	13.01
8"	S2@.172	Ea	10.80	6.20	—	17.00
9"	S2@.212	Ea	13.20	7.64	—	20.84
10"	S2@.300	Ea	18.70	10.80	—	29.50
12"	S2@.400	Ea	25.00	14.40	—	39.40
14" - 16"	S2@.576	Ea	35.80	20.80	—	56.60
18" - 20"	S2@.850	Ea	53.40	30.60	—	84.00
22" - 24"	S2@1.19	Ea	74.40	42.90	—	117.30
26" - 28"	S2@1.61	Ea	101.00	58.00	—	159.00
30" - 32"	S2@2.06	Ea	130.00	74.20	—	204.20
34" - 36"	S2@3.01	Ea	188.00	108.00	—	296.00

Description	Craft@Hrs	Unit	Material $	Labor $	Equipment $	Total $

Galvanized steel spiral duct coupling

Description	Craft@Hrs	Unit	Material $	Labor $	Equipment $	Total $
3"	S2@.020	Ea	2.49	.72	—	3.21
4"	S2@.024	Ea	3.06	.86	—	3.92
5"	S2@.028	Ea	3.48	1.01	—	4.49
6"	S2@.036	Ea	4.49	1.30	—	5.79
7"	S2@.040	Ea	5.02	1.44	—	6.46
8"	S2@.048	Ea	6.03	1.73	—	7.76
9"	S2@.104	Ea	13.00	3.75	—	16.75
10"	S2@.116	Ea	14.70	4.18	—	18.88
12"	S2@.140	Ea	17.60	5.05	—	22.65
14" - 16"	S2@.174	Ea	22.00	6.27	—	28.27
18" - 20"	S2@.220	Ea	27.30	7.93	—	35.23
22" - 24"	S2@.266	Ea	33.40	9.59	—	42.99
26" - 28"	S2@.312	Ea	39.00	11.20	—	50.20
30" - 32"	S2@.357	Ea	44.90	12.90	—	57.80
34" - 36"	S2@.403	Ea	50.60	14.50	—	65.10

Galvanized steel spiral duct with reducer

Description	Craft@Hrs	Unit	Material $	Labor $	Equipment $	Total $
3"	S2@.040	Ea	5.02	1.44	—	6.46
4"	S2@.048	Ea	6.03	1.73	—	7.76
5"	S2@.056	Ea	7.03	2.02	—	9.05
6"	S2@.072	Ea	9.07	2.59	—	11.66
7"	S2@.080	Ea	10.00	2.88	—	12.88
8"	S2@.096	Ea	11.90	3.46	—	15.36
9"	S2@.132	Ea	16.50	4.76	—	21.26
10"	S2@.176	Ea	22.00	6.34	—	28.34
12"	S2@.212	Ea	26.70	7.64	—	34.34
14" - 16"	S2@.262	Ea	32.90	9.44	—	42.34
18" - 20"	S2@.330	Ea	41.50	11.90	—	53.40
22" - 24"	S2@.398	Ea	49.80	14.30	—	64.10
26" - 28"	S2@.465	Ea	58.30	16.80	—	75.10
30" - 32"	S2@.536	Ea	66.80	19.30	—	86.10
34" - 36"	S2@.606	Ea	75.70	21.80	—	97.50

Galvanized Steel Spiral Tees

Costs for Steel Spiral Tees with Reducing Branch. Costs per tee, based on quantities less than 1,000 pounds. Deduct 15% from the material cost for quantities over 1,000 pounds. Add 50% to the material cost for lined duct. These costs include delivery, typical bracing, cleats, scrap, hangers, end closures, sealants and miscellaneous hardware.

Description	Craft@Hrs	Unit	Material $	Labor $	Equipment $	Total $

Galvanized steel spiral duct tees with 3" reducing branch

Description	Craft@Hrs	Unit	Material $	Labor $	Equipment $	Total $
4"	S2@.109	Ea	11.20	3.93	—	15.13
5"	S2@.112	Ea	13.10	4.04	—	17.14
6"	S2@.128	Ea	15.10	4.61	—	19.71
7"	S2@.148	Ea	17.00	5.33	—	22.33
8"	S2@.164	Ea	19.10	5.91	—	25.01
9"	S2@.180	Ea	20.50	6.49	—	26.99
10"	S2@.256	Ea	32.30	9.23	—	41.53
12"	S2@.308	Ea	38.20	11.10	—	49.30
14" - 16"	S2@.382	Ea	48.40	13.80	—	62.20
18" - 20"	S2@.485	Ea	60.80	17.50	—	78.30
22" - 24"	S2@.585	Ea	73.30	21.10	—	94.40
26" - 28"	S2@.683	Ea	85.40	24.60	—	110.00
30" - 32"	S2@.786	Ea	98.80	28.30	—	127.10
34" - 36"	S2@1.02	Ea	128.00	36.80	—	164.80

Galvanized steel spiral duct tees with 4" reducing branch

Description	Craft@Hrs	Unit	Material $	Labor $	Equipment $	Total $
5"	S2@.120	Ea	14.60	4.32	—	18.92
6"	S2@.140	Ea	17.00	5.05	—	22.05
7"	S2@.156	Ea	18.50	5.62	—	24.12
8"	S2@.176	Ea	21.00	6.34	—	27.34
9"	S2@.192	Ea	22.30	6.92	—	29.22
10"	S2@.268	Ea	34.20	9.66	—	43.86
12"	S2@.324	Ea	41.50	11.70	—	53.20
14" - 16"	S2@.403	Ea	51.50	14.50	—	66.00
18" - 20"	S2@.508	Ea	65.10	18.30	—	83.40
22" - 24"	S2@.619	Ea	78.90	22.30	—	101.20
26" - 28"	S2@.724	Ea	92.40	26.10	—	118.50
30" - 32"	S2@.830	Ea	106.00	29.90	—	135.90
34" - 36"	S2@1.08	Ea	137.00	38.90	—	175.90

Galvanized steel spiral duct tees with 5" reducing branch

Description	Craft@Hrs	Unit	Material $	Labor $	Equipment $	Total $
6"	S2@.148	Ea	18.50	5.33	—	23.83
7"	S2@.164	Ea	20.50	5.91	—	26.41
8"	S2@.188	Ea	22.90	6.78	—	29.68
9"	S2@.204	Ea	24.90	7.35	—	32.25
10"	S2@.284	Ea	37.00	10.20	—	47.20
12"	S2@.340	Ea	44.10	12.30	—	56.40
14" - 16"	S2@.428	Ea	55.60	15.40	—	71.00
18" - 20"	S2@.538	Ea	70.50	19.40	—	89.90
22" - 24"	S2@.651	Ea	84.20	23.50	—	107.70
26" - 28"	S2@.764	Ea	98.90	27.50	—	126.40
30" - 32"	S2@.873	Ea	114.00	31.50	—	145.50
34" - 36"	S2@1.15	Ea	149.00	41.40	—	190.40

Description	Craft@Hrs	Unit	Material $	Labor $	Equipment $	Total $

Galvanized steel spiral duct tees with 6" reducing branch

Description	Craft@Hrs	Unit	Material $	Labor $	Equipment $	Total $
7"	S2@.180	Ea	22.90	6.49	—	29.39
8"	S2@.200	Ea	24.90	7.21	—	32.11
9"	S2@.220	Ea	27.10	7.93	—	35.03
10"	S2@.300	Ea	39.60	10.80	—	50.40
12"	S2@.356	Ea	47.30	12.80	—	60.10
14" - 16"	S2@.449	Ea	59.20	16.20	—	75.40
18" - 20"	S2@.566	Ea	74.80	20.40	—	95.20
22" - 24"	S2@.683	Ea	90.00	24.60	—	114.60
26" - 28"	S2@.807	Ea	105.00	29.10	—	134.10
30" - 32"	S2@.924	Ea	121.00	33.30	—	154.30
34" - 36"	S2@1.20	Ea	158.00	43.20	—	201.20

Galvanized steel spiral duct tees with 7" reducing branch

Description	Craft@Hrs	Unit	Material $	Labor $	Equipment $	Total $
8"	S2@.164	Ea	27.10	5.91	—	33.01
9"	S2@.180	Ea	29.30	6.49	—	35.79
10"	S2@.252	Ea	41.50	9.08	—	50.58
12"	S2@.304	Ea	50.20	11.00	—	61.20
14" - 16"	S2@.381	Ea	63.00	13.70	—	76.70
18" - 20"	S2@.482	Ea	78.90	17.40	—	96.30
22" - 24"	S2@.579	Ea	95.10	20.90	—	116.00
26" - 28"	S2@.683	Ea	113.00	24.60	—	137.60
30" - 32"	S2@.784	Ea	128.00	28.30	—	156.30
34" - 36"	S2@1.02	Ea	168.00	36.80	—	204.80

Galvanized steel spiral duct tees with 8" reducing branch

Description	Craft@Hrs	Unit	Material $	Labor $	Equipment $	Total $
9"	S2@.228	Ea	37.70	8.22	—	45.92
10"	S2@.268	Ea	44.10	9.66	—	53.76
12"	S2@.324	Ea	53.10	11.70	—	64.80
14" - 16"	S2@.401	Ea	66.20	14.50	—	80.70
18" - 20"	S2@.508	Ea	83.80	18.30	—	102.10
22" - 24"	S2@.617	Ea	101.00	22.20	—	123.20
26" - 28"	S2@.722	Ea	118.00	26.00	—	144.00
30" - 32"	S2@.830	Ea	135.00	29.90	—	164.90
34" - 36"	S2@1.08	Ea	177.00	38.90	—	215.90

Galvanized steel spiral duct tees with 9" reducing branch

Description	Craft@Hrs	Unit	Material $	Labor $	Equipment $	Total $
10"	S2@.284	Ea	47.00	10.20	—	57.20
12"	S2@.340	Ea	55.90	12.30	—	68.20
14" - 16"	S2@.426	Ea	70.50	15.40	—	85.90
18" - 20"	S2@.540	Ea	88.40	19.50	—	107.90
22" - 24"	S2@.649	Ea	106.00	23.40	—	129.40
26" - 28"	S2@.760	Ea	124.00	27.40	—	151.40
30" - 32"	S2@.875	Ea	146.00	31.50	—	177.50
34" - 36"	S2@1.15	Ea	187.00	41.40	—	228.40

Galvanized Steel Spiral Tees

Description	Craft@Hrs	Unit	Material $	Labor $	Equipment $	Total $
Galvanized steel spiral duct tees with 10" reducing branch						
12"	S2@.360	Ea	59.20	13.00	—	72.20
14" - 16"	S2@.447	Ea	73.30	16.10	—	89.40
18" - 20"	S2@.564	Ea	93.10	20.30	—	113.40
22" - 24"	S2@.683	Ea	113.00	24.60	—	137.60
26" - 28"	S2@.805	Ea	132.00	29.00	—	161.00
30" - 32"	S2@.924	Ea	151.00	33.30	—	184.30
34" - 36"	S2@1.20	Ea	199.00	43.20	—	242.20
Galvanized steel spiral duct tees with 12" reducing branch						
14" - 16"	S2@.572	Ea	80.90	20.60	—	101.50
18" - 20"	S2@.681	Ea	102.00	24.50	—	126.50
22" - 24"	S2@.820	Ea	122.00	29.60	—	151.60
26" - 28"	S2@.965	Ea	146.00	34.80	—	180.80
30" - 32"	S2@1.11	Ea	167.00	40.00	—	207.00
34" - 36"	S2@1.44	Ea	220.00	51.90	—	271.90
Galvanized steel spiral duct tees with 14" reducing branch						
14" - 16"	S2@.572	Ea	93.80	20.60	—	114.40
18" - 20"	S2@.681	Ea	112.00	24.50	—	136.50
22" - 24"	S2@.820	Ea	134.00	29.60	—	163.60
26" - 28"	S2@.965	Ea	158.00	34.80	—	192.80
30" - 32"	S2@1.11	Ea	183.00	40.00	—	223.00
34" - 36"	S2@1.44	Ea	240.00	51.90	—	291.90
Galvanized steel spiral duct tees with 16" reducing branch						
18" - 20"	S2@.739	Ea	121.00	26.60	—	147.60
22" - 24"	S2@.888	Ea	148.00	32.00	—	180.00
26" - 28"	S2@1.04	Ea	171.00	37.50	—	208.50
30" - 32"	S2@1.20	Ea	197.00	43.20	—	240.20
34" - 36"	S2@1.56	Ea	259.00	56.20	—	315.20
Galvanized steel spiral duct tees with 18" reducing branch						
18" - 20"	S2@.837	Ea	136.00	30.20	—	166.20
22" - 24"	S2@.954	Ea	157.00	34.40	—	191.40
26" - 28"	S2@1.12	Ea	185.00	40.40	—	225.40
30" - 32"	S2@1.29	Ea	212.00	46.50	—	258.50
34" - 36"	S2@1.68	Ea	280.00	60.50	—	340.50
Galvanized steel spiral duct tees with 20" reducing branch						
22" - 24"	S2@1.02	Ea	169.00	36.80	—	205.80
26" - 28"	S2@1.20	Ea	199.00	43.20	—	242.20
30" - 32"	S2@1.38	Ea	226.00	49.70	—	275.70
34" - 36"	S2@1.81	Ea	296.00	65.20	—	361.20
Galvanized steel spiral duct tees with 22" reducing branch						
22" - 24"	S2@1.57	Ea	187.00	56.60	—	243.60
26" - 28"	S2@1.32	Ea	220.00	47.60	—	267.60
30" - 32"	S2@1.48	Ea	244.00	53.30	—	297.30
34" - 36"	S2@1.92	Ea	320.00	69.20	—	389.20

Costs for Steel Spiral Tees with Reducing Run and Branch. Costs per tee, based on quantities less than 1,000 pounds. Deduct 15% from the material cost for quantities over 1,000 pounds. Add 50% to the material cost for lined duct. These costs include delivery, typical bracing, cleats, scrap, hangers, end closures, sealants and miscellaneous hardware.

Description	Craft@Hrs	Unit	Material $	Labor $	Equipment $	Total $

Galvanized steel spiral duct tees with 3" run and branch

Description	Craft@Hrs	Unit	Material $	Labor $	Equipment $	Total $
4"	S2@.109	Ea	18.20	3.93	—	22.13
5"	S2@.112	Ea	18.50	4.04	—	22.54
6"	S2@.128	Ea	21.00	4.61	—	25.61
7"	S2@.148	Ea	24.40	5.33	—	29.73
8"	S2@.164	Ea	27.10	5.91	—	33.01
9"	S2@.180	Ea	29.30	6.49	—	35.79
10"	S2@.256	Ea	42.30	9.23	—	51.53
12"	S2@.308	Ea	50.40	11.10	—	61.50
14" - 16"	S2@.382	Ea	63.00	13.80	—	76.80
18" - 20"	S2@.485	Ea	79.40	17.50	—	96.90
22" - 24"	S2@.585	Ea	96.40	21.10	—	117.50
26" - 28"	S2@.683	Ea	113.00	24.60	—	137.60
30" - 32"	S2@.786	Ea	130.00	28.30	—	158.30
34" - 36"	S2@1.02	Ea	168.00	36.80	—	204.80

Galvanized steel spiral duct tees with 4" run and branch

Description	Craft@Hrs	Unit	Material $	Labor $	Equipment $	Total $
5"	S2@.120	Ea	19.70	4.32	—	24.02
6"	S2@.140	Ea	22.90	5.05	—	27.95
7"	S2@.156	Ea	25.90	5.62	—	31.52
8"	S2@.176	Ea	28.70	6.34	—	35.04
9"	S2@.192	Ea	31.40	6.92	—	38.32
10"	S2@.268	Ea	44.10	9.66	—	53.76
12"	S2@.324	Ea	53.10	11.70	—	64.80
14" - 16"	S2@.403	Ea	66.50	14.50	—	81.00
18" - 20"	S2@.508	Ea	83.80	18.30	—	102.10
22" - 24"	S2@.619	Ea	101.00	22.30	—	123.30
26" - 28"	S2@.724	Ea	118.00	26.10	—	144.10
30" - 32"	S2@.830	Ea	135.00	29.90	—	164.90
34" - 36"	S2@1.08	Ea	178.00	38.90	—	216.90

Galvanized steel spiral duct tees with 5" run and branch

Description	Craft@Hrs	Unit	Material $	Labor $	Equipment $	Total $
6"	S2@.148	Ea	24.40	5.33	—	29.73
7"	S2@.164	Ea	27.10	5.91	—	33.01
8"	S2@.188	Ea	31.00	6.78	—	37.78
9"	S2@.204	Ea	33.60	7.35	—	40.95
10"	S2@.284	Ea	47.00	10.20	—	57.20
12"	S2@.340	Ea	55.90	12.30	—	68.20
14" - 16"	S2@.428	Ea	70.50	15.40	—	85.90
18" - 20"	S2@.538	Ea	88.20	19.40	—	107.60
22" - 24"	S2@.651	Ea	107.00	23.50	—	130.50
26" - 28"	S2@.764	Ea	125.00	27.50	—	152.50
30" - 32"	S2@.873	Ea	146.00	31.50	—	177.50
34" - 36"	S2@1.15	Ea	187.00	41.40	—	228.40

Galvanized Steel Spiral Tees

Description	Craft@Hrs	Unit	Material $	Labor $	Equipment $	Total $

Galvanized steel spiral duct tees with 6" run and branch

Description	Craft@Hrs	Unit	Material $	Labor $	Equipment $	Total $
7"	S2@.180	Ea	29.30	6.49	—	35.79
8"	S2@.200	Ea	32.80	7.21	—	40.01
9"	S2@.220	Ea	35.70	7.93	—	43.63
10"	S2@.300	Ea	49.40	10.80	—	60.20
12"	S2@.356	Ea	58.80	12.80	—	71.60
14" - 16"	S2@.449	Ea	73.70	16.20	—	89.90
18" - 20"	S2@.566	Ea	93.10	20.40	—	113.50
22" - 24"	S2@.683	Ea	113.00	24.60	—	137.60
26" - 28"	S2@.807	Ea	132.00	29.10	—	161.10
30" - 32"	S2@.924	Ea	151.00	33.30	—	184.30
34" - 36"	S2@1.20	Ea	199.00	43.20	—	242.20

Galvanized steel spiral duct tees with 7" run and branch

Description	Craft@Hrs	Unit	Material $	Labor $	Equipment $	Total $
8"	S2@.212	Ea	34.80	7.64	—	42.44
9"	S2@.232	Ea	38.20	8.36	—	46.56
10"	S2@.312	Ea	51.00	11.20	—	62.20
12"	S2@.376	Ea	61.90	13.60	—	75.50
14" - 16"	S2@.470	Ea	77.50	16.90	—	94.40
18" - 20"	S2@.636	Ea	104.00	22.90	—	126.90
22" - 24"	S2@.717	Ea	118.00	25.80	—	143.80
26" - 28"	S2@.845	Ea	137.00	30.50	—	167.50
30" - 32"	S2@.969	Ea	162.00	34.90	—	196.90
34" - 36"	S2@1.27	Ea	209.00	45.80	—	254.80

Galvanized steel spiral duct tees with 8" run and branch

Description	Craft@Hrs	Unit	Material $	Labor $	Equipment $	Total $
9"	S2@.248	Ea	40.90	8.94	—	49.84
10"	S2@.328	Ea	54.30	11.80	—	66.10
12"	S2@.396	Ea	65.00	14.30	—	79.30
14" - 16"	S2@.493	Ea	80.50	17.80	—	98.30
18" - 20"	S2@.623	Ea	102.00	22.50	—	124.50
22" - 24"	S2@.751	Ea	122.00	27.10	—	149.10
26" - 28"	S2@.882	Ea	147.00	31.80	—	178.80
30" - 32"	S2@1.02	Ea	167.00	36.80	—	203.80
34" - 36"	S2@1.33	Ea	220.00	47.90	—	267.90

Galvanized steel spiral duct tees with 9" run and branch

Description	Craft@Hrs	Unit	Material $	Labor $	Equipment $	Total $
10"	S2@.344	Ea	56.40	12.40	—	68.80
12"	S2@.412	Ea	67.50	14.80	—	82.30
14" - 16"	S2@.515	Ea	84.80	18.60	—	103.40
18" - 20"	S2@.653	Ea	107.00	23.50	—	130.50
22" - 24"	S2@.779	Ea	130.00	28.10	—	158.10
26" - 28"	S2@.924	Ea	151.00	33.30	—	184.30
30" - 32"	S2@1.06	Ea	175.00	38.20	—	213.20
34" - 36"	S2@1.39	Ea	226.00	50.10	—	276.10

Description	Craft@Hrs	Unit	Material $	Labor $	Equipment $	Total $

Galvanized steel spiral duct tees with 10" run and branch

Description	Craft@Hrs	Unit	Material $	Labor $	Equipment $	Total $
12"	S2@.427	Ea	35.30	15.40	—	50.70
14" - 16"	S2@.538	Ea	44.50	19.40	—	63.90
18" - 20"	S2@.681	Ea	111.00	24.50	—	135.50
22" - 24"	S2@.820	Ea	134.00	29.60	—	163.60
26" - 28"	S2@.965	Ea	158.00	34.80	—	192.80
30" - 32"	S2@1.11	Ea	183.00	40.00	—	223.00
34" - 36"	S2@1.44	Ea	238.00	51.90	—	289.90

Galvanized steel spiral duct tees with 12" run and branch

Description	Craft@Hrs	Unit	Material $	Labor $	Equipment $	Total $
14" - 16"	S2@.581	Ea	95.20	20.90	—	116.10
18" - 20"	S2@.739	Ea	121.00	26.60	—	147.60
22" - 24"	S2@.888	Ea	148.00	32.00	—	180.00
26" - 28"	S2@1.04	Ea	171.00	37.50	—	208.50
30" - 32"	S2@1.20	Ea	197.00	43.20	—	240.20
34" - 36"	S2@1.57	Ea	259.00	56.60	—	315.60

Galvanized steel spiral duct tees with 14" run and branch

Description	Craft@Hrs	Unit	Material $	Labor $	Equipment $	Total $
14" - 16"	S2@.666	Ea	109.00	24.00	—	133.00
18" - 20"	S2@.794	Ea	131.00	28.60	—	159.60
22" - 24"	S2@.959	Ea	157.00	34.60	—	191.60
26" - 28"	S2@1.11	Ea	183.00	40.00	—	223.00
30" - 32"	S2@1.29	Ea	212.00	46.50	—	258.50
34" - 36"	S2@1.69	Ea	280.00	60.90	—	340.90

Galvanized steel spiral duct tees with 16" run and branch

Description	Craft@Hrs	Unit	Material $	Labor $	Equipment $	Total $
18" - 20"	S2@.833	Ea	135.00	30.00	—	165.00
22" - 24"	S2@1.03	Ea	169.00	37.10	—	206.10
26" - 28"	S2@1.20	Ea	199.00	43.20	—	242.20
30" - 32"	S2@1.39	Ea	226.00	50.10	—	276.10
34" - 36"	S2@1.81	Ea	296.00	65.20	—	361.20

Galvanized steel spiral duct tees with 18" run and branch

Description	Craft@Hrs	Unit	Material $	Labor $	Equipment $	Total $
18" - 20"	S2@.956	Ea	157.00	34.50	—	191.50
22" - 24"	S2@1.09	Ea	178.00	39.30	—	217.30
26" - 28"	S2@1.29	Ea	211.00	46.50	—	257.50
30" - 32"	S2@1.48	Ea	244.00	53.30	—	297.30
34" - 36"	S2@1.93	Ea	320.00	69.60	—	389.60

Galvanized steel spiral duct tees with 20" run and branch

Description	Craft@Hrs	Unit	Material $	Labor $	Equipment $	Total $
22" - 24"	S2@1.16	Ea	191.00	41.80	—	232.80
26" - 28"	S2@1.37	Ea	225.00	49.40	—	274.40
30" - 32"	S2@1.57	Ea	259.00	56.60	—	315.60
34" - 36"	S2@2.03	Ea	336.00	73.20	—	409.20

Galvanized steel spiral duct tees with 22" run and branch

Description	Craft@Hrs	Unit	Material $	Labor $	Equipment $	Total $
22" - 24"	S2@1.28	Ea	211.00	46.10	—	257.10
26" - 28"	S2@1.45	Ea	240.00	52.30	—	292.30
30" - 32"	S2@1.64	Ea	272.00	59.10	—	331.10
34" - 36"	S2@2.16	Ea	354.00	77.80	—	431.80

Galvanized Steel Spiral Crosses

Costs for Steel Spiral Crosses with Reducing Branches. Costs per cross, based on quantities less than 1,000 pounds. Deduct 15% from the material cost for quantities over 1,000 pounds. Add 50% to the material cost for lined duct. These costs include delivery, typical bracing, cleats, scrap, hangers, end closures, sealants and miscellaneous hardware.

Description	Craft@Hrs	Unit	Material $	Labor $	Equipment $	Total $

Galvanized steel spiral duct crosses with 3" branch

Description	Craft@Hrs	Unit	Material $	Labor $	Equipment $	Total $
4"	S2@.104	Ea	17.00	3.75	—	20.75
5"	S2@.128	Ea	21.00	4.61	—	25.61
6"	S2@.152	Ea	24.90	5.48	—	30.38
7"	S2@.176	Ea	28.70	6.34	—	35.04
8"	S2@.196	Ea	32.30	7.06	—	39.36
9"	S2@.216	Ea	35.30	7.78	—	43.08
10"	S2@.304	Ea	50.20	11.00	—	61.20
12"	S2@.352	Ea	58.30	12.70	—	71.00
14" - 16"	S2@.428	Ea	70.90	15.40	—	86.30
18" - 20"	S2@.525	Ea	86.30	18.90	—	105.20
22" - 24"	S2@.634	Ea	104.00	22.80	—	126.80
26" - 28"	S2@.745	Ea	121.00	26.80	—	147.80
30" - 32"	S2@.856	Ea	139.00	30.90	—	169.90
34" - 36"	S2@1.11	Ea	184.00	40.00	—	224.00

Galvanized steel spiral duct crosses with 4" branch

Description	Craft@Hrs	Unit	Material $	Labor $	Equipment $	Total $
5"	S2@.144	Ea	23.80	5.19	—	28.99
6"	S2@.168	Ea	28.00	6.05	—	34.05
7"	S2@.196	Ea	32.30	7.06	—	39.36
8"	S2@.216	Ea	35.30	7.78	—	43.08
9"	S2@.240	Ea	39.60	8.65	—	48.25
10"	S2@.328	Ea	54.30	11.80	—	66.10
12"	S2@.384	Ea	63.70	13.80	—	77.50
14" - 16"	S2@.461	Ea	75.90	16.60	—	92.50
18" - 20"	S2@.568	Ea	93.60	20.50	—	114.10
22" - 24"	S2@.685	Ea	113.00	24.70	—	137.70
26" - 28"	S2@.805	Ea	132.00	29.00	—	161.00
30" - 32"	S2@.927	Ea	153.00	33.40	—	186.40
34" - 36"	S2@1.20	Ea	199.00	43.20	—	242.20

Galvanized steel spiral duct crosses with 5" branch

Description	Craft@Hrs	Unit	Material $	Labor $	Equipment $	Total $
6"	S2@.184	Ea	30.50	6.63	—	37.13
7"	S2@.216	Ea	35.30	7.78	—	43.08
8"	S2@.240	Ea	39.60	8.65	—	48.25
9"	S2@.264	Ea	43.70	9.51	—	53.21
10"	S2@.356	Ea	58.80	12.80	—	71.60
12"	S2@.412	Ea	67.50	14.80	—	82.30
14" - 16"	S2@.497	Ea	81.20	17.90	—	99.10
18" - 20"	S2@.613	Ea	100.00	22.10	—	122.10
22" - 24"	S2@.741	Ea	121.00	26.70	—	147.70
26" - 28"	S2@.867	Ea	145.00	31.20	—	176.20
30" - 32"	S2@.993	Ea	164.00	35.80	—	199.80
34" - 36"	S2@1.29	Ea	212.00	46.50	—	258.50

Description	Craft@Hrs	Unit	Material $	Labor $	Equipment $	Total $

Galvanized steel spiral duct crosses with 6" branch

Description	Craft@Hrs	Unit	Material $	Labor $	Equipment $	Total $
7"	S2@.232	Ea	38.20	8.36	—	46.56
8"	S2@.260	Ea	43.00	9.37	—	52.37
9"	S2@.288	Ea	47.30	10.40	—	57.70
10"	S2@.380	Ea	62.90	13.70	—	76.60
12"	S2@.440	Ea	72.40	15.90	—	88.30
14" - 16"	S2@.532	Ea	87.40	19.20	—	106.60
18" - 20"	S2@.653	Ea	107.00	23.50	—	130.50
22" - 24"	S2@.790	Ea	131.00	28.50	—	159.50
26" - 28"	S2@.927	Ea	153.00	33.40	—	186.40
30" - 32"	S2@1.06	Ea	175.00	38.20	—	213.20
34" - 36"	S2@1.38	Ea	226.00	49.70	—	275.70

Galvanized steel spiral duct crosses with 7" branch

Description	Craft@Hrs	Unit	Material $	Labor $	Equipment $	Total $
8"	S2@.280	Ea	45.80	10.10	—	55.90
9"	S2@.308	Ea	50.40	11.10	—	61.50
10"	S2@.408	Ea	67.10	14.70	—	81.80
12"	S2@.470	Ea	76.80	16.90	—	93.70
14" - 16"	S2@.566	Ea	93.10	20.40	—	113.50
18" - 20"	S2@.692	Ea	114.00	24.90	—	138.90
22" - 24"	S2@.841	Ea	136.00	30.30	—	166.30
26" - 28"	S2@.986	Ea	163.00	35.50	—	198.50
30" - 32"	S2@1.13	Ea	186.00	40.70	—	226.70
34" - 36"	S2@1.48	Ea	244.00	53.30	—	297.30

Galvanized steel spiral duct crosses with 8" branch

Description	Craft@Hrs	Unit	Material $	Labor $	Equipment $	Total $
9"	S2@.310	Ea	54.60	11.20	—	65.80
10"	S2@.403	Ea	71.60	14.50	—	86.10
12"	S2@.470	Ea	81.70	16.90	—	98.60
14" - 16"	S2@.561	Ea	98.90	20.20	—	119.10
18" - 20"	S2@.690	Ea	121.00	24.90	—	145.90
22" - 24"	S2@.835	Ea	148.00	30.10	—	178.10
26" - 28"	S2@.980	Ea	171.00	35.30	—	206.30
30" - 32"	S2@1.12	Ea	199.00	40.40	—	239.40
34" - 36"	S2@1.47	Ea	259.00	53.00	—	312.00

Galvanized steel spiral duct crosses with 9" branch

Description	Craft@Hrs	Unit	Material $	Labor $	Equipment $	Total $
10"	S2@.427	Ea	74.80	15.40	—	90.20
12"	S2@.495	Ea	86.50	17.80	—	104.30
14" - 16"	S2@.596	Ea	105.00	21.50	—	126.50
18" - 20"	S2@.732	Ea	128.00	26.40	—	154.40
22" - 24"	S2@.882	Ea	156.00	31.80	—	187.80
26" - 28"	S2@1.04	Ea	183.00	37.50	—	220.50
30" - 32"	S2@1.19	Ea	209.00	42.90	—	251.90
34" - 36"	S2@1.55	Ea	272.00	55.90	—	327.90

Galvanized Steel Spiral Crosses

Description	Craft@Hrs	Unit	Material $	Labor $	Equipment $	Total $
Galvanized steel spiral duct crosses with 12" branch						
14" - 16"	S2@.692	Ea	121.00	24.90	—	145.90
18" - 20"	S2@.850	Ea	150.00	30.60	—	180.60
22" - 24"	S2@1.03	Ea	182.00	37.10	—	219.10
26" - 28"	S2@1.18	Ea	209.00	42.50	—	251.50
30" - 32"	S2@1.38	Ea	244.00	49.70	—	293.70
34" - 36"	S2@1.80	Ea	320.00	64.90	—	384.90
Galvanized steel spiral duct crosses with 14" branch						
14" - 16"	S2@1.81	Ea	61.90	65.20	—	127.10
18" - 20"	S2@2.11	Ea	71.90	76.00	—	147.90
22" - 24"	S2@2.55	Ea	87.40	91.90	—	179.30
26" - 28"	S2@2.98	Ea	102.00	107.00	—	209.00
30" - 32"	S2@3.43	Ea	116.00	124.00	—	240.00
34" - 36"	S2@4.47	Ea	153.00	161.00	—	314.00
Galvanized steel spiral duct crosses with 16" branch						
18" - 20"	S2@1.08	Ea	177.00	38.90	—	215.90
22" - 24"	S2@1.30	Ea	214.00	46.90	—	260.90
26" - 28"	S2@1.53	Ea	251.00	55.10	—	306.10
30" - 32"	S2@1.76	Ea	287.00	63.40	—	350.40
34" - 36"	S2@2.31	Ea	380.00	83.30	—	463.30
Galvanized steel spiral duct crosses with 18" branch						
18" - 20"	S2@1.22	Ea	201.00	44.00	—	245.00
22" - 24"	S2@1.40	Ea	232.00	50.50	—	282.50
26" - 28"	S2@1.65	Ea	272.00	59.50	—	331.50
30" - 32"	S2@1.89	Ea	311.00	68.10	—	379.10
34" - 36"	S2@2.47	Ea	409.00	89.00	—	498.00
Galvanized steel spiral duct crosses with 20" branch						
22" - 24"	S2@1.51	Ea	248.00	54.40	—	302.40
26" - 28"	S2@1.77	Ea	291.00	63.80	—	354.80
30" - 32"	S2@2.05	Ea	341.00	73.90	—	414.90
34" - 36"	S2@2.65	Ea	437.00	95.50	—	532.50
Galvanized steel spiral duct crosses with 22" branch						
22" - 24"	S2@1.68	Ea	280.00	60.50	—	340.50
26" - 28"	S2@1.89	Ea	311.00	68.10	—	379.10
30" - 32"	S2@2.17	Ea	354.00	78.20	—	432.20
34" - 36"	S2@2.84	Ea	470.00	102.00	—	572.00

Costs for Steel 26 Gauge Rectangular Duct. Costs per linear foot of duct, based on quantities less than 1,000 pounds of prefabricated duct. Deduct 15% from the material cost for quantities over 1,000 pounds. Add 50% to the material cost for lined duct. These costs include delivery, typical bracing, cleats, scrap, hangers, end closures, sealants and miscellaneous hardware.

Description	Craft@Hrs	Unit	Material $	Labor $	Equipment $	Total $

Galvanized steel 4" rectangular duct 26 gauge

Description	Craft@Hrs	Unit	Material $	Labor $	Equipment $	Total $
4"	S2@.048	LF	4.06	1.73	—	5.79
5"	S2@.054	LF	4.49	1.95	—	6.44
6"	S2@.060	LF	5.02	2.16	—	7.18
7"	S2@.066	LF	5.55	2.38	—	7.93
8"	S2@.072	LF	6.05	2.59	—	8.64
9"	S2@.078	LF	6.50	2.81	—	9.31
10"	S2@.084	LF	7.02	3.03	—	10.05
11"	S2@.091	LF	7.53	3.28	—	10.81
12"	S2@.097	LF	8.04	3.50	—	11.54

Galvanized steel 5" rectangular duct 26 gauge

Description	Craft@Hrs	Unit	Material $	Labor $	Equipment $	Total $
5"	S2@.060	LF	5.02	2.16	—	7.18
6"	S2@.066	LF	5.55	2.38	—	7.93
7"	S2@.072	LF	6.05	2.59	—	8.64
8"	S2@.078	LF	6.50	2.81	—	9.31
9"	S2@.084	LF	7.02	3.03	—	10.05
10"	S2@.091	LF	7.53	3.28	—	10.81
11"	S2@.097	LF	8.04	3.50	—	11.54
12"	S2@.103	LF	8.51	3.71	—	12.22

Galvanized steel 6" rectangular duct 26 gauge

Description	Craft@Hrs	Unit	Material $	Labor $	Equipment $	Total $
6"	S2@.072	LF	6.05	2.59	—	8.64
7"	S2@.078	LF	6.50	2.81	—	9.31
8"	S2@.084	LF	7.02	3.03	—	10.05
9"	S2@.091	LF	7.53	3.28	—	10.81
10"	S2@.097	LF	8.04	3.50	—	11.54
11"	S2@.103	LF	8.51	3.71	—	12.22
12"	S2@.109	LF	9.01	3.93	—	12.94

Galvanized steel 7" rectangular duct 26 gauge

Description	Craft@Hrs	Unit	Material $	Labor $	Equipment $	Total $
7"	S2@.084	LF	7.02	3.03	—	10.05
8"	S2@.091	LF	7.53	3.28	—	10.81
9"	S2@.097	LF	8.04	3.50	—	11.54
10"	S2@.103	LF	8.51	3.71	—	12.22
11"	S2@.109	LF	9.01	3.93	—	12.94
12"	S2@.115	LF	9.51	4.14	—	13.65

Galvanized Steel Rectangular Ductwork

Description	Craft@Hrs	Unit	Material $	Labor $	Equipment $	Total $

Galvanized steel 8" rectangular duct 26 gauge

Description	Craft@Hrs	Unit	Material $	Labor $	Equipment $	Total $
8"	S2@.097	LF	8.04	3.50	—	11.54
9"	S2@.103	LF	8.51	3.71	—	12.22
10"	S2@.109	LF	9.01	3.93	—	12.94
11"	S2@.115	LF	9.51	4.14	—	13.65
12"	S2@.121	LF	10.00	4.36	—	14.36

Galvanized steel 9" rectangular duct 26 gauge

Description	Craft@Hrs	Unit	Material $	Labor $	Equipment $	Total $
9"	S2@.109	LF	9.01	3.93	—	12.94
10"	S2@.115	LF	9.51	4.14	—	13.65
11"	S2@.121	LF	10.00	4.36	—	14.36
12"	S2@.127	LF	10.50	4.58	—	15.08

Galvanized steel 10" rectangular duct 26 gauge

Description	Craft@Hrs	Unit	Material $	Labor $	Equipment $	Total $
10"	S2@.129	LF	10.00	4.65	—	14.65
11"	S2@.127	LF	10.50	4.58	—	15.08
12"	S2@.133	LF	11.00	4.79	—	15.79

Galvanized steel 11" rectangular duct 26 gauge

Description	Craft@Hrs	Unit	Material $	Labor $	Equipment $	Total $
11"	S2@.133	LF	11.00	4.79	—	15.79
12"	S2@.139	LF	11.40	5.01	—	16.41

Galvanized steel 12" rectangular duct 26 gauge

Description	Craft@Hrs	Unit	Material $	Labor $	Equipment $	Total $
12"	S2@.145	LF	11.90	5.23	—	17.13

Galvanized Steel Rectangular Ductwork

Costs for Galvanized 24 Gauge Rectangular Duct. Costs per linear foot of duct, based on quantities less than 1,000 pounds of prefabricated duct. Deduct 15% from the material cost for over 1,000 pounds. Add 50% to the material cost for lined duct. These costs include delivery, typical bracing, cleats, scrap, hangers, end closures.

Description	Craft@Hrs	Unit	Material $	Labor $	Equipment $	Total $

Galvanized steel 14" rectangular duct 24 gauge

Description	Craft@Hrs	Unit	Material $	Labor $	Equipment $	Total $
6"	S2@.154	LF	12.80	5.55	—	18.35
8"	S2@.170	LF	14.10	6.13	—	20.23
10"	S2@.184	LF	15.50	6.63	—	22.13
12"	S2@.200	LF	16.80	7.21	—	24.01
14" - 16"	S2@.223	LF	18.60	8.04	—	26.64
18" - 20"	S2@.254	LF	21.20	9.15	—	30.35
22" - 24"	S2@.285	LF	24.00	10.30	—	34.30
26" - 28"	S2@.316	LF	26.70	11.40	—	38.10
30"	S2@.339	LF	28.40	12.20	—	40.60

Galvanized steel 16" rectangular duct 24 gauge

Description	Craft@Hrs	Unit	Material $	Labor $	Equipment $	Total $
6"	S2@.170	LF	14.10	6.13	—	20.23
8"	S2@.184	LF	15.50	6.63	—	22.13
10"	S2@.200	LF	16.80	7.21	—	24.01
12"	S2@.216	LF	18.20	7.78	—	25.98
14" - 16"	S2@.239	LF	20.00	8.61	—	28.61
18" - 20"	S2@.269	LF	22.50	9.69	—	32.19
22" - 24"	S2@.300	LF	24.90	10.80	—	35.70
26" - 28"	S2@.331	LF	28.00	11.90	—	39.90
30"	S2@.354	LF	30.00	12.80	—	42.80

Galvanized steel 18" rectangular duct 24 gauge

Description	Craft@Hrs	Unit	Material $	Labor $	Equipment $	Total $
6"	S2@.184	LF	15.50	6.63	—	22.13
8"	S2@.200	LF	16.80	7.21	—	24.01
10"	S2@.216	LF	18.20	7.78	—	25.98
12"	S2@.231	LF	19.30	8.33	—	27.63
14" - 16"	S2@.254	LF	21.20	9.15	—	30.35
18" - 20"	S2@.285	LF	24.00	10.30	—	34.30
22" - 24"	S2@.316	LF	26.70	11.40	—	38.10
26" - 28"	S2@.346	LF	28.70	12.50	—	41.20
30"	S2@.370	LF	31.10	13.30	—	44.40

Galvanized Steel Rectangular Ductwork

Description	Craft@Hrs	Unit	Material $	Labor $	Equipment $	Total $

Galvanized steel 20" rectangular duct 24 gauge

Description	Craft@Hrs	Unit	Material $	Labor $	Equipment $	Total $
6"	S2@.200	LF	16.80	7.21	—	24.01
8"	S2@.216	LF	18.20	7.78	—	25.98
10"	S2@.231	LF	19.30	8.33	—	27.63
12"	S2@.247	LF	20.70	8.90	—	29.60
14" - 16"	S2@.269	LF	22.50	9.69	—	32.19
18" - 20"	S2@.300	LF	24.90	10.80	—	35.70
22" - 24"	S2@.331	LF	28.00	11.90	—	39.90
26" - 28"	S2@.362	LF	30.50	13.00	—	43.50
30"	S2@.385	LF	32.30	13.90	—	46.20

Galvanized steel 22" rectangular duct 24 gauge

Description	Craft@Hrs	Unit	Material $	Labor $	Equipment $	Total $
6"	S2@.216	LF	18.20	7.78	—	25.98
8"	S2@.231	LF	19.30	8.33	—	27.63
10"	S2@.247	LF	20.70	8.90	—	29.60
12"	S2@.262	LF	22.00	9.44	—	31.44
14" - 16"	S2@.285	LF	24.00	10.30	—	34.30
18" - 20"	S2@.316	LF	26.70	11.40	—	38.10
22" - 24"	S2@.346	LF	28.70	12.50	—	41.20
26" - 28"	S2@.377	LF	32.00	13.60	—	45.60
30"	S2@.400	LF	33.60	14.40	—	48.00

Galvanized steel 24" rectangular duct 24 gauge

Description	Craft@Hrs	Unit	Material $	Labor $	Equipment $	Total $
6"	S2@.231	LF	19.30	8.33	—	27.63
8"	S2@.247	LF	20.70	8.90	—	29.60
10"	S2@.262	LF	22.00	9.44	—	31.44
12"	S2@.278	LF	23.20	10.00	—	33.20
14" - 16"	S2@.300	LF	24.90	10.80	—	35.70
18" - 20"	S2@.331	LF	28.00	11.90	—	39.90
22" - 24"	S2@.362	LF	30.50	13.00	—	43.50
26" - 28"	S2@.392	LF	32.80	14.10	—	46.90
30"	S2@.415	LF	34.80	15.00	—	49.80

Galvanized steel 26" rectangular duct 24 gauge

Description	Craft@Hrs	Unit	Material $	Labor $	Equipment $	Total $
6"	S2@.247	LF	20.70	8.90	—	29.60
8"	S2@.262	LF	22.00	9.44	—	31.44
10"	S2@.278	LF	23.20	10.00	—	33.20
12"	S2@.292	LF	24.50	10.50	—	35.00
14" - 16"	S2@.316	LF	26.70	11.40	—	38.10
18" - 20"	S2@.346	LF	28.70	12.50	—	41.20
22" - 24"	S2@.377	LF	32.00	13.60	—	45.60
26" - 28"	S2@.407	LF	34.20	14.70	—	48.90
30"	S2@.431	LF	35.70	15.50	—	51.20

Description	Craft@Hrs	Unit	Material $	Labor $	Equipment $	Total $

Galvanized steel 28" rectangular duct 24 gauge

Description	Craft@Hrs	Unit	Material $	Labor $	Equipment $	Total $
6"	S2@.262	LF	22.00	9.44	—	31.44
8"	S2@.278	LF	23.20	10.00	—	33.20
10"	S2@.292	LF	24.50	10.50	—	35.00
12"	S2@.308	LF	25.90	11.10	—	37.00
14" - 16"	S2@.331	LF	28.00	11.90	—	39.90
18" - 20"	S2@.362	LF	30.50	13.00	—	43.50
22" - 24"	S2@.392	LF	32.80	14.10	—	46.90
26" - 28"	S2@.423	LF	35.40	15.20	—	50.60
30"	S2@.448	LF	37.70	16.10	—	53.80

Galvanized steel 30" rectangular duct 24 gauge

Description	Craft@Hrs	Unit	Material $	Labor $	Equipment $	Total $
6"	S2@.278	LF	23.20	10.00	—	33.20
8"	S2@.292	LF	24.50	10.50	—	35.00
10"	S2@.308	LF	25.90	11.10	—	37.00
12"	S2@.323	LF	27.20	11.60	—	38.80
14" - 16"	S2@.346	LF	28.70	12.50	—	41.20
18" - 20"	S2@.377	LF	32.00	13.60	—	45.60
22" - 24"	S2@.407	LF	34.20	14.70	—	48.90
26" - 28"	S2@.440	LF	36.70	15.90	—	52.60
30"	S2@.465	LF	38.70	16.80	—	55.50

Costs for Galvanized 22 Gauge Rectangular Duct. Costs per linear foot of duct, based on quantities less than 1,000 pounds of prefabricated duct. Deduct 15% from the material cost for over 1,000 pounds. Add 50% to the material cost for lined duct. These costs include delivery, typical bracing, cleats, scrap, hangers, end closures.

Galvanized steel 32" rectangular duct 22 gauge

Description	Craft@Hrs	Unit	Material $	Labor $	Equipment $	Total $
8"	S2@.374	LF	31.40	13.50	—	44.90
10"	S2@.393	LF	32.80	14.20	—	47.00
12"	S2@.412	LF	34.40	14.80	—	49.20
14" - 16"	S2@.442	LF	37.00	15.90	—	52.90
18" - 20"	S2@.478	LF	40.20	17.20	—	57.40
22" - 24"	S2@.517	LF	43.30	18.60	—	61.90
26" - 28"	S2@.553	LF	46.30	19.90	—	66.20
30" - 32"	S2@.589	LF	49.60	21.20	—	70.80
34" - 36"	S2@.628	LF	52.60	22.60	—	75.20
38" - 40"	S2@.664	LF	55.90	23.90	—	79.80
42" - 44"	S2@.702	LF	58.50	25.30	—	83.80
46" - 48"	S2@.739	LF	61.90	26.60	—	88.50
50" - 52"	S2@.777	LF	65.00	28.00	—	93.00
54"	S2@.807	LF	67.30	29.10	—	96.40

Galvanized Steel Rectangular Ductwork

Description	Craft@Hrs	Unit	Material $	Labor $	Equipment $	Total $

Galvanized steel 34" rectangular duct 22 gauge

Description	Craft@Hrs	Unit	Material $	Labor $	Equipment $	Total $
8"	S2@.393	LF	32.80	14.20	—	47.00
10"	S2@.412	LF	34.40	14.80	—	49.20
12"	S2@.431	LF	35.70	15.50	—	51.20
14" - 16"	S2@.461	LF	38.40	16.60	—	55.00
18" - 20"	S2@.497	LF	42.20	17.90	—	60.10
22" - 24"	S2@.534	LF	44.50	19.20	—	63.70
26" - 28"	S2@.572	LF	48.10	20.60	—	68.70
30" - 32"	S2@.608	LF	50.90	21.90	—	72.80
34" - 36"	S2@.645	LF	54.30	23.20	—	77.50
38" - 40"	S2@.683	LF	57.20	24.60	—	81.80
42" - 44"	S2@.722	LF	60.70	26.00	—	86.70
46" - 48"	S2@.758	LF	63.70	27.30	—	91.00
50" - 52"	S2@.796	LF	66.50	28.70	—	95.20
54"	S2@.824	LF	69.20	29.70	—	98.90

Galvanized steel 36" rectangular duct 22 gauge

Description	Craft@Hrs	Unit	Material $	Labor $	Equipment $	Total $
8"	S2@.412	LF	34.40	14.80	—	49.20
10"	S2@.431	LF	35.70	15.50	—	51.20
12"	S2@.453	LF	38.00	16.30	—	54.30
14" - 16"	S2@.478	LF	40.20	17.20	—	57.40
18" - 20"	S2@.517	LF	43.30	18.60	—	61.90
22" - 24"	S2@.553	LF	46.30	19.90	—	66.20
26" - 28"	S2@.589	LF	49.60	21.20	—	70.80
30" - 32"	S2@.628	LF	52.60	22.60	—	75.20
34" - 36"	S2@.664	LF	55.90	23.90	—	79.80
38" - 40"	S2@.702	LF	58.50	25.30	—	83.80
42" - 44"	S2@.739	LF	61.90	26.60	—	88.50
46" - 48"	S2@.777	LF	65.00	28.00	—	93.00
50" - 52"	S2@.816	LF	67.70	29.40	—	97.10
54"	S2@.841	LF	70.90	30.30	—	101.20

Galvanized steel 38" rectangular duct 22 gauge

Description	Craft@Hrs	Unit	Material $	Labor $	Equipment $	Total $
8"	S2@.431	LF	35.70	15.50	—	51.20
10"	S2@.453	LF	38.00	16.30	—	54.30
12"	S2@.470	LF	39.10	16.90	—	56.00
14" - 16"	S2@.497	LF	42.20	17.90	—	60.10
18" - 20"	S2@.534	LF	44.50	19.20	—	63.70
22" - 24"	S2@.572	LF	48.10	20.60	—	68.70
26" - 28"	S2@.608	LF	50.90	21.90	—	72.80
30" - 32"	S2@.645	LF	54.30	23.20	—	77.50
34" - 36"	S2@.683	LF	57.20	24.60	—	81.80
38" - 40"	S2@.722	LF	60.70	26.00	—	86.70
42" - 44"	S2@.758	LF	63.70	27.30	—	91.00
46" - 48"	S2@.796	LF	66.50	28.70	—	95.20
50" - 52"	S2@.833	LF	69.90	30.00	—	99.90
54"	S2@.863	LF	72.00	31.10	—	103.10

Description	Craft@Hrs	Unit	Material $	Labor $	Equipment $	Total $

Galvanized steel 40" rectangular duct 22 gauge

Description	Craft@Hrs	Unit	Material $	Labor $	Equipment $	Total $
8"	S2@.453	LF	38.00	16.30	—	54.30
10"	S2@.470	LF	39.10	16.90	—	56.00
12"	S2@.487	LF	41.00	17.60	—	58.60
14" - 16"	S2@.517	LF	43.30	18.60	—	61.90
18" - 20"	S2@.553	LF	46.30	19.90	—	66.20
22" - 24"	S2@.589	LF	49.60	21.20	—	70.80
26" - 28"	S2@.628	LF	52.60	22.60	—	75.20
30" - 32"	S2@.664	LF	55.90	23.90	—	79.80
34" - 36"	S2@.702	LF	58.50	25.30	—	83.80
38" - 40"	S2@.739	LF	61.90	26.60	—	88.50
42" - 44"	S2@.777	LF	65.00	28.00	—	93.00
46" - 48"	S2@.816	LF	67.70	29.40	—	97.10
50" - 52"	S2@.852	LF	71.60	30.70	—	102.30
54"	S2@.880	LF	73.30	31.70	—	105.00

Galvanized steel 42" rectangular duct 22 gauge

Description	Craft@Hrs	Unit	Material $	Labor $	Equipment $	Total $
8"	S2@.470	LF	39.10	16.90	—	56.00
10"	S2@.487	LF	41.00	17.60	—	58.60
12"	S2@.508	LF	42.80	18.30	—	61.10
14" - 16"	S2@.534	LF	44.50	19.20	—	63.70
18" - 20"	S2@.572	LF	48.10	20.60	—	68.70
22" - 24"	S2@.611	LF	50.90	22.00	—	72.90
26" - 28"	S2@.645	LF	54.30	23.20	—	77.50
30" - 32"	S2@.683	LF	57.20	24.60	—	81.80
34" - 36"	S2@.722	LF	60.70	26.00	—	86.70
38" - 40"	S2@.758	LF	63.70	27.30	—	91.00
42" - 44"	S2@.798	LF	66.50	28.80	—	95.30
46" - 48"	S2@.833	LF	69.90	30.00	—	99.90
50" - 52"	S2@.871	LF	73.00	31.40	—	104.40
54"	S2@.901	LF	75.00	32.50	—	107.50

Galvanized steel 44" rectangular duct 22 gauge

Description	Craft@Hrs	Unit	Material $	Labor $	Equipment $	Total $
8"	S2@.487	LF	40.90	17.60	—	58.50
10"	S2@.508	LF	42.30	18.30	—	60.60
12"	S2@.525	LF	43.70	18.90	—	62.60
14" - 16"	S2@.555	LF	45.80	20.00	—	65.80
18" - 20"	S2@.589	LF	49.40	21.20	—	70.60
22" - 24"	S2@.628	LF	52.10	22.60	—	74.70
26" - 28"	S2@.666	LF	55.60	24.00	—	79.60
30" - 32"	S2@.705	LF	58.30	25.40	—	83.70
34" - 36"	S2@.739	LF	61.70	26.60	—	88.30
38" - 40"	S2@.777	LF	64.70	28.00	—	92.70
42" - 44"	S2@.816	LF	67.50	29.40	—	96.90
46" - 48"	S2@.852	LF	71.20	30.70	—	101.90
50" - 52"	S2@.890	LF	73.80	32.10	—	105.90
54"	S2@.918	LF	76.30	33.10	—	109.40

Galvanized Steel Rectangular Ductwork

Description	Craft@Hrs	Unit	Material $	Labor $	Equipment $	Total $

Galvanized steel 46" rectangular duct 22 gauge

Description	Craft@Hrs	Unit	Material $	Labor $	Equipment $	Total $
8"	S2@.508	LF	42.80	18.30	—	61.10
10"	S2@.525	LF	43.70	18.90	—	62.60
12"	S2@.542	LF	45.40	19.50	—	64.90
14" - 16"	S2@.572	LF	48.10	20.60	—	68.70
18" - 20"	S2@.611	LF	50.90	22.00	—	72.90
22" - 24"	S2@.649	LF	54.30	23.40	—	77.70
26" - 28"	S2@.683	LF	57.20	24.60	—	81.80
30" - 32"	S2@.722	LF	60.70	26.00	—	86.70
34" - 36"	S2@.758	LF	63.70	27.30	—	91.00
38" - 40"	S2@.796	LF	66.50	28.70	—	95.20
42" - 44"	S2@.833	LF	69.90	30.00	—	99.90
46" - 48"	S2@.871	LF	73.00	31.40	—	104.40
50" - 52"	S2@.909	LF	75.90	32.80	—	108.70
54"	S2@.935	LF	78.20	33.70	—	111.90

Galvanized steel 48" rectangular duct 22 gauge

Description	Craft@Hrs	Unit	Material $	Labor $	Equipment $	Total $
8"	S2@.525	LF	43.70	18.90	—	62.60
10"	S2@.542	LF	45.40	19.50	—	64.90
12"	S2@.564	LF	47.10	20.30	—	67.40
14" - 16"	S2@.589	LF	49.60	21.20	—	70.80
18" - 20"	S2@.628	LF	52.60	22.60	—	75.20
22" - 24"	S2@.664	LF	55.90	23.90	—	79.80
26" - 28"	S2@.702	LF	58.50	25.30	—	83.80
30" - 32"	S2@.739	LF	61.90	26.60	—	88.50
34" - 36"	S2@.777	LF	65.00	28.00	—	93.00
38" - 40"	S2@.816	LF	67.70	29.40	—	97.10
42" - 44"	S2@.852	LF	71.60	30.70	—	102.30
46" - 48"	S2@.890	LF	74.30	32.10	—	106.40
50" - 52"	S2@.927	LF	77.80	33.40	—	111.20
54"	S2@.956	LF	79.40	34.50	—	113.90

Galvanized steel 50" rectangular duct 22 gauge

Description	Craft@Hrs	Unit	Material $	Labor $	Equipment $	Total $
8"	S2@.542	LF	45.40	19.50	—	64.90
10"	S2@.564	LF	47.10	20.30	—	67.40
12"	S2@.581	LF	48.60	20.90	—	69.50
14" - 16"	S2@.611	LF	50.90	22.00	—	72.90
18" - 20"	S2@.645	LF	54.30	23.20	—	77.50
22" - 24"	S2@.683	LF	57.20	24.60	—	81.80
26" - 28"	S2@.722	LF	60.70	26.00	—	86.70
30" - 32"	S2@.758	LF	63.70	27.30	—	91.00
34" - 36"	S2@.798	LF	66.50	28.80	—	95.30
38" - 40"	S2@.833	LF	69.90	30.00	—	99.90
42" - 44"	S2@.871	LF	73.00	31.40	—	104.40
46" - 48"	S2@.909	LF	75.90	32.80	—	108.70
50" - 52"	S2@.946	LF	78.90	34.10	—	113.00
54"	S2@.974	LF	81.20	35.10	—	116.30

Description	Craft@Hrs	Unit	Material $	Labor $	Equipment $	Total $

Galvanized steel 52" rectangular duct 22 gauge

Description	Craft@Hrs	Unit	Material $	Labor $	Equipment $	Total $
8"	S2@.564	LF	47.10	20.30	—	67.40
10"	S2@.581	LF	48.60	20.90	—	69.50
12"	S2@.598	LF	50.20	21.60	—	71.80
14" - 16"	S2@.628	LF	52.60	22.60	—	75.20
18" - 20"	S2@.662	LF	55.90	23.90	—	79.80
22" - 24"	S2@.700	LF	58.50	25.20	—	83.70
26" - 28"	S2@.739	LF	61.90	26.60	—	88.50
30" - 32"	S2@.777	LF	65.00	28.00	—	93.00
34" - 36"	S2@.816	LF	67.70	29.40	—	97.10
38" - 40"	S2@.852	LF	71.60	30.70	—	102.30
42" - 44"	S2@.890	LF	74.30	32.10	—	106.40
46" - 48"	S2@.927	LF	77.80	33.40	—	111.20
50" - 52"	S2@.976	LF	80.50	35.20	—	115.70
54"	S2@.991	LF	82.50	35.70	—	118.20

Galvanized steel 54" rectangular duct 22 gauge

Description	Craft@Hrs	Unit	Material $	Labor $	Equipment $	Total $
8"	S2@.581	LF	48.60	20.90	—	69.50
10"	S2@.598	LF	50.20	21.60	—	71.80
12"	S2@.619	LF	51.70	22.30	—	74.00
14" - 16"	S2@.645	LF	54.30	23.20	—	77.50
18" - 20"	S2@.683	LF	57.20	24.60	—	81.80
22" - 24"	S2@.722	LF	60.70	26.00	—	86.70
26" - 28"	S2@.758	LF	63.70	27.30	—	91.00
30" - 32"	S2@.796	LF	66.50	28.70	—	95.20
34" - 36"	S2@.833	LF	69.90	30.00	—	99.90
38" - 40"	S2@.871	LF	73.00	31.40	—	104.40
42" - 44"	S2@.909	LF	75.90	32.80	—	108.70
46" - 48"	S2@.946	LF	78.90	34.10	—	113.00
50" - 52"	S2@.982	LF	81.70	35.40	—	117.10
54"	S2@1.01	LF	84.80	36.40	—	121.20

Galvanized Steel Rectangular Ductwork

Costs for Galvanized 20 Gauge Rectangular Duct. Costs per linear foot of duct, based on quantities less than 1,000 pounds of prefabricated duct. Deduct 15% from the material cost for over 1,000 pounds. Add 50% to the material cost for lined duct. These costs include delivery, typical bracing, cleats, scrap, hangers, end closures, sealants and miscellaneous hardware.

Description	Craft@Hrs	Unit	Material $	Labor $	Equipment $	Total $
Galvanized steel 56" rectangular duct 20 gauge						
18" - 20"	S2@.828	LF	69.90	29.80	—	99.70
22" - 24"	S2@.871	LF	73.30	31.40	—	104.70
26" - 28"	S2@.916	LF	76.80	33.00	—	109.80
30" - 32"	S2@.959	LF	80.90	34.60	—	115.50
34" - 36"	S2@1.00	LF	84.80	36.00	—	120.80
38" - 40"	S2@1.05	LF	88.40	37.80	—	126.20
42" - 44"	S2@1.09	LF	92.20	39.30	—	131.50
46" - 48"	S2@1.14	LF	95.80	41.10	—	136.90
50" - 52"	S2@1.18	LF	99.80	42.50	—	142.30
54" - 56"	S2@1.23	LF	103.00	44.30	—	147.30
58" - 60"	S2@1.27	LF	107.00	45.80	—	152.80
62" - 64"	S2@1.31	LF	111.00	47.20	—	158.20
66" - 68"	S2@1.36	LF	114.00	49.00	—	163.00
70" - 72"	S2@1.40	LF	118.00	50.50	—	168.50
Galvanized steel 58" rectangular duct 20 gauge						
18" - 20"	S2@.850	LF	71.70	30.60	—	102.30
22" - 24"	S2@.895	LF	75.30	32.30	—	107.60
26" - 28"	S2@.937	LF	78.90	33.80	—	112.70
30" - 32"	S2@.980	LF	82.80	35.30	—	118.10
34" - 36"	S2@1.02	LF	86.30	36.80	—	123.10
38" - 40"	S2@1.07	LF	90.10	38.60	—	128.70
42" - 44"	S2@1.11	LF	94.00	40.00	—	134.00
46" - 48"	S2@1.16	LF	97.40	41.80	—	139.20
50" - 52"	S2@1.02	LF	102.00	36.80	—	138.80
54" - 56"	S2@1.25	LF	105.00	45.10	—	150.10
58" - 60"	S2@1.29	LF	108.00	46.50	—	154.50
62" - 64"	S2@1.33	LF	113.00	47.90	—	160.90
66" - 68"	S2@1.38	LF	116.00	49.70	—	165.70
70" - 72"	S2@1.42	LF	119.00	51.20	—	170.20
Galvanized steel 60" rectangular duct 20 gauge						
18" - 20"	S2@.871	LF	73.30	31.40	—	104.70
22" - 24"	S2@.916	LF	76.80	33.00	—	109.80
26" - 28"	S2@.959	LF	80.90	34.60	—	115.50
30" - 32"	S2@1.00	LF	84.80	36.00	—	120.80
34" - 36"	S2@1.05	LF	88.40	37.80	—	126.20
38" - 40"	S2@1.09	LF	92.20	39.30	—	131.50
42" - 44"	S2@1.14	LF	95.80	41.10	—	136.90
46" - 48"	S2@1.18	LF	99.80	42.50	—	142.30
50" - 52"	S2@1.23	LF	103.00	44.30	—	147.30
54" - 56"	S2@1.27	LF	107.00	45.80	—	152.80
58" - 60"	S2@1.31	LF	111.00	47.20	—	158.20
62" - 64"	S2@1.36	LF	114.00	49.00	—	163.00
66" - 68"	S2@1.40	LF	118.00	50.50	—	168.50
70" - 72"	S2@1.45	LF	121.00	52.30	—	173.30

Description	Craft@Hrs	Unit	Material $	Labor $	Equipment $	Total $

Galvanized steel 62" rectangular duct 20 gauge

Description	Craft@Hrs	Unit	Material $	Labor $	Equipment $	Total $
18" - 20"	S2@.895	LF	75.30	32.30	—	107.60
22" - 24"	S2@.937	LF	78.90	33.80	—	112.70
26" - 28"	S2@.980	LF	82.80	35.30	—	118.10
30" - 32"	S2@1.02	LF	86.30	36.80	—	123.10
34" - 36"	S2@1.07	LF	90.10	38.60	—	128.70
38" - 40"	S2@1.11	LF	94.00	40.00	—	134.00
42" - 44"	S2@1.16	LF	97.40	41.80	—	139.20
46" - 48"	S2@1.20	LF	102.00	43.20	—	145.20
50" - 52"	S2@1.25	LF	105.00	45.10	—	150.10
54" - 56"	S2@1.29	LF	108.00	46.50	—	154.50
58" - 60"	S2@1.33	LF	113.00	47.90	—	160.90
62" - 64"	S2@1.38	LF	116.00	49.70	—	165.70
66" - 68"	S2@1.42	LF	119.00	51.20	—	170.20
70" - 72"	S2@1.47	LF	122.00	53.00	—	175.00

Galvanized steel 64" rectangular duct 20 gauge

Description	Craft@Hrs	Unit	Material $	Labor $	Equipment $	Total $
18" - 20"	S2@.916	LF	76.80	33.00	—	109.80
22" - 24"	S2@.959	LF	80.90	34.60	—	115.50
26" - 28"	S2@1.00	LF	84.80	36.00	—	120.80
30" - 32"	S2@1.05	LF	88.40	37.80	—	126.20
34" - 36"	S2@1.09	LF	92.20	39.30	—	131.50
38" - 40"	S2@1.14	LF	95.80	41.10	—	136.90
42" - 44"	S2@1.18	LF	99.80	42.50	—	142.30
46" - 48"	S2@1.23	LF	103.00	44.30	—	147.30
50" - 52"	S2@1.27	LF	107.00	45.80	—	152.80
54" - 56"	S2@1.31	LF	111.00	47.20	—	158.20
58" - 60"	S2@1.36	LF	114.00	49.00	—	163.00
62" - 64"	S2@1.40	LF	118.00	50.50	—	168.50
66" - 68"	S2@1.45	LF	121.00	52.30	—	173.30
70" - 72"	S2@1.49	LF	125.00	53.70	—	178.70

Galvanized steel 66" rectangular duct 20 gauge

Description	Craft@Hrs	Unit	Material $	Labor $	Equipment $	Total $
18" - 20"	S2@.937	LF	78.90	33.80	—	112.70
22" - 24"	S2@.980	LF	82.80	35.30	—	118.10
26" - 28"	S2@1.02	LF	86.30	36.80	—	123.10
30" - 32"	S2@1.07	LF	90.10	38.60	—	128.70
34" - 36"	S2@1.11	LF	94.00	40.00	—	134.00
38" - 40"	S2@1.16	LF	97.40	41.80	—	139.20
42" - 44"	S2@1.20	LF	102.00	43.20	—	145.20
46" - 48"	S2@1.25	LF	105.00	45.10	—	150.10
50" - 52"	S2@1.29	LF	108.00	46.50	—	154.50
54" - 56"	S2@1.33	LF	113.00	47.90	—	160.90
58" - 60"	S2@1.38	LF	116.00	49.70	—	165.70
62" - 64"	S2@1.42	LF	119.00	51.20	—	170.20
66" - 68"	S2@1.47	LF	122.00	53.00	—	175.00
70" - 72"	S2@1.51	LF	126.00	54.40	—	180.40

Galvanized Steel Rectangular Ductwork

Description	Craft@Hrs	Unit	Material $	Labor $	Equipment $	Total $

Galvanized steel 68" rectangular duct 20 gauge

Description	Craft@Hrs	Unit	Material $	Labor $	Equipment $	Total $
18" - 20"	S2@.959	LF	80.90	34.60	—	115.50
22" - 24"	S2@1.00	LF	84.80	36.00	—	120.80
26" - 28"	S2@1.05	LF	88.40	37.80	—	126.20
30" - 32"	S2@1.09	LF	92.20	39.30	—	131.50
34" - 36"	S2@1.14	LF	95.80	41.10	—	136.90
38" - 40"	S2@1.18	LF	99.80	42.50	—	142.30
42" - 44"	S2@1.23	LF	103.00	44.30	—	147.30
46" - 48"	S2@1.27	LF	107.00	45.80	—	152.80
50" - 52"	S2@1.31	LF	111.00	47.20	—	158.20
54" - 56"	S2@1.36	LF	114.00	49.00	—	163.00
58" - 60"	S2@1.40	LF	118.00	50.50	—	168.50
62" - 64"	S2@1.45	LF	121.00	52.30	—	173.30
66" - 68"	S2@1.49	LF	125.00	53.70	—	178.70
70" - 72"	S2@1.53	LF	130.00	55.10	—	185.10

Galvanized steel 70" rectangular duct 20 gauge

Description	Craft@Hrs	Unit	Material $	Labor $	Equipment $	Total $
18" - 20"	S2@.980	LF	82.80	35.30	—	118.10
22" - 24"	S2@1.02	LF	86.30	36.80	—	123.10
26" - 28"	S2@1.07	LF	90.10	38.60	—	128.70
30" - 32"	S2@1.11	LF	94.00	40.00	—	134.00
34" - 36"	S2@1.16	LF	97.40	41.80	—	139.20
38" - 40"	S2@1.20	LF	102.00	43.20	—	145.20
42" - 44"	S2@1.25	LF	105.00	45.10	—	150.10
46" - 48"	S2@1.29	LF	108.00	46.50	—	154.50
50" - 52"	S2@1.33	LF	113.00	47.90	—	160.90
54" - 56"	S2@1.38	LF	116.00	49.70	—	165.70
58" - 60"	S2@1.42	LF	119.00	51.20	—	170.20
62" - 64"	S2@1.47	LF	122.00	53.00	—	175.00
66" - 68"	S2@1.51	LF	126.00	54.40	—	180.40
70" - 72"	S2@1.55	LF	131.00	55.90	—	186.90

Galvanized steel 72" rectangular duct 20 gauge

Description	Craft@Hrs	Unit	Material $	Labor $	Equipment $	Total $
18" - 20"	S2@1.00	LF	84.80	36.00	—	120.80
22" - 24"	S2@1.05	LF	88.40	37.80	—	126.20
26" - 28"	S2@1.09	LF	92.20	39.30	—	131.50
30" - 32"	S2@1.14	LF	95.80	41.10	—	136.90
34" - 36"	S2@1.18	LF	99.80	42.50	—	142.30
38" - 40"	S2@1.23	LF	103.00	44.30	—	147.30
42" - 44"	S2@1.27	LF	107.00	45.80	—	152.80
46" - 48"	S2@1.31	LF	111.00	47.20	—	158.20
50" - 52"	S2@1.36	LF	114.00	49.00	—	163.00
54" - 56"	S2@1.40	LF	118.00	50.50	—	168.50
58" - 60"	S2@1.45	LF	121.00	52.30	—	173.30
62" - 64"	S2@1.49	LF	125.00	53.70	—	178.70
66" - 68"	S2@1.53	LF	130.00	55.10	—	185.10
70" - 72"	S2@1.58	LF	132.00	56.90	—	188.90

Costs for Galvanized Rectangular 90-Degree Elbows. Costs per elbow, based on quantities less than 1,000 pounds. Deduct 15% from the material cost for over 1,000 pounds. Add 50% to the material cost for lined duct. These costs include delivery, typical bracing, cleats, scrap, hangers, end closures, sealants and miscellaneous hardware. For turning vanes, please see page 342.

Description	Craft@Hrs	Unit	Material $	Labor $	Equipment $	Total $

Galvanized steel 4" rectangular duct 90-degree elbow

Description	Craft@Hrs	Unit	Material $	Labor $	Equipment $	Total $
4"	S2@.042	Ea	7.09	1.51	—	8.60
6"	S2@.058	Ea	9.73	2.09	—	11.82
8"	S2@.078	Ea	13.10	2.81	—	15.91
10"	S2@.104	Ea	17.50	3.75	—	21.25
12"	S2@.132	Ea	22.20	4.76	—	26.96
14" - 16"	S2@.236	Ea	39.60	8.51	—	48.11
18" - 20"	S2@.339	Ea	57.20	12.20	—	69.40
22" - 24"	S2@.468	Ea	78.20	16.90	—	95.10
26" - 28"	S2@.615	Ea	102.00	22.20	—	124.20
30" - 32"	S2@.854	Ea	146.00	30.80	—	176.80
34" - 36"	S2@1.19	Ea	200.00	42.90	—	242.90
38" - 40"	S2@1.45	Ea	244.00	52.30	—	296.30
42" - 44"	S2@1.73	Ea	287.00	62.30	—	349.30
46" - 48"	S2@2.04	Ea	342.00	73.50	—	415.50
50"	S2@2.28	Ea	384.00	82.20	—	466.20

Galvanized steel 6" rectangular duct 90-degree elbow

Description	Craft@Hrs	Unit	Material $	Labor $	Equipment $	Total $
4"	S2@.058	Ea	9.73	2.09	—	11.82
6"	S2@.078	Ea	13.10	2.81	—	15.91
8"	S2@.104	Ea	17.50	3.75	—	21.25
10"	S2@.132	Ea	22.20	4.76	—	26.96
12"	S2@.148	Ea	24.80	5.33	—	30.13
14" - 16"	S2@.266	Ea	44.90	9.59	—	54.49
18" - 20"	S2@.370	Ea	61.90	13.30	—	75.20
22" - 24"	S2@.506	Ea	85.00	18.20	—	103.20
26" - 28"	S2@.655	Ea	109.00	23.60	—	132.60
30" - 32"	S2@.905	Ea	151.00	32.60	—	183.60
34" - 36"	S2@1.24	Ea	209.00	44.70	—	253.70
38" - 40"	S2@1.49	Ea	249.00	53.70	—	302.70
42" - 44"	S2@1.78	Ea	296.00	64.20	—	360.20
46" - 48"	S2@2.09	Ea	350.00	75.30	—	425.30
50"	S2@2.34	Ea	394.00	84.30	—	478.30

Galvanized Steel Rectangular Elbows

Description	Craft@Hrs	Unit	Material $	Labor $	Equipment $	Total $

Galvanized steel 8" rectangular duct 90-degree elbow

Description	Craft@Hrs	Unit	Material $	Labor $	Equipment $	Total $
4"	S2@.078	Ea	13.10	2.81	—	15.91
6"	S2@.104	Ea	17.50	3.75	—	21.25
8"	S2@.132	Ea	22.20	4.76	—	26.96
10"	S2@.148	Ea	24.80	5.33	—	30.13
12"	S2@.170	Ea	28.60	6.13	—	34.73
14" - 16"	S2@.300	Ea	50.20	10.80	—	61.00
18" - 20"	S2@.403	Ea	67.50	14.50	—	82.00
22" - 24"	S2@.549	Ea	92.10	19.80	—	111.90
26" - 28"	S2@.694	Ea	116.00	25.00	—	141.00
30" - 32"	S2@.950	Ea	162.00	34.20	—	196.20
34" - 36"	S2@1.28	Ea	215.00	46.10	—	261.10
38" - 40"	S2@1.54	Ea	261.00	55.50	—	316.50
42" - 44"	S2@1.83	Ea	307.00	66.00	—	373.00
46" - 48"	S2@2.13	Ea	357.00	76.80	—	433.80
50"	S2@2.39	Ea	406.00	86.10	—	492.10

Galvanized steel 10" rectangular duct 90-degree elbow

Description	Craft@Hrs	Unit	Material $	Labor $	Equipment $	Total $
4"	S2@.104	Ea	17.50	3.75	—	21.25
6"	S2@.132	Ea	22.20	4.76	—	26.96
8"	S2@.148	Ea	24.80	5.33	—	30.13
10"	S2@.170	Ea	28.60	6.13	—	34.73
12"	S2@.194	Ea	32.40	6.99	—	39.39
14" - 16"	S2@.336	Ea	56.00	12.10	—	68.10
18" - 20"	S2@.438	Ea	73.30	15.80	—	89.10
22" - 24"	S2@.587	Ea	98.80	21.20	—	120.00
26" - 28"	S2@.737	Ea	122.00	26.60	—	148.60
30" - 32"	S2@.997	Ea	168.00	35.90	—	203.90
34" - 36"	S2@1.33	Ea	223.00	47.90	—	270.90
38" - 40"	S2@1.59	Ea	269.00	57.30	—	326.30
42" - 44"	S2@1.87	Ea	314.00	67.40	—	381.40
46" - 48"	S2@2.18	Ea	367.00	78.60	—	445.60
50"	S2@2.43	Ea	410.00	87.60	—	497.60

Galvanized steel 12" rectangular duct 90-degree elbow

Description	Craft@Hrs	Unit	Material $	Labor $	Equipment $	Total $
4"	S2@.132	Ea	22.20	4.76	—	26.96
6"	S2@.148	Ea	24.80	5.33	—	30.13
8"	S2@.170	Ea	28.60	6.13	—	34.73
10"	S2@.194	Ea	32.40	6.99	—	39.39
12"	S2@.224	Ea	37.70	8.07	—	45.77
18" - 20"	S2@.478	Ea	79.40	17.20	—	96.60
22" - 24"	S2@.628	Ea	105.00	22.60	—	127.60
26" - 28"	S2@.775	Ea	131.00	27.90	—	158.90
30" - 32"	S2@1.04	Ea	175.00	37.50	—	212.50
34" - 36"	S2@1.38	Ea	232.00	49.70	—	281.70
38" - 40"	S2@1.64	Ea	274.00	59.10	—	333.10
42" - 44"	S2@1.92	Ea	323.00	69.20	—	392.20
46" - 48"	S2@2.23	Ea	377.00	80.40	—	457.40
50"	S2@2.48	Ea	415.00	89.40	—	504.40

Galvanized Steel Rectangular Elbows

Description	Craft@Hrs	Unit	Material $	Labor $	Equipment $	Total $

Galvanized steel 14" rectangular duct 90-degree elbow

Description	Craft@Hrs	Unit	Material $	Labor $	Equipment $	Total $
4"	S2@.212	Ea	35.40	7.64	—	43.04
6"	S2@.232	Ea	39.10	8.36	—	47.46
8"	S2@.260	Ea	43.60	9.37	—	52.97
10"	S2@.292	Ea	49.00	10.50	—	59.50
12"	S2@.328	Ea	55.10	11.80	—	66.90
14" - 16"	S2@.415	Ea	69.90	15.00	—	84.90
18" - 20"	S2@.527	Ea	88.40	19.00	—	107.40
22" - 24"	S2@.867	Ea	112.00	31.20	—	143.20
26" - 28"	S2@.816	Ea	135.00	29.40	—	164.40
30" - 32"	S2@1.09	Ea	184.00	39.30	—	223.30
34" - 36"	S2@1.43	Ea	242.00	51.50	—	293.50
38" - 40"	S2@1.69	Ea	283.00	60.90	—	343.90
42" - 44"	S2@1.97	Ea	333.00	71.00	—	404.00
46" - 48"	S2@2.28	Ea	382.00	82.20	—	464.20
50"	S2@2.53	Ea	428.00	91.20	—	519.20

Galvanized steel 16" rectangular duct 90-degree elbow

Description	Craft@Hrs	Unit	Material $	Labor $	Equipment $	Total $
4"	S2@.260	Ea	43.60	9.37	—	52.97
6"	S2@.310	Ea	52.10	11.20	—	63.30
8"	S2@.340	Ea	57.10	12.30	—	69.40
10"	S2@.380	Ea	64.10	13.70	—	77.80
12"	S2@.420	Ea	70.50	15.10	—	85.60
14" - 16"	S2@.480	Ea	80.40	17.30	—	97.70
18" - 20"	S2@.555	Ea	93.10	20.00	—	113.10
22" - 24"	S2@.707	Ea	118.00	25.50	—	143.50
26" - 28"	S2@.856	Ea	145.00	30.90	—	175.90
30" - 32"	S2@1.14	Ea	191.00	41.10	—	232.10
34" - 36"	S2@1.48	Ea	248.00	53.30	—	301.30
38" - 40"	S2@1.73	Ea	291.00	62.30	—	353.30
42" - 44"	S2@2.02	Ea	341.00	72.80	—	413.80
46" - 48"	S2@2.32	Ea	394.00	83.60	—	477.60
50"	S2@2.58	Ea	433.00	93.00	—	526.00

Galvanized steel 18" rectangular duct 90-degree elbow

Description	Craft@Hrs	Unit	Material $	Labor $	Equipment $	Total $
4"	S2@.312	Ea	52.60	11.20	—	63.80
6"	S2@.350	Ea	59.00	12.60	—	71.60
8"	S2@.390	Ea	65.80	14.10	—	79.90
10"	S2@.427	Ea	71.90	15.40	—	87.30
12"	S2@.470	Ea	78.20	16.90	—	95.10
14" - 16"	S2@.519	Ea	87.40	18.70	—	106.10
18" - 20"	S2@.604	Ea	101.00	21.80	—	122.80
22" - 24"	S2@.747	Ea	125.00	26.90	—	151.90
26" - 28"	S2@.895	Ea	150.00	32.30	—	182.30
30" - 32"	S2@1.18	Ea	199.00	42.50	—	241.50
34" - 36"	S2@1.53	Ea	259.00	55.10	—	314.10
38" - 40"	S2@1.78	Ea	296.00	64.20	—	360.20
42" - 44"	S2@2.07	Ea	348.00	74.60	—	422.60
46" - 48"	S2@2.37	Ea	402.00	85.40	—	487.40
50"	S2@2.63	Ea	441.00	94.80	—	535.80

383

Galvanized Steel Rectangular Elbows

Description	Craft@Hrs	Unit	Material $	Labor $	Equipment $	Total $

Galvanized steel 20" rectangular duct 90-degree elbow

Description	Craft@Hrs	Unit	Material $	Labor $	Equipment $	Total $
4"	S2@.370	Ea	61.90	13.30	—	75.20
6"	S2@.390	Ea	65.80	14.10	—	79.90
8"	S2@.415	Ea	69.90	15.00	—	84.90
10"	S2@.448	Ea	75.10	16.10	—	91.20
12"	S2@.487	Ea	81.30	17.60	—	98.90
14" - 16"	S2@.553	Ea	92.60	19.90	—	112.50
18" - 20"	S2@.658	Ea	110.00	23.70	—	133.70
22" - 24"	S2@.786	Ea	132.00	28.30	—	160.30
26" - 28"	S2@.931	Ea	156.00	33.60	—	189.60
30" - 32"	S2@1.23	Ea	206.00	44.30	—	250.30
34" - 36"	S2@1.58	Ea	267.00	56.90	—	323.90
38" - 40"	S2@1.83	Ea	310.00	66.00	—	376.00
42" - 44"	S2@2.11	Ea	354.00	76.00	—	430.00
46" - 48"	S2@2.42	Ea	409.00	87.20	—	496.20
50"	S2@2.67	Ea	453.00	96.20	—	549.20

Galvanized steel rectangular 22" duct 90-degree elbow

Description	Craft@Hrs	Unit	Material $	Labor $	Equipment $	Total $
22" - 24"	S2@.828	Ea	137.00	29.80	—	166.80
26" - 28"	S2@1.01	Ea	169.00	36.40	—	205.40
30" - 32"	S2@1.31	Ea	220.00	47.20	—	267.20
34" - 36"	S2@1.71	Ea	286.00	61.60	—	347.60
38" - 40"	S2@2.01	Ea	341.00	72.40	—	413.40
42" - 44"	S2@2.31	Ea	384.00	83.30	—	467.30
46" - 48"	S2@2.61	Ea	437.00	94.10	—	531.10
50" - 52"	S2@2.91	Ea	490.00	105.00	—	595.00
54" - 56"	S2@3.51	Ea	590.00	127.00	—	717.00
58" - 60"	S2@4.26	Ea	717.00	154.00	—	871.00
62" - 64"	S2@4.76	Ea	793.00	172.00	—	965.00
66" - 68"	S2@5.25	Ea	882.00	189.00	—	1,071.00
70" - 72"	S2@5.76	Ea	964.00	208.00	—	1,172.00

Galvanized steel 24" rectangular duct 90-degree elbow

Description	Craft@Hrs	Unit	Material $	Labor $	Equipment $	Total $
22" - 24"	S2@.922	Ea	155.00	33.20	—	188.20
26" - 28"	S2@1.10	Ea	185.00	39.60	—	224.60
30" - 32"	S2@1.41	Ea	238.00	50.80	—	288.80
34" - 36"	S2@1.81	Ea	305.00	65.20	—	370.20
38" - 40"	S2@2.10	Ea	354.00	75.70	—	429.70
42" - 44"	S2@2.41	Ea	406.00	86.90	—	492.90
46" - 48"	S2@2.71	Ea	454.00	97.70	—	551.70
50" - 52"	S2@3.01	Ea	502.00	108.00	—	610.00
54" - 56"	S2@3.62	Ea	608.00	130.00	—	738.00
58" - 60"	S2@4.38	Ea	733.00	158.00	—	891.00
62" - 64"	S2@4.87	Ea	813.00	176.00	—	989.00
66" - 68"	S2@5.38	Ea	900.00	194.00	—	1,094.00
70" - 72"	S2@5.87	Ea	988.00	212.00	—	1,200.00

Description	Craft@Hrs	Unit	Material $	Labor $	Equipment $	Total $

Galvanized steel 26" rectangular duct 90-degree elbow

Description	Craft@Hrs	Unit	Material $	Labor $	Equipment $	Total $
22" - 24"	S2@1.01	Ea	169.00	36.40	—	205.40
26" - 28"	S2@1.19	Ea	200.00	42.90	—	242.90
30" - 32"	S2@1.50	Ea	251.00	54.10	—	305.10
34" - 36"	S2@1.92	Ea	323.00	69.20	—	392.20
38" - 40"	S2@2.21	Ea	370.00	79.60	—	449.60
42" - 44"	S2@2.51	Ea	423.00	90.50	—	513.50
46" - 48"	S2@2.81	Ea	470.00	101.00	—	571.00
50" - 52"	S2@3.11	Ea	526.00	112.00	—	638.00
54" - 56"	S2@3.73	Ea	630.00	134.00	—	764.00
58" - 60"	S2@4.50	Ea	753.00	162.00	—	915.00
62" - 64"	S2@5.02	Ea	838.00	181.00	—	1,019.00
66" - 68"	S2@5.53	Ea	931.00	199.00	—	1,130.00
70" - 72"	S2@6.04	Ea	1,010.00	218.00	—	1,228.00

Galvanized steel 28" rectangular duct 90-degree elbow

Description	Craft@Hrs	Unit	Material $	Labor $	Equipment $	Total $
22" - 24"	S2@1.10	Ea	184.00	39.60	—	223.60
26" - 28"	S2@1.28	Ea	215.00	46.10	—	261.10
30" - 32"	S2@1.60	Ea	271.00	57.70	—	328.70
34" - 36"	S2@2.03	Ea	342.00	73.20	—	415.20
38" - 40"	S2@2.31	Ea	384.00	83.30	—	467.30
42" - 44"	S2@2.61	Ea	437.00	94.10	—	531.10
46" - 48"	S2@2.91	Ea	490.00	105.00	—	595.00
50" - 52"	S2@3.21	Ea	543.00	116.00	—	659.00
54" - 56"	S2@3.84	Ea	647.00	138.00	—	785.00
58" - 60"	S2@4.61	Ea	777.00	166.00	—	943.00
62" - 64"	S2@5.15	Ea	854.00	186.00	—	1,040.00
66" - 68"	S2@5.66	Ea	944.00	204.00	—	1,148.00
70" - 72"	S2@6.15	Ea	1,020.00	222.00	—	1,242.00

Galvanized steel 30" rectangular duct 90-degree elbow

Description	Craft@Hrs	Unit	Material $	Labor $	Equipment $	Total $
22" - 24"	S2@1.19	Ea	199.00	42.90	—	241.90
26" - 28"	S2@1.37	Ea	232.00	49.40	—	281.40
30" - 32"	S2@1.70	Ea	284.00	61.30	—	345.30
34" - 36"	S2@2.14	Ea	357.00	77.10	—	434.10
38" - 40"	S2@2.41	Ea	406.00	86.90	—	492.90
42" - 44"	S2@2.71	Ea	454.00	97.70	—	551.70
46" - 48"	S2@3.01	Ea	502.00	108.00	—	610.00
50" - 52"	S2@3.31	Ea	556.00	119.00	—	675.00
54" - 56"	S2@3.95	Ea	662.00	142.00	—	804.00
58" - 60"	S2@4.74	Ea	791.00	171.00	—	962.00
62" - 64"	S2@5.25	Ea	882.00	189.00	—	1,071.00
66" - 68"	S2@5.79	Ea	966.00	209.00	—	1,175.00
70" - 72"	S2@6.28	Ea	1,050.00	226.00	—	1,276.00

Galvanized Steel Rectangular Elbows

Description	Craft@Hrs	Unit	Material $	Labor $	Equipment $	Total $

Galvanized steel 32" rectangular duct 90-degree elbow

Description	Craft@Hrs	Unit	Material $	Labor $	Equipment $	Total $
22" - 24"	S2@1.53	Ea	259.00	55.10	—	314.10
26" - 28"	S2@1.73	Ea	287.00	62.30	—	349.30
30" - 32"	S2@1.93	Ea	324.00	69.60	—	393.60
34" - 36"	S2@2.25	Ea	380.00	81.10	—	461.10
38" - 40"	S2@2.51	Ea	423.00	90.50	—	513.50
42" - 44"	S2@2.81	Ea	470.00	101.00	—	571.00
46" - 48"	S2@3.11	Ea	521.00	112.00	—	633.00
50" - 52"	S2@3.41	Ea	572.00	123.00	—	695.00
54" - 56"	S2@4.06	Ea	677.00	146.00	—	823.00
58" - 60"	S2@4.85	Ea	812.00	175.00	—	987.00
62" - 64"	S2@5.38	Ea	900.00	194.00	—	1,094.00
66" - 68"	S2@5.89	Ea	989.00	212.00	—	1,201.00
70" - 72"	S2@6.40	Ea	1,070.00	231.00	—	1,301.00

Galvanized steel 34" rectangular duct 90-degree elbow

Description	Craft@Hrs	Unit	Material $	Labor $	Equipment $	Total $
22" - 24"	S2@1.67	Ea	281.00	60.20	—	341.20
26" - 28"	S2@1.91	Ea	323.00	68.80	—	391.80
30" - 32"	S2@2.12	Ea	354.00	76.40	—	430.40
34" - 36"	S2@2.36	Ea	396.00	85.10	—	481.10
38" - 40"	S2@2.60	Ea	437.00	93.70	—	530.70
42" - 44"	S2@2.91	Ea	490.00	105.00	—	595.00
46" - 48"	S2@3.21	Ea	532.00	116.00	—	648.00
50" - 52"	S2@3.51	Ea	590.00	127.00	—	717.00
54" - 56"	S2@4.17	Ea	702.00	150.00	—	852.00
58" - 60"	S2@4.97	Ea	836.00	179.00	—	1,015.00
62" - 64"	S2@5.49	Ea	922.00	198.00	—	1,120.00
66" - 68"	S2@6.04	Ea	1,010.00	218.00	—	1,228.00
70" - 72"	S2@6.51	Ea	1,080.00	235.00	—	1,315.00

Galvanized steel 36" rectangular duct 90-degree elbow

Description	Craft@Hrs	Unit	Material $	Labor $	Equipment $	Total $
22" - 24"	S2@1.83	Ea	310.00	66.00	—	376.00
26" - 28"	S2@2.04	Ea	344.00	73.50	—	417.50
30" - 32"	S2@2.26	Ea	380.00	81.50	—	461.50
34" - 36"	S2@2.47	Ea	415.00	89.00	—	504.00
38" - 40"	S2@2.70	Ea	454.00	97.30	—	551.30
42" - 44"	S2@3.01	Ea	502.00	108.00	—	610.00
46" - 48"	S2@3.31	Ea	556.00	119.00	—	675.00
50" - 52"	S2@3.61	Ea	607.00	130.00	—	737.00
54" - 56"	S2@4.27	Ea	717.00	154.00	—	871.00
58" - 60"	S2@5.10	Ea	851.00	184.00	—	1,035.00
62" - 64"	S2@5.61	Ea	940.00	202.00	—	1,142.00
66" - 68"	S2@6.15	Ea	1,020.00	222.00	—	1,242.00
70" - 72"	S2@6.64	Ea	1,110.00	239.00	—	1,349.00

Description	Craft@Hrs	Unit	Material $	Labor $	Equipment $	Total $

Galvanized steel 38" rectangular duct 90-degree elbow

Description	Craft@Hrs	Unit	Material $	Labor $	Equipment $	Total $
22" - 24"	S2@1.98	Ea	333.00	71.40	—	404.40
26" - 28"	S2@2.18	Ea	367.00	78.60	—	445.60
30" - 32"	S2@2.38	Ea	402.00	85.80	—	487.80
34" - 36"	S2@2.58	Ea	436.00	93.00	—	529.00
38" - 40"	S2@2.81	Ea	470.00	101.00	—	571.00
42" - 44"	S2@3.11	Ea	521.00	112.00	—	633.00
46" - 48"	S2@3.41	Ea	572.00	123.00	—	695.00
50" - 52"	S2@3.71	Ea	620.00	134.00	—	754.00
54" - 56"	S2@4.38	Ea	733.00	158.00	—	891.00
58" - 60"	S2@5.21	Ea	878.00	188.00	—	1,066.00
62" - 64"	S2@5.74	Ea	958.00	207.00	—	1,165.00
66" - 68"	S2@6.28	Ea	1,050.00	226.00	—	1,276.00
70" - 72"	S2@6.75	Ea	1,130.00	243.00	—	1,373.00

Galvanized steel 40" rectangular duct 90-degree elbow

Description	Craft@Hrs	Unit	Material $	Labor $	Equipment $	Total $
40" - 42"	S2@3.06	Ea	515.00	110.00	—	625.00
44" - 46"	S2@3.36	Ea	560.00	121.00	—	681.00
48" - 50"	S2@3.66	Ea	616.00	132.00	—	748.00
52" - 54"	S2@3.98	Ea	667.00	143.00	—	810.00
56" - 58"	S2@5.17	Ea	863.00	186.00	—	1,049.00
60" - 62"	S2@5.59	Ea	938.00	201.00	—	1,139.00
64" - 66"	S2@6.11	Ea	1,020.00	220.00	—	1,240.00
68" - 70"	S2@6.66	Ea	1,120.00	240.00	—	1,360.00
72"	S2@7.05	Ea	1,170.00	254.00	—	1,424.00

Galvanized steel 42" rectangular duct 90-degree elbow

Description	Craft@Hrs	Unit	Material $	Labor $	Equipment $	Total $
40" - 42"	S2@3.21	Ea	543.00	116.00	—	659.00
44" - 46"	S2@3.52	Ea	592.00	127.00	—	719.00
48" - 50"	S2@3.83	Ea	647.00	138.00	—	785.00
52" - 54"	S2@4.16	Ea	699.00	150.00	—	849.00
56" - 58"	S2@5.34	Ea	893.00	192.00	—	1,085.00
60" - 62"	S2@5.81	Ea	970.00	209.00	—	1,179.00
64" - 66"	S2@6.32	Ea	1,060.00	228.00	—	1,288.00
68" - 70"	S2@6.85	Ea	1,140.00	247.00	—	1,387.00
72"	S2@7.22	Ea	1,210.00	260.00	—	1,470.00

Galvanized steel 44" rectangular duct 90-degree elbow

Description	Craft@Hrs	Unit	Material $	Labor $	Equipment $	Total $
40" - 42"	S2@3.36	Ea	560.00	121.00	—	681.00
44" - 46"	S2@3.69	Ea	619.00	133.00	—	752.00
48" - 50"	S2@4.01	Ea	671.00	145.00	—	816.00
52" - 54"	S2@4.32	Ea	724.00	156.00	—	880.00
56" - 58"	S2@5.51	Ea	926.00	199.00	—	1,125.00
60" - 62"	S2@6.00	Ea	1,000.00	216.00	—	1,216.00
64" - 66"	S2@6.51	Ea	1,080.00	235.00	—	1,315.00
68" - 70"	S2@7.05	Ea	1,180.00	254.00	—	1,434.00
72"	S2@7.39	Ea	1,220.00	266.00	—	1,486.00

Galvanized Steel Rectangular Elbows

Description	Craft@Hrs	Unit	Material $	Labor $	Equipment $	Total $

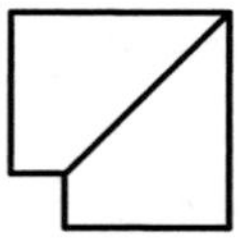

Galvanized steel 46" rectangular duct 90-degree elbow

Description	Craft@Hrs	Unit	Material $	Labor $	Equipment $	Total $
40" - 42"	S2@3.52	Ea	592.00	127.00	—	719.00
44" - 46"	S2@3.86	Ea	650.00	139.00	—	789.00
48" - 50"	S2@4.16	Ea	699.00	150.00	—	849.00
52" - 54"	S2@4.32	Ea	750.00	156.00	—	906.00
56" - 58"	S2@5.72	Ea	955.00	206.00	—	1,161.00
60" - 62"	S2@6.19	Ea	1,030.00	223.00	—	1,253.00
64" - 66"	S2@6.70	Ea	1,130.00	241.00	—	1,371.00
68" - 70"	S2@7.26	Ea	1,210.00	262.00	—	1,472.00
72"	S2@7.60	Ea	1,260.00	274.00	—	1,534.00

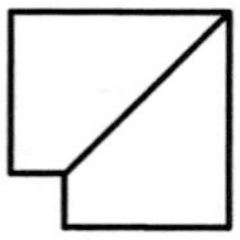

Galvanized steel 48" rectangular duct 90-degree elbow

Description	Craft@Hrs	Unit	Material $	Labor $	Equipment $	Total $
40" - 42"	S2@3.67	Ea	617.00	132.00	—	749.00
44" - 46"	S2@4.01	Ea	671.00	145.00	—	816.00
48" - 50"	S2@4.32	Ea	724.00	156.00	—	880.00
52" - 54"	S2@4.61	Ea	768.00	166.00	—	934.00
56" - 58"	S2@5.91	Ea	997.00	213.00	—	1,210.00
60" - 62"	S2@6.40	Ea	1,070.00	231.00	—	1,301.00
64" - 66"	S2@6.90	Ea	1,150.00	249.00	—	1,399.00
68" - 70"	S2@7.41	Ea	1,240.00	267.00	—	1,507.00
72"	S2@7.77	Ea	1,310.00	280.00	—	1,590.00

Galvanized steel 50" rectangular duct 90-degree elbow

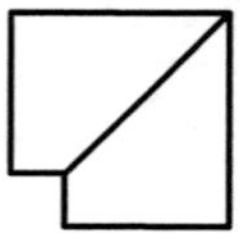

Description	Craft@Hrs	Unit	Material $	Labor $	Equipment $	Total $
40" - 42"	S2@3.82	Ea	647.00	138.00	—	785.00
44" - 46"	S2@4.16	Ea	699.00	150.00	—	849.00
48" - 50"	S2@4.46	Ea	750.00	161.00	—	911.00
52" - 54"	S2@4.78	Ea	794.00	172.00	—	966.00
56" - 58"	S2@6.11	Ea	102.00	220.00	—	322.00
60" - 62"	S2@6.60	Ea	1,110.00	238.00	—	1,348.00
64" - 66"	S2@7.09	Ea	1,180.00	256.00	—	1,436.00
68" - 70"	S2@7.62	Ea	1,280.00	275.00	—	1,555.00
72"	S2@7.94	Ea	1,330.00	286.00	—	1,616.00

Galvanized steel 52" rectangular duct 90-degree elbow

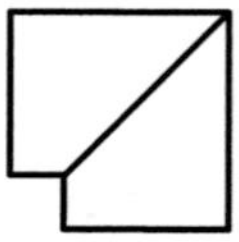

Description	Craft@Hrs	Unit	Material $	Labor $	Equipment $	Total $
40" - 42"	S2@3.98	Ea	667.00	143.00	—	810.00
44" - 46"	S2@4.32	Ea	724.00	156.00	—	880.00
48" - 50"	S2@4.61	Ea	768.00	166.00	—	934.00
52" - 54"	S2@4.95	Ea	828.00	178.00	—	1,006.00
56" - 58"	S2@6.32	Ea	1,060.00	228.00	—	1,288.00
60" - 62"	S2@6.79	Ea	1,140.00	245.00	—	1,385.00
64" - 66"	S2@7.30	Ea	1,210.00	263.00	—	1,473.00
68" - 70"	S2@7.81	Ea	1,310.00	281.00	—	1,591.00
72"	S2@8.16	Ea	1,350.00	294.00	—	1,644.00

Description	Craft@Hrs	Unit	Material $	Labor $	Equipment $	Total $

Galvanized steel 54" rectangular duct 90-degree elbow

Description	Craft@Hrs	Unit	Material $	Labor $	Equipment $	Total $
40" - 42"	S2@4.16	Ea	699.00	150.00	—	849.00
44" - 46"	S2@4.46	Ea	750.00	161.00	—	911.00
48" - 50"	S2@4.78	Ea	794.00	172.00	—	966.00
52" - 54"	S2@5.15	Ea	854.00	186.00	—	1,040.00
56" - 58"	S2@6.51	Ea	1,080.00	235.00	—	1,315.00
60" - 62"	S2@7.00	Ea	1,170.00	252.00	—	1,422.00
64" - 66"	S2@7.49	Ea	1,250.00	270.00	—	1,520.00
68" - 70"	S2@8.01	Ea	1,340.00	289.00	—	1,629.00
72"	S2@8.41	Ea	1,390.00	303.00	—	1,693.00

Galvanized steel 56" rectangular duct 90-degree elbow

Description	Craft@Hrs	Unit	Material $	Labor $	Equipment $	Total $
40" - 42"	S2@5.10	Ea	851.00	184.00	—	1,035.00
44" - 46"	S2@5.49	Ea	922.00	198.00	—	1,120.00
48" - 50"	S2@5.85	Ea	985.00	211.00	—	1,196.00
52" - 54"	S2@6.26	Ea	1,050.00	226.00	—	1,276.00
56" - 58"	S2@6.68	Ea	1,120.00	241.00	—	1,361.00
60" - 62"	S2@7.15	Ea	1,190.00	258.00	—	1,448.00
64" - 66"	S2@7.66	Ea	1,280.00	276.00	—	1,556.00
68" - 70"	S2@8.18	Ea	1,350.00	295.00	—	1,645.00
72"	S2@8.58	Ea	1,460.00	309.00	—	1,769.00

Galvanized steel 58" rectangular duct 90-degree elbow

Description	Craft@Hrs	Unit	Material $	Labor $	Equipment $	Total $
58" - 60"	S2@7.15	Ea	1,210.00	258.00	—	1,468.00
62" - 64"	S2@7.62	Ea	1,280.00	275.00	—	1,555.00
66" - 68"	S2@8.13	Ea	1,360.00	293.00	—	1,653.00
70" - 72"	S2@8.69	Ea	1,480.00	313.00	—	1,793.00

Galvanized steel 60" rectangular duct 90-degree elbow

Description	Craft@Hrs	Unit	Material $	Labor $	Equipment $	Total $
58" - 60"	S2@7.37	Ea	1,240.00	266.00	—	1,506.00
62" - 64"	S2@7.88	Ea	1,330.00	284.00	—	1,614.00
66" - 68"	S2@8.41	Ea	1,450.00	303.00	—	1,753.00
70" - 72"	S2@8.95	Ea	1,510.00	323.00	—	1,833.00

Galvanized steel 62" rectangular duct 90-degree elbow

Description	Craft@Hrs	Unit	Material $	Labor $	Equipment $	Total $
58" - 60"	S2@7.62	Ea	1,280.00	275.00	—	1,555.00
62" - 64"	S2@8.13	Ea	1,350.00	293.00	—	1,643.00
66" - 68"	S2@8.69	Ea	1,470.00	313.00	—	1,783.00
70" - 72"	S2@9.18	Ea	1,550.00	331.00	—	1,881.00

Galvanized steel 64" rectangular duct 90-degree elbow

Description	Craft@Hrs	Unit	Material $	Labor $	Equipment $	Total $
58" - 60"	S2@7.88	Ea	1,320.00	284.00	—	1,604.00
62" - 64"	S2@8.41	Ea	1,410.00	303.00	—	1,713.00
66" - 68"	S2@8.95	Ea	1,500.00	323.00	—	1,823.00
70" - 72"	S2@9.44	Ea	1,580.00	340.00	—	1,920.00

Galvanized Steel Rectangular Elbows

Description	Craft@Hrs	Unit	Material $	Labor $	Equipment $	Total $

Galvanized steel 66" rectangular duct 90-degree elbow

Description	Craft@Hrs	Unit	Material $	Labor $	Equipment $	Total $
58" - 60"	S2@8.13	Ea	1,350.00	293.00	—	1,643.00
62" - 64"	S2@8.69	Ea	1,470.00	313.00	—	1,783.00
66" - 68"	S2@9.18	Ea	1,550.00	331.00	—	1,881.00
70" - 72"	S2@9.69	Ea	1,640.00	349.00	—	1,989.00

Galvanized steel 68" rectangular duct 90-degree elbow

Description	Craft@Hrs	Unit	Material $	Labor $	Equipment $	Total $
58" - 60"	S2@8.41	Ea	1,410.00	303.00	—	1,713.00
62" - 64"	S2@8.95	Ea	1,500.00	323.00	—	1,823.00
66" - 68"	S2@9.44	Ea	1,580.00	340.00	—	1,920.00
70" - 72"	S2@9.95	Ea	1,680.00	359.00	—	2,039.00

Galvanized steel 70" rectangular duct 90-degree elbow

Description	Craft@Hrs	Unit	Material $	Labor $	Equipment $	Total $
58" - 60"	S2@8.69	Ea	1,470.00	313.00	—	1,783.00
62" - 64"	S2@9.18	Ea	1,550.00	331.00	—	1,881.00
66" - 68"	S2@9.69	Ea	1,640.00	349.00	—	1,989.00
70" - 72"	S2@10.2	Ea	1,710.00	368.00	—	2,078.00

Galvanized steel 72" rectangular duct 90-degree elbow

Description	Craft@Hrs	Unit	Material $	Labor $	Equipment $	Total $
58" - 60"	S2@8.95	Ea	1,500.00	323.00	—	1,823.00
62" - 64"	S2@9.44	Ea	1,580.00	340.00	—	1,920.00
66" - 68"	S2@9.95	Ea	1,680.00	359.00	—	2,039.00
70" - 72"	S2@10.5	Ea	1,760.00	378.00	—	2,138.00

Costs for Galvanized Rectangular Drops and Tap-in Tees. Costs per drop or tap-in tee, based on quantities less than 1,000 pounds. Deduct 15% from the material cost for over 1,000 pounds. Add 50% to the material cost for the lined duct. These costs include delivery, typical bracing, cleats, scrap, hangers, end closures, sealants and miscellaneous hardware. For turning vanes, please see page 342.

Description	Craft@Hrs	Unit	Material $	Labor $	Equipment $	Total $

Galvanized steel 6" rectangular duct drops and tap-in tees

Description	Craft@Hrs	Unit	Material $	Labor $	Equipment $	Total $
6"	S2@.072	Ea	13.20	2.59	—	15.79
8"	S2@.084	Ea	15.50	3.03	—	18.53
10"	S2@.097	Ea	17.60	3.50	—	21.10
12" - 16"	S2@.139	Ea	25.40	5.01	—	30.41
20" - 24"	S2@.216	Ea	39.40	7.78	—	47.18
28" - 32"	S2@.309	Ea	56.00	11.10	—	67.10
36" - 40"	S2@.412	Ea	75.10	14.80	—	89.90
44" - 48"	S2@.489	Ea	89.10	17.60	—	106.70
52" - 56"	S2@.591	Ea	107.00	21.30	—	128.30
60"	S2@.747	Ea	135.00	26.90	—	161.90

Galvanized steel 8" rectangular duct drops and tap-in tees

Description	Craft@Hrs	Unit	Material $	Labor $	Equipment $	Total $
6"	S2@.084	Ea	15.50	3.03	—	18.53
8"	S2@.097	Ea	17.60	3.50	—	21.10
10"	S2@.109	Ea	19.90	3.93	—	23.83
12" - 16"	S2@.152	Ea	28.10	5.48	—	33.58
20" - 24"	S2@.231	Ea	42.30	8.33	—	50.63
28" - 32"	S2@.326	Ea	60.00	11.70	—	71.70
36" - 40"	S2@.432	Ea	78.90	15.60	—	94.50
44" - 48"	S2@.506	Ea	92.20	18.20	—	110.40
52" - 56"	S2@.634	Ea	115.00	22.80	—	137.80
60"	S2@.760	Ea	137.00	27.40	—	164.40

Galvanized steel 10" rectangular duct drops and tap-in tees

Description	Craft@Hrs	Unit	Material $	Labor $	Equipment $	Total $
6"	S2@.097	Ea	17.60	3.50	—	21.10
8"	S2@.109	Ea	19.90	3.93	—	23.83
10"	S2@.121	Ea	22.20	4.36	—	26.56
12" - 16"	S2@.167	Ea	30.50	6.02	—	36.52
20" - 24"	S2@.246	Ea	45.30	8.87	—	54.17
28" - 32"	S2@.343	Ea	63.00	12.40	—	75.40
36" - 40"	S2@.450	Ea	81.70	16.20	—	97.90
44" - 48"	S2@.525	Ea	95.50	18.90	—	114.40
52" - 56"	S2@.653	Ea	118.00	23.50	—	141.50
60"	S2@.773	Ea	139.00	27.90	—	166.90

Galvanized Steel Rectangular Drops and Tees

Description	Craft@Hrs	Unit	Material $	Labor $	Equipment $	Total $

Galvanized steel 12" rectangular duct drops and tap-in tees

Description	Craft@Hrs	Unit	Material $	Labor $	Equipment $	Total $
6"	S2@.109	Ea	19.90	3.93	—	23.83
8"	S2@.121	Ea	22.20	4.36	—	26.56
10"	S2@.133	Ea	24.40	4.79	—	29.19
12" - 16"	S2@.180	Ea	32.80	6.49	—	39.29
20" - 24"	S2@.262	Ea	48.10	9.44	—	57.54
28" - 32"	S2@.360	Ea	65.80	13.00	—	78.80
36" - 40"	S2@.470	Ea	85.10	16.90	—	102.00
44" - 48"	S2@.544	Ea	98.90	19.60	—	118.50
52" - 56"	S2@.675	Ea	122.00	24.30	—	146.30
60"	S2@.794	Ea	146.00	28.60	—	174.60

Galvanized steel 16" rectangular duct drops and tap-in tees

Description	Craft@Hrs	Unit	Material $	Labor $	Equipment $	Total $
6"	S2@.170	Ea	31.00	6.13	—	37.13
8"	S2@.184	Ea	34.10	6.63	—	40.73
10"	S2@.200	Ea	36.70	7.21	—	43.91
12" - 16"	S2@.231	Ea	42.30	8.33	—	50.63
20" - 24"	S2@.293	Ea	53.10	10.60	—	63.70
28" - 32"	S2@.396	Ea	71.90	14.30	—	86.20
36" - 40"	S2@.506	Ea	92.20	18.20	—	110.40
44" - 48"	S2@.581	Ea	106.00	20.90	—	126.90
52" - 56"	S2@.709	Ea	130.00	25.60	—	155.60
60"	S2@.841	Ea	153.00	30.30	—	183.30

Galvanized steel 20" rectangular duct drops and tap-in tees

Description	Craft@Hrs	Unit	Material $	Labor $	Equipment $	Total $
6"	S2@.200	Ea	36.70	7.21	—	43.91
8"	S2@.216	Ea	39.40	7.78	—	47.18
10"	S2@.231	Ea	42.30	8.33	—	50.63
12" - 16"	S2@.262	Ea	48.10	9.44	—	57.54
20" - 24"	S2@.324	Ea	59.20	11.70	—	70.90
28" - 32"	S2@.428	Ea	78.00	15.40	—	93.40
36" - 40"	S2@.544	Ea	98.90	19.60	—	118.50
44" - 48"	S2@.617	Ea	112.00	22.20	—	134.20
52" - 56"	S2@.758	Ea	136.00	27.30	—	163.30
60"	S2@.884	Ea	162.00	31.90	—	193.90

Galvanized Steel Rectangular Drops and Tees

Description	Craft@Hrs	Unit	Material $	Labor $	Equipment $	Total $

Galvanized steel 24" rectangular duct drops and tap-in tees

Description	Craft@Hrs	Unit	Material $	Labor $	Equipment $	Total $
6"	S2@.231	Ea	42.30	8.33	—	50.63
8"	S2@.247	Ea	45.30	8.90	—	54.20
10"	S2@.262	Ea	48.10	9.44	—	57.54
12" - 16"	S2@.293	Ea	53.10	10.60	—	63.70
20" - 24"	S2@.354	Ea	64.70	12.80	—	77.50
28" - 32"	S2@.462	Ea	83.60	16.70	—	100.30
36" - 40"	S2@.581	Ea	106.00	20.90	—	126.90
44" - 48"	S2@.655	Ea	118.00	23.60	—	141.60
52" - 56"	S2@.798	Ea	147.00	28.80	—	175.80
60"	S2@.927	Ea	169.00	33.40	—	202.40

Galvanized steel 28" rectangular duct drops and tap-in tees

Description	Craft@Hrs	Unit	Material $	Labor $	Equipment $	Total $
6"	S2@.262	Ea	48.10	9.44	—	57.54
8"	S2@.278	Ea	50.20	10.00	—	60.20
10"	S2@.292	Ea	53.10	10.50	—	63.60
12" - 16"	S2@.324	Ea	59.20	11.70	—	70.90
20" - 24"	S2@.385	Ea	70.20	13.90	—	84.10
28" - 32"	S2@.497	Ea	90.60	17.90	—	108.50
36" - 40"	S2@.617	Ea	112.00	22.20	—	134.20
44" - 48"	S2@.694	Ea	125.00	25.00	—	150.00
52" - 56"	S2@.837	Ea	153.00	30.20	—	183.20
60"	S2@.969	Ea	177.00	34.90	—	211.90

Galvanized steel 32" rectangular duct drops and tap-in tees

Description	Craft@Hrs	Unit	Material $	Labor $	Equipment $	Total $
6"	S2@.356	Ea	65.00	12.80	—	77.80
8"	S2@.374	Ea	69.00	13.50	—	82.50
10"	S2@.393	Ea	71.70	14.20	—	85.90
12" - 16"	S2@.432	Ea	78.90	15.60	—	94.50
20" - 24"	S2@.506	Ea	92.20	18.20	—	110.40
28" - 32"	S2@.581	Ea	106.00	20.90	—	126.90
36" - 40"	S2@.655	Ea	118.00	23.60	—	141.60
44" - 48"	S2@.730	Ea	132.00	26.30	—	158.30
52" - 56"	S2@.877	Ea	162.00	31.60	—	193.60
60"	S2@1.02	Ea	185.00	36.80	—	221.80

Galvanized Steel Round Ductwork

Description	Craft@Hrs	Unit	Material $	Labor $	Equipment $	Total $

Galvanized steel round duct snap-lock with slip joints

Description	Craft@Hrs	Unit	Material $	Labor $	Equipment $	Total $
3"	S2@.080	LF	3.00	2.88	—	5.88
4"	S2@.090	LF	3.10	3.24	—	6.34
5"	S2@.095	LF	3.32	3.42	—	6.74
6"	S2@.100	LF	3.87	3.60	—	7.47
7"	S2@.110	LF	5.95	3.96	—	9.91
8"	S2@.120	LF	6.15	4.32	—	10.47
9"	S2@.140	LF	6.43	5.05	—	11.48
10"	S2@.160	LF	6.70	5.77	—	12.47

Galvanized steel round duct 90-degree adjustable elbow

Description	Craft@Hrs	Unit	Material $	Labor $	Equipment $	Total $
3"	S2@.250	Ea	5.80	9.01	—	14.81
4"	S2@.300	Ea	8.00	10.80	—	18.80
5"	S2@.350	Ea	9.67	12.60	—	22.27
6"	S2@.400	Ea	12.50	14.40	—	26.90
7"	S2@.450	Ea	43.00	16.20	—	59.20
8"	S2@.500	Ea	48.10	18.00	—	66.10
9"	S2@.600	Ea	52.10	21.60	—	73.70
10"	S2@.700	Ea	56.90	25.20	—	82.10

Galvanized steel round duct 45-degree adjustable elbow

Description	Craft@Hrs	Unit	Material $	Labor $	Equipment $	Total $
3"	S2@.250	Ea	5.80	9.01	—	14.81
4"	S2@.300	Ea	8.00	10.80	—	18.80
5"	S2@.350	Ea	9.67	12.60	—	22.27
6"	S2@.400	Ea	12.50	14.40	—	26.90
7"	S2@.450	Ea	43.00	16.20	—	59.20
8"	S2@.500	Ea	48.10	18.00	—	66.10
9"	S2@.600	Ea	52.10	21.60	—	73.70
10"	S2@.700	Ea	56.90	25.20	—	82.10

Fiberglass Ductwork

Standard fiberglass ductwork, or duct board, should not be used in systems where air velocities exceed 2,500 feet per minute, or where static pressures are higher than 2 inches of water gauge.

Additionally, it should not be used for the following applications:

1) Equipment rooms

2) Underground or outdoors

3) Final connections to air handling equipment

4) Plenums or casings

5) Fume, heat or moisture exhaust systems

The costs for manufacturing and installing fiberglass ductwork are based on the net areas of the duct and fittings.

Here are approximate current material costs, per square foot:

Fiberglass Type	Up to 2,500 SF Material $/SF	Over 2,500 SF Material $/SF
475	4.47	4.08
800	4.90	4.45
1400	6.01	5.78

Additional costs: Approximately 25 cents per square foot should be added for the cost of tie rods, hangers, staples, tape and waste.

Type 475 fiberglass board is used almost exclusively for residential and small commercial air systems, while Types 800 and 1400 are used in larger commercial and industrial systems.

	Fiberglass Ductwork Fabrication Labor*					
	Size ranges (inches)					
	Up through 16		17 through 32		33 through 48	
Activity	SF/Hour Output	Labor $/SF	SF/Hour Output	Labor $/SF	SF/Hour Output	Labor $/SF
Duct only	60	.59	58	.61	56	.64
Fittings only	40	.89	38	.94	35	1.02
Duct & 15% fittings	55	.65	53	.67	50	.71
Duct & 25% fittings	53	.67	51	.70	48	.74
Duct & 35% fittings	51	.70	49	.73	46	.77

*Labor rates are based on using a standard grooving machine and include all tie-rod reinforcing.

Labor correction factors:

1) Hand grooving in lieu of machine grooving: 1.58

2) Use of an auto-closer: .82

3) Tie-rods enclosed in conduit: 1.36

4) T-bar or channel reinforcing in lieu of tie-rods: 1.25

In the following installation table the measurements refer to the "semi-perimeter", the width plus depth of the ductwork. Example – 12" duct (8" wide plus 4" high) has 2.00 SF of area for each linear foot.

Fiberglass Ductwork

Description	Craft@Hrs	Unit	Material $	Labor $	Equipment $	Total $
Fiberglass ductwork installation						
12" (2.00 SF/LF)	S2@.070	LF	—	2.52	—	2.52
14" (2.35 SF/LF)	S2@.082	LF	—	2.96	—	2.96
16" (2.65 SF/LF)	S2@.093	LF	—	3.35	—	3.35
18" (3.00 SF/LF)	S2@.105	LF	—	3.78	—	3.78
20" (3.35 SF/LF)	S2@.117	LF	—	4.22	—	4.22
22" (3.65 SF/LF)	S2@.128	LF	—	4.61	—	4.61
24" (4.00 SF/LF)	S2@.140	LF	—	5.05	—	5.05
26" (4.35 SF/LF)	S2@.152	LF	—	5.48	—	5.48
28" (4.65 SF/LF)	S2@.163	LF	—	5.87	—	5.87
30" (5.00 SF/LF)	S2@.175	LF	—	6.31	—	6.31
32" (5.35 SF/LF)	S2@.187	LF	—	6.74	—	6.74
34" (5.65 SF/LF)	S2@.198	LF	—	7.14	—	7.14
36" (6.00 SF/LF)	S2@.210	LF	—	7.57	—	7.57
38" (6.35 SF/LF)	S2@.222	LF	—	8.00	—	8.00
40" (6.65 SF/LF)	S2@.232	LF	—	8.36	—	8.36
42" (7.00 SF/LF)	S2@.245	LF	—	8.83	—	8.83
44" (7.35 SF/LF)	S2@.257	LF	—	9.26	—	9.26
46" (7.65 SF/LF)	S2@.268	LF	—	9.66	—	9.66
48" (8.00 SF/LF)	S2@.280	LF	—	10.10	—	10.10
50" (8.35 SF/LF)	S2@.292	LF	—	10.50	—	10.50
52" (8.65 SF/LF)	S2@.303	LF	—	10.90	—	10.90
54" (9.00 SF/LF)	S2@.315	LF	—	11.40	—	11.40
56" (9.35 SF/LF)	S2@.327	LF	—	11.80	—	11.80
58" (9.65 SF/LF)	S2@.338	LF	—	12.20	—	12.20
60" (10.0 SF/LF)	S2@.350	LF	—	12.60	—	12.60
64" (10.6 SF/LF)	S2@.373	LF	—	13.40	—	13.40
68" (11.4 SF/LF)	S2@.397	LF	—	14.30	—	14.30
72" (12.0 SF/LF)	S2@.420	LF	—	15.10	—	15.10
76" (12.6 SF/LF)	S2@.443	LF	—	16.00	—	16.00
80" (13.4 SF/LF)	S2@.467	LF	—	16.80	—	16.80
84" (14.0 SF/LF)	S2@.490	LF	—	17.70	—	17.70
88" (14.6 SF/LF)	S2@.513	LF	—	18.50	—	18.50
92" (15.4 SF/LF)	S2@.537	LF	—	19.40	—	19.40
96" (16.0 SF/LF)	S2@.560	LF	—	20.20	—	20.20

For estimating purposes, when measuring the length of ductwork, include the fitting.
Then apply the labor per run foot to the total.

1¼" thick wire reinforced flexible fiberglass duct with vinyl cover

Description	Craft@Hrs	Unit	Material $	Labor $	Equipment $	Total $
4"	S2@.060	LF	8.16	2.16	—	10.32
5"	S2@.070	LF	8.45	2.52	—	10.97
6"	S2@.080	LF	9.25	2.88	—	12.13
8"	S2@.090	LF	11.80	3.24	—	15.04
9"	S2@.100	LF	12.50	3.60	—	16.10
10"	S2@.120	LF	13.20	4.32	—	17.52
12"	S2@.140	LF	17.00	5.05	—	22.05
14"	S2@.160	LF	20.30	5.77	—	26.07
16"	S2@.180	LF	23.60	6.49	—	30.09
18"	S2@.200	LF	26.80	7.21	—	34.01

Fiberglass Pipe Insulation with AP-T Plus (all-purpose self-sealing) Jacket. By nominal pipe diameter. R factor equals 2.56 at 300 degrees F. These costs do not include scaffolding; add for same, if required. Also see fittings, flanges and valves at the end of this section.

Description	Craft@Hrs	Unit	Material $	Labor $	Equipment $	Total $

Jacketed fiberglass insulation for ½" pipe

Description	Craft@Hrs	Unit	Material $	Labor $	Equipment $	Total $
½" thick	P1@.038	LF	2.54	1.39	—	3.93
1" thick	P1@.038	LF	2.98	1.39	—	4.37
1½" thick	P1@.040	LF	6.08	1.47	—	7.55

Jacketed fiberglass insulation for ¾" pipe

Description	Craft@Hrs	Unit	Material $	Labor $	Equipment $	Total $
½" thick	P1@.038	LF	2.75	1.39	—	4.14
1" thick	P1@.038	LF	3.39	1.39	—	4.78
1½" thick	P1@.040	LF	6.30	1.47	—	7.77

Jacketed fiberglass insulation for 1" pipe

Description	Craft@Hrs	Unit	Material $	Labor $	Equipment $	Total $
½" thick	P1@.040	LF	2.92	1.47	—	4.39
1" thick	P1@.040	LF	3.88	1.47	—	5.35
1½" thick	P1@.042	LF	6.67	1.54	—	8.21

Jacketed fiberglass insulation for 1¼" pipe

Description	Craft@Hrs	Unit	Material $	Labor $	Equipment $	Total $
½" thick	P1@.040	LF	3.24	1.47	—	4.71
1" thick	P1@.040	LF	3.92	1.47	—	5.39
1½" thick	P1@.042	LF	7.10	1.54	—	8.64

Jacketed fiberglass insulation for 1½" pipe

Description	Craft@Hrs	Unit	Material $	Labor $	Equipment $	Total $
½" thick	P1@.042	LF	3.48	1.54	—	5.02
1" thick	P1@.042	LF	4.42	1.54	—	5.96
1½" thick	P1@.044	LF	7.64	1.61	—	9.25

Jacketed fiberglass insulation for 2" pipe

Description	Craft@Hrs	Unit	Material $	Labor $	Equipment $	Total $
½" thick	P1@.044	LF	3.64	1.61	—	5.25
1" thick	P1@.044	LF	5.07	1.61	—	6.68
1½" thick	P1@.046	LF	8.20	1.69	—	9.89

Jacketed fiberglass insulation for 2½" pipe

Description	Craft@Hrs	Unit	Material $	Labor $	Equipment $	Total $
1" thick	P1@.047	LF	5.23	1.72	—	6.95
1½" thick	P1@.048	LF	7.42	1.76	—	9.18
2" thick	P1@.050	LF	11.80	1.84	—	13.64

Jacketed fiberglass insulation for 3" pipe

Description	Craft@Hrs	Unit	Material $	Labor $	Equipment $	Total $
1" thick	P1@.051	LF	5.76	1.87	—	7.63
1½" thick	P1@.054	LF	8.11	1.98	—	10.09
2" thick	P1@.056	LF	13.00	2.06	—	15.06

Fiberglass Pipe Insulation

Description	Craft@Hrs	Unit	Material $	Labor $	Equipment $	Total $

Jacketed fiberglass insulation for 4" pipe

Description	Craft@Hrs	Unit	Material $	Labor $	Equipment $	Total $
1" thick	P1@.063	LF	6.78	2.31	—	9.09
1½" thick	P1@.066	LF	9.04	2.42	—	11.46
2" thick	P1@.069	LF	14.50	2.53	—	17.03

Jacketed fiberglass insulation for 6" pipe

Description	Craft@Hrs	Unit	Material $	Labor $	Equipment $	Total $
1" thick	P1@.077	LF	7.89	2.83	—	10.72
1½" thick	P1@.080	LF	10.90	2.94	—	13.84
2" thick	P1@.085	LF	18.40	3.12	—	21.52

Jacketed fiberglass insulation for 8" pipe

Description	Craft@Hrs	Unit	Material $	Labor $	Equipment $	Total $
1" thick	P1@.117	LF	10.90	4.29	—	15.19
1½" thick	P1@.123	LF	13.20	4.51	—	17.71
2" thick	P1@.129	LF	21.60	4.73	—	26.33

Jacketed fiberglass insulation for 10" pipe

Description	Craft@Hrs	Unit	Material $	Labor $	Equipment $	Total $
1" thick	P1@.140	LF	12.90	5.14	—	18.04
1½" thick	P1@.147	LF	16.30	5.39	—	21.69
2" thick	P1@.154	LF	25.80	5.65	—	31.45

Jacketed fiberglass insulation for 12" pipe

Description	Craft@Hrs	Unit	Material $	Labor $	Equipment $	Total $
1" thick	P1@.155	LF	14.90	5.69	—	20.59
1½" thick	P1@.163	LF	18.40	5.98	—	24.38
2" thick	P1@.171	LF	28.20	6.28	—	34.48
Add for .016" aluminum jacket						
Per SF of surface	P1@.035	SF	2.13	1.28	—	3.41

Pipe fittings and flanges

For each fitting or flange use the cost for 3 LF of pipe of the same pipe size

Valves

Body only: for each valve body use the cost for 5 LF of pipe of the same pipe size

Body and bonnet or yoke: for each valve use the cost for 10 LF of pipe of the same pipe size

Flanged valves: add the cost for flanges per above

Calcium Silicate Pipe Insulation with Aluminum Jacket

Calcium Silicate Pipe Insulation with .016" Aluminum Jacket. By nominal pipe diameter. R factor equals 2.20 at 300 degrees F. Manufactured by Pabco. These costs do not include scaffolding; add for same, if required. Also see fittings, flanges and valves below.

Description	Craft@Hrs	Unit	Material $	Labor $	Equipment $	Total $

Jacketed calcium silicate insulation for 6" pipe

Description	Craft@Hrs	Unit	Material $	Labor $	Equipment $	Total $
2" thick	P1@.183	LF	34.60	6.72	—	41.32
4" thick	P1@.218	LF	74.90	8.00	—	82.90
6" thick	P1@.288	LF	106.00	10.60	—	116.60

Jacketed calcium silicate insulation for 8" pipe

Description	Craft@Hrs	Unit	Material $	Labor $	Equipment $	Total $
2" thick	P1@.194	LF	41.50	7.12	—	48.62
4" thick	P1@.228	LF	87.80	8.37	—	96.17
6" thick	P1@.308	LF	133.00	11.30	—	144.30

Jacketed calcium silicate insulation for 10" pipe

Description	Craft@Hrs	Unit	Material $	Labor $	Equipment $	Total $
2" thick	P1@.202	LF	48.40	7.41	—	55.81
4" thick	P1@.239	LF	103.00	8.77	—	111.77
6" thick	P1@.337	LF	161.00	12.40	—	173.40

Jacketed calcium silicate insulation for 12" pipe

Description	Craft@Hrs	Unit	Material $	Labor $	Equipment $	Total $
2" thick	P1@.208	LF	56.00	7.63	—	63.63
4" thick	P1@.261	LF	115.00	9.58	—	124.58
6" thick	P1@.388	LF	182.00	14.20	—	196.20

Pipe fittings and flanges
For each fitting or flange use the cost for 3 LF of pipe of the same pipe size

Valves
Body only: for each valve body use the cost for 5 LF of pipe of the same pipe size

Body and bonnet or yoke: for each valve use the cost for 10 LF of pipe of the same pipe size

Flanged valves: add the cost for flanges per above

Closed Cell Elastomeric Pipe Insulation

Description	Craft@Hrs	Unit	Material $	Labor $	Equipment $	Total $

Closed Cell Elastomeric Pipe and Tubing Insulation. Semi split, by nominal pipe or tube diameter with insulation wall thickness as shown, no cover. R factor equals 3.58 at 220 degrees F. Manufactured by Rubatex. These costs do not include scaffolding; add for same, if required. Also see fittings, flanges and valves below.

Description	Craft@Hrs	Unit	Material $	Labor $	Equipment $	Total $
¼", ½" thick	P1@.039	LF	1.05	1.43	—	2.48
3/8", ½" thick	P1@.039	LF	1.08	1.43	—	2.51
½", ½" thick	P1@.039	LF	1.20	1.43	—	2.63
¾", ½" thick	P1@.039	LF	1.37	1.43	—	2.80
1", ½" thick	P1@.042	LF	1.53	1.54	—	3.07
1¼", ½" thick	P1@.042	LF	1.71	1.54	—	3.25
1½", ½" thick	P1@.042	LF	2.00	1.54	—	3.54
2", ½" thick	P1@.046	LF	2.50	1.69	—	4.19
2½", 3/8" thick	P1@.056	LF	3.22	2.06	—	5.28
2½", ½" thick	P1@.056	LF	4.31	2.06	—	6.37
2½", ¾" thick	P1@.056	LF	5.56	2.06	—	7.62
3", 3/8" thick	P1@.062	LF	3.67	2.28	—	5.95
3", ½" thick	P1@.062	LF	4.84	2.28	—	7.12
3", ¾" thick	P1@.062	LF	6.30	2.28	—	8.58
4", 3/8" thick	P1@.068	LF	4.31	2.50	—	6.81
4", ½" thick	P1@.068	LF	5.68	2.50	—	8.18
4", ¾" thick	P1@.068	LF	7.39	2.50	—	9.89

Pipe fittings and flanges
For each fitting or flange use the cost for 3 LF of pipe of the same pipe size

Valves
Body only: for each valve body use the cost for 5 LF of pipe of the same pipe size

Body and bonnet or yoke: for each valve use the cost for 10 LF of pipe of the same pipe size

Flanged valves: add the cost for flanges per above

Description	Craft@Hrs	Unit	Material $	Labor $	Equipment $	Total $

Fiberglass flexible wrap duct insulation with foil facing vapor barrier to 250 degrees F

Description	Craft@Hrs	Unit	Material $	Labor $	Equipment $	Total $
1"	S2@.023	SF	1.10	.83	—	1.93
1½"	S2@.023	SF	1.32	.83	—	2.15
2"	S2@.025	SF	1.80	.90	—	2.70
Deduct foil facing	—	%	−30.0	—	—	—

Fiberglass rigid board duct insulation plain, *without* foil facing vapor barrier to 450 degrees F

Description	Craft@Hrs	Unit	Material $	Labor $	Equipment $	Total $
1"	S2@.067	SF	1.35	2.41	—	3.76
1½"	S2@.067	SF	2.29	2.41	—	4.70
2"	S2@.075	SF	2.70	2.70	—	5.40
3"	S2@.085	SF	4.03	3.06	—	7.09

Fiberglass rigid board duct insulation with FSK (foil) facing vapor barrier to 250 degrees F

Description	Craft@Hrs	Unit	Material $	Labor $	Equipment $	Total $
1"	S2@.067	SF	1.80	2.41	—	4.21
1½"	S2@.067	SF	2.19	2.41	—	4.60
2"	S2@.075	SF	2.60	2.70	—	5.30
3"	S2@.085	SF	3.71	3.06	—	6.77

Fiberglass flexible blanket duct lining to 250 degrees F

Description	Craft@Hrs	Unit	Material $	Labor $	Equipment $	Total $
½"	S2@.032	SF	.82	1.15	—	1.97
1"	S2@.032	SF	1.07	1.15	—	2.22
1½"	S2@.035	SF	1.53	1.26	—	2.79
2"	S2@.035	SF	2.00	1.26	—	3.26

Fiberglass rigid duct lining to 250 degrees F

Description	Craft@Hrs	Unit	Material $	Labor $	Equipment $	Total $
5/8"	S2@.032	SF	1.79	1.15	—	2.94
1"	S2@.032	SF	2.31	1.15	—	3.46
1½"	S2@.035	SF	3.04	1.26	—	4.30
2"	S2@.035	SF	3.87	1.26	—	5.13

Note: Material prices include all necessary tape, pins, mastic and 15% waste.

Balancing of HVAC Systems

Description	Craft@Hrs	Unit	Material $	Labor $	Equipment $	Total $

Air balancing – air handling units

Description	Craft@Hrs	Unit	Material $	Labor $	Equipment $	Total $
Central station	S2@6.00	Ea	—	216.00	—	216.00
Multi-zone A/C	S2@4.00	Ea	—	144.00	—	144.00
Single-zone A/C	S2@3.00	Ea	—	108.00	—	108.00
Packaged A/C	S2@2.50	Ea	—	90.10	—	90.10
Built-up low pressure	S2@4.50	Ea	—	162.00	—	162.00
Built-up high pres.	S2@6.00	Ea	—	216.00	—	216.00
Built-up VAV	S2@6.00	Ea	—	216.00	—	216.00
Built-up dual duct	S2@6.00	Ea	—	216.00	—	216.00

Air balancing – terminal boxes

Description	Craft@Hrs	Unit	Material $	Labor $	Equipment $	Total $
Variable air volume	S2@0.40	Ea	—	14.40	—	14.40
Constant volume	S2@0.40	Ea	—	14.40	—	14.40
Dual duct box	S2@0.70	Ea	—	25.20	—	25.20
Terminal box and all downstream diffusers cost per zone	S2@1.75	Ea	—	63.10	—	63.10

Air balancing – diffusers

Description	Craft@Hrs	Unit	Material $	Labor $	Equipment $	Total $
Ceiling diffusers to 500 CFM	S2@.250	Ea	—	9.01	—	9.01
Ceiling diffusers over 500 CFM	S2@.350	Ea	—	12.60	—	12.60
Ceiling diffusers perforated	S2@.370	Ea	—	13.30	—	13.30
Sidewall diffusers double deflection	S2@.250	Ea	—	9.01	—	9.01
Linear slot diffusers 2 slot	S2@.060	LF	—	2.16	—	2.16
Linear slot diffusers 3 slot	S2@.070	LF	—	2.52	—	2.52
Linear slot diffusers 4 slot	S2@.080	LF	—	2.88	—	2.88
Plug-in diffuser	S2@.250	Ea	—	9.01	—	9.01
Light troffers 1 slot	S2@.195	Ea	—	7.03	—	7.03
Light troffers 2 slot	S2@.220	Ea	—	7.93	—	7.93
Laminar flow diff.	S2@.750	Ea	—	27.00	—	27.00
Laminar flow diffusers with HEPA filter	S2@3.00	Ea	—	108.00	—	108.00

Description	Craft@Hrs	Unit	Material $	Labor $	Equipment $	Total $

Air balancing – grilles

Description	Craft@Hrs	Unit	Material $	Labor $	Equipment $	Total $
Return air grilles to 800 CFM	S2@.250	Ea	—	9.01	—	9.01
Return air grilles over 800 CFM	S2@.350	Ea	—	12.60	—	12.60
Bar type return air grilles 2 slot	S2@.060	LF	—	2.16	—	2.16
Bar type return air grilles 4 slot	S2@.070	LF	—	2.52	—	2.52
Bar type return air grilles 6 slot	S2@.080	LF	—	2.88	—	2.88

Air balancing – centrifugal fans

Description	Craft@Hrs	Unit	Material $	Labor $	Equipment $	Total $
Supply and return air to 3,000 CFM	S2@1.00	Ea	—	36.00	—	36.00
Supply and return air to 5,000 CFM	S2@1.50	Ea	—	54.10	—	54.10
Supply and return air to 10,000 CFM	S2@2.00	Ea	—	72.10	—	72.10
Supply and return air to 20,000 CFM	S2@2.75	Ea	—	99.10	—	99.10
Roof exhaust to 6,000 CFM	S2@.725	Ea	—	26.10	—	26.10
Roof exhaust to 10,000 CFM	S2@.925	Ea	—	33.30	—	33.30
Roof exhaust to 20,000 CFM	S2@1.50	Ea	—	54.10	—	54.10
Range hood exhaust to 10,000 CFM	S2@1.75	Ea	—	63.10	—	63.10
Range hood exhaust to 20,000 CFM	S2@2.50	Ea	—	90.10	—	90.10

Air balancing – fan coil units

Description	Craft@Hrs	Unit	Material $	Labor $	Equipment $	Total $
Vertical floor model	S2@.500	Ea	—	18.00	—	18.00
Horiz. ceiling mtd	S2@.650	Ea	—	23.40	—	23.40

Air balancing – fume hoods

Description	Craft@Hrs	Unit	Material $	Labor $	Equipment $	Total $
Lab fume hood	S2@2.00	Ea	—	72.10	—	72.10
General purpose	S2@.450	Ea	—	16.20	—	16.20
Dust collectors	S2@1.50	Ea	—	54.10	—	54.10

Balancing of HVAC Systems

Description	Craft@Hrs	Unit	Material $	Labor $	Equipment $	Total $
Wet balancing						
Boiler	S2@1.65	Ea	—	59.50	—	59.50
Chiller	S2@3.50	Ea	—	126.00	—	126.00
Cooling tower	S2@2.00	Ea	—	72.10	—	72.10
Circulating pump	S2@1.65	Ea	—	59.50	—	59.50
Heat exchanger	S2@1.75	Ea	—	63.10	—	63.10
AHU water coil	S2@2.50	Ea	—	90.10	—	90.10
Reheat coil	S2@.450	Ea	—	16.20	—	16.20
Wall fin radiation	S2@.350	Ea	—	12.60	—	12.60
Radiant heat panel	S2@.450	Ea	—	16.20	—	16.20
Fan coil unit, 2 pipe	S2@.350	Ea	—	12.60	—	12.60
Fan coil unit, 4 pipe	S2@.450	Ea	—	16.20	—	16.20
Cabinet heater	S2@.350	Ea	—	12.60	—	12.60
Unit ventilator	S2@.450	Ea	—	16.20	—	16.20
Unit heater	S2@.350	Ea	—	12.60	—	12.60

Commercial HVAC controls. Temperature, pressure and humidity control for any device. 0-10 Volts DC or 4-20 milliamp input. Honeywell T775U or equal. Complies with ANSI/ACCA standard 5QI-2007. With two modulating and two relay outputs, each with their own setpoint.

Description	Craft@Hrs	Unit	Material $	Labor $	Equipment $	Total $
Install control module enclosure	BE@1.00	Ea	—	40.40	—	40.40
Twisted pair Category 6 PVC cable	BE@.008	LF	.54	.32	—	.86
Panel wiring	BE@.250	Ea	2.87	10.10	—	12.97
Relay and damper actuator wiring	BE@.250	Ea	2.57	10.10	—	12.67
Sensor wiring	BE@.300	Ea	2.57	12.10	—	14.67
Zone control module wiring	BE@.400	Ea	2.40	16.20	—	18.60
Control module and wiring	BE@2.00	Ea	458.00	80.80	—	538.80
Ground fault protection wiring	BE@1.25	Ea	143.00	50.50	—	193.50
Programming and system test	BE@3.00	Ea	—	121.00	—	121.00
Commissioning and final test	BE@8.00	Ea	—	323.00	—	323.00

Sensor and zone control modules. Honeywell Webstat TR23 or equal. 20K non-linear sensor with setpoint adjust 55-85, and override LED. With LonWorks interface card and power supply. Add the cost of control wiring.

Description	Craft@Hrs	Unit	Material $	Labor $	Equipment $	Total $
Room sensor	BE@.500	Ea	87.20	20.20	—	107.40

Damper actuators. Modutrol IV or equal. For rectangular volume control dampers. Add the cost of wiring and the damper from the section Ductwork Specialties.

Description	Craft@Hrs	Unit	Material $	Labor $	Equipment $	Total $
Lever arm actuator	BE@.500	Ea	543.00	20.20	—	563.20

Description	Craft@Hrs	Unit	Material $	Labor $	Equipment $	Total $

Remote CO_2 sensors. Honeywell C72321 or equal, with LonWorks interface card and power supply. Add the cost of wiring.

Description	Craft@Hrs	Unit	Material $	Labor $	Equipment $	Total $
Panel mount	BE@.500	Ea	92.90	20.20	—	113.10

Monitoring and recording equipment. Honeywell DR4500A digital circular chart recorder or equal, with LonWorks card and power supply. Add the cost of wiring.

Description	Craft@Hrs	Unit	Material $	Labor $	Equipment $	Total $
Panel mount	BE@.500	Ea	4,410.00	20.20	—	4,430.20

Thermostat, electric. Install and connect. Make additional allowances for sub-subcontractor markups and control wiring as required.

Description	Craft@Hrs	Unit	Material $	Labor $	Equipment $	Total $
Heat/cool						
low voltage	BE@.575	Ea	46.40	23.20	—	69.60
Heat only						
low voltage	BE@.500	Ea	38.70	20.20	—	58.90
Heat /cool						
line voltage	BE@.650	Ea	140.00	26.30	—	166.30
Programmable – residential						
low voltage	BE@.850	LF	213.00	34.30	—	247.30
Programmable – commercial w/subbase						
low voltage	BE@1.10	LF	438.00	44.40	—	482.40

Thermostat, pneumatic. Install and connect. Make additional allowances for sub-subcontractor markups and pneumatic tubing as required.

Description	Craft@Hrs	Unit	Material $	Labor $	Equipment $	Total $
Direct acting						
55 to 85 Deg. F.	BE@.675	Ea	120.00	27.30	—	147.30
Reverse acting						
55 to 85 Deg. F.	BE@.675	Ea	100.00	27.30	—	127.30
Floating	BE@.675	Ea	216.00	27.30	—	243.30

Control valve, 2-way, electric. 6.9 Volt Input. Proportional Spring Return. Connect only. N.O. means: "normally open." Make additional allowances for sub-subcontractor markups, control wiring and valve installation by piping trades as required. (See Plumbing and Piping Specialties)

Description	Craft@Hrs	Unit	Material $	Labor $	Equipment $	Total $
½", 2-way, N.O.	BE@.400	Ea	550.00	16.20	—	566.20
¾", 2-way, N.O.	BE@.400	Ea	584.00	16.20	—	600.20
1", 2-way, N.O.	BE@.400	Ea	630.00	16.20	—	646.20
1¼" 2-way, N.O.	BE@.400	Ea	688.00	16.20	—	704.20
1½" 2-way, N.O.	BE@.400	Ea	1,480.00	16.20	—	1,496.20
2", 2-way, N.O.	BE@.400	Ea	1,620.00	16.20	—	1,636.20

Control valve, 3-way, electric. 6.9 Volt Input, Proportional Spring Return. Connect only. Make additional allowances for sub-subcontractor markups, control wiring and valve installation by piping trades as required. (See Plumbing and Piping Specialties)

Description	Craft@Hrs	Unit	Material $	Labor $	Equipment $	Total $
½", 3-way	BE@.450	Ea	573.00	18.20	—	591.20
¾", 3-way	BE@.450	Ea	624.00	18.20	—	642.20
1", 3-way	BE@.450	Ea	747.00	18.20	—	765.20

Temperature Controls

Description	Craft@Hrs	Unit	Material $	Labor $	Equipment $	Total $

Control valve, 2-way, pneumatic. Proportional Spring Return, 3 to 7 psi Input, Connect only. N.O. means "normally open." Make additional allowances for sub-subcontractor markups, pneumatic tubing and valve installation by piping trades as required. (See Plumbing and Piping Specialties)

Description	Craft@Hrs	Unit	Material $	Labor $	Equipment $	Total $
½", 2-way, N.O.	BE@.400	Ea	226.00	16.20	—	242.20
¾", 2-way, N.O.	BE@.400	Ea	264.00	16.20	—	280.20
1", 2-way, N.O.	BE@.400	Ea	311.00	16.20	—	327.20
1¼" 2-way, N.O.	BE@.400	Ea	368.00	16.20	—	384.20
1½" 2-way, N.O.	BE@.400	Ea	656.00	16.20	—	672.20
2", 2-way, N.O.	BE@.400	Ea	783.00	16.20	—	799.20

Control valve, 3-way, pneumatic. Proportional Spring Return, 3 to 7 psi Input, Connect only. Make additional allowances for sub-subcontractor markups, pneumatic tubing and valve installation by piping trades as required. (See Plumbing and Piping Specialties)

Description	Craft@Hrs	Unit	Material $	Labor $	Equipment $	Total $
½", 3-way	BE@.450	Ea	253.00	18.20	—	271.20
¾", 3-way	BE@.450	Ea	304.00	18.20	—	322.20
1", 3-way	BE@.450	Ea	422.00	18.20	—	440.20
1¼", 3-way	BE@.450	Ea	489.00	18.20	—	507.20
1½", 3-way	BE@.450	Ea	701.00	18.20	—	719.20
2", 3-way	BE@.450	Ea	820.00	18.20	—	838.20

Air compressor. Reciprocating piston-type. Larger systems include HVAC actuator controls. Underwriters Laboratories and CSA rated. With ASME U-stamped pressure tank, pressure safety relief valve, silencer, lubrication system with lube cooling radiator and fan. Single-phase 110 volt electric motor, drive belt and OSHA-rated safety belt guard. Add the cost of compressor hose and fittings and hose reel, if required. Set in place only. Add installation costs from the table below.

Description	Craft@Hrs	Unit	Material $	Labor $	Equipment $	Total $
½ hp	SL@3.00	Ea	417.00	109.00	—	526.00
¾ hp	SL@3.00	Ea	457.00	109.00	—	566.00
1 hp	SL@3.50	Ea	531.00	128.00	—	659.00
1½ hp	SL@3.50	Ea	588.00	128.00	—	716.00
2 hp	SL@3.50	Ea	625.00	128.00	—	753.00

Installation of air compressors.

Description	Craft@Hrs	Unit	Material $	Labor $	Equipment $	Total $
Cut, frame support and gasket hole in sidewall	S2@1.00	Ea	31.00	36.00	—	67.00
Install piping for compressor line	P1@.100	LF	4.16	3.67	—	7.83
Install electrical wiring	BE@.150	LF	1.93	6.06	—	7.99
Install compressor controls	BE@.500	Ea	—	20.20	—	20.20
Commission and test	P1@2.00	Ea	—	73.40	—	73.40
Fine-tune system	P1@1.00	Ea	—	36.70	—	36.70

Ductile Iron, Class 153, Cement-Lined with Mechanical Joints

Description	Craft@Hrs	Unit	Material $	Labor $	Equipment $	Total $

Cement-lined Class 153 ductile iron pipe with mechanical joints

Description	Craft@Hrs	Unit	Material $	Labor $	Equipment $	Total $
4"	P1@.140	LF	13.40	5.14	—	18.54
6"	ER@.180	LF	19.40	7.09	3.40	29.89
8"	ER@.200	LF	26.60	7.88	3.78	38.26
10"	ER@.230	LF	39.40	9.06	4.34	52.80
12"	ER@.270	LF	53.80	10.60	5.10	69.50
14"	ER@.310	LF	81.70	12.20	5.85	99.75
16"	ER@.350	LF	105.00	13.80	6.61	125.41

Cement-lined Class 153 ductile iron 1/16 bend with mechanical joints

Description	Craft@Hrs	Unit	Material $	Labor $	Equipment $	Total $
4"	P1@.900	Ea	53.30	33.00	—	86.30
6"	ER@1.30	Ea	82.20	51.20	24.50	157.90
8"	ER@1.70	Ea	129.00	67.00	32.10	228.10
10"	ER@2.15	Ea	176.00	84.70	40.60	301.30
12"	ER@2.40	Ea	244.00	94.60	45.30	383.90
14"	ER@2.70	Ea	536.00	106.00	51.00	693.00
16"	ER@3.00	Ea	605.00	118.00	56.60	779.60

Cement-lined Class 153 ductile iron 1/8 bend with mechanical joints

Description	Craft@Hrs	Unit	Material $	Labor $	Equipment $	Total $
4"	P1@.900	Ea	55.00	33.00	—	88.00
6"	ER@1.30	Ea	81.70	51.20	24.50	157.40
8"	ER@1.70	Ea	131.00	67.00	32.10	230.10
10"	ER@2.15	Ea	179.00	84.70	40.60	304.30
12"	ER@2.40	Ea	248.00	94.60	45.30	387.90
14"	ER@2.70	Ea	533.00	106.00	51.00	690.00
16"	ER@3.00	Ea	623.00	118.00	56.60	797.60

Cement-lined Class 153 ductile iron 1/4 bend with mechanical joints

Description	Craft@Hrs	Unit	Material $	Labor $	Equipment $	Total $
4"	P1@.900	Ea	58.90	33.00	—	91.90
6"	ER@1.30	Ea	95.70	51.20	24.50	171.40
8"	ER@1.70	Ea	201.00	67.00	32.10	300.10
10"	ER@2.15	Ea	301.00	84.70	40.60	426.30
12"	ER@2.40	Ea	412.00	94.60	45.30	551.90
14"	ER@2.70	Ea	1,060.00	106.00	51.00	1,217.00
16"	ER@3.00	Ea	1,110.00	118.00	56.60	1,284.60

Cement-lined Class 153 ductile iron tee with mechanical joints

Description	Craft@Hrs	Unit	Material $	Labor $	Equipment $	Total $
4"	P1@1.65	Ea	89.00	60.60	—	149.60
6"	ER@2.35	Ea	131.00	92.60	44.40	268.00
8"	ER@3.15	Ea	201.00	124.00	59.50	384.50
10"	ER@4.00	Ea	301.00	158.00	75.50	534.50
12"	ER@4.45	Ea	412.00	175.00	84.00	671.00
14"	ER@4.90	Ea	1,060.00	193.00	92.50	1,345.50
16"	ER@5.30	Ea	1,110.00	209.00	100.00	1,419.00

Ductile Iron, Class 153, Cement-Lined with Mechanical Joints

Description	Craft@Hrs	Unit	Material $	Labor $	Equipment $	Total $

Cement-lined Class 153 ductile iron wye with mechanical joints

Description	Craft@Hrs	Unit	Material $	Labor $	Equipment $	Total $
4"	P1@1.65	Ea	201.00	60.60	—	261.60
6"	ER@2.35	Ea	248.00	92.60	44.40	385.00
8"	ER@3.15	Ea	376.00	124.00	59.50	559.50
10"	ER@4.00	Ea	699.00	158.00	75.50	932.50
12"	ER@4.45	Ea	1,050.00	175.00	84.00	1,309.00
14"	ER@4.90	Ea	2,160.00	193.00	92.50	2,445.50
16"	ER@5.30	Ea	2,700.00	209.00	100.00	3,009.00

Cement-lined Class 153 ductile iron reducer with mechanical joints

Description	Craft@Hrs	Unit	Material $	Labor $	Equipment $	Total $
6"	ER@1.45	Ea	65.40	57.10	27.40	149.90
8"	ER@2.00	Ea	103.00	78.80	37.80	219.60
10"	ER@2.55	Ea	137.00	100.00	48.10	285.10
12"	ER@3.05	Ea	176.00	120.00	57.60	353.60
14"	ER@3.60	Ea	394.00	142.00	68.00	604.00
16"	ER@4.20	Ea	480.00	165.00	79.30	724.30

Ductile Iron, 153, Double Cement-Lined with Mechanical Joints

Description	Craft@Hrs	Unit	Material $	Labor $	Equipment $	Total $

Double cement-lined Class 153 ductile iron pipe with mechanical joints

Description	Craft@Hrs	Unit	Material $	Labor $	Equipment $	Total $
4"	P1@.140	LF	14.30	5.14	—	19.44
6"	ER@.180	LF	20.60	7.09	3.40	31.09
8"	ER@.200	LF	28.10	7.88	3.78	39.76
10"	ER@.230	LF	42.00	9.06	4.34	55.40
12"	ER@.270	LF	57.20	10.60	5.10	72.90
14"	ER@.310	LF	87.00	12.20	5.85	105.05
16"	ER@.350	LF	112.00	13.80	6.61	132.41

Double cement-lined Class 153 ductile iron 1/16 bend with mechanical joints

Description	Craft@Hrs	Unit	Material $	Labor $	Equipment $	Total $
4"	P1@.900	Ea	62.70	33.00	—	95.70
6"	ER@1.30	Ea	97.00	51.20	24.50	172.70
8"	ER@1.70	Ea	155.00	67.00	32.10	254.10
10"	ER@2.15	Ea	208.00	84.70	40.60	333.30
12"	ER@2.40	Ea	284.00	94.60	45.30	423.90
14"	ER@2.70	Ea	636.00	106.00	51.00	793.00
16"	ER@3.00	Ea	721.00	118.00	56.60	895.60

Double cement-lined Class 153 ductile iron 1/8 bend with mechanical joints

Description	Craft@Hrs	Unit	Material $	Labor $	Equipment $	Total $
4"	P1@.900	Ea	62.70	33.00	—	95.70
6"	ER@1.30	Ea	94.40	51.20	24.50	170.10
8"	ER@1.70	Ea	151.00	67.00	32.10	250.10
10"	ER@2.15	Ea	207.00	84.70	40.60	332.30
12"	ER@2.40	Ea	281.00	94.60	45.30	420.90
14"	ER@2.70	Ea	619.00	106.00	51.00	776.00
16"	ER@3.00	Ea	721.00	118.00	56.60	895.60

Double cement-lined Class 153 ductile iron 1/4 bend with mechanical joints

Description	Craft@Hrs	Unit	Material $	Labor $	Equipment $	Total $
4"	P1@.900	Ea	67.80	33.00	—	100.80
6"	ER@1.30	Ea	110.00	51.20	24.50	185.70
8"	ER@1.70	Ea	171.00	67.00	32.10	270.10
10"	ER@2.15	Ea	266.00	84.70	40.60	391.30
12"	ER@2.40	Ea	345.00	94.60	45.30	484.90
14"	ER@2.70	Ea	822.00	106.00	51.00	979.00
16"	ER@3.00	Ea	933.00	118.00	56.60	1,107.60

Ductile Iron, Class 153, Double Cement-Lined with Mechanical Joints

Description	Craft@Hrs	Unit	Material $	Labor $	Equipment $	Total $

Double cement-lined Class 153 ductile iron tee with mechanical joints

Description	Craft@Hrs	Unit	Material $	Labor $	Equipment $	Total $
4"	P1@1.65	Ea	104.00	60.60	—	164.60
6"	ER@2.35	Ea	151.00	92.60	44.40	288.00
8"	ER@3.15	Ea	230.00	124.00	59.50	413.50
10"	ER@4.00	Ea	345.00	158.00	75.50	578.50
12"	ER@4.45	Ea	473.00	175.00	84.00	732.00
14"	ER@4.90	Ea	1,220.00	193.00	92.50	1,505.50
16"	ER@5.30	Ea	1,270.00	209.00	100.00	1,579.00

Double cement-lined Class 153 ductile iron wye with mechanical joints

Description	Craft@Hrs	Unit	Material $	Labor $	Equipment $	Total $
4"	P1@1.65	Ea	230.00	60.60	—	290.60
6"	ER@2.35	Ea	293.00	92.60	44.40	430.00
8"	ER@3.15	Ea	432.00	124.00	59.50	615.50
10"	ER@4.00	Ea	815.00	158.00	75.50	1,048.50
12"	ER@4.45	Ea	1,210.00	175.00	84.00	1,469.00
14"	ER@4.90	Ea	2,480.00	193.00	92.50	2,765.50
16"	ER@5.30	Ea	3,070.00	209.00	100.00	3,379.00

Double cement-lined Class 153 ductile iron reducer with mechanical joints

Description	Craft@Hrs	Unit	Material $	Labor $	Equipment $	Total $
6"	ER@1.45	Ea	75.40	57.10	27.40	159.90
8"	ER@2.00	Ea	107.00	78.80	37.80	223.60
10"	ER@2.55	Ea	140.00	100.00	48.10	288.10
12"	ER@3.05	Ea	205.00	120.00	57.60	382.60
14"	ER@3.60	Ea	438.00	142.00	68.00	648.00
16"	ER@4.20	Ea	551.00	165.00	79.30	795.30

Ductile Iron, Class 110, Cement-Lined with Mechanical Joints

Description	Craft@Hrs	Unit	Material $	Labor $	Equipment $	Total $

Cement-lined Class 110 ductile iron pipe with mechanical joints

Description	Craft@Hrs	Unit	Material $	Labor $	Equipment $	Total $
4"	P1@.180	LF	15.60	6.61	—	22.21
6"	ER@.220	LF	22.10	8.67	4.15	34.92
8"	ER@.240	LF	30.60	9.46	4.53	44.59
10"	ER@.270	LF	45.60	10.60	5.10	61.30
12"	ER@.320	LF	62.00	12.60	6.04	80.64
16"	ER@.400	LF	121.00	15.80	7.55	144.35

Cement-lined Class 110 ductile iron 1/16 bend with mechanical joints

Description	Craft@Hrs	Unit	Material $	Labor $	Equipment $	Total $
4"	P1@1.40	Ea	156.00	51.40	—	207.40
6"	ER@2.00	Ea	208.00	78.80	37.80	324.60
8"	ER@2.50	Ea	285.00	98.50	47.20	430.70
10"	ER@3.20	Ea	435.00	126.00	60.40	621.40
12"	ER@3.60	Ea	553.00	142.00	68.00	763.00
16"	ER@4.50	Ea	1,440.00	177.00	85.00	1,702.00

Cement-lined Class 110 ductile iron 1/8 bend with mechanical joints

Description	Craft@Hrs	Unit	Material $	Labor $	Equipment $	Total $
4"	P1@1.40	Ea	154.00	51.40	—	205.40
6"	ER@2.00	Ea	206.00	78.80	37.80	322.60
8"	ER@2.50	Ea	285.00	98.50	47.20	430.70
10"	ER@3.20	Ea	419.00	126.00	60.40	605.40
12"	ER@3.60	Ea	552.00	142.00	68.00	762.00
16"	ER@4.50	Ea	1,420.00	177.00	85.00	1,682.00

Cement-lined Class 110 ductile iron 1/4 bend with mechanical joints

Description	Craft@Hrs	Unit	Material $	Labor $	Equipment $	Total $
4"	P1@1.40	Ea	169.00	51.40	—	220.40
6"	ER@2.00	Ea	227.00	78.80	37.80	343.60
8"	ER@2.50	Ea	326.00	98.50	47.20	471.70
10"	ER@3.20	Ea	505.00	126.00	60.40	691.40
12"	ER@3.60	Ea	757.00	142.00	68.00	967.00
16"	ER@4.50	Ea	1,840.00	177.00	85.00	2,102.00

Cement-lined Class 110 ductile iron tee with mechanical joints

Description	Craft@Hrs	Unit	Material $	Labor $	Equipment $	Total $
4"	P1@2.50	Ea	256.00	91.80	—	347.80
6"	ER@3.50	Ea	338.00	138.00	66.10	542.10
8"	ER@4.70	Ea	484.00	185.00	88.70	757.70
10"	ER@6.00	Ea	835.00	236.00	113.00	1,184.00
12"	ER@6.70	Ea	1,060.00	264.00	126.00	1,450.00
16"	ER@8.00	Ea	2,790.00	315.00	151.00	3,256.00

Cement-lined Class 110 ductile iron reducer with mechanical joints

Description	Craft@Hrs	Unit	Material $	Labor $	Equipment $	Total $
6"	ER@2.20	Ea	169.00	86.70	41.50	297.20
8"	ER@3.00	Ea	233.00	118.00	56.60	407.60
10"	ER@3.80	Ea	382.00	150.00	71.70	603.70
12"	ER@4.60	Ea	451.00	181.00	86.80	718.80
16"	ER@6.30	Ea	1,230.00	248.00	119.00	1,597.00

Cast Iron, Class 150 with Mechanical Joints

Description	Craft@Hrs	Unit	Material $	Labor $	Equipment $	Total $

Class 150 cast iron pipe with mechanical joints

Description	Craft@Hrs	Unit	Material $	Labor $	Equipment $	Total $
1½"	P1@.080	LF	7.58	2.94	—	10.52
2"	P1@.100	LF	7.41	3.67	—	11.08
3"	P1@.120	LF	10.20	4.40	—	14.60
4"	P1@.140	LF	13.30	5.14	—	18.44
6"	ER@.180	LF	22.90	7.09	3.40	33.39
8"	ER@.200	LF	36.60	7.88	3.78	48.26

Class 150 cast iron 1/8 bend with mechanical joints

Description	Craft@Hrs	Unit	Material $	Labor $	Equipment $	Total $
1½"	P1@.750	Ea	8.35	27.50	—	35.85
2"	P1@.800	Ea	9.30	29.40	—	38.70
3"	P1@1.00	Ea	12.50	36.70	—	49.20
4"	P1@1.20	Ea	15.90	44.00	—	59.90
6"	ER@1.70	Ea	37.40	67.00	32.10	136.50
8"	ER@2.25	Ea	107.00	88.70	42.50	238.20

Class 150 cast iron 1/4 bend with mechanical joints

Description	Craft@Hrs	Unit	Material $	Labor $	Equipment $	Total $
2"	P1@.800	Ea	13.40	29.40	—	42.80
3"	P1@1.00	Ea	19.30	36.70	—	56.00
4"	P1@1.20	Ea	30.20	44.00	—	74.20
6"	ER@1.70	Ea	62.70	67.00	32.10	161.80
8"	ER@2.25	Ea	170.00	88.70	42.50	301.20

Class 150 cast iron sanitary tee with mechanical joints

Description	Craft@Hrs	Unit	Material $	Labor $	Equipment $	Total $
1½"	P1@.750	Ea	14.00	27.50	—	41.50
2"	P1@1.00	Ea	15.40	36.70	—	52.10
3"	P1@1.20	Ea	18.50	44.00	—	62.50
4"	P1@1.65	Ea	28.40	60.60	—	89.00
6"	ER@2.35	Ea	81.20	92.60	44.40	218.20
8"	ER@3.15	Ea	260.00	124.00	59.50	443.50

Class 150 cast iron wye with mechanical joints

Description	Craft@Hrs	Unit	Material $	Labor $	Equipment $	Total $
1½"	P1@.750	Ea	14.30	27.50	—	41.80
2"	P1@1.00	Ea	14.30	36.70	—	51.00
3"	P1@1.20	Ea	20.10	44.00	—	64.10
4"	P1@1.65	Ea	32.60	60.60	—	93.20
6"	ER@2.35	Ea	87.30	92.60	44.40	224.30
8"	ER@3.15	Ea	206.00	124.00	59.50	389.50

Class 150 cast iron reducer

Description	Craft@Hrs	Unit	Material $	Labor $	Equipment $	Total $
2" x 1½"	P1@.600	Ea	7.63	22.00	—	29.63
3" x 2"	P1@.800	Ea	7.57	29.40	—	36.97
4" x 2"	P1@1.00	Ea	11.90	36.70	—	48.60
6" x 4"	ER@1.45	Ea	31.40	57.10	27.40	115.90
8" x 6"	ER@2.00	Ea	53.80	78.80	37.80	170.40

Asbestos-Cement, Class 2400 or 3000 with Mechanical Joints

Description	Craft@Hrs	Unit	Material $	Labor $	Equipment $	Total $

Asbestos-cement pipe with mechanical joints

Description	Craft@Hrs	Unit	Material $	Labor $	Equipment $	Total $
4"	ER@.080	LF	11.60	3.15	1.51	16.26
6"	ER@.100	LF	7.56	3.94	1.89	13.39
8"	ER@.140	LF	11.80	5.52	2.64	19.96
10"	ER@.200	LF	18.60	7.88	3.78	30.26
12"	ER@.260	LF	24.70	10.20	4.91	39.81
14"	ER@.300	LF	33.20	11.80	5.66	50.66
16"	ER@.350	LF	34.90	13.80	6.61	55.31

Asbestos-cement 1/16 bend with mechanical joints

Description	Craft@Hrs	Unit	Material $	Labor $	Equipment $	Total $
4"	ER@1.00	Ea	170.00	39.40	18.90	228.30
6"	ER@1.25	Ea	223.00	49.30	23.60	295.90
8"	ER@1.70	Ea	307.00	67.00	32.10	406.10
10"	ER@2.20	Ea	471.00	86.70	41.50	599.20
12"	ER@2.70	Ea	539.00	106.00	51.00	696.00
14"	ER@3.40	Ea	1,120.00	134.00	64.20	1,318.20
16"	ER@3.90	Ea	1,560.00	154.00	73.60	1,787.60

Asbestos-cement 1/8 bend with mechanical joints

Description	Craft@Hrs	Unit	Material $	Labor $	Equipment $	Total $
4"	ER@1.00	Ea	162.00	39.40	18.90	220.30
6"	ER@1.25	Ea	222.00	49.30	23.60	294.90
8"	ER@1.70	Ea	307.00	67.00	32.10	406.10
10"	ER@2.20	Ea	452.00	86.70	41.50	580.20
12"	ER@2.70	Ea	602.00	106.00	51.00	759.00
14"	ER@3.40	Ea	1,070.00	134.00	64.20	1,268.20
16"	ER@3.90	Ea	1,550.00	154.00	73.60	1,777.60

Asbestos-cement 1/4 bend with mechanical joints

Description	Craft@Hrs	Unit	Material $	Labor $	Equipment $	Total $
4"	ER@1.00	Ea	183.00	39.40	18.90	241.30
6"	ER@1.25	Ea	247.00	49.30	23.60	319.90
8"	ER@1.70	Ea	353.00	67.00	32.10	452.10
10"	ER@2.20	Ea	538.00	86.70	41.50	666.20
12"	ER@2.70	Ea	713.00	106.00	51.00	870.00
14"	ER@3.40	Ea	1,320.00	134.00	64.20	1,518.20
16"	ER@3.90	Ea	1,950.00	154.00	73.60	2,177.60

Asbestos-Cement, Class 2400 or 3000 with Mechanical Joints

Description	Craft@Hrs	Unit	Material $	Labor $	Equipment $	Total $
Asbestos-cement tee or wye with mechanical joints						
4"	ER@1.40	Ea	275.00	55.20	26.40	356.60
6"	ER@1.75	Ea	364.00	69.00	33.00	466.00
8"	ER@2.40	Ea	520.00	94.60	45.30	659.90
10"	ER@3.10	Ea	898.00	122.00	58.50	1,078.50
12"	ER@3.80	Ea	1,130.00	150.00	71.70	1,351.70
14"	ER@4.75	Ea	2,060.00	187.00	89.70	2,336.70
16"	ER@5.45	Ea	2,970.00	215.00	103.00	3,288.00
Asbestos-cement reducer with mechanical joints						
6" x 4"	ER@1.20	Ea	192.00	47.30	22.70	262.00
8" x 6"	ER@1.50	Ea	284.00	59.10	28.30	371.40
10" x 8"	ER@1.95	Ea	413.00	76.80	36.80	526.60
12" x 10"	ER@2.45	Ea	574.00	96.50	46.30	716.80

Description	Craft@Hrs	Unit	Material $	Labor $	Equipment $	Total $

Fiberglass septic tank. 2 compartments. Shallow burial (2 feet of earth cover). 32" manway entrance. One-piece construction. Including excavation, backfill and equipment costs.

Description	Craft@Hrs	Unit	Material $	Labor $	Equipment $	Total $
750 gallon	P1@8.00	Ea	1,540.00	294.00	529.00	2,363.00
1,000 gallon	P1@9.00	Ea	2,040.00	330.00	595.00	2,965.00
1,250 gallon	P1@10.0	Ea	2,620.00	367.00	661.00	3,648.00
1,500 gallon	P1@11.5	Ea	2,960.00	422.00	760.00	4,142.00
1,750 gallon	P1@12.0	Ea	3,310.00	440.00	793.00	4,543.00
2,000 gallon	P1@12.5	Ea	3,640.00	459.00	826.00	4,925.00

Fiberglass septic tank. 2 compartments. Deep burial (7 feet of earth cover). 32" x 40" manway extension. One-piece construction. Including excavation, backfill and equipment costs.

Description	Craft@Hrs	Unit	Material $	Labor $	Equipment $	Total $
750 gallon	P1@8.00	Ea	1,740.00	294.00	529.00	2,563.00
1,000 gallon	P1@9.00	Ea	2,280.00	330.00	595.00	3,205.00
1,250 gallon	P1@10.0	Ea	2,990.00	367.00	661.00	4,018.00
1,500 gallon	P1@11.5	Ea	3,360.00	422.00	760.00	4,542.00
1,750 gallon	P1@12.0	Ea	3,720.00	440.00	792.00	4,952.00
2,000 gallon	P1@12.5	Ea	4,100.00	459.00	826.00	5,385.00
Manway extension	P1@.350	LF	51.50	12.80	—	64.30

Fiberglass holding tank. Single compartment. Shallow burial (2 feet of earth cover). 32" manway entrance. One-piece construction. Including excavation, backfill and equipment costs. Can be used for drink water storage or sewage holding.

Description	Craft@Hrs	Unit	Material $	Labor $	Equipment $	Total $
750 gallon	P1@8.00	Ea	1,340.00	294.00	529.00	2,163.00
1,000 gallon	P1@9.00	Ea	1,750.00	330.00	595.00	2,675.00
1,250 gallon	P1@10.0	Ea	2,280.00	367.00	661.00	3,308.00
1,500 gallon	P1@11.5	Ea	2,620.00	422.00	760.00	3,802.00
1,750 gallon	P1@12.0	Ea	2,960.00	440.00	792.00	4,192.00
2,000 gallon	P1@12.5	Ea	3,310.00	459.00	826.00	4,595.00

Fiberglass holding tank. Single compartment. Deep burial (7 feet of earth cover). 32" x 40" manway extension. One-piece construction. Including excavation, backfill and equipment costs. Can be used for drink water storage or sewage holding.

Description	Craft@Hrs	Unit	Material $	Labor $	Equipment $	Total $
750 gallon	P1@8.00	Ea	1,570.00	294.00	529.00	2,393.00
1,000 gallon	P1@9.00	Ea	1,980.00	330.00	595.00	2,905.00
1,250 gallon	P1@10.0	Ea	2,660.00	367.00	661.00	3,688.00
1,500 gallon	P1@11.5	Ea	3,020.00	422.00	760.00	4,202.00
1,750 gallon	P1@12.0	Ea	3,380.00	440.00	793.00	4,613.00
2,000 gallon	P1@12.5	Ea	3,760.00	459.00	826.00	5,045.00
Manway extension	P1@.350	LF	51.50	12.80	—	64.30

Plastic Tanks

Description	Craft@Hrs	Unit	Material $	Labor $	Equipment $	Total $

Plastic water tank. Suitable for potable water. Labor includes setting in place only. Freight, hoisting & rigging excluded.

Description	Craft@Hrs	Unit	Material $	Labor $	Equipment $	Total $
Vertical tank, D31" x H55"						
165 gallon	P1@2.00	Ea	632.00	73.40	—	705.40
Vertical tank, D48" x H102"						
500 gallon	P1@4.00	Ea	1,170.00	147.00	—	1,317.00
Vertical tank, D64" x H79"						
1,000 gallon	P1@4.00	Ea	1,670.00	147.00	—	1,817.00
Vertical tank, D87" x H65"						
1,500 gallon	P1@5.00	Ea	2,190.00	184.00	—	2,374.00
Vertical tank, D95" x H89"						
2,500 gallon	P1@5.00	Ea	2,270.00	184.00	—	2,454.00
Vertical tank, D95" x H107"						
3,000 gallon	P1@6.00	Ea	3,570.00	220.00	—	3,790.00
Vertical tank, D102" x H125"						
4,000 gallon	P1@6.00	Ea	3,800.00	220.00	—	4,020.00
Vertical tank, D102" x H150"						
5,000 gallon	P1@8.00	Ea	7,140.00	294.00	—	7,434.00
Vertical tank, D141" x H192"						
12,000 gallon	P1@9.00	Ea	11,100.00	330.00	—	11,430.00

Plastic holding tank. Single compartment. One-piece polyethylene construction. Shallow burial (2 feet of earth cover). 28" manway entrance. Including excavation, backfill and equipment costs. Used for drink water storage.

Description	Craft@Hrs	Unit	Material $	Labor $	Equipment $	Total $
30 gallon	P1@3.00	Ea	211.00	110.00	—	321.00
40 gallon	P1@3.00	Ea	241.00	110.00	—	351.00
75 gallon	P1@3.00	Ea	268.00	110.00	198.00	576.00
300 gallon	P1@5.00	Ea	1,180.00	184.00	330.00	1,694.00
525 gallon	P1@6.00	Ea	1,700.00	220.00	396.00	2,316.00
1,150 gallon	P1@8.00	Ea	2,970.00	294.00	529.00	3,793.00
1,700 gallon	P1@11.0	Ea	3,830.00	404.00	727.00	4,961.00
1,950 gallon	P1@12.0	Ea	4,080.00	440.00	793.00	5,313.00

Plastic holding tank. Single compartment. One-piece polyethylene construction. Above ground applications only. Used for drink water storage.

Description	Craft@Hrs	Unit	Material $	Labor $	Equipment $	Total $
180 gallon	P1@3.00	Ea	551.00	110.00	198.00	859.00
200 gallon	P1@3.75	Ea	816.00	138.00	248.00	1,202.00
300 gallon	P1@4.25	Ea	1,060.00	156.00	281.00	1,497.00
525 gallon	P1@5.00	Ea	1,570.00	184.00	330.00	2,084.00
1,000 gallon	P1@5.75	Ea	2,110.00	211.00	380.00	2,701.00
1,150 gallon	P1@6.25	Ea	2,600.00	229.00	413.00	3,242.00
1,700 gallon	P1@7.00	Ea	3,440.00	257.00	462.00	4,159.00

Plastic septic tank. 750 gallon. Single compartment. One-piece polyethylene construction. Including excavation, backfill and equipment costs.

Description	Craft@Hrs	Unit	Material $	Labor $	Equipment $	Total $
Trickle tank	P1@8.00	Ea	2,410.00	294.00	529.00	3,233.00
Shallow buried	P1@8.00	Ea	2,410.00	294.00	529.00	3,233.00
Deep buried	P1@8.00	Ea	2,540.00	294.00	529.00	3,363.00
Manway extension						
	P1@.350	LF	96.00	12.80	—	108.80

Description	Craft@Hrs	Unit	Material $	Labor $	Equipment $	Total $

Plastic sump pit. One-piece polyethylene construction. Complete with radon gasket electrical and discharge grommets. Including excavation, backfill and equipment costs.

Description	Craft@Hrs	Unit	Material $	Labor $	Equipment $	Total $
30 gallon	P1@3.00	Ea	197.00	110.00	—	307.00
40 gallon	P1@3.00	Ea	241.00	110.00	—	351.00
75 gallon	P1@3.00	Ea	284.00	110.00	198.00	592.00
110 gallon	P1@3.25	Ea	450.00	119.00	215.00	784.00
150 gallon	P1@3.75	Ea	607.00	138.00	215.00	960.00
250 gallon	P1@4.75	Ea	960.00	174.00	314.00	1,448.00
380 gallon	P1@5.50	Ea	1,500.00	202.00	363.00	2,065.00

Plastic sewage lift tank. One-piece polyethylene construction. Complete with radon gasket, inlet, vent, electrical and discharge grommets. Including excavation, backfill and equipment costs.

Description	Craft@Hrs	Unit	Material $	Labor $	Equipment $	Total $
30 gallon	P1@3.00	Ea	248.00	110.00	—	358.00
40 gallon	P1@3.00	Ea	273.00	110.00	—	383.00
75 gallon	P1@3.00	Ea	307.00	110.00	198.00	615.00
110 gallon	P1@3.25	Ea	521.00	119.00	215.00	855.00
150 gallon	P1@3.75	Ea	633.00	138.00	248.00	1,019.00
250 gallon	P1@4.75	Ea	1,040.00	174.00	314.00	1,528.00
380 gallon	P1@5.50	Ea	1,550.00	202.00	363.00	2,115.00

Plastic grey water treatment tank. 75 gallon. One-piece polyethylene construction. Including excavation, backfill and equipment costs.

Description	Craft@Hrs	Unit	Material $	Labor $	Equipment $	Total $
Buried	P1@3.00	Ea	494.00	110.00	198.00	802.00
Non-buried	P1@1.50	Ea	450.00	55.10	—	505.10

Trenching

Because of varying soil compositions and densities, trenching costs can sometimes be difficult to accurately estimate. The tables in this section are based on average soil conditions, which are loosely defined as dirt, soft clay or gravel mixed with rocks. The costs for trenching through limestone, sandstone or shale can increase by as much as 1,000 percent. For these conditions, it is best to consult with experts or to subcontract the work to specialists.

The primary consideration in excavations is to ensure the safety of the pipe layers. Current OSHA statutes require all excavations over 5 feet deep to be sloped, shored, sheeted or braced to prevent cave-ins. The use of shoring minimizes trenching costs by eliminating the need to slope the trench walls, but increases pipe-laying costs because of the difficulty in working between cross-bracing.

The estimator should carefully compare the costs for installing and removing shoring, and reduced pipe installation productivity, against the additional excavation costs for sloping the walls of the trench. For average soil conditions, not exceeding 6 to 8 feet in depth, it's usually more economical to slope the trench walls.

The current average cost for a backhoe with operator is $105.00 per hour. Some trenching contractors additionally charge $25.00 move-on and $25.00 move-off costs, particularly for small projects.

Total trenching, backfill and compaction costs are calculated by multiplying the backhoe hours from the following charts by $105.00 per hour, and adding move-on and move-off charges, if applicable.

Backhoe Trenching						
Trench walls at a 90-degree angle, based on average soil, including trenching, backfill and compaction. Hours per 100 linear feet of trench						
Trench depth (feet)	Trench width (feet)					
	1	2	3	4	5	6
2	2.6	5.1	7.7	10.2	12.8	15.3
3	3.8	7.7	11.5	15.3	19.2	23.0
4	5.1	10.2	15.3	20.4	25.6	30.7
5	6.4	12.8	19.2	25.6	32.0	38.3
6	7.7	15.3	23.0	30.7	38.3	46.0
7	8.9	17.9	26.8	35.8	44.7	53.7
8	10.2	20.4	30.7	40.9	51.1	61.3
9	11.5	23.0	34.5	46.0	57.5	69.0
10	12.8	25.5	38.3	51.1	63.9	76.6

Notes:
1) Compaction is based on the use of a compactor wheel.
2) Additional labor costs must be added if watering-in, stone removal or excess spoil removal is required.

	Backhoe Trenching Trench walls at a 45-degree angle, based on average soil including trenching, backfill and compaction. Hours per 100 linear feet of trench					
	Trench width (feet)					
Trench depth (feet)	**1**	**2**	**3**	**4**	**5**	**6**
2	7.6	10.3	12.6	15.5	17.9	20.3
3	15.2	22.6	22.6	26.6	30.7	34.5
4	25.5	30.7	35.5	40.7	45.9	51.0
5	38.3	44.5	51.0	57.6	63.8	70.0
6	53.4	61.0	69.0	76.6	84.1	91.7
7	71.4	80.3	89.3	98.3	107.2	115.9
8	92.1	102.1	112.4	122.4	132.8	143.1
9	114.8	126.6	137.9	149.3	161.0	172.4
10	140.3	153.1	165.9	178.6	191.4	204.5

Notes:
1) Compaction is based on the use of a compactor wheel.
2) Additional labor costs must be added if watering-in, stone removal or excess spoil removal is required.

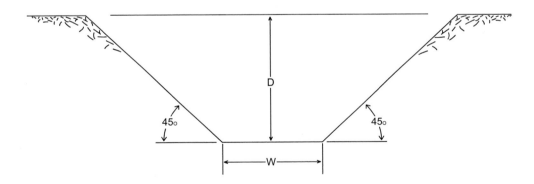

Equipment Rental			
Description	**Per day ($)**	**Per week ($)**	**Per month ($)**
Air compressor, gas or diesel, 100-125 CFM with 100' hose, air tool and 4 bits	248.00	748.00	1,900.00
Air compressor, gas or diesel, 165-185 CFM with 100' hose, air tool and 4 bits	287.00	868.00	2,250.00
Air compressor, gas or diesel, 100-125 CFM, no accessories	120.00	358.00	1,080.00
Air compressor, gas or diesel, 165-185 CFM, no accessories	159.00	478.00	1,430.00
Backhoe with bucket, 55 HP	388.00	1,410.00	3,280.00
Backhoe with bucket, 100 HP	432.00	1,360.00	4,050.00
Backhoe hydraulic compactor, 5-10 ton	186.00	479.00	1,200.00
Bender, hydraulic	87.00	322.00	909.00
Bevel machine with torch	28.00	93.00	230.00
Boom lift, 31' - 40'	471.00	1,170.00	2,390.00
Boom lift, 51' - 60'	556.00	1,200.00	2,460.00
Chain hoist, to 2 tons	31.00	101.00	221.00
Chain hoist, 10 tons	166.00	331.00	552.00
Come-along, ½ to 1 ton	26.00	35.00	208.00
Come-along, 1½ to 2 tons	45.00	120.00	364.00
Come-along, 4-ton	65.00	188.00	583.00
Compactor, soil, gas or diesel, 125 lb	111.00	337.00	716.00
Crane, truck-mounted, 13,000 lb capacity	404.00	1,480.00	3,670.00
Crane, truck-mounted, 30,000 lb capacity	506.00	1,560.00	4,410.00
Forklift, 2,000 pound capacity	208.00	551.00	1,490.00
Forklift, 4,000 pound capacity	199.00	541.00	1,640.00
Forklift, 6,000 pound capacity	261.00	1,170.00	1,840.00
Front-end loader, 1 yard capacity	347.00	1,010.00	2,830.00
Front-end loader, 2 yard capacity	544.00	1,600.00	4,230.00
Front-end loader, 3¼ yard capacity	841.00	2,400.00	6,550.00
Pipe machine, up to 2" capacity	66.00	195.00	543.00
Pipe machine, 2½" to 4" capacity	72.00	253.00	774.00
Ramset (nail gun)	29.00	116.00	354.00
Scissor-lift, to 20', diesel, self-propelled	108.00	269.00	699.00
Scissor-lift, 21' – 30'	229.00	637.00	1,290.00
Skid steer loader 1,750 lb capacity	301.00	875.00	1,980.00
Trailer, office, 25'	62.00	199.00	525.00
Truck, ¾-ton pickup	95.00	369.00	1,490.00
Truck, 1-ton pickup	143.00	452.00	1,430.00
Truck, 1-ton stake bed	194.00	584.00	1,720.00
Van, storage, 26'	—	—	369.00
Van, storage, 40'	—	—	450.00
Welding machine, gas, 200 amp, trailer-mounted with 50' bare lead	77.00	199.00	435.00
Welding machine, diesel, 400 amp, trailer-mounted with 50' bare lead	133.00	307.00	712.00

Description	Craft@Hrs	Unit	Material $	Labor $	Equipment $	Total $

1½" brass valve tags with black filled numbers

Description	Craft@Hrs	Unit	Material $	Labor $	Equipment $	Total $
Typical costs	P1@.150	Ea	3.75	5.51	—	9.26

1½" blank brass valve tags

Description	Craft@Hrs	Unit	Material $	Labor $	Equipment $	Total $
Typical costs	P1@.150	Ea	1.85	5.51	—	7.36

1½" plastic valve tags with engraved numbers

Description	Craft@Hrs	Unit	Material $	Labor $	Equipment $	Total $
Typical costs	P1@.150	Ea	3.11	5.51	—	8.62

Engraved plastic equipment nameplates

Description	Craft@Hrs	Unit	Material $	Labor $	Equipment $	Total $
Typical costs	P1@.400	Ea	8.06	14.70	—	22.76

Pipe and duct markers

Description	Craft@Hrs	Unit	Material $	Labor $	Equipment $	Total $
Typical costs	P1@.250	Ea	4.61	9.18	—	13.79

As-built drawings

Description	Craft@Hrs	Unit	Material $	Labor $	Equipment $	Total $
Typical costs	P1@.500	Ea	—	18.40	—	18.40

Instructing owner's operating personnel

Description	Craft@Hrs	Unit	Material $	Labor $	Equipment $	Total $
Typical costs	P1@8.00	Ea	—	294.00	—	294.00

HVAC & Plumbing Demolition

Description	Craft@Hrs	Unit	Material $	Labor $	Equipment $	Total $

Remove rectangular galvanized ductwork. 22 to 30 gauge galvanized sheet metal (dimension/gauge per SMACNA standards). Per linear foot of sheet metal ductwork. The cost of removing ductwork includes the cost of removing the support hangers. Add the cost of disconnecting and removing heating coils, air volume zone boxes, air handing units, condensate drains, motorized dampers or other associated accessories, piping or wiring. These figures assume demolished materials are piled on site. Disposal costs or salvage values excluded. (SMACNA = Sheetmetal & Air Conditioning National Association.) Equipment may be needed for some sizes and locations. Refer to the equipment rental costs at the front of this section.

Description	Craft@Hrs	Unit	Material $	Labor $	Equipment $	Total $
4" to 12" rectangular/square duct, remove	S2@.015	LF	—	.54	—	.54
14" to 24" rectangular/square duct, remove	S2@.020	LF	—	.72	—	.72
26" to 36" rectangular/square duct, remove	S2@.065	LF	—	2.34	—	2.34
38" to 60" rectangular/square duct, remove	S2@.090	LF	—	3.24	—	3.24

Remove round galvanized ductwork. 22 to 30 gauge galvanized sheet metal (dimension/gauge per SMACNA standards). Per linear foot of sheet metal ductwork. The cost of removing ductwork includes the cost of removing the support hangers. Add the cost of disconnecting and removing heating coils, air volume zone boxes, air handing units, condensate drains, motorized dampers or other associated accessories, piping or wiring. These figures assume demolished materials are piled on site. Disposal costs or salvage values excluded. (SMACNA = Sheetmetal & Air Conditioning National Association.) Equipment may be needed for some sizes and locations. Refer to the equipment rental costs at the front of this section.

Description	Craft@Hrs	Unit	Material $	Labor $	Equipment $	Total $
4" to 12" round duct, remove	S2@.012	LF	—	.43	—	.43
14" to 24" round duct, remove	S2@.016	LF	—	.58	—	.58
26" to 36" round duct, remove	S2@.052	LF	—	1.87	—	1.87
38" to 60" round duct, remove	S2@.072	LF	—	2.59	—	2.59

Description	Craft@Hrs	Unit	Material $	Labor $	Equipment $	Total $

Remove welded steel ductwork. 10 to 16 gauge mild steel. Per linear foot of welded milled steel ductwork. The cost of removing ductwork includes the cost of removing the support hangers. Add the cost of disconnecting and removing exhaust fans, condensate drains, sprinkler heads or other associated accessories, piping or wiring. These figures assume demolished materials are piled on site. Disposal costs or salvage values excluded. Equipment may be needed for some sizes and locations. Refer to the equipment rental costs at the front of this section.

Description	Craft@Hrs	Unit	Material $	Labor $	Equipment $	Total $
4" to 12", square or round duct, remove	S2@.045	LF	—	1.62	—	1.62
14" to 24", square or round duct, remove	S2@.060	LF	—	2.16	—	2.16
26" to 36", square or round duct, remove	S2@.095	LF	—	3.42	—	3.42
38" to 60", square or round duct, remove	S2@.150	LF	—	5.41	—	5.41

Remove diffuser or grille. These figures assume demolished materials are piled on site. Disposal costs or salvage values excluded. Equipment may be needed for some sizes and locations. Refer to the equipment rental costs at the front of this section.

Description	Craft@Hrs	Unit	Material $	Labor $	Equipment $	Total $
6" to 12", square or round diffuser/grille, remove	S2@.150	Ea	—	5.41	—	5.41
14" to 24", square or round diffuser/grille, remove	S2@.180	Ea	—	6.49	—	6.49
26" to 36", square or round diffuser/grille, remove	S2@.400	Ea	—	14.40	—	14.40
38" to 60" square or round diffuser/grille, remove	S2@.800	Ea	—	28.80	—	28.80

Remove air mixing box. These figures include isolating and disconnecting hot water, chilled water, steam, refrigerant and/or pneumatic control tube or control wiring. These figures assume demolished materials are piled on site. Disposal costs or salvage values excluded. Equipment may be needed for some sizes and locations. Refer to the equipment rental costs at the front of this section.

Description	Craft@Hrs	Unit	Material $	Labor $	Equipment $	Total $
Air box, 50 to 150 CFM, remove	S2@.500	Ea	—	18.00	—	18.00
Air box, 160 to 500 CFM, remove	S2@.650	Ea	—	23.40	—	23.40
Air box, 550 to 750 CFM, remove	S2@.900	Ea	—	32.40	—	32.40

HVAC & Plumbing Demolition

Description	Craft@Hrs	Unit	Material $	Labor $	Equipment $	Total $

Remove air handling unit. These figures include isolating and disconnecting hot water, chilled water, steam or gas and/or ductwork, electrical connections and unit disassembly when required for removal. Disposal costs or salvage values excluded. Equipment may be needed for some sizes and locations. Refer to the equipment rental costs at the front of this section.

Description	Craft@Hrs	Unit	Material $	Labor $	Equipment $	Total $
AHU, 2,000-5,000 CFM, remove	P1@8.00	Ea	—	294.00	—	294.00
AHU, 5,000-10,000 CFM, remove	P1@12.0	Ea	—	440.00	—	440.00
AHU,10,000-20,000 CFM, remove	P1@18.0	Ea	—	661.00	—	661.00
AHU, over 20,000 CFM, remove	P1@28.0	Ea	—	1,030.00	—	1,030.00

Remove rooftop unit. These figures include isolating and disconnecting hot water, chilled water, steam or gas and/or ductwork, electrical connections and unit disassembly when required for removal. Disposal costs or salvage values excluded. Make additional hoisting allowances for roof heights above 40'. Equipment may be needed for some sizes and locations. Refer to the equipment rental costs at the front of this section.

Description	Craft@Hrs	Unit	Material $	Labor $	Equipment $	Total $
RTU, 2,000-5,000 CFM, remove	P1@6.00	Ea	—	220.00	—	220.00
RTU, 5,000-10,000 CFM, remove	P1@10.0	Ea	—	367.00	—	367.00
RTU. 10,000-20,000 CFM, remove	P1@16.0	Ea	—	587.00	—	587.00
RTU, over 20,000 CFM, remove	P1@26.0	Ea	—	954.00	—	954.00

Description	Craft@Hrs	Unit	Material $	Labor $	Equipment $	Total $

Remove boiler. These figures include isolating and disconnecting hot water, steam or gas and/or electrical connections and unit disassembly when required for removal. Disposal costs or salvage values excluded. Equipment may be needed for some sizes and locations. Refer to the equipment rental costs at the front of this section.

Description	Craft@Hrs	Unit	Material $	Labor $	Equipment $	Total $
10-50 KW electric boiler, remove	P1@8.00	Ea	—	294.00	—	294.00
50-100 KW electric boiler, remove	P1@12.0	Ea	—	440.00	—	440.00
100-200 KW electric boiler, remove	P1@26.0	Ea	—	954.00	—	954.00
200-500 KW electric boiler, remove	P1@36.0	Ea	—	1,320.00	—	1,320.00
over 500 KW electric boiler, remove	P1@44.0	Ea	—	1,610.00	—	1,610.00
to 150 Mbh gas-fired boiler, remove	P1@9.00	Ea	—	330.00	—	330.00
150-300 Mbh gas-fired boiler, remove	P1@14.0	Ea	—	514.00	—	514.00
300-750 Mbh gas-fired boiler, remove	P1@30.0	Ea	—	1,100.00	—	1,100.00
750-1,500 Mbh gas-fired boiler, remove	P1@46.0	Ea	—	1,690.00	—	1,690.00
1,500-2,000 Mbh gas-fired boiler, remove	P1@65.0	Ea	—	2,390.00	—	2,390.00
over 2,000 Mbh gas-fired boiler, remove	P1@80.0	Ea	—	2,940.00	—	2,940.00

Remove furnace. These figures include isolating and disconnecting plenums, fuel and/or electrical connections. Disposal costs or salvage values excluded. Equipment may be needed for some sizes and locations. Refer to the equipment rental costs at the front of this section.

Description	Craft@Hrs	Unit	Material $	Labor $	Equipment $	Total $
5-30 KW electric furnace, remove	P1@2.50	Ea	—	91.80	—	91.80
30-60 KW electric furnace, remove	P1@3.75	Ea	—	138.00	—	138.00
30-150 Mbh gas-fired furnace, remove	P1@5.00	Ea	—	184.00	—	184.00
150-225 Mbh gas-fired furnace, remove	P1@7.00	Ea	—	257.00	—	257.00

HVAC & Plumbing Demolition

Description	Craft@Hrs	Unit	Material $	Labor $	Equipment $	Total $

Remove chiller. These figures include isolating and disconnecting chilled water and electrical connections, evacuating refrigerant to storage vessels and unit disassembly when required for removal. Disposal costs or salvage values excluded. Equipment may be needed for some sizes and locations. Refer to the equipment rental costs at the front of this section.

Description	Craft@Hrs	Unit	Material $	Labor $	Equipment $	Total $
Chiller, 20-50 tons,						
remove	P1@34.0	Ea	—	1,250.00	—	1,250.00
Chiller, 50-100 tons,						
remove	P1@48.0	Ea	—	1,760.00	—	1,760.00
Chiller, 100-150 tons,						
remove	P1@56.0	Ea	—	2,060.00	—	2,060.00
Chiller, 150-200 tons,						
remove	P1@68.0	Ea	—	2,500.00	—	2,500.00
Chiller, 200-250 tons,						
remove	P1@80.0	Ea	—	2,940.00	—	2,940.00
Chiller, over 250 tons,						
remove	P1@100	Ea	—	3,670.00	—	3,670.00

Remove air-cooled condenser. These figures include isolating and disconnecting refrigerant lines and electrical connections, evacuating refrigerant to storage vessels and unit disassembly when required for removal. Disposal costs or salvage values excluded. Make additional hoisting allowances for roof heights above 40'. Equipment may be needed for some sizes and locations. Refer to the equipment rental costs at the front of this section.

Description	Craft@Hrs	Unit	Material $	Labor $	Equipment $	Total $
Air-cooled condenser, 2-4 tons,						
remove	P1@4.00	Ea	—	147.00	—	147.00
Air-cooled condenser, 5-10 tons,						
remove	P1@8.00	Ea	—	294.00	—	294.00
Air-cooled condenser, 10-20 tons,						
remove	P1@20.0	Ea	—	734.00	—	734.00
Air-cooled condenser, 20-30 tons,						
remove	P1@28.0	Ea	—	1,030.00	—	1,030.00
Air-cooled condenser, 30-50 tons,						
remove	P1@36.0	Ea	—	1,320.00	—	1,320.00
Air-cooled condenser, 50-150 tons,						
remove	P1@44.0	Ea	—	1,610.00	—	1,610.00
Air-cooled condenser, 150-250 tons,						
remove	P1@52.0	Ea	—	1,910.00	—	1,910.00

Description	Craft@Hrs	Unit	Material $	Labor $	Equipment $	Total $

Remove cooling tower. These figures include isolating and disconnecting chilled water lines and electrical connections, and unit disassembly when required for removal. Disposal costs or salvage values excluded. Make additional hoisting allowances for roof heights above 40'. Equipment may be needed for some sizes and locations. Refer to the equipment rental costs at the front of this section.

Description	Craft@Hrs	Unit	Material $	Labor $	Equipment $	Total $
Cooling tower, 5-10 tons, remove	P1@8.00	Ea	—	294.00	—	294.00
Cooling tower, 10-20 tons, remove	P1@16.0	Ea	—	587.00	—	587.00
Cooling tower, 20-30 tons, remove	P1@24.0	Ea	—	881.00	—	881.00
Cooling tower, 30-50 tons, remove	P1@32.0	Ea	—	1,170.00	—	1,170.00
Cooling tower, 50-150 tons, remove	P1@40.0	Ea	—	1,470.00	—	1,470.00
Cooling tower, 150-250 tons, remove	P1@48.0	Ea	—	1,760.00	—	1,760.00
Cooling tower, 250-500 tons, remove	P1@70.0	Ea	—	2,570.00	—	2,570.00

Remove fan coil. These figures include isolating and disconnecting chilled water and/or hot water lines and/or duct and electrical connections, and unit disassembly when required for removal. Disposal costs or salvage values excluded. Equipment may be needed for some sizes and locations. Refer to the equipment rental costs at the front of this section.

Description	Craft@Hrs	Unit	Material $	Labor $	Equipment $	Total $
2-pipe fan coil, 300-500 CFM, remove	P1@4.00	Ea	—	147.00	—	147.00
2-pipe fan coil, 550-750 CFM, remove	P1@4.50	Ea	—	165.00	—	165.00
2-pipe fan coil, 800-1,200 CFM, remove	P1@6.00	Ea	—	220.00	—	220.00
4-pipe fan coil, 300-500 CFM, remove	P1@4.50	Ea	—	165.00	—	165.00
4-pipe fan coil, 550-750 CFM, remove	P1@5.00	Ea	—	184.00	—	184.00
4-pipe fan coil, 800-1,200 CFM, remove	P1@6.50	Ea	—	239.00	—	239.00

HVAC & Plumbing Demolition

Description	Craft@Hrs	Unit	Material $	Labor $	Equipment $	Total $

Remove unit heater. These figures include isolating and disconnecting hot water, steam lines or gas and/or electrical connections. Disposal costs or salvage values excluded. Equipment may be needed for some sizes and locations. Refer to the equipment rental costs at the front of this section.

Description	Craft@Hrs	Unit	Material $	Labor $	Equipment $	Total $
Gas-fired unit heater, 50-150 Mbh,						
remove	P1@2.25	Ea	—	82.60	—	82.60
Gas-fired unit heater, 200-350 Mbh,						
remove	P1@3.50	Ea	—	128.00	—	128.00
Steam unit heater, 75-200 Mbh,						
remove	P1@3.25	Ea	—	119.00	—	119.00
Steam unit heater, 250-350 Mbh,						
remove	P1@4.75	Ea	—	174.00	—	174.00
Electric unit heater, 5-25 KW,						
remove	P1@2.00	Ea	—	73.40	—	73.40
Electric unit heater, 30-50 KW,						
remove	P1@2.75	Ea	—	101.00	—	101.00

Remove heat pump. These figures include isolating and disconnecting pipe, duct and electrical connections and evacuating refrigerant to storage vessels when required for removal. Disposal costs or salvage values excluded. Equipment may be needed for some sizes and locations. Refer to the equipment rental costs at the front of this section.

Description	Craft@Hrs	Unit	Material $	Labor $	Equipment $	Total $
Heat pump, split system, 2-5 ton,						
remove	P1@4.75	Ea	—	174.00	—	174.00
Heat pump, split system, 7.5-10 ton,						
remove	P1@8.50	Ea	—	312.00	—	312.00
Heat pump, packaged system, 2-5 ton,						
remove	P1@3.25	Ea	—	119.00	—	119.00
Heat pump, pkg. system, 7.5-10 ton,						
remove	P1@5.50	Ea	—	202.00	—	202.00

Description	Craft@Hrs	Unit	Material $	Labor $	Equipment $	Total $

Remove heat exchanger. These figures include isolating and disconnecting chilled water, hot water and/or steam line connections. Disposal costs or salvage values excluded. Equipment may be needed for some sizes and locations. Refer to the equipment rental costs at the front of this section.

Description	Craft@Hrs	Unit	Material $	Labor $	Equipment $	Total $
Heat exchanger, 6" x 30", remove	P1@2.75	Ea	—	101.00	—	101.00
Heat exchanger, 10" x 48", remove	P1@4.00	Ea	—	147.00	—	147.00
Heat exchanger, 12" x 72", remove	P1@5.25	Ea	—	193.00	—	193.00
Heat exchanger, 16" x 90", remove	P1@6.00	Ea	—	220.00	—	220.00

Remove duct-mounted coil. These figures include isolating and disconnecting steam, chilled water, hot water, condensate, refrigerant or electrical and duct connections. Disposal costs or salvage values excluded. Refrigerant coil removal costs include system refrigerant evacuation and storage. Equipment may be needed for some sizes and locations. Refer to the equipment rental costs at the front of this section.

Description	Craft@Hrs	Unit	Material $	Labor $	Equipment $	Total $
Single-row water coil, 12" to 18", remove	P1@2.00	Ea	—	73.40	—	73.40
Single-row water coil, 20" to 30", remove	P1@3.50	Ea	—	128.00	—	128.00
Single-row water coil, 32" to 60", remove	P1@8.00	Ea	—	294.00	—	294.00
Single-row steam coil, 12" to 18", remove	P1@2.25	Ea	—	82.60	—	82.60
Single-row steam coil, 20" to 30", remove	P1@3.75	Ea	—	138.00	—	138.00
Single-row steam coil, 32" to 60", remove	P1@9.00	Ea	—	330.00	—	330.00
Electric resistance, 12" to 18", remove	P1@1.25	Ea	—	45.90	—	45.90
Electric resistance, 20" to 30", remove	P1@1.75	Ea	—	64.20	—	64.20
Electric resistance, 32" to 60", remove	P1@4.00	Ea	—	147.00	—	147.00
Refrigerant coil, 2-3 ton, remove	P1@4.00	Ea	—	147.00	—	147.00
Refrigerant coil, 4-5 ton, remove	P1@6.50	Ea	—	239.00	—	239.00
Refrigerant coil, 6-10 ton, remove	P1@12.0	Ea	—	440.00	—	440.00
Refrigerant coil, 12-15 ton, remove	P1@16.0	Ea	—	587.00	—	587.00

HVAC & Plumbing Demolition

Description	Craft@Hrs	Unit	Material $	Labor $	Equipment $	Total $

Remove steel piping. These figures include cutting out and removing threaded or welded steel piping. These figures assume demolished materials are piled on site. Equipment, fixture or apparatus disconnection costs and salvage values excluded. Equipment may be needed for some sizes and locations. Refer to the equipment rental costs at the front of this section.

Description	Craft@Hrs	Unit	Material $	Labor $	Equipment $	Total $
Steel pipe, Sch. 40, ½"-1"						
remove	P1@.038	LF	—	1.39	—	1.39
Steel pipe, Sch. 40, 1½"-2"						
remove	P1@.068	LF	—	2.50	—	2.50
Steel pipe, Sch. 40, 2½"-4"						
remove	P1@.108	LF	—	3.96	—	3.96
Steel pipe, Sch. 40, 6"-8"						
remove	P1@.230	LF	—	8.44	—	8.44
Steel pipe, Sch. 40, 10"-12"						
remove	P1@.330	LF	—	12.10	—	12.10
Steel pipe, Sch. 80, ½"-1"						
remove	P1@.042	LF	—	1.54	—	1.54
Steel pipe, Sch. 80, 1½"-2"						
remove	P1@.075	LF	—	2.75	—	2.75
Steel pipe, Sch. 80, 2½"-4"						
remove	P1@.118	LF	—	4.33	—	4.33
Steel pipe, Sch. 80, 6"-8"						
remove	P1@.290	LF	—	10.60	—	10.60
Steel pipe, Sch. 80, 10"-12"						
remove	P1@.360	LF	—	13.20	—	13.20
Remove steel pipe, Sch. 160,						
add	—	%	—	30.00	—	—

Remove copper piping. These figures include cutting out and removing copper piping. These figures assume demolished materials are piled on site. Equipment, fixture or apparatus disconnection costs and salvage values excluded.

Description	Craft@Hrs	Unit	Material $	Labor $	Equipment $	Total $
Copper pipe, ½"-1"						
remove	P1@.022	LF	—	.81	.08	.89
Copper pipe, 1½"-2"						
remove	P1@.028	LF	—	1.03	.09	1.12
Copper pipe, 2½"-4"						
remove	P1@.044	LF	—	1.61	.12	1.73

Remove plastic piping. These figures include cutting out and removing plastic piping. These figures assume demolished materials are piled on site. Equipment, fixture or apparatus disconnection costs and salvage values excluded. Equipment may be needed for some sizes and locations. Refer to the equipment rental costs at the front of this section.

Description	Craft@Hrs	Unit	Material $	Labor $	Equipment $	Total $
Plastic pipe, ½"-1"						
remove	P1@.015	LF	—	.55	—	.55
Plastic pipe, 1½"-2"						
remove	P1@.030	LF	—	1.10	—	1.10
Plastic pipe, 2½"-4"						
remove	P1@.038	LF	—	1.39	—	1.39
Plastic pipe, 6"-8"						
remove	P1@.080	LF	—	2.91	—	2.91

Description	Craft@Hrs	Unit	Material $	Labor $	Equipment $	Total $

Remove pump. These figures include isolating and disconnecting pipe and electrical connections. Disposal costs or salvage values excluded. Equipment may be needed for some sizes and locations. Refer to the equipment rental costs at the front of this section.

Description	Craft@Hrs	Unit	Material $	Labor $	Equipment $	Total $
Base-mounted pump, ½-1 HP, remove	P1@1.00	Ea	—	36.70	—	36.70
Base-mounted pump, 1½-3 HP, remove	P1@1.75	Ea	—	64.20	—	64.20
Base-mounted pump, 5-10 HP, remove	P1@2.75	Ea	—	101.00	—	101.00
Base-mounted pump, 15-25 HP, remove	P1@3.75	Ea	—	138.00	—	138.00
Base-mounted pump, 30-50 HP, remove	P1@4.55	Ea	—	167.00	—	167.00
In-line mounted pump, 1/12-1/3 HP, remove	P1@.650	Ea	—	23.90	—	23.90
In-line mounted pump, ½-1 HP, remove	P1@.750	Ea	—	27.50	—	27.50
In-line mounted pump, 3-5 HP, remove	P1@1.85	Ea	—	67.90	—	67.90
In-line mounted pump, 7.5-10 HP, remove	P1@2.55	Ea	—	93.60	—	93.60
In-line mounted pump, 15-25 HP, remove	P1@3.90	Ea	—	143.00	—	143.00
In-line mounted pump, 30-50 HP, remove	P1@4.85	Ea	—	178.00	—	178.00

Remove valve. These figures include isolation and disconnection. Disposal costs or salvage values excluded. Equipment may be needed for some sizes and locations. Refer to the equipment rental costs at the front of this section.

Description	Craft@Hrs	Unit	Material $	Labor $	Equipment $	Total $
Soldered valve, ½"-1", remove	P1@.165	Ea	—	6.06	—	6.06
Soldered valve, 1½"-2", remove	P1@.225	Ea	—	8.26	—	8.26
Soldered valve, 2½"-4", remove	P1@.650	Ea	—	23.90	—	23.90
Threaded valve, ½"-1", remove	P1@.150	Ea	—	5.51	—	5.51
Threaded valve, 1½"-2", remove	P1@.235	Ea	—	8.62	—	8.62
Threaded valve, 2½"-4", remove	P1@.680	Ea	—	25.00	—	25.00
Flanged valve, ½"-1", remove	P1@.200	Ea	—	7.34	—	7.34
Flanged valve, 1½"-2", remove	P1@.280	Ea	—	10.30	—	10.30
Flanged valve, 2½"-4", remove	P1@.625	Ea	—	22.90	—	22.90
Flanged valve, 6"-8", remove	P1@1.70	Ea	—	62.40	—	62.40
Flanged valve, 10"-12", remove	P1@2.65	Ea	—	97.30	—	97.30

HVAC & Plumbing Demolition

Description	Craft@Hrs	Unit	Material $	Labor $	Equipment $	Total $

Remove hot water tank. These figures include isolation and disconnection. Disposal costs or salvage values excluded. Equipment may be needed for some sizes and locations. Refer to the equipment rental costs at the front of this section.

Description	Craft@Hrs	Unit	Material $	Labor $	Equipment $	Total $
Gas-fired hot water tank, 40 gallon, remove	P1@2.00	Ea	—	73.40	—	73.40
Gas-fired hot water tank, 60 gallon, remove	P1@2.50	Ea	—	91.80	—	91.80
Gas-fired hot water tank, 100 gallon, remove	P1@3.00	Ea	—	110.00	—	110.00
Electric hot water tank, 40 gallon, remove	P1@1.60	Ea	—	58.70	—	58.70
Electric hot water tank, 60 gallon, remove	P1@2.00	Ea	—	73.40	—	73.40
Electric hot water tank, 100 gallon, remove	P1@2.60	Ea	—	95.40	—	95.40

Remove plumbing fixture. These figures include isolation and disconnection. Disposal costs or salvage values excluded. Equipment may be needed for some sizes and locations. Refer to the equipment rental costs at the front of this section.

Description	Craft@Hrs	Unit	Material $	Labor $	Equipment $	Total $
Floor-mount, tank-type water closet, remove	P1@.850	ea	—	31.20	—	31.20
Wall-mount, tank-type water closet, remove	P1@1.35	Ea	—	49.50	—	49.50
Floor-mount, flush valve water closet, remove	P1@.950	Ea	—	34.90	—	34.90
Wal- mount, flush valve water closet, remove	P1@1.55	Ea	—	56.90	—	56.90
Wall-mount, flush valve urinal, remove	P1@1.15	Ea	—	42.20	—	42.20
Wall-mount, tank-type urinal, remove	P1@1.65	Ea	—	60.60	—	60.60
Kitchen sink, remove	P1@.850	Ea	—	31.20	—	31.20
Bathroom sink, remove	P1@.850	Ea	—	31.20	—	31.20
Bathtub, remove	P1@2.25	Ea	—	82.60	—	82.60
Shower stall, remove	P1@2.00	Ea	—	73.40	—	73.40
Tub and shower combination, remove	P1@2.85	Ea	—	105.00	—	105.00
Laundry sink, remove	P1@.950	Ea	—	34.90	—	34.90
Mop sink, remove	P1@1.50	Ea	—	55.10	—	55.10
Slop sink, remove	P1@1.75	Ea	—	64.20	—	64.20
Drinking fountain, remove	P1@1.40	Ea	—	51.40	—	51.40

Description	Craft@Hrs	Unit	Material $	Labor $	Equipment $	Total $

Disconnect plumbing fixture. These figures include isolation and disconnection only. Fixtures removed by others. Disposal costs or salvage values excluded. Equipment may be needed for some sizes and locations. Refer to the equipment rental costs at the front of this section.

Description	Craft@Hrs	Unit	Material $	Labor $	Equipment $	Total $
Floor-mount, tank-type water closet, disconnect	P1@.450	Ea	—	16.50	—	16.50
Wall-mount, tank-type water closet, disconnect	P1@.650	Ea	—	23.90	—	23.90
Floor-mount, flush valve water closet, disconnect	P1@.550	Ea	—	20.20	—	20.20
Wall-mount, flush valve water closet, disconnect	P1@.650	Ea	—	23.90	—	23.90
Wall-mount, flush valve urinal, disconnect	P1@.650	Ea	—	23.90	—	23.90
Wall-mount, tank-type urinal, disconnect	P1@.950	Ea	—	34.90	—	34.90
Kitchen sink, disconnect	P1@.450	Ea	—	16.50	—	16.50
Bathroom sink, disconnect	P1@.450	Ea	—	16.50	—	16.50
Bathtub, disconnect	P1@.650	Ea	—	23.90	—	23.90
Shower stall, disconnect	P1@.650	Ea	—	23.90	—	23.90
Tub and shower combination, disconnect	P1@.650	Ea	—	23.90	—	23.90
Laundry sink, disconnect	P1@.450	Ea	—	16.50	—	16.50
Mop sink, disconnect	P1@.450	Ea	—	16.50	—	16.50
Slop sink, disconnect	P1@.750	Ea	—	27.50	—	27.50
Drinking fountain, disconnect	P1@.650	Ea	—	23.90	—	23.90
Gas-fired hot water tank, 40 gallon, disconnect	P1@.900	Ea	—	33.00	—	33.00
Gas-fired hot water tank, 60 gallon, disconnect	P1@1.15	Ea	—	42.20	—	42.20
Gas-fired hot water tank, 100 gallon, disconnect	P1@1.35	Ea	—	49.50	—	49.50
Electric hot water tank, 40 gallon, disconnect	P1@.600	Ea	—	22.00	—	22.00
Electric hot water tank, 60 gallon, disconnect	P1@.900	Ea	—	33.00	—	33.00
Electric hot water tank, 100 gallon, disconnect	P1@1.00	Ea	—	36.70	—	36.70

Description	Craft@Hrs	Unit	Material $	Labor $	Equipment $	Total $

Remove pipe insulation. These figures assume demolished materials are piled on site. Disposal costs or salvage values excluded.

Description	Craft@Hrs	Unit	Material $	Labor $	Equipment $	Total $
Fiberglass pipe insulation, ½"-1", remove	P1@.020	LF	—	.73	—	.73
Fiberglass pipe insulation, 1½"-2", remove	P1@.025	LF	—	.92	—	.92
Fiberglass pipe insulation, 2½"-4", remove	P1@.035	LF	—	1.28	—	1.28
Fiberglass pipe insulation, 6"-8", remove	P1@.060	LF	—	2.20	—	2.20
Fiberglass pipe insulation, 10"-12", remove	P1@.090	LF	—	3.30	—	3.30
Pipe insul. with canvas jacket – add	—	%	—	10.00	—	—
Pipe insul. with alum. jacket – add	—	%	—	20.00	—	—
Calcium silicate pipe insul. – add	—	%	—	15.00	—	—

Remove duct insulation. These figures assume demolished materials are piled on site. Disposal costs or salvage values excluded.

Description	Craft@Hrs	Unit	Material $	Labor $	Equipment $	Total $
Fiberglass blanket insulation, remove	S2@.020	SF	—	.72	—	.72
Fiberglass rigid board insulation, remove	S2@.025	SF	—	.90	—	.90
Insul. with canvas jacket – add	—	%	—	8.00	—	—
Insul. with alum jacket – add	—	%	—	10.00	—	—

Budget Estimates for Plumbing and HVAC Work

General contractors and owners often ask mechanical subcontractors to prepare budget cost estimates for use in planning projects. A budget estimate isn't a bid for the contract and you probably won't be paid for preparing the estimate. That's why you won't want to spend too much time on these estimates. Accuracy within 8 to 10 percent of the final bid price will usually be acceptable. Normally you'll have only preliminary architectural drawings and possibly a guide specification to work from. The tables that follow will help you prepare budget estimates for nearly any plumbing or HVAC job.

The costs in these tables are averages and will vary considerably with design requirements and by geographic area. Make it clear to the owner or architect requesting the figures that these are budget estimates and will be revised when final plans and specifications are available.

Budget Estimates for Plumbing: First, count the number of plumbing fixtures. Multiply the number of fixtures of each type by the price per fixture shown in the table below. These prices include all the usual labor and material for a complete average plumbing installation. Prices are based on branch piping lengths averaging 35 feet between the fixture and the main. Typical main piping is also included in the prices.

Description	Craft@Hrs	Unit	Material $	Labor $	Equipment $	Total $

Budget plumbing estimate, cast iron DWV, copper supply pipe

Description	Craft@Hrs	Unit	Material $	Labor $	Equipment $	Total $
Bathtubs	P1@15.0	Ea	2,460.00	551.00	—	3,011.00
Showers	P1@9.00	Ea	1,340.00	330.00	—	1,670.00
Lavatories	P1@11.0	Ea	2,300.00	404.00	—	2,704.00
Kitchen sinks	P1@12.0	Ea	2,410.00	440.00	—	2,850.00
Service sinks	P1@10.0	Ea	2,070.00	367.00	—	2,437.00
Bar sinks	P1@9.00	Ea	1,740.00	330.00	—	2,070.00
Floor sinks	P1@7.00	Ea	1,070.00	257.00	—	1,327.00
Water closets						
Tank-type	P1@11.0	Ea	1,740.00	404.00	—	2,144.00
Flush valve	P1@12.0	Ea	1,960.00	440.00	—	2,400.00
Urinals	P1@8.00	Ea	2,520.00	294.00	—	2,814.00
Drinking fountains						
(refrigerated)	P1@12.0	Ea	2,910.00	440.00	—	3,350.00
Wash fountains	P1@25.0	Ea	6,260.00	918.00	—	7,178.00
Can washers	P1@18.0	Ea	3,190.00	661.00	—	3,851.00
Floor drains	P1@8.00	Ea	985.00	294.00	—	1,279.00
Area drains	P1@6.00	Ea	846.00	220.00	—	1,066.00
Roof drains	P1@18.0	Ea	2,240.00	661.00	—	2,901.00
Overflow drains	P1@18.0	Ea	2,240.00	661.00	—	2,901.00
Deck drains	P1@4.00	Ea	697.00	147.00	—	844.00
Cleanouts	P1@5.00	Ea	753.00	184.00	—	937.00
Trap primers	P1@2.00	Ea	382.00	73.40	—	455.40
Sump pumps	P1@6.00	Ea	3,720.00	220.00	—	3,940.00
Water heaters						
Gas, 40 gal.	P1@16.0	Ea	4,390.00	587.00	—	4,977.00
Gas, 80 gal.	P1@17.0	Ea	4,870.00	624.00	—	5,494.00
Gas, 120 gal.	P1@18.0	Ea	5,690.00	661.00	—	6,351.00

Budget Estimating

Description	Craft@Hrs	Unit	Material $	Labor $	Equipment $	Total $

Budget plumbing estimate, plastic DWV and plastic supply pipe

Description	Craft@Hrs	Unit	Material $	Labor $	Equipment $	Total $
Bathtubs	P1@12.0	Ea	1,960.00	440.00	—	2,400.00
Showers	P1@7.00	Ea	950.00	257.00	—	1,207.00
Lavatories	P1@9.00	Ea	1,810.00	330.00	—	2,140.00
Kitchen sinks	P1@10.0	Ea	2,020.00	367.00	—	2,387.00
Service sinks	P1@8.00	Ea	1,630.00	294.00	—	1,924.00
Bar sinks	P1@7.00	Ea	1,380.00	257.00	—	1,637.00
Floor sinks	P1@6.00	Ea	826.00	220.00	—	1,046.00
Water closets						
Tank-type	P1@8.00	Ea	1,270.00	294.00	—	1,564.00
Flush valve	P1@9.00	Ea	1,530.00	330.00	—	1,860.00
Urinals	P1@6.00	Ea	2,060.00	220.00	—	2,280.00
Drinking fountains						
(refrigerated)	P1@10.0	Ea	2,400.00	367.00	—	2,767.00
Wash fountains	P1@22.0	Ea	5,460.00	807.00	—	6,267.00
Can washers	P1@15.0	Ea	2,680.00	551.00	—	3,231.00
Floor drains	P1@7.00	Ea	683.00	257.00	—	940.00
Area drains	P1@5.00	Ea	559.00	184.00	—	743.00
Roof drains	P1@16.0	Ea	1,910.00	587.00	—	2,497.00
Overflow drains	P1@16.0	Ea	1,910.00	587.00	—	2,497.00
Deck drains	P1@3.00	Ea	475.00	110.00	—	585.00
Cleanouts	P1@4.00	Ea	580.00	147.00	—	727.00
Trap primers	P1@2.00	Ea	288.00	73.40	—	361.40
Sump pumps	P1@5.00	Ea	3,350.00	184.00	—	3,534.00
Water heaters						
Gas, 40 gal.	P1@15.0	Ea	4,050.00	551.00	—	4,601.00
Gas, 80 gal.	P1@16.0	Ea	4,430.00	587.00	—	5,017.00
Gas, 120 gal.	P1@17.0	Ea	5,140.00	624.00	—	5,764.00

Budget Estimates for HVAC Work: Budget estimates for heating, ventilating and air conditioning work are figured by the square foot of floor. Multiply the area of conditioned space by the cost per square foot in the table that follows. The result is the estimated cost including a complete heating, ventilating and air conditioning system and 20 percent overhead and profit. The table also shows average heating and cooling requirements for various types of buildings.

Description	Cooling (Btu/SF)	Heating (Btu/SF)	Total Selling Price ($) Per SF	Per Ton
Apartments and condominiums	25	25	14.10	5,180.00
Auditoriums	40	40	36.00	8,280.00
Banks	45	30	37.30	8,260.00
Barber shops	40	40	36.80	8,430.00
Beauty shops	60	30	43.70	6,680.00
Bowling alleys	40	40	28.70	6,590.00
Churches	35	35	31.20	8.130.00
Classrooms	45	40	56.70	11,500.00
Cocktail lounges	70	30	39.40	5,180.00
Computer rooms	145	20	80.20	5,080.00
Department stores	35	30	25.20	6,640.00
Laboratories	60	45	69.80	10,700.00
Libraries	45	35	41.70	8,500.00
Manufacturing plants	40	35	16.20	3,720.00
Medical buildings	35	35	27.50	7,260.00
Motels	30	30	19.30	5,890.00
Museums	35	40	30.70	8.070.00
Nursing homes	45	35	35.40	7,290.00
Office buildings, low-rise	35	35	28.00	7,370.00
Office buildings, high-rise	40	30	38.20	8,730.00
Residences	20	30	8.61	3,980.00
Retail shops	45	30	28.70	5,850.00
Supermarkets	30	30	15.00	4,570.00
Theaters	40	40	28.00	6,480.00

Note that these are budget estimates and can vary widely with conditions which may be unique to the job site. Many of these conditions are identified below. Avoid submitting even a budget estimate without either a site walk-through, a conference with the consulting engineer or architect or a careful review of the plans.

Be alert to any of the following conditions. Any of these can have a major impact on cost:

1. Do HVAC unit dimensions exceed the size of available openings?

2. If a crane is required, is there ample clearance for crane operations?

3. Will any doors, permanent ladders, scaffolding or load-bearing supports obstruct installation?

4. Do the plans show all vents, drains and cleanouts required by the code? Many items like these are not specified by the engineer but will be required by the inspector.

5. For retrofit work, consider carefully problems associated with demolition, including removal of the existing HVAC system, structural modifications and any encased asbestos that may be present.

6. In retrofit jobs, the type and size of HVAC equipment is limited by the window area and installed insulation.

7. How will available power, gas and water lines be extended to the new HVAC equipment?

Plumbing & HVAC Forms and Letters

The following pages contain business forms and letters most plumbing and HVAC contractors can use.

Change Orders
Change Estimate Takeoff
Change Estimate Worksheet
Change Estimate Summary
Change Order Log

Subcontract Forms
Standard Form Subcontract
Subcontract Change Order

Purchase Order

Construction Schedule

Letter of Intent

Submittal Data Cover Sheet

Submittal Index

Billing Breakdown Worksheet

An explanation is provided on how the forms function in the contractor's office, how to fill them out, and a sample provided. Where appropriate, a blank copy of the form is provided for your use.

Permission is hereby given by the publisher for the duplication of any of the blank forms on a copy machine or at a printer, provided they are for your own use and not to be sold at a cost exceeding the actual printing and paper cost.

Space has been provided at the top of the forms so you can add your own letterhead or logo.

Change Estimates

Very few projects are completed without some changes to the original plans. Most changes in the scope of work are a result of either errors by the designer or requests by the owner.

Here's the usual sequence in handling changes.

1) The project architect usually initiates a change by sending a Request for Quotation (R.F.Q.) to the general contractor with a detailed description of the work to be modified.

2) The general contractor prepares a Change Estimate (C.E.) if the work is to be done by the general.

3) The subcontractor prepares a change estimate which shows the work to be added to or deleted from the contract and the cost of changes, including markup. The general contractor reviews the change estimate and forwards a summary to the architect for decision.

4) If the decision is to go ahead with the change, the architect issues a Change Order (C.O.) to the general contractor. This change order is authorization to proceed with the change at the price quoted.

Your costs for making changes will normally be higher than your costs for similar items on the original estimate. Here's why:

Equipment and materials for changes are almost always purchased in smaller quantities. Volume discounts are seldom available.

Special handling and delivery of small orders will further increase equipment and material costs.

Scheduled work will be disrupted.

Overhead costs will be higher because special purchase orders or subcontracts will be required.

Vendor restocking charges for deleted items will range from 20 to 50 percent of the original cost.

Large change orders may require additional manpower, which usually results in decreased efficiency.

The cumulative effect of change orders will be to increase overhead costs if the project completion date is extended.

To compensate for these higher costs, most estimators use a higher markup on change estimates:

Add an extra 10 percent markup to equipment and material costs to cover increased overhead and handling costs, prior to adding final Overhead and Profit markup.

Increase competitive labor units (manhours required to perform a specific task) by 15 percent to compensate for inefficiencies such as disruptions in the work schedule, extra handling of equipment and material, and additional manpower.

Mark up subcontractor prices by 10 percent.

Figure overhead and profit markup at 15/10 or 10/10.

By following these suggestions, the mathematical gross profit will appear to be 40 to 45 percent, but the true gross will be closer to 25 percent because your costs for change order work are higher, as listed above.

Change Estimate Example

Let's assume that the general contractor on your job has sent a R.F.Q. to your attention. The extra work is a run of hot water piping that will connect to equipment being furnished and set by others. The first step is to list all material and labor required to do the work. Notice the items entered on the Change Estimate Take-off form on page 440.

When the take-off is complete, transfer the manhour total to the Change Estimate Worksheet on page 441. To complete this form:

1) Multiply manhours by the cost per hour.

2) Add the cost of supervision.

3) Total direct labor cost.

4) Add the cost of preparing the change order estimate (C.E.).

5) Add equipment and tool costs.

6) Add subcontract costs.

Then bring totals forward to the Change Estimate Summary on page 442:

1) Enter the material cost total from the Change Estimate Take-off form on page 444.

2) Enter direct and indirect labor cost totals from the Change Estimate Worksheet.

3) Enter equipment, tool and subcontract cost totals from the Change Estimate Worksheet.

Notice the additional bond charge on line M of the Change Estimate Summary. This is the extra premium charged by the bonding company to cover the increase in contract value. The next line, Service Reserve, is the allowance for service work (if any is required) during the warranty period. A sample Change Order Log, used to record the status of changes, is also included on page 443.

ACME
Mechanical Contractors

7600 Oak Avenue - Smallville, U.S.A. 12345-6789
Voice 1-234-5678 / FAX 1-234-5680

Change Estimate Take-off

Project CBC Bank Building

Take-off by DCO

Date 2-2-20

C. E. Number One

Job No 7777

Item	Equipment and Material — Description	Quantity	Material $ — Unit	Material $ — Total	Labor Manhours — Unit	Labor Manhours — Total
1	1" Type L hard copper pipe	40 LF	3.87	154.80	.038	1.52
2	1" 90 degree copper ell	6 Ea	4.21	25.26	.23	1.38
3	1" copper coupling	4 Ea	2.54	10.16	.24	.96
4	1" bronze ball valve	2 Ea	17.30	34.60	.36	.72
5	2" x 1" reducing copper tee	2 Ea	20.60	41.20	.40	.80
6	1" copper union	2 Ea	19.10	38.20	.26	.52
7	1" pipe hanger	8 Ea	4.33	34.64	.25	2.00
8	Miscellaneous material	1 Ea	LS	50.00	1.78	1.78
		Totals		388.86		9.68

440

ACME
Mechanical Contractors

7600 Oak Avenue - Smallville, U.S.A. 12345-6789
Voice 1-234-5678 / FAX 1-234-5680

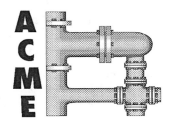

Change Estimate Worksheet

Estimated by   DCO   **C. E. Number**   One

Date   2-3-20   **Job No**   7777

Direct Labor Expense (including benefits)

Plumber or pipefitter journeyman	9.68	manhours at $	41.85	per hour =	$ 405.11
Plumber or pipefitter supervisor	1.00	manhours at $	45.00	per hour =	$ 45.00
Sheet metal journeyman		manhours at $		per hour =	$
Sheet metal supervisor		manhours at $		per hour =	$
Laborer journeyman		manhours at $		per hour =	$
		manhours at $		per hour =	$

Total Direct Labor Cost $ 450.11

Indirect Costs

Travel		$
Subsistence		$
Labor	Cost Estimate Preparation (1 Hour)	$ 40.00

Total Indirect Cost $ 40.00

Equipment and Tool Costs

Ramset rental	$ 35.00
Propane	$ 4.50

Total Equipment and Tool Cost $ 39.50

Subcontract Costs

Quikrap Insulation Co.	$ 100.00
	$

Total Subcontract Cost $ 100.00

441

ACME
Mechanical Contractors

7600 Oak Avenue - Smallville, U.S.A. 12345-6789
Voice 1-234-5678 / FAX 1-234-5680

Change Estimate Summary

Quotation to ABC General Contractors, Inc **Date** 2-3-20

Address 380 First Street **Change Order No.** One

Address **Job Number** 7777

City / State / Zip Smallville, CA 63876

Attention: Mr. J.H. Smith

Job Name CBC Bank Building

Reference: Your R.F.O. No. 7 Labor and Material to furnish and install additional hot water piping.

A.	Materials and equipment:	$ 388.86		
B.	Sales tax: 7% .	$ 27.22		
C.	Direct labor:	$ 450.11		
D.	Indirect costs:	$ 40.00		
E.	Equipment and tools:	$ 39.50		
F.			Subtotal $	945.69
G.	Overhead at : 15% of line F:	$ 141.85		
H.	Subcontracts:	$ 100.00		
I.	Overhead at : 10% of line H:	$ 10.00		
J.			Subtotal $	1,197.54
K.	Profit at 10% of line J:	$ 119.75		
L.			Subtotal $	1,317.29
M.	Bond premium at 1% of line L:	$ 13.17		
N.	Service reserve at 0.5% of line L:	$ 6.59		
O.	**Total cost estimate, lines L thru N:**	Add Deduct	$	1,337.05

P.   Exclusions from this estimate: Per Base

Q.   This quotation is valid for 30 days

R.   We require 1 day(s) extension of the contract time.

S.   We are proceeding with this work per your authorization.

T.   Please forward your confirming change order.

Signed by: _____
Project Manager

442

Change Order Log

Job Name _____ **Job Number** _____

Change Number	Customer Number	Date	Amount	Date Approved	Amount Approved	Remarks

Use this space to insert your company logo and letterhead.

Change Estimate Take-off

Project _____   **C. E. Number** _____

Take-off by _____   **Job No** _____

Date _____

Item	Description	Quantity	Unit	Total	Unit	Total
	Equipment and Material		**Material $**		**Labor Manhours**	
		Totals				

Use this space to insert your company logo and letterhead.

Change Estimate Worksheet

Estimated by _____   **C. E. Number** _____

Date _____   **Job No** _____

Direct Labor Expense (including benefits)

Plumber or pipefitter journeyman	_____ manhours at $ _____	per hour =	$ _____
Plumber or pipefitter supervisor	_____ manhours at $ _____	per hour =	$ _____
Sheet metal journeyman	_____ manhours at $ _____	per hour =	$ _____
Sheet metal supervisor	_____ manhours at $ _____	per hour =	$ _____
Laborer journeyman	_____ manhours at $ _____	per hour =	$ _____
_____	_____ manhours at $ _____	per hour =	$ _____
		Total Direct Labor Cost	$ _____

Indirect Costs

Travel	_____	$ _____
Subsistence	_____	$ _____
Labor	_____	$ _____
	Total Indirect Cost	$ _____

Equipment and Tool Costs

_____	$ _____
_____	$ _____
Total Equipment and Tool Cost	$ _____

Subcontract Costs

_____	$ _____
_____	$ _____
Total Subcontract Cost	$ _____

Use this space to insert your company logo and letterhead.

_____

Change Estimate Summary

Quotation to _____   **Date** _____

Address _____   **Change Order No.** _____

Address _____   **Job Number** _____

City / State / Zip _____

Attention: _____

Job Name _____

Reference: _____

_____

A. Materials and equipment: $ _____

B. Sales tax: _____ . $ _____

C. Direct labor: $ _____

D. Indirect costs: $ _____

E. Equipment and tools: $ _____

F.   **Subtotal $** _____

G. Overhead at _____ of line F: $ _____

H. Subcontracts: $ _____

I. Overhead at _____ of line H: $ _____

J.   **Subtotal $** _____

K. Profit at _____ of line J: $ _____

L.   **Subtotal $** _____

M. Bond premium at 1% of line L: $ _____

N. Service reserve at 0.5% of line L: $ _____

O. **Total cost estimate, lines L thru N:**   Add   Deduct $ _____

P. Exclusions from this estimate: _____

Q. This quotation is valid for _____ days

R. We require _____ day(s) extension of the contract time.

S. We are proceeding with this work per your authorization.

T. Please forward your confirming change order.

Signed by: _____

Project Manager

Another term for the general contractor is *prime contractor*. Subcontractors working directly for the prime contractor are called *second-tier subcontractors*. Contractors working for these second-tier subs are usually known as *third-tier subcontractors*. On a large job, many second-tier subs may hire third-tier subs to perform specialized work such as insulation, fireproofing, test and balance, water treatment, instrumentation and trenching.

If a portion of your work will be subcontracted to others, I recommend drawing up a subcontract agreement for that work. Any subcontractors who furnish both material and labor for a job should have a written contract.

Most larger plumbing, heating and A/C subcontractors use a two-part *Standard Form Subcontract* published by the Associated General Contractors of America. This two-part form is used to write the base subcontract. The A.G.C. also publishes a *Subcontract Change Order* form which is used when adding or deleting items after the base subcontract is signed. These forms are sold at modest cost at most A.G.C. offices.

The first section of page one of the Standard Form Subcontract identifies the two contractors entering into the agreement. The party issuing the subcontract is the *contractor*. Recipient of the agreement is the *subcontractor*.

Under *Recitals,* the contractor must affirm the right to issue a subcontract by declaring that a legal contract exists with the *owner* (general contractor).

Section 1 - Entire Contract describes the contract documents: the project plans, specifications and addenda. This list of documents should always be the same as the documents identified in the second-tier subcontractor's contract with the general contractor.

Section 2 - Scope describes the specific scope of work to be performed.

Section 3 - Contract Price. Complete this section by inserting the agreed-upon price for doing the work described.

Section 4 - Payment Schedule on page two should show the same payment terms as your contract with the general contractor.

Section 7 - Special Provisions is used to list any special agreements or exclusions.

When the contract is typed, do not sign it. Instead, forward it to the third-tier subcontractor with instructions to sign it and return it intact. When the signed contract is received, sign it and send a copy to the third-tier subcontractor.

For contracts that comply with your state's regulations, please visit www.constructioncontractwriter.com

Standard Form Subcontract

Subcontract No. 7777-1

THIS AGREEMENT, made and entered into at Smallville, CA. this 3rd day of May, 2020, by and between Acme Mechanical Contractors hereinafter called **CONTRACTOR,** with principal office at 7600 Oak Avenue, Smallville, California, and Quikrap Insulation Co., 320 Maple Blvd., Smallville, CA hereinafter called a **SUBCONTRACTOR.**

RECITALS

On or about the ___2nd___ day of ___January___, 20_20_, CONTRACTOR entered into a prime contract with

ABC General Contractors, Inc. _____ hereinafter called OWNER, whose address is 380 First Street, Smallville, California ___ to perform the following construction work: The installation of all mechanical and plumbing systems for the proposed Mile-Hi office building to be located at 2600 Second, North, Smallville, California.

Said work is to be performed in accordance with the prime contract and the plans and specifications. Said plans and specifications have been prepared by or on behalf of Quik-Draw and Associates, Smallville, California ___, **ARCHITECT**

SECTION 1 - ENTIRE CONTRACT

SUBCONTRACTOR certifies and agrees that he is fully familiar with all of the terms, conditions and obligations of the Contract Documents, as hereinafter defined, the location of the job site, and the conditions under which the work is to be performed, and that he enters into this Agreement based upon his investigation of all of such matters and is in no way relying upon any opinions or representations of CONTRACTOR. It is agreed that this Agreement represents the entire agreement. It is further agreed that the Contract Documents are incorporated in this Agreement by this reference, with the same force and effect as if the same were set forth at length herein, and that SUBCONTRACTOR and his subcontractors will be and are bound by any and all of said Contract Documents insofar as they relate in any part or in any way, directly or indirectly to the work covered by this Agreement. SUBCONTRACTOR agrees to be bound to CONTRACTOR in the same manner and to the same extent as CONTRACTOR is bound to OWNER under the Contract Documents, to the extent of the work provided for in this Agreement, and that where, in the Contract Documents, reference is made to CONTRACTOR and the work or specification therein pertains to SUBCONTRACTOR'S trade, craft, or type of work, then such work or specification shall be interpreted to apply to SUBCONTRACTOR instead of CONTRACTOR. The phrase "Contract Documents" is defined to mean and include:

Drawing Nos. A-1 thru A-17, C-1 thru C-4, S-1 thru S-7, E-1 thru E-6 and M-1 thru M-7, all dated
10-4-2019, Specifications dated 10-2-19 and Addendum No. 1 dated 10-18-19.

SECTION 2 - SCOPE

SUBCONTRACTOR agrees to furnish all labor, services, materials, installation, cartage, hoisting, supplies, insurance, equipment, scaffolding, tools and other facilities of every kind and description required for the prompt and efficient execution of the work described herein and to perform the work necessary or incidental to complete thermal insulation for all plumbing, HHW, CHW and condensate piping systems, including pumps P-1, P-2 and P-3 for the project in strict accordance with the Contract Documents and as more particularly, though not exclusively, specified in:

Division 15, Section 400 entitled "Thermal Insulation systems" and General Conditions, as applicable.

SECTION 3 - CONTRACT PRICE

CONTRACTOR agrees to pay SUBCONTRACTOR for the strict performance of his work, the sum of:

Forty-two thousand, three hundred dollars ($42,300.00), subject to additions and deductions for changes in the work as may be agreed upon, and to make payment in accordance with the Payment Schedule, Section 4.

SECTION 4 - PAYMENT SCHEDULE

CONTRACTOR agrees to pay SUBCONTRACTOR in monthly payments of ___90___ % of labor and materials which have been placed in position and for which payment has been made by OWNER to CONTRACTOR. The remaining ___10___ % shall be retained by CONTRACTOR until he receives final payment from OWNER, but not less than thirty-five days after the entire work required by the prime contract has been fully completed in conformity with the Contract Documents and has been delivered and accepted by OWNER, ARCHITECT, and CONTRACTOR. Subject to the provisions of the next sentence, the retained percentage shall be paid SUBCONTRACTOR promptly after CONTRACTOR receives his final payment from OWNER. SUBCONTRACTOR agrees to furnish, if and when required by CONTRACTOR, payroll affidavits, receipts, vouchers, release of claims for labor, material and subcontractors performing work or furnishing materials under this Agreement, all in form satisfactory to CONTRACTOR, and it is agreed that no payment hereunder shall be made, except at CONTRACTOR'S option, until and unless such payroll affidavits, receipts, vouchers or release; or any or all of them, have been furnished. And payment made hereunder prior to completion and acceptance of the work, as referred to above, shall not be construed as evidence of acceptance of any part of SUBCONTRACTOR'S work.

SECTION 5 - GENERAL SUBCONTRACT PROVISIONS

General Subcontract Provisions on back of Pages 1 and 2 are an integral part of this Agreement.

SECTION 6 - GENERAL PROVISIONS

1. *SUBCONTRACTOR* agrees to begin work as soon as instructed by the CONTRACTOR, and shall carry on said work promptly, efficiently and at a speed that will not cause delay in the progress of the CONTRACTOR'S work or work of other subcontractors. If, in the opinion of the CONTRACTOR, the SUBCONTRACTOR falls behind in the progress of the work, the CONTRACTOR may direct the SUBCONTRACTOR to take such steps as the CONTRACTOR deems necessary to improve the rate of progress, including, without limitation, requiring the SUBCONTRACTOR to increase the number of shifts, personnel, overtime operations, days of work, equipment, amount of plant, or other remedies and to submit to CONTRACTOR for CONTRACTOR'S approval an outline schedule demonstrating the manner in which the required rate of progress will be regained, without additional cost to the CONTRACTOR. CONTRACTOR may require SUBCONTRACTOR to prosecute, in preference to other parts of the work, such part or parts of the work as CONTRACTOR may specify.

The SUBCONTRACTOR shall complete the work as required by the progress schedule prepared by the CONTRACTOR, which may be amended from time to time. The progress schedule may be reviewed in the office of the CONTRACTOR and sequence of construction will be as directed by the CONTRACTOR.

The SUBCONTRACTOR agrees to have an acceptable representative (an officer of SUBCONTRACTOR if requested by the CONTRACTOR) present at all job meetings and to submit weekly progress reports in writing if requested by the CONTRACTOR. Any job progress schedules are hereby made a part of and incorporated herein by reference.

2. *Reserved Gate Usage*

SUBCONTRACTOR shall notify in writing, and assign its employees, material men and suppliers, to such gates or entrances as may be established for their use by CONTRACTOR and in accordance with such conditions and at such times as may be imposed by CONTRACTOR. Strict compliance with CONTRACTOR'S gate usage procedures shall be required by the SUBCONTRACTOR who shall be responsible for such gate usage by its employees, material men, suppliers, subcontractors, and their material men and suppliers.

3. *Staggered Days and Hours of Work and for Deliveries*

SUBCONTRACTOR shall schedule the work and the presence of its employees at the jobsite and any deliveries of supplies or materials by its material men and suppliers to the jobsite on such days, and at such times and during such hours, as may be directed by CONTRACTOR. SUBCONTRACTOR shall assume responsibility for such schedule compliance not only for its employees but for all its material men, suppliers and subcontractors, and their material men and suppliers.

SECTION 7 - SPECIAL PROVISIONS

```
1. Pipe hanger inserts will be furnished and installed by Contractor.
2. All trash pickup by Subcontractor; haul-away will be by General Contractor.
```

Contractors are required by law to be licensed and regulated by the Contractors' State License Board. Any questions concerning a contractor may be referred to the registrar of the board whose address is:

Contractors' State License Board -- P. O. Box 26000, Sacramento, California 95826

IN WITNESS WHEREOF: The parties hereto have executed this Agreement for themselves, their heirs, executors, successors, administrators, and assignees on the day and year first above written.

SUBCONTRACTOR	CONTRACTOR
Quikrap Insulation Company	Acme Mechanical Contractors
By _____	By _____
Name Title	Name Title
☐ Corporation ☐ Partnership ☐ Proprietorship	
(Seal)	
Contractor's State License No. _____	Contractor's State License No. _____

Revised 5-09   Published by AGC San Diego
Page 2 of 2

For contracts that comply with your state's regulations, please visit www.constructioncontractwriter.com

Subcontract Change Order

Number   One

Date   6-11-20   Subcontract Number   7777-1

To   Quickrap Insulation Co.   Our Job Number _____

_____   Our Proposal Number _____

_____   Architect's C.O. Number _____

                                                     Effective Date of Charge   6-17-20

Subject to all the provisions of this Change Order, you are hereby directed to make the following change(s)

Delete HHW and CHW piping insulation located above Room No. 322 per ABC General

Contractor's Change Order No. 18, dated 6-5-20.

_____

_____

_____

_____

_____

_____

_____

_____

The following change(s) will alter the price provided in your subcontract by the   deduction   *of $*   465.00

Surety Consent		
This Change is Approved Date _____	Adjusted Subcontract Price Through C.O. Number 0	$ 42,300.00
Name: _____		
	Amount this C.O. Number:	$ (465.00)
By: _____		
Title: _____	Current Adjusted Subcontract Price:	$ 41,835.00

When this Change Order is signed by both parties (and by Subcontractor's surety is subcontract is bonded), it constitutes their agreement:

(A) That the subcontract price is adjusted as shown above and that no further adjustment in that price by reason of the change(s) provided herein shall be made; and (B) That all the terms and conditions of the subcontract, except as modified by this and any previous changes, shall remain in full force and effect and apply to the work as so changed.

Accepted and Agreed:	Date: _____
By _____	By _____
Authorized Signature – Subcontractor	Authorized Signature – Contractor
Title	Title

* Subcontractor: Sign pink copy and return immediately. If subcontract is bonded, obtain consent of surety endorsed thereon.
No payments on account of this Change Order will be made until you have complied with the foregoing.

Published by San Diego Chapter A.G.C.

A purchase order is a legal document used to order equipment and materials from a vendor, with the implied promise to pay for these goods within a specified time after receipt of the items. Purchase orders are sometimes used to write subcontracts whose value does not exceed one or two thousand dollars. Refer to the sample purchase order on page 452.

While most of the purchase order is self-explanatory, certain items need to be emphasized:

"Description" should include all pertinent data such as model numbers, arrangements, performance requirements, applicable standards, electrical characteristics and so on.

"Tag" is used to identify a particular piece of equipment to facilitate unloading it at the proper location at the jobsite. Tag numbers for all major equipment can usually be found on the contract drawings.

"48 hour delivery notice": This can be very important if the material or equipment being delivered is too heavy to be unloaded by hand. Time will be required to arrange for a crane or forklift to unload the truck.

"Sales tax" is the current state sales tax which must be included unless the goods are to be delivered and installed at a project located out-of-state.

"Terms": The usual payment terms are "Net 30" which means the entire amount of the purchase order must be paid within thirty days after receipt of all goods, undamaged. If items are received damaged, payment should be withheld until satisfactory repair or replacement is made.

ACME
Mechanical Contractors
7600 Oak Avenue - Smallville, U.S.A. 12345-6789
Voice 1-234-5678 / FAX 1-234-5680

Purchase Order

P.O. Number: 7777-36   **Date:** 1-6-20

Job: CBC Bank Building

Job No: 7777

Seller: Smallville Industrial Sales Co.	**Ship to:** Acme Mechanical Contractors
3657 Third Avenue Northwest	c/o CBC Bank Building
Smallville, CA 63876	465 Commercial Street
	Smallville, CA 63876

Attn: Mr. L.H. Seller   **By:** RJS

Delivery date: 2-28-20   **24 Hr. Delivery Notice Required**   **Payment terms:** Net 30

Quantity	Description	Unit	Total
Two	Big-Blo Model No. 1JN 86-13 centrifugal	$443.00	$886.00
	fans with 480/3/60 one-half H.P., O.D.P.		
	motor. 3250 CFM @ 0.50 T.S.P.		
	Tag one EF-13 and one EF-23		

Number of copies of operation and maintenance
manuals required 2.

All shipments are to be F.O.B. jobsite or shop unless
otherwise noted.

Total before tax	$886.00
Sales tax	$62.02
Freight	Included
Total Order	$948.02

Buyer: RJS

Acceptance by seller _____ Date _____

The general contractor's construction schedule may range from a simple bar chart to a complicated computer-generated CPM (Critical Path Method) diagram. More complex jobs require more complete and detailed schedules. On larger jobs, lower-tier subcontractors may be required to prepare a bar chart showing their construction schedule.

If you have to prepare a construction schedule, make conservative estimates of the time required to complete each part of the job. A slower schedule makes it possible to use smaller crews which are usually more efficient.

You can't schedule your own work until the general contractor has supplied a schedule for the balance of the project. When you have that schedule, prepare a list of tasks your crews will perform. Show this list to the general contractor to be certain there are no conflicts with the master schedule.

When the list of tasks to be performed has been approved, begin recording the manhours required for each task. Take these manhours from your estimate. When the duration for each task has been determined, figure the crew sizes required.

For example, suppose the estimate shows that it will take 1,040 manhours to install the underground plumbing for a project. The work must be completed in two months. Assuming one worker averages 173 hours of work in one month, how many workers will be required to meet the schedule? Here's the solution:

$$\frac{1,040 \text{ manhours}}{173 \text{ MH}/\text{Mo. x 2 months}} = 3 \text{ workers}$$

The first of the two bar charts on the following pages shows the schedule a general contractor might expect to receive from a plumbing and HVAC subcontractor. The chart on page 454 is intended for use by the subcontractor. It includes information a mechanical subcontractor needs to determine proper crew sizes for each task. The *Cumulative Manhours* line shows the budgeted manhours for each month of the project. If labor costs are to be kept within budget, actual hours should not exceed estimated hours. Monitor labor costs each month by comparing the line *Actual MH Used* (actual manhours used) with the line *Cumulative MH* (estimated cumulative manhours).

Construction Schedule

Company: Acme Mechanical Contractors   **Project:** Mile-Hi Office Bldg   **Job Number:** 7777   **Date:** 6/8/20

Activity	2020						2021					Crew Sizes
	Jul	Aug	Sep	Oct	Nov	Dec	Jan	Feb	Mar	Apr	May	
Submittals	▨											
Purchasing		▨	▨									
U/G Plumbing				▨								
A/G Plumbing					▨	▨	▨					
Fin. Plumbing								▨				
HVAC Piping					▨	▨	▨	▨				
HVAC Ducting					▨	▨	▨	▨				
Equipment									▨			
Insulation								▨	▨			
Temp. Controls					▨	▨	▨	▨	▨			
Start-up										▨		

Construction Schedule

Company: Acme Mechanical Contractors   **Project:** Mile-Hi Office Bldg   **Job Number:** 7777   **Date:** 6/8/20

Activity	2020						2021					Crew Sizes
	Jul	Aug	Sep	Oct	Nov	Dec	Jan	Feb	Mar	Apr	May	
Submittals	▓	▓										N/A
Purchasing		▓	▓									N/A
U/G Plumbing			▓	▓								1040 MH ÷ (173 x 2) = 3
A/G Plumbing					▓	▓	▓					2075 MH ÷ (173 x 4) = 3
Fin. Plumbing								▓				130 MH ÷ (173 x .5) = 1.5
HVAC Piping					▓	▓	▓	▓				2760 MH ÷ (173 x 4) = 4
HVAC Ducting												2790 MH ÷ (173 x 4) = 4
Equipment								▓	▓			690 MH ÷ (173 x 2) = 2
Insulation									▓			Subcontract
Temp. Controls					▓	▓	▓	▓	▓			Subcontract
Start-up										▓		85 MH ÷ (173 x .5) = 1
Crew size/month			3	6	11	11	11	11.5	2	1		
Cumulative MH*			520	1159	3466	5373	7280	9143	9448	9573		MH Estimated = 9573
Actual MH used												
*Per estimate												
*Rounded												

455

Letter of Intent

On most large projects the contractor is required to submit drawings and technical data to the owner's representative for approval before work actually begins. Most of these submittals will be prepared by your suppliers and third-tier subs. Vendors and subs may be reluctant to prepare these documents without some assurance that they've been selected to do the work. That's the purpose of a letter of intent. It notifies a proposed vendor or subcontractor that you plan to contract with them when the owner approves the information submitted.

A letter of intent isn't a contract. It's an expressed intention to make a contract, which isn't a contract at all. But a vendor who gets your letter of intent and supplies all the information requested should get the contract. If you place the order with a different vendor, obviously, your letter of intent was worthless. That first vendor may be very reluctant to cooperate in the future.

A sample letter of intent follows this page.

ACME
Mechanical Contractors

7600 Oak Avenue - Smallville, U.S.A. 12345-6789
Voice 1-234-5678 / FAX 1-234-5680

December 20, 2019

Quikrap Insulation Company
320 Maple Boulevard
Smallville, U.S.A. 12345-6789

Attn: Mr. R. S. Quikrap

Dear Mr. Quikrap:

Acme Mechanical Contractors has been issued a subcontract to furnish and install the mechanical and plumbing systems for the proposed Mile-Hi Office Building to be located at 2600 Second Street in Smallville.

It is my intention to award your firm a subcontract to provide the thermal insulation for these systems in accordance with your proposal dated 12-13-19 in the amount of $42,300.00.

Please prepare ten copies of your submittal data and forward them to me no later than January 15, 2020. Upon approval of your submittal, I will mail you the subcontract for your review and signature.

Thank you,

Bobby Thompson
Project Manager

CC: Project file

Submittal Data

The construction specifications may require that you submit manufacturer's technical data on certain equipment you plan to install. You may have to submit six to ten copies of this technical data for approval before buying the equipment or contracting with third-tier subcontractors.

First, find out how many submittal copies will be required. Request that number of copies plus at least two additional sets for your use. A subcontractor who provides only services (no materials) may have to submit a detailed written explanation of the work to be done.

The second step is to label each submittal with the paragraph number of the specification where that equipment, material or service is described. Type or write this number in the upper right corner of each submittal sheet. From this set of numbered submittals, prepare a submittal index, a list of submittals identified by title and paragraph number. This index should be arranged in specification paragraph order. A blank submittal index form follows this section.

I prefer to use a three-ring binder to hold submittals. That makes it easier to add and delete sections as needed. Be sure to include a cover sheet, usually on company letterhead, which identifies the project, job number and scope of work being submitted. A sample submittal cover sheet follows this section.

Purchase orders and subcontracts should not be written until an approved copy of the submittal brochure has been received.

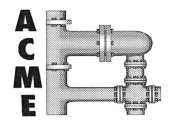

ACME
Mechanical Contractors

7600 Oak Avenue - Smallville, U.S.A. 12345-6789
Voice 1-234-5678 / FAX 1-234-5680

Division 15

Heating, Ventilating, Air Conditioning and Plumbing Systems

Submittal Data for

Mile-Hi Office Building
2600 Second Street
Smallville, U.S.A.

Submittal Index

Specification Section	Specification Paragraph	Item Description	Proposed Vendor or Sub	As Specified	See Submittal Data Page

The format to be used for monthly progress billings is seldom specified in the contract documents. In any case, your bill has to include enough detail to satisfy the owner and the lender.

Before preparing your bill, ask the general contractor how much detail is required. Some general contractors will want detailed cost breakdowns showing costs for each system on each building floor. Others will require much less detail.

Find out if billings can include the cost of equipment and materials delivered and stored either on site or off site but not yet installed. If billings can include materials stored off site, do these materials have to be stored in a bonded warehouse? Do vendor's invoices have to accompany progress billings?

When you understand what's required for monthly progress billings, prepare a billing breakdown worksheet. A sample worksheet follows this section.

Column one, Activity Lists each work item in the order work will be performed. The first item listed will usually be mobilization. This includes the cost of the job site trailer, electrical and telephone hookups, office furniture and supplies and initial labor costs. Billings for mobilization costs are sometimes denied by the lender. Many contractors minimize these costs to increase the chance of approval. It's better to receive partial payment for these costs than none at all.

Column two, Cost ($) Lists the actual contractor's costs for each activity, including labor, material, equipment and sales taxes. Markup for overhead and profit are not included.

Column three, Factor This column shows the markup assigned to each activity. Activities scheduled to be completed first are usually assigned the highest markups. This is called front loading and is common in the construction industry. Most bills won't be paid for 30 to 60 days. When the bill is paid, the amount received probably won't include the percentage allowed for retention. Generally, retention is 10 percent and isn't released until the project is complete and accepted by the owner.

Front loading helps contractors carry the financial burden of work in progress. Few subcontractors have enough cash to pay all their bills when due and still wait months to collect for work that's been completed. Employees have to be paid in full and on time. Assigning a higher markup to work completed first accelerates the payment schedule and helps spread payments more evenly over the entire project. Front loading is routine with subcontractors and unpopular with owners and lenders. In practice, subcontractors have no choice but to place a higher price or higher markup on work completed first.

Note that figures in the Factor column are less than 1.0 for work done near the end of the project. That's because front loading doesn't change the contract price. It changes only when that money is received.

Column four, Sell ($) Values for this column are the product of the cost and factor columns. Use the figures in this column when submitting monthly progress invoices.

ACME
Mechanical Contractors

7600 Oak Avenue - Smallville, U.S.A. 12345-6789
Voice 1-234-5678 / FAX 1-234-5680

Billing Breakdown Worksheet
(CBC Bank Building - Job No. 7777)

Activity	Cost ($)	Factor	Sell ($)
Mobilization	2,000	1.50	3,000
Underground chilled and heated water piping	120,000	1.35	162,000
Underground plumbing & piping	87,000	1.35	117,450
Above ground chilled and heating water piping	115,000	1.20	138,000
Above ground plumbing & piping	87,000	1.20	104,400
HVAC ducting	96,000	1.20	115,200
Equipment	312,000	1.10	343,200
Duct and pipe insulation	119,000	1.00	119,000
Temperature controls	53,000	.90	47,700
Plumbing fixtures	43,000	.65	27,950
Water treatment	1,800	.55	990
Test and balance	13,000	.55	7,150
Validation	500	.50	250
	1,049,300		1,186,290

$$\text{Estimated gross profit} = \frac{\$1,186,290 - \$1,049,300}{\$1,049,300} = 13\%$$

Index

Practical References for Builders

Basic Plumbing with Illustrations, Revised

This completely revised edition brings this comprehensive manual fully up-to-date with all the latest plumbing codes. It is the journeyman's and apprentice's guide to installing plumbing, piping, and fixtures in residential and light commercial buildings: how to select the right materials, lay out the job and do professional-quality plumbing work, use essential tools and materials, make repairs, maintain plumbing systems, install fixtures, and add to existing systems. Includes extensive study questions at the end of each chapter, and a section with all the correct answers.
384 pages, 8^1/$_2$ x 11, $44.75
eBook (PDF) also available; $22.37 at www.craftsman-book.com

Construction Contract Writer

Relying on a "one-size-fits-all" boilerplate construction contract to fit your jobs can be dangerous — almost as dangerous as a handshake agreement. *Construction Contract Writer* lets you draft a contract in minutes that precisely fits your needs and the particular job, and meets both state and federal requirements. You just answer a series of questions — like an interview — to construct a legal contract for each project you take on. Anticipate where disputes could arise and settle them in the contract before they happen. Include the warranty protection you intend, the payment schedule, and create subcontracts from the prime contract by just clicking a box. Includes a feedback button to an attorney on the Craftsman staff to help should you get stumped — *No extra charge.*
$149.95. *Download the Construction Contract Writer at*
http://www.constructioncontractwriter.com

CD Estimator

If your computer has *Windows*™ and a CD-ROM drive, CD Estimator puts at your fingertips over 150,000 construction costs for new construction, remodeling, renovation & insurance repair, home improvement, framing & finish carpentry, electrical, concrete & masonry, painting, earthwork and heavy equipment and plumbing & HVAC. Quarterly cost updates are available at no charge on the Internet. You'll also have the *National Estimator* program — a stand-alone estimating program for *Windows*™ that *Remodeling* magazine called a "computer wiz," and *Job Cost Wizard*, a program that lets you export your estimates to QuickBooks Pro for actual job costing. A 60-minute interactive video teaches you how to use this CD-ROM to estimate construction costs. And to top it off, to help you create professional-looking estimates, the disk includes over 40 construction estimating and bidding forms in a format that's perfect for nearly any *Windows*™ word processing or spreadsheet program. **CD Estimator is $149.50**

Plumber's Handbook Revised

This edition shows what will and won't pass inspection in drainage, vent, and waste piping, septic tanks, water supply, graywater recycling systems, pools and spas, fire protection, and gas piping systems. All tables, standards, and specifications are completely up-to-date with recent plumbing code changes. Covers common layouts for residential work, how to size piping, select and hang fixtures, practical recommendations, and trade tips. It's the approved reference for the plumbing contractor's exam in many states. Includes an extensive set of multiple-choice questions after each chapter, with answers and explanations in the back of the book, along with a complete sample plumber's exam. **352 pages, 8^1/$_2$ x 11, $44.50**
eBook (PDF) also available; $20.75 at www.craftsman-book.com

Building Code Compliance for Contractors & Inspectors

Have you ever failed a construction inspection? Have you ever dealt with an inspector who has his own interpretation of the Code and forces you to comply with it? This new book explains what it takes to pass inspections under the 2009 *International Residential Code*. It includes a Code checklist — with explanations and the Code section number — for every trade, covering some of the most common reasons why inspectors reject residential work. The author uses his 30 years' experience as a building code official to provide you with little-known information on what code officials look for during inspections. **232 pages, 8^1/$_2$ x 11, $32.50**
eBook (PDF) also available; $16.25 at www.craftsman-book.com

Uniform Plumbing Code Quick-Card

This 6-page, laminated, fold-out guide provides the basic numbers, flow rates and formulas the plumber needs based on the *2015 Uniform Plumbing Code (UPC)* and *2015 International Plumbing Code (IPC)*. Like a Cliffs Notes to the Plumbing Code. **6 pages, 8^1/$_2$ x 11, $7.95**

Plumbing & HVAC Manhour Estimates

Hundreds of tested and proven manhours for installing just about any plumbing and HVAC component you're likely to use in residential, commercial, and industrial work. You'll find manhours for installing piping systems, specialties, fixtures and accessories, ducting systems, and HVAC equipment. If you estimate the price of plumbing, you shouldn't be without the reliable, proven manhours in this unique book.
224 pages, 5^1/$_2$ x 8^1/$_2$, $28.25

Paper Contracting: The How-To of Construction Management Contracting

Risk, and the headaches that go with it, have always been a major part of any construction project — risk of loss, negative cash flow, construction claims, regulations, excessive changes, disputes, slow pay — sometimes you'll make money, and often you won't. But many contractors today are avoiding almost all of that risk by working under a construction management contract, where they are simply a paid consultant to the owner, running the job, but leaving him the risk. This manual is the how-to of construction management contracting. You'll learn how the process works, how to get started as a CM contractor, what the job entails, how to deal with the issues that come up, when to step back, and how to get the job completed on time and on budget. Includes a link to free downloads of CM contracts legal in each state.
272 pages, 8^1/$_2$ x 11, $55.50
eBook (PDF) also available; $27.75 at www.craftsman-book.com

Planning Drain, Waste & Vent Systems

How to design plumbing systems in residential, commercial, and industrial buildings. Covers designing systems that meet code requirements for homes, commercial buildings, private sewage disposal systems, and even mobile home parks. Includes relevant code sections and many illustrations to guide you through what the code requires in designing drainage, waste, and vent systems. **192 pages, 8^1/$_2$ x 11, $29.95**

Code Check Plumbing & Mechanical, 4th Edition

Save time, money, and potential delays for code violations. Code Check Plumbing & Mechanical is an essential reference guide, providing reliable information on up-to-date residential plumbing and mechanical codes. This handy guide, cross-referenced to the current International Residential Code, Uniform Plumbing Code, and Uniform Mechanical Code, provides answers to hundreds of commonly-asked plumbing questions. This spiral-bound book, in easy-to-use flip chart format, and with durable laminated pages, easily withstands abuse on the jobsite. It's specifically designed for quick on-site reference, and summarizes national code specifications. It includes 116 drawings and 45 tables to answer questions on the plumbing code in force anywhere in the U.S. **49 pages, 8^1/$_2$ x 11, $24.95**

Construction Forms for Contractors

This practical guide contains 78 practical forms, letters and checklists, guaranteed to help you streamline your office, organize your jobsites, gather and organize records and documents, keep a handle on your subs, reduce estimating errors, administer change orders and lien issues, monitor crew productivity, track your equipment use, and more. Includes accounting forms, change order forms, forms for customers, estimating forms, field work forms, HR forms, lien forms, office forms, bids and proposals, subcontracts, and more. All are also on the CD-ROM included, in Excel spreadsheets, as formatted Rich Text that you can fill out on your computer, and as PDFs. **360 pages, 8^1/$_2$ x 11, $48.50**
eBook (PDF) also available; $24.25 at www.craftsman-book.com

National Construction Estimator

Current building costs for residential, commercial, and industrial construction. Estimated prices for every common building material. Provides manhours, recommended crew, and gives the labor cost for installation. Includes a free download of an electronic version of the book with *National Estimator*, a stand-alone *Windows*™ estimating program. Additional information and *National Estimator* ShowMe tutorial video is available on our website under the "Support" dropdown tab.
672 pages, 8^1/$_2$ x 11, $97.50. Revised annually
eBook (PDF) also available; $48.75 at www.craftsman-book.com

National Appraisal Estimator

An Online Appraisal Estimating Service. Produce credible single-family residence appraisals – in as little as five minutes. A smart resource for appraisers using the cost approach. Reports consider all significant cost variables and both physical and functional depreciation. For more information, visit www.craftsman-book.com/national-appraisal-estimator-online-software

Plumber's Exam Preparation Guide

Hundreds of questions and answers to help you pass the apprentice, journeyman, or master plumber's exam. Questions are in the style of the actual exam. Gives answers for both the Standard and Uniform plumbing codes. Includes tips on studying for the exam and the best way to prepare yourself for examination day. **320 pages, 8¹/₂ x 11, $34.00**

Insurance Restoration Contracting: Startup to Success

Insurance restoration — the repair of buildings damaged by water, fire, smoke, storms, vandalism and other disasters — is an exciting field of construction that provides lucrative work that's immune to economic downturns. And, with insurance companies funding the repairs, your payment is virtually guaranteed. But this type of work requires special knowledge and equipment, and that's what you'll learn about in this book. It covers fire repairs and smoke damage, water losses and specialized drying methods, mold remediation, content restoration, even damage to mobile and manufactured homes. You'll also find information on equipment needs, training classes, estimating books and software, and how restoration leads to lucrative remodeling jobs. It covers all you need to know to start and succeed as the restoration contractor that both homeowners and insurance companies call on first for the best jobs. **640 pages, 8¹/₂ x 11, $69.00**

eBook (PDF) also available; **$34.50** at www.craftsman-book.com

Insurance Replacement Estimator

Insurance underwriters demand detailed, accurate valuation data. There's no better authority on replacement cost for single-family homes than the *Insurance Replacement Estimator*. In minutes you get an insurance-to-value report showing the cost of re-construction based on your specification. You can generate and save unlimited reports. For more details, visit www.craftsman-book.com/insurance-replacement-estimator-online-software

Craftsman eLibrary

Craftsman's eLibrary license gives you immediate access to 60+ PDF eBooks in our bookstore for 12 full months! **You pay only one low price. $129.99. Visit** www.craftsman-book.com **for more details.**

Home Building Mistakes & Fixes

This is an encyclopedia of practical fixes for real-world home building and repair problems. There's never an end to "surprises" when you're in the business of building and fixing homes, yet there's little published on how to deal with construction that went wrong - where out-of-square or non-standard or jerry-rigged turns what should be a simple job into a nightmare. This manual describes jaw-dropping building mistakes that actually occurred, from disastrous misunderstandings over property lines, through basement floors leveled with an out-of-level instrument, to a house collapse when a siding crew removed the old siding. You'll learn the pitfalls the painless way, and real-world working solutions for the problems every contractor finds in a home building or repair jobsite. Includes dozens of those "surprises" and the author's step-by-step, clearly illustrated tips, tricks and workarounds for dealing with them. **384 pages, 8¹/₂ x 11, $52.50**

eBook (PDF) also available; **$26.25** at www.craftsman-book.com